T0238901

Computer Algebra
with SymbolicC++

Yorick Hardy
University of Johannesburg
South Africa

Kiat Shi Tan
Ilog Co., Ltd., Singapore

Willi-Hans Steeb
University of Johannesburg
South Africa

Computer Algebra
with SymbolicC++

World Scientific

NEW JERSEY · LONDON · SINGAPORE · BEIJING · SHANGHAI · HONG KONG · TAIPEI · CHENNAI

Published by

World Scientific Publishing Co. Pte. Ltd.

5 Toh Tuck Link, Singapore 596224

USA office: 27 Warren Street, Suite 401-402, Hackensack, NJ 07601

UK office: 57 Shelton Street, Covent Garden, London WC2H 9HE

British Library Cataloguing-in-Publication Data
A catalogue record for this book is available from the British Library.

COMPUTER ALGEBRA WITH SYMBOLICC++

Copyright © 2008 by World Scientific Publishing Co. Pte. Ltd.

All rights reserved. This book, or parts thereof, may not be reproduced in any form or by any means, electronic or mechanical, including photocopying, recording or any information storage and retrieval system now known or to be invented, without written permission from the Publisher.

For photocopying of material in this volume, please pay a copying fee through the Copyright Clearance Center, Inc., 222 Rosewood Drive, Danvers, MA 01923, USA. In this case permission to photocopy is not required from the publisher.

ISBN-13 978-981-283-360-0
ISBN-10 981-283-360-9
ISBN-13 978-981-283-361-7 (pbk)
ISBN-10 981-283-361-7 (pbk)

Printed in Singapore.

Preface

In this text we show how object-oriented programming can be used to implement a symbolic algebra system and how the system is applied to different areas in mathematics and physics.

In the most restrictive sense, computer algebra is used for the manipulation of scientific and engineering formulae. Usually, a mathematical formula described in the programming languages such as C, C++ and Java can only be evaluated numerically, by assigning the respective values to each variable. However, the same formula may be treated as a mathematical object in a symbolic algebra system, which allows formal transformation, such as differentiation, integration and series expansion, in addition to the numerical manipulations. This is therefore an indispensable tool for research and scientific computation.

Object-oriented programming has created a new era for programming in computer science as it has been suggested as a possible solution to software development. Basically, object-oriented programming is an important approach to analyzing problems, designing systems and building solutions. By applying this method effectively, the software products become less error prone, easier to maintain, more reusable and extensible.

The purpose of this book is to demonstrate how the features of object-oriented programming may be applied to the development of a computer algebra system. Among the many object-oriented programming languages available nowadays, we have selected C++ as our programming language. It is the most widely used object-oriented programming language, which has been successfully utilized by many programmers in various application areas. The design is based partly on acknowledged principles and partly on solid experience and feedback from actual use. Many experienced individuals and organizations in the industry and academia use C++. In addition to the reasons stated above, we have selected C++ over other object-oriented languages because of its efficiency in execution speed and its utilization of pointers and templates. The Standard Template Library provided by C++ is very helpful for the implementation of a computer algebra system.

Chapter 1 introduces the general notion of Computer Algebra. We discuss the essential properties and requirements of a computer algebra system. Some pitfalls and limitations are also listed for reference. Finally, we present a computer algebra system — **SymbolicC++**. This new system has many advantages over existing computer algebra systems.

Chapter 2 presents the general mathematics for a computer algebra system. We describe how fundamental mathematical quantities are built up to form more complex mathematical structures.

Chapter 3 gives a brief introduction to some computer algebra systems available in the market place, such as Reduce, Maple, Axiom, Mathematica, MuPAD and Maxima. The basic operations are described for each system. Examples are used to demonstrate the features of these systems.

In Chapter 4, we introduce the language tools in C++ such as the `this` pointer, classes, constructors, and templates. We describe error handling techniques and introduce the concept of exception handling. Examples are also given for demonstration purposes. We also describe recursion. A number of programs illustrate the concepts.

String classes are discussed in Chapter 5. We construct the `String` data type, which serves as a vehicle for introducing the facilities available in C++. The built-in `string` class of C++ is also described in detail. A number of examples show the use of this class. This `string` class of C++ will be used in **SymbolicC++**.

The Standard Template Library (STL) is introduced in Chapter 6 together with a large number of examples. At the core of the Standard Template Library are the three foundational items: containers, algorithm, and iterators. These items work in conjunction with one another. The built-in class in C++ to deal with complex numbers is also introduced. The built-in classes `list`, `vector`, `map`, `complex` will be used in **SymbolicC++**.

Chapter 7 gives a collection of useful classes for computer algebra. We investigate very long integers, rational numbers, quaternions, exact derivatives, vectors, matrices, arrays, bit vectors, finite sets and polynomials. They are the building blocks of mathematics as described in Chapter 2. The internal structures and external interfaces of these classes are described in great detail.

In Chapter 8, we describe how a mathematical expression can be constructed using object-oriented techniques. The computer algebra system **SymbolicC++** is introduced and its internal representations and public interfaces are described. Several examples are also presented to demonstrate the functionalities of the system. A symbolic numeric interface is also described.

In Chapter 9, we apply the classes developed in Chapters 7 and 8 to problems in mathematics and physics. Applications are categorized according to classes. Several classes may be used simultaneously to solve a particular problem. Many interesting problems are presented, such as ghost solutions, Padé approximant, Lie series techniques, Picard's method, Mandelbrot set, etc.

In Chapter 10, we discuss how the programming language Lisp can be used to implement a computer algebra system. We implement an algebraic simplification and differentiation program.

We develop a Lisp system using the object-oriented language C++ in Chapter 11. λ-calculus and its implementation in C++ are also be introduced.

Gene expression programming and its use for numerical and symbolic manipulations is studied in Chapter 12. A number of programs are given to illustrate the technique.

The header files of the classes (abstract data type) introduced in Chapters 8 and 9 are listed in Chapter 13.

The level of presentation is such that one can study the subject early on in one's education in science. There is a balance between practical programming and the underlying language. The book is ideally suited for use in lectures on symbolic computation and object-oriented programming. The beginner will also benefit from the book.

The reference list gives a collection of textbooks useful in the study of the computer language C++ [9], [16], [28], [34], [39], [47], [59]. For data structures we refer to Budd (1994) [11]. For applications in science we refer to Steeb et al. (1993) [49], Steeb (1994) [50], Steeb (2005) [54] and Steeb et al. (2004) [55].

The C++ programs have been tested with all newer C++ compilers which comply with the C++ Standard and include an implementation of the Standard Template Library.

All programs and header files of **SymbolicC++** fall under the GNU General Public License. We omit the following comment (or its equivalent) in all header file and program file listings in the interest of brevity:

```
/*
    SymbolicC++ : An object oriented computer algebra system written in C++

    Copyright (C) 2008 Yorick Hardy and Willi-Hans Steeb

    This program is free software; you can redistribute it and/or modify
    it under the terms of the GNU General Public License as published by
    the Free Software Foundation; either version 2 of the License, or
    (at your option) any later version.

    This program is distributed in the hope that it will be useful,
    but WITHOUT ANY WARRANTY; without even the implied warranty of
    MERCHANTABILITY or FITNESS FOR A PARTICULAR PURPOSE.  See the
    GNU General Public License for more details.
```

```
    You should have received a copy of the GNU General Public License along
    with this program; if not, write to the Free Software Foundation, Inc.,
    51 Franklin Street, Fifth Floor, Boston, MA 02110-1301 USA.
*/
```

The license is included with **SymbolicC++** which is available from the web site of the International School for Scientific Computing as described below.

Without doubt, this book can be extended. If you have comments or suggestions, we would be pleased to have them. The email addresses of the authors are:

```
Yorick Hardy:      yhardy@uj.ac.za
                   yorickhardy@gmail.com
Willi-Hans Steeb:  whsteeb@uj.ac.za
                   steebwilli@gmail.com
```

SymbolicC++ was developed by the International School for Scientific Computing. The web pages of the International School for Scientific Computing are

```
    http://issc.uj.ac.za/
```

The web page also provides the header files for **SymbolicC++**.

Johannesburg, Singapore,   *Yorick Hardy*
March 2008   *Kiat Shi Tan*
                                                         Willi-Hans Steeb

Contents

Chapter 1

Introduction

1.1   What is Computer Algebra?

Computer algebra [15], [38], [45] is the name of the technology for manipulating mathematical formulae symbolically by digital computers. For example an expression such as

$$x - 2 * x + \frac{d}{dx}(x - a)^2$$

should evaluate symbolically to

$$x - 2 * a.$$

Symbolic simplifications of algebraic expressions are the basic properties of computer algebra. Symbolic differentiation using the sum rule, product rule and division rule has to be part of a computer algebra system. Symbolic integration should also be included in a computer algebra system. Furthermore expressions such as $\sin^2(x) + \cos^2(x)$ and $\cosh^2(x) - \sinh^2(x)$ should simplify to 1. Thus another important ingredient of a computer algebra system is that it should allow one to define rules. Examples are the implementations of the exterior product and Lie algebras. Another important part of a computer algebra system is the symbolic manipulation of polynomials, for example to find the greatest common divisor of two polynomials or finding the coefficients of a polynomial.

The name of this discipline has long hesitated between symbolic and algebraic calculation, symbolic and algebraic manipulations and finally settled down as Computer Algebra. Algebraic computation programs have already been applied to a large number of areas in science and engineering. It is most extensively used in the fields where the algebraic calculations are extremely tedious and time consuming, such as general relativity, celestial mechanics and quantum chromodynamics. One of the first applications was the calculation of the curvature of a given metric tensor field. This involves mainly symbolic differentiation.

1.2   Properties of Computer Algebra Systems

What should a computer algebra system be able to do? First of all it should be able to handle the data types such as very long integers, rational numbers, complex numbers, quaternions, etc. The basic properties of the symbolic part should be simplifications of expressions, for example

$$a + 0 = a, \qquad 0 + a = a$$

$$a - a = 0, \qquad -a + a = 0$$

$$a * 0 = 0, \qquad 0 * a = 0$$

$$a * 1 = a, \qquad 1 * a = a$$

$$a^0 = 1.$$

In most systems it is assumed that the symbols are *commutative*, i.e.,

$$a * b = b * a.$$

Thus an expression such as

$$(a + b) * (a - b)$$

should be evaluated to

$$a * a - b * b.$$

If the symbols are not commutative then a special command should be given to indicate so. Furthermore, a computer algebra system should do simplifications of trigonometric and hyperbolic functions such as

$$\sin(0) = 0, \qquad \cos(0) = 1$$

$$\cosh(0) = 1, \qquad \sinh(0) = 0.$$

The expression $\exp(0)$ should simplify to 1 and $\ln 1$ should simplify to 0. Expressions such as

$$\sin^2(x) + \cos^2(x), \qquad \cosh^2(x) - \sinh^2(x)$$

should simplify to 1. Besides symbolic differentiation and integration a computer algebra system should also allow the symbolic manipulation of vectors, matrices and arrays. Thus the scalar product and vector product must be calculated symbolically. For square matrices the trace and determinant have to be evaluated symbolically. Furthermore, the system should also allow numerical manipulations. Thus it should be able to switch from symbolic to numerical manipulations. The computer algebra system should also be a programming language. For example, it should allow if-conditions and for-loops. Moreover, it must allow functions and procedures.

1.3 Pitfalls in Computer Algebra Systems

Although computer algebra systems have been around for many years there are still bugs and limitations in these systems. Here we list a number of typical pitfalls.

One of the typical pitfalls is the evaluation of

$$\sqrt{a^2 + b^2 - 2ab}.$$

Some computer algebra systems indicate that $a - b$ is the solution. Obviously, the result should be

$$\pm|a - b|.$$

As another example consider the *rank* of the matrix

$$A = \begin{pmatrix} 0 & 0 \\ x & 0 \end{pmatrix}.$$

The rank of a matrix is the number of linearly independent columns (which is equal to the number of linearly independent rows). If $x = 0$, then the rank is equal to 0. On the other hand if $x \neq 0$, then the rank of A is 1. Thus computer algebra systems face an ambiguity in the determination of the rank of the matrix. A similar problem arises when we consider the inverse of the matrix

$$B = \begin{pmatrix} 1 & 1 \\ x & 0 \end{pmatrix}.$$

It only exists when $x \neq 0$.

Another problem arises when we ask computer algebra systems to integrate

$$\int x^n dx$$

where n is an integer. If $n \neq -1$ then the integral is given by

$$\frac{x^{n+1}}{n+1}.$$

If $n = -1$, the integral is

$$\ln(x).$$

Another ambiguity arises when we consider

$$0^0.$$

Consider for example

$$f(x) = x^x \equiv \exp(x \ln(x))$$

for $x > 0$. Applying *L'Hospital's rule* we find $0^0 = 1$ as a possible definition of 0^0. Many computer algebra systems have problems finding

$$\frac{x}{x + \sin(x)}$$

at $x = 0$ using L'Hospital's rule. The result is $1/2$.

We must also be aware that when we solve the equation

$$a * x = 0$$

the computer algebra system has to distinguish between the cases $a = 0$ and $x = 0$.

A large number of pitfalls can arise when we consider complex numbers and branch points in complex analysis. Complex numbers and functions should satisfy the Aslaksen test [3]. Thus

$$\exp(\ln(z))$$

should simplify to z, but

$$\ln(\exp(z))$$

should not simplify for complex numbers. We have to take care of the branch cuts when we consider multiple-valued complex functions. Most computer algebra systems assume by default that the argument is real-valued.

Example. Consider the equation

$$\operatorname{Log}(zw) = \operatorname{Log} z + \operatorname{Log} w$$

where $z \neq 0$, $w \neq 0$ and Log is the principal logarithm. The left hand side of the equation is equivalent to

$$\operatorname{Log}(zw) = \operatorname{Log}|r| + \operatorname{Log}|z| + i\operatorname{Arg}(zw)$$

where $\operatorname{Arg}(zw) \in (-\pi, \pi]$ is the principle argument of zw. The right hand side of the equation can be written as

$$\operatorname{Log} z + \operatorname{Log} w = \operatorname{Log}|r| + \operatorname{Log}|w| + i(\operatorname{Arg}(z) + \operatorname{Arg}(w)).$$

For the equation to hold we must have

$$\operatorname{Arg}(z) + \operatorname{Arg}(w) \in (-\pi, \pi].$$

For a more in-depth survey of the pitfalls in computer algebra systems Stoutemeyer [58] may be perused.

1.4 Design of a Computer Algebra System

Most computer algebra systems are based on Lisp. The computer language Lisp takes its name from list processing. The main task of Lisp is the manipulation of quantities called lists, which are enclosed in parentheses. A number of powerful computer algebra systems are based in Lisp, for example Reduce, Maxima, Derive, Axiom and MuPAD. The design of Axiom is based on object-oriented programming using Lisp. The computer algebra systems Maple and Mathematica are based on C. All of these systems are powerful software systems which can perform symbolic calculations. However, these software systems are independent systems and the transfer of expressions from one of them to another programming environment such as C is rather tedious, time consuming and error prone. It would therefore be helpful to enable a higher level language to manipulate symbolic expressions. On the other hand, the object-oriented programming languages provide all the necessary tools to perform this task elegantly.

Here we show that object-oriented programming using C++ can be used to develop a computer algebra system. Object-oriented programming is an approach to software design that is based on classes rather than procedures. This approach maximizes modularity and information hiding. Object-oriented design provides many advantages. For example, it combines both the data and the functions that operate on that data into a single unit. Such a unit (*abstract data type*) is called a *class*.

We use C++ as our object-oriented programming language for the following reasons. C++ allows the introduction of abstract data types. Thus we can introduce the data types used in the computer algebra system as abstract data types. The language C++ supports the central concepts of object-oriented programming: encapsulation, inheritance, polymorphism (including dynamic binding) and operator overloading. It has good support for dynamic memory management and supports both procedural and object-oriented programming. A less abstract form of polymorphism is provided via template support. We overload the operators

$$+, \quad -, \quad *, \quad /$$

for our abstract data types, such as verylong integers, rational numbers, complex numbers or symbolic data types. The vector and matrix classes are implemented on a template basis so that they can be used with the other abstract data types.

Another advantage of this approach is that, since the system of symbolic manipulations itself is written in C++, it is easy to enlarge it and to fit it to the special problem at hand. The classes (abstract data types) are included in a header file and can be provided in any C++ program by giving the command `#include "ADT.h"` at the beginning of the program.

For the realization of this concept we need to apply the following features of C++

(1) the class concept
(2) overloading of operators
(3) overloading of functions
(4) inheritance of classes
(5) virtual functions
(6) function templates
(7) class templates
(8) Standard Template Library.

The developed system **SymbolicC++** includes the following abstract data types (classes)

`Verylong`	handles very long integers
`Rational`	template class that handles rational numbers
`Quaternion`	template class that handles quaternions
`Derive`	template class to handle exact differentiation
`Vector`	template class that handles vectors
`Matrix`	template class that handles matrices
`Array`	template class that handles arrays
`Polynomial`	template class that handles polynomials
`Symbolic`	class that handles symbolic manipulations, such as rules, simplifications, differentiation, integration, commutativity and non-commutativity
`Type/Pair`	handles the atom and dotted pair for a Lisp system All the standard Lisp functions are included.

Suitable type conversions between the abstract data types and between abstract data types and basic data types are also provided.

The advantages of SymbolicC++ are as follows

(1) Object-oriented design and proper handling of basic and abstract data types.
(2) The system is operating system independent, i.e. for all operating systems powerful C++ compilers are available.
(3) The user is provided with the source code.
(4) New classes (abstract data types) can easily be added.
(5) The ANSI C++ standard is taken into account.
(6) The user only needs to learn C++ to apply the computer algebra system.
(7) Assembler code can easily be added to run on a specific CPU.
(8) Member functions and type conversion operators provide a symbolic-numeric interface.
(9) The classes (abstract data types) are included in a header file.
(10) Standard Template Library is used with SymbolicC++.

Chapter 2

Mathematics for Computer Algebra

2.1 Sets

A *set* can be thought of as a collection of objects. However, a collection of objects described by some property is not always a set.

Let X be a set, then $x \in X$ states that x is a member (or element) of the set X, while $x \notin X$ states that x is not a member of the set X.

The empty set $\emptyset$ is the set which has no members, i.e. $\forall x : x \notin \emptyset$.

A *subset* Y of X ($Y \subseteq X$) is a set Y where $\forall y \in Y : y \in X$. If $Y \subseteq X$ and $\exists x \in X : x \notin Y$ then Y is a *proper subset* of X ($Y \subset X$).

If $Y \subseteq X$ and $X \subseteq Y$ then $X = Y$.

The characteristic function $\chi_Y : X \to \{0, 1\}$ of a subset $Y \subseteq X$ is a defined by

$$\chi_Y(x) := \begin{cases} 1 & x \in Y \\ 0 & x \notin Y \end{cases}.$$

A number of operations defined on sets are

- The union $X \cup Y$ of sets X and Y is

$$X \cup Y := \{\, x : x \in X \text{ or } x \in Y \,\}.$$

- The intersection $X \cap Y$ of sets X and Y is

$$X \cap Y := \{\, x : x \in X \text{ and } x \in Y \,\}.$$

  If $X \cap Y = \emptyset$ then X and Y are said to be disjoint.

- The complement of set Y under the set X is

$$Y \backslash X := \{ \, x \, : \, x \in X \text{ and } x \notin Y \, \}.$$

- The difference $Y - X$ between sets $X \subseteq Z$ and $Y \subseteq Z$ is

$$X - Y := X \cap (Y \backslash Z).$$

- The symmetric difference $X \triangle Y$ between sets X and Y is

$$X \triangle Y := (X - Y) \cup (Y - X).$$

- The cartesian product $X \times Y$ between sets X and Y is

$$X \cup Y := \{ \, (x, y) \, : \, x \in X \text{ and } y \in Y \, \}.$$

- The cardinality $|X|$ of a finite set X is the number of elements in the set.

The *power set* of a set $P(X)$ is the set of all subsets of X. For a finite set X the power set has $|P(X)| = 2^{|X|}$ elements.

The Standard Template Library (STL) provides a template `set` class, however only data of the same type can be stored in the template set class. Below is a C++ program which introduces a data type `AnyData` that wraps any C++ data and allows the STL `set` class to contain data of any type.

```
// anyset.cpp

#include <iostream>
#include <set>
#include <string>
#include <typeinfo>
using namespace std;

class Data
{
 public: virtual const type_info &type() const = 0;
         virtual ostream &print(ostream&) const = 0;
         virtual Data *copy() const = 0;
         virtual int operator<(const Data&) const = 0;
         virtual ~Data() {}
};

ostream &operator<<(ostream &o,const Data &d)
{ return d.print(o); }

template <class T>
class DataT: public Data
{
 private: T data;
 public: DataT(const T &t) : data(t) {}
         DataT(const DataT<T> &t) : data(t.data) {}
         virtual ~DataT() {}
```

```
            virtual const type_info &type() const { return typeid(T); }
            virtual ostream &print(ostream &o) const { return (o << data); }
            virtual Data *copy() const { return new DataT(*this); };
            virtual int operator<(const Data &d) const
            {
              if(type() == d.type()) return (data < ((const DataT *)&d)->data);
              return (type().name() < d.type().name());
            }
            const T &getdata() const { return data; }
};

class AnyData
{
 private: Data *data;
 public: template <class T> AnyData(const T &t) { data = new DataT<T>(t); }
         AnyData(const AnyData &a) { data = a.data->copy(); }
         ~AnyData() { delete data; }
         const type_info &type() const { return data->type(); }
         ostream &print(ostream &o) const { return (o << *data); }
         // for set implementation
         int operator<(const AnyData &a) const { return *data < *(a.data); }
         template <class T> operator T() const
         {
           if(type() != typeid(T)) throw string("Wrong type");
           return ((DataT<T> *)data)->getdata();
         }
};

ostream &operator<<(ostream &o,const AnyData &d)
{ return d.print(o); }

int main(void)
{
 set<AnyData> s;
 s.insert(2);
 s.insert(string("string"));
 s.insert(3.5);
 s.insert(2);
 set<AnyData>::iterator si;
 for(si=s.begin();si!=s.end();si++)
 {
   cout << *si << " of type " << si->type().name() << endl;
   if(si->type() == typeid(int) && int(*si) == 2)
     cout << "Found the integer value 2" << endl;
 }

 return 0;
}
```

2.2 Rings and Fields

The data types in a computer algebra system include integers, rational numbers, real numbers, complex numbers, quaternions and symbols. On the other hand the basic data types in computer languages such as C and C++ include integers, characters, float (double) numbers and pointers. From these basic data types we can form arrays, for example arrays of integers or arrays of characters (strings). The mathematical structure of these data types (except pointers) are rings and fields [4], [27].

A *ring* is an ordered triple $(R, +, *)$ consisting of a set R with two binary operations $+$ (addition) and $*$ (multiplication), satisfying the following conditions
(a) the pair $(R, +)$ is a commutative group;
(b) multiplication is associative;
(c) it admits an identity (or unit) element, denoted by I;
(d) multiplication is distributive (on both sides) over addition, i.e.

$$x * (y + z) = x * y + x * z \quad \text{and} \quad (x + y) * z = x * z + y * z, \quad \text{for all } x, y, z \in R.$$

It is common not to include the existence of a multiplicative identity element, amongst the ring axioms. One then distinguishes between rings and rings with an identity (unit) element. Similar differences also apply to the definitions of subring and integral domain given below. If, moreover, multiplication is commutative, then R is said to be a *commutative ring*.

A *subring* of R is a subset S of R satisfying
(a) S is a subgroup of the additive group R;
(b) $x \in S$ and $y \in S$ together imply $x * y \in S$;
(c) $I \in S$.

Thus a subring is a ring.

If an element $a \in R$ possesses an inverse element with respect to multiplication, i.e. if there exists a (unique) $a^{-1} \in R$ such that

$$a * a^{-1} = a^{-1} * a = I,$$

then we say that a is an *invertible element* of R.

The set of invertible elements of a ring R is denoted by R^*. If every non-zero element of R is invertible, then R is said to be a *division ring*. A commutative division ring is called a *field*. S is a *subfield* of F if S is a subring of F and $x \in S$, $x \neq 0$ together imply $x^{-1} \in S$.

A ring is ordered when there is a non-empty subset $P \subset R$, called the set of positive elements of R, satisfying
(a) $a \in P$ and $b \in P$ together imply $a + b \in P$ and $a * b \in P$;
(b) for each $a \in R$ exactly one of the following holds

$$a \in P, \quad a = 0, \quad (-a) \in P.$$

A commutative ring is said to be an *integral domain* if $x \in K$, $y \in K$ and $x * y = 0$, together imply $x = 0$ or $y = 0$. If $x * y = 0$, $x \neq 0$ and $y \neq 0$, then x and y are said to be *zero divisors* in K.

The *characteristic* of a ring is defined as the additive order of its (multiplicative) identity element, i.e. R has finite characteristic m if m is the least positive integer for which $m * I = 0$. It has characteristic 0 (or ∞, for usage differs) if no such multiple is zero.

Examples of rings are $(\mathbf{Z}, +, *)$ and $(\mathbf{Q}, +, *)$. $\mathbf{Z}$ is a subring of $\mathbf{Q}$, but it can be shown that $\mathbf{Z}$ has no proper subrings. The only invertible elements of $\mathbf{Z}$ are 1 and -1 whereas all non-zero elements of $\mathbf{Q}$ are invertible. Since $\mathbf{Q}$ is a commutative ring, it follows that $\mathbf{Q}$ is a field. $\mathbf{Q}$ is a subfield of $\mathbf{R}$. $\mathbf{Z}$, $\mathbf{Q}$ and $\mathbf{R}$ can be ordered by $<$ used in its usual sense. They are all integral domains and have characteristic 0.

A two-sided *ideal*, I, of a ring R is a non-empty subset of R satisfying
(a) I is a subgroup of the additive group R;
(b) for all $x \in I$ and all $a, b \in R$ we have

$$\text{(i)} \quad a * x \in I \quad \text{and} \quad \text{(ii)} \quad x * b \in I.$$

I is a left ideal if it satisfies axioms (a) and (b)(i) and a right ideal if it satisfies (a) and (b)(ii). When R is commutative all ideals are two-sided.

An ideal, $I \neq K$, of a commutative ring K is said to be a maximal ideal of K if, whenever M is an ideal of K satisfying $I \subset M \subset K$, either $M = I$ or $M = K$.

Example. The sets

$$I_2 = \{\ldots, -4, -2, 0, 2, 4, \ldots\} \quad \text{and} \quad I_3 = \{\ldots, -6, -3, 0, 3, 6, \ldots\}$$

are maximal ideals of $\mathbf{Z}$.

$$I_{12} = \{\ldots, -24, -12, 0, 12, 24, \ldots\} \quad \text{and} \quad I_8 = \{\ldots, -16, -8, 0, 8, 16, \ldots\}$$

are ideals but not maximal ideals.

Given two ideals I and J of a commutative ring K, we define their sum, $I + J$, to be the set of all elements of K of the form $x + y$ where $x \in I$ and $y \in J$, and their product, IJ, to be all elements of K which can be written in the form

$$x_1 * y_1 + x_2 * y_2 + \cdots + x_n * y_n$$

where $n \in \mathbf{N}$, $x_i \in I$ and $y_i \in J$ for all $i = 1, \ldots, n$. $I + J$ and IJ are again ideals.

If K is a commutative ring and $I = \{x * b \mid x \in K\}$ where b is a fixed element of K, then I is an ideal, called a principal ideal. b is then said to generate I. An integral domain all of whose ideals are principal is called a principal ideal domain (or principal ideal ring). If I is an ideal of a principal ideal domain D and is generated by the elements $x_1, \ldots, x_n$ i.e. I is the set of all elements of the form

$$u_1 * x_1 + u_2 * x_2 + \cdots + u_n * x_n$$

where $u_i \in D$, then I is generated by some single element d. We say that d is a *greatest common divisor* (g.c.d.) of $x_1, \ldots, x_n$. $x_1, \ldots, x_n$ are said to be mutually or relatively prime if they have g.c.d. 1.

The intersection of the n ideals generated by $x_1, \ldots, x_n$ taken one at a time is also an ideal and so will be generated by a single element m. m is called a least common multiple (l.c.m.) of $x_1, \ldots, x_n$. An element p of D is said to be prime or irreducible if it is not an invertible element of D and if $p = a * b$ $(a, b \in D)$ implies that a or b is invertible.

An integral domain D is a unique factorization domain if

(a) for all $a \in D \setminus \{0\}$, a is either invertible or can be written as the product of a finite number of irreducible elements of D, and
(b) the decomposition in (a) is unique up to the ordering of the irreducible elements and substitution by associates.

2.3   Integers

In this section we introduce integers. For the proof of the following theorems we refer to the literature [4], [27]. Let $\mathbf{N}$ be the set of natural numbers

$$\mathbf{N} := \{\, 0,\, 1,\, 2,\, \ldots \,\}.$$

The set of *integers*

$$\mathbf{Z} := \{\ldots, -3, -2, -1, 0, 1, 2, 3, \ldots\}$$

can be constructed from $\mathbf{N}$ in the following way: define an equivalence relation E on $\mathbf{N} \times \mathbf{N}$ by

$$(x, y)E(x', y') \Leftrightarrow x + y' = x' + y.$$

We want the formula $x - y = x' - y'$ to hold once 'minus' has been defined. The set $\mathbf{Z}$ is then defined as $(\mathbf{N} \times \mathbf{N})/E$. Given elements z and z' of $\mathbf{Z}$ such that

$$z = p_E(x, y), \qquad z' = p_E(x', y')$$

where p_E denotes the canonical mapping from $\mathbf{N} \times \mathbf{N}$ onto $\mathbf{Z}$, we define the sum and product of z and z' by

$$z + z' = p_E(x + x', y + y'), \qquad z * z' = p_E(x * x' + y * y', x * y' + x' * y).$$

This definition is chosen because we want the formulae

$$(x - y) + (x' - y') = (x + x') - (y + y')$$

and

$$(x - y) * (x' - y') = (x * x' + y * y') - (x * y' + x' * y)$$

to hold. With these two operations $\mathbf{Z}$ is a *commutative ring*. The mapping $n \mapsto p_E(n, 0)$ is an injection of $\mathbf{N}$ into $\mathbf{Z}$ which preserves addition and multiplication.

We can, therefore, identify $\mathbf{N}$ with a subset of $\mathbf{Z}$. If we define the negative of z, written $-z$, as the inverse element of z under addition, then it can be shown that either $z \in \mathbf{N}$ or $-z \in \mathbf{N}$. If $z \in \mathbf{N} \setminus \{0\}$ we say that z is a positive integer, if $-z \in \mathbf{N} - \{0\}$ we say that z is a negative integer. Note that $a - b = a + (-b)$ if $a, b \in \mathbf{N}$. An integer $a \neq 0$ is called a *divisor* of an integer b (written $a|b$) if there exists an integer c such that $b = a * c$. When $a|b$ we also say that b is an integral multiple of a. To show that the restriction $a \neq 0$ is necessary, suppose $0|b$. If $b \neq 0$, we must have $b = 0 * c$ for some $c \in \mathbf{Z}$, which is impossible; while if $b = 0$, we would have $0 = 0 * c$, which is true for every $c \in \mathbf{Z}$. When $b, c, x, y \in \mathbf{Z}$, the integer $b * x + c * y$ is called a linear combination of b and c. We have

Theorem. If $a|b$ and $a|c$ then $a|(b * x + c * y)$ for all $x, y \in \mathbf{Z}$.

An integer $p \neq 0, \pm 1$ is called a *prime* if and only if its only divisors are ± 1 and $\pm p$. It is clear that $-p$ is a prime if and only if p is a prime. Hereafter, we restrict our attention mainly to positive primes. The number of positive primes is infinite. When $a = b * c$ with $|b| > 1$ and $|c| > 1$, we call the integer a composite. Thus every integer $a \neq 0, \pm 1$ is either a prime or a composite. When $a|b$ and $a|c$, we call a a *common divisor* of b and c. When, in addition, every common divisor of b and c is also a divisor of a, we call a a *greatest common divisor* of b and c. Suppose c and d are two different greatest common divisors of $a \neq 0$ and $b \neq 0$. Then $c|d$ and $d|c$; hence c and d differ only in sign. We limit our attention to the positive greatest common divisor of two integers a and b and use either d or (a, b) to denote it. Thus, d is truly the largest (greatest) integer which divides both a and b.

We have assumed (a) that every two non-zero integers have a positive greatest common divisor and (b) that any integer $a > 1$ has a unique factorization, except for the order of the factors, as a product of positive primes. Of course, in (b) it must be understood that when a is itself a prime, "a product of positive primes" consists of a single prime.

Division Algorithm. For any two non-zero integers a and b, there exist unique integers q and r, called respectively *quotient* and *remainder*, such that

$$a = b * q + r, \qquad 0 \leq r < |b|.$$

It follows that $b|a$ and $(a, b) = b$ if and only if $r = 0$. When $r \neq 0$ it is easy to show that a common divisor of a and b also divides r and a common divisor of b and r also divides a. Then $(a, b)|(b, r)$ and $(b, r)|(a, b)$ so that $(a, b) = (b, r)$. Now either $r|b$ or $r \nmid b$. In the latter case, we use the division algorithm to obtain

$$b = r * q_1 + r_1, \qquad 0 < r_1 < r.$$

Again, either $r_1|r$ and $(a, b) = r_1$ or, using the division algorithm,

$$r = r_1 * q_2 + r_2, \qquad q < r_2 < r_1$$

and $(a, b) = (b, r) = (r, r_1) = (r_1, r_2)$. Since the remainders $r_1, r_2, \ldots$, assuming the process to continue, constitute a set of decreasing non-negative integers there must eventually be one which is zero. Suppose the process terminates with

$$(k) \qquad r_{k-3} \;=\; r_{k-2} * q_{k-1} + r_{k-1} \qquad 0 < r_{k-1} < r_{k-2}$$
$$(k+1) \quad r_{k-2} \;=\; r_{k-1} * q_k + r_k \qquad\quad 0 < r_k < r_{k-1}$$
$$(k+2) \quad r_{k-1} \;=\; r_k * q_{k+1} + 0.$$

Then

$$(a,b) = (b,r) = (r,r_1) = \cdots = (r_{k-2}, r_{k-1}) = (r_{k-1}, r_k) = r_k.$$

Since $r = a - b * q = a + (-q) * b = m_1 * a + n_1 * b$ we find

$$r_1 = b - r * q_1$$
$$= b - (m_1 * a + n_1 * b) * q_1$$
$$= -m_1 * q_1 * a + (1 - n_1 * q_1) * b$$
$$= m_2 * a + n_2 * b$$

and

$$r_2 = r - r_1 * q_2$$
$$= (m_1 * a + n_1 * b) - (m_2 * a + n_2 * b) * q_2$$
$$= (m_1 - q_2 * m_2) * a + (n_1 - q_2 * n_2) * b$$
$$= m_3 * a + n_3 * b$$

and continuing, we obtain finally

$$r_k = m_{k+1} * a + n_{k+1} * b.$$

Thus, we have

Theorem. When $d = (a,b)$, there exist $m, n \in \mathbf{Z}$ such that $d = (a,b) = m*a+n*b$.

In $(a,b) = m * a + n * b$, the integers m and n are not unique; in fact, $(a,b) = (m + k * b) * a + (n - k * a) * b$ for every $k \in \mathbf{N}$. The importance of this theorem is as follows: if $a|c$, $b|c$, and if $(a,b) = d$, then $a * b|c * d$. Since $a|c$ and $b|c$, there exist integers s and t such that $c = a * s = b * t$. There exist $m, n \in \mathbf{Z}$ such that $d = m * a + n * b$. Then

$$c * d = c * m * a + c * n * b = b * t * m * a + a * s * n * b = a * b * (t * m + s * n)$$

and $a * b | c * d$.

A second consequence of the Division Algorithm is

Theorem. Any non-empty set K of integers which is closed under the binary operations addition and subtraction is either $\{0\}$ or consists of all multiples of its least positive element.

For given $a, b \in \mathbf{Z}$, suppose there exist $m, n \in \mathbf{Z}$ such that $a * m + b * n = 1$. Now every common factor of a and b is a factor of the right member 1; hence, $(a,b) = 1$. Two integers a and b for which $(a,b) = 1$ are said to be *relatively prime*.

Unique Factorization Theorem. Every integer $a > 1$ has a unique factorization, except for order,

$$a = p_1 * p_2 * \cdots * p_n$$

into a product of positive primes. Evidently,

$$-a = -(p_1 * p_2 * \cdots * p_n).$$

Moreover, since the p_i's are not necessarily distinct, we may write

$$a = p_1^{\alpha_1} * p_2^{\alpha_2} * \cdots * p_s^{\alpha_s}$$

where each $\alpha_i \geq 1$ and the primes $p_1, p_2, \ldots, p_s$ are distinct.

Let m be a positive integer. The relation congruent modulo m ($\equiv (\bmod\ m)$) is defined on all $a, b \in \mathbf{Z}$ by $a \equiv b \,(\bmod\ m)$ if and only if $m|(a - b)$. An alternate definition, often more useful than the original, is $a \equiv b \,(\bmod\ m)$ if and only if a and b have the same remainder when divided by m. As immediate consequences of these definitions, we have

Theorem. If $a \equiv b \,(\bmod\ m)$, then for any $n \in \mathbf{Z}$, $m * n + a \equiv b \,(\bmod\ m)$ and conversely.

Theorem. If $a \equiv b \,(\bmod\ m)$, then for all $x \in \mathbf{Z}$, $a + x \equiv b + x \,(\bmod\ m)$ and $a * x \equiv b * x \,(\bmod\ m)$.

Theorem. If $a \equiv b \,(\bmod\ m)$ and $c \equiv e \,(\bmod\ m)$, then $a + c \equiv b + e \,(\bmod\ m)$, $a - c \equiv b - c \,(\bmod\ m)$, $a * c \equiv b * e \,(\bmod\ m)$.

Theorem. Let $(c, m) = d$ and write $m = m_1 * d$. If $c * a \equiv c * b \,(\bmod\ m)$, then $a \equiv b \,(\bmod\ m_1)$ and conversely.

The relation $\equiv (\bmod\ m)$ on $\mathbf{Z}$ is an equivalence relation and separates the integers into m equivalence classes, $[0], [1], [2], \ldots, [m-1]$, called *residue classes modulo m*, where

$$[r] := \{\, a : \ a \in \mathbf{Z},\ a \equiv r \,(\bmod\ m)\,\}.$$

We denote the set of all residue classes modulo m by $\mathbf{Z}/(m)$. Two basic properties of the residue classes modulo m are: If a and b are elements of the same residue class $[s]$, then $a \equiv b \,(\bmod\ m)$. If $[s]$ and $[t]$ are distinct residue classes with $a \in [s]$ and $b \in [t]$, then $a \not\equiv b \,(\bmod\ m)$.

Consider the linear congruence

$$a * x \equiv b \,(\bmod\ m)$$

in which a, b, m are fixed integers with $m > 0$. By a solution of the congruence we mean an integer $x = x_1$ for which $m|(a * x_1 - b)$. Now if x_1 is a solution so that

$m|(a*x_1 - b)$, then for any $k \in \mathbf{Z}$, $m|(a*(x_1 + k*m) - b)$ and $x_1 + k*m$ is another solution. Thus, if x_1 is a solution so is every other element of the residue class $[x_1]$ modulo m. If the linear congruence has solutions, they consists of all the elements of one or more of the residue classes of $\mathbf{Z}/(m)$.

Suppose $(a, m) = 1 = s*a + t*m$. Then $b = b*s*a + b*t*m$ and $x_1 = b*s$ is a solution. Now assume $x_2 \not\equiv x_1 \pmod{m}$ to be another solution. Since $a*x_1 \equiv b \pmod{m}$ and $a*x_2 \equiv b \pmod{m}$, it follows from the transitive property of $\equiv \pmod{m}$ that $a*x_1 \equiv a*x_2 \pmod{m}$. Then $m|a*(x_1 - x_2)$ and $x_1 \equiv x_2 \pmod{m}$, contrary to the assumption. Thus, one has just one incongruent solution, say x_1, and the residue class $[x_1] \in \mathbf{Z}/(m)$, also called a *congruence class*, includes all solutions.

Theorem. The congruence $a*x \equiv b \pmod{m}$ has a solution if and only if $d = (a, m)$ is a divisor of b. When $d|b$, the congruence has exactly d incongruent solutions (d congruence classes of solutions).

The number 827 016 can be written as

$$827016 = 8*10^5 + 2*10^4 + 7*10^3 + 0*10^2 + 1*10 + 6.$$

This representation is an application of the congruence properties of integers. For suppose a is a positive integer. By the division algorithm,

$$a = 10*q_0 + r_0, \qquad 0 \leq r_0 < 10.$$

If $q_0 = 0$, we write $a = r_0$. If $q_0 > 0$, then $q_0 = 10*q_1 + r_1$, $0 \leq r_1 < 10$. Now if $q_1 = 0$, then $a = 10*r_1 + r_2$ and we write $a = r_1 r_0$; if $q_1 > 0$, then $q_1 = 10*q_2 + r_2$, $0 \leq r_2 < 10$. Again, if $q_2 = 0$, then $a = 10^2*r_2 + 10*r_1 + r_0$ and we write $a = r_2 r_1 r_0$; if $q_2 > 0$, we repeat the process. This must end eventually and we have

$$a = 10^s*r_s + 10^{s-1}*r_{s-1} + \cdots + 10*r_1 + r_0 = r_s r_{s-1} \cdots r_1 r_0.$$

This follows from the fact that the q_i's constitute a set of decreasing non-negative integers. In this representation the symbols r_i used are from the set $\{0, 1, 2, 3, \ldots, 9\}$ of remainders modulo 10. The representation is unique. The process is independent of the base and any other positive integer may be used. Thus, if 4 is taken as base, any positive integer will be represented by a sequence of the symbols $0, 1, 2, 3$. For example, the integer (base 10) 155 is given in base 4 as

$$155 = 4^3*2 + 4^2*1 + 4*2 + 3 = 2123 \quad \text{base 4.}$$

Next, we describe an algorithm which generates the prime number sequence called the "sieve of Eratosthenes". It was worked out in the third century B.C. This algorithm discovers all the prime numbers less than a given integer N. It works by removing all the non-prime numbers, leaving the prime number sequence.

Consider the following sequence of integers > 1

$$2\ 3\ 4\ 5\ 6\ 7\ 8\ 9\ 10\ 11\ 12\ 13\ 14\ 15\ 16\ 17\ 18\ 19\ 20\ 21 \ldots$$

The first prime number is 2, and all the multiples of 2 are not prime. Thus we cross out all the even numbers greater than 2

$$2 \mid 3 \; \cancel{4} \; 5 \; \cancel{6} \; 7 \; \cancel{8} \; 9 \; \cancel{10} \; 11 \; \cancel{12} \; 13 \; \cancel{14} \; 15 \; \cancel{16} \; 17 \; \cancel{18} \; 19 \; \cancel{20} \; 21 \ldots$$

The next number on the sequence which has not been crossed out is 3. Thus we know it is the next prime number. Similarly, we cross out all the multiples of 3. We attempt to cross out the numbers $3 \times 2 = 6$, $3 \times 4 = 12$, $3 \times 6 = 18$ which have already been removed as they are also multiples of 2. In fact, we could save some operations by just removing $3(3 + 0) = 9$, $3(3 + 2) = 15$, $3(3 + 4) = 21$, ...

$$2 \, 3 \mid 5 \; 7 \; \cancel{9} \; 11 \; 13 \; \cancel{15} \; 17 \; 19 \; \cancel{21} \; 23 \; 25 \; \cancel{27} \; 29 \; 31 \; \cancel{33} \; 35 \; 37 \ldots$$

In general, starting with a prime number p, we successively cross out the multiples p^2, $p(p + 2)$, $p(p + 4)$, ... We start the crossing out process from p^2 because all the multiples smaller than that would have been removed in the earlier stages of the process. For example, starting with the prime number 5, we cross out $5(5 + 0) = 25$, $5(5 + 2) = 35$, $5(5 + 4) = 45$, ... We do not need to cross out 5×2 or 5×3 as they have been removed for $p = 2$ or $p = 3$, respectively.

With this process, we may still end up crossing out numbers more than once. For example, $5(5 + 4) = 45$ has already been crossed out as a multiple of 3. The sequence for $p = 5$ looks like

$$2 \, 3 \, 5 \mid 7 \; 11 \; 13 \; 17 \; 19 \; 23 \; \cancel{25} \; 29 \; 31 \; \cancel{35} \; 37 \; 41 \; 43 \; 47 \; 53 \; \cancel{55} \ldots$$

We continue this process until we reach a prime p with $p^2 > N$, where N is the largest number we wish to consider. Then all the non-prime numbers $\leq N$ would have been crossed out. What remains is the prime number sequence $\leq N$. Below we list the prime numbers less than 100 that were generated using the algorithm described above

$$2 \, 3 \, 5 \, 7 \, 11 \, 13 \, 17 \, 19 \, 23 \, 29 \, 31 \, 37 \, 41 \, 43 \, 47 \, 53 \, 59 \, 61 \, 67 \, 71 \, 73 \, 79 \, 83 \, 89 \, 97.$$

2.4 Rational Numbers

The set of *rational numbers*, $\mathbf{Q}$, can be constructed in a similar manner to $\mathbf{Z}$, as follows. Let E be the equivalence relation on $\mathbf{Z} \times (\mathbf{Z} \setminus \{0\})$ defined by

$$(a, b)E(c, d) \Leftrightarrow a * d = b * c,$$

and define $\mathbf{Q}$ as

$$\mathbf{Z} \times (\mathbf{Z} \setminus \{0\})/E.$$

Addition and multiplication are defined on $\mathbf{Q}$ in terms of the canonical mapping, p_E, by

$$p_E(a, b) + p_E(c, d) = p_E(a * d + b * c, b * d)$$
$$p_E(a, b) \times p_E(c, d) = p_E(a * c, b * d).$$

With these operations $\mathbf{Q}$ is a *field* and there is a natural injection which maps $\mathbf{Z}$ into $\mathbf{Q}$ and preserves the operations of multiplication and addition. We can, therefore, consider $\mathbf{Z}$ to be a subset of $\mathbf{Q}$.

Corresponding to each ordered pair (a, b) of $\mathbf{Z} \times (\mathbf{Z} \setminus \{0\})$ is the *fraction a/b* with *numerator a* and non-zero *denominator b* (the need to make b non-zero accounts for the use of $\mathbf{Z} \setminus \{0\}$, rather than $\mathbf{Z}$, in the definition). Two fractions are then equivalent if the corresponding ordered pairs are equivalent in the sense defined above, and a rational number is an equivalence class of fractions.

We now define two special rational numbers

$$\text{zero} \to p_E(0, b), \qquad \text{one} \to p_E(a, a)$$

and the inverse

$$(\text{additive}): \quad -p_E(a, b) = p_E(-a, b)$$

$$(\text{multiplicative}): \quad p_E(a, b)^{-1} = p_E(b, a).$$

In the following we use the notation a/b. Addition and multiplication obey the distributive and associative laws. Moreover, for every a/b $(a, b \neq 0)$ there exists a multiplicative inverse b/a such that $(a/b) * (b/a) = 1$.

Subtraction is defined by

$$\frac{a}{b} - \frac{c}{d} := \frac{a * d - b * c}{b * d}, \qquad b, d \neq 0.$$

Division is defined by

$$\frac{\dfrac{a}{b}}{\dfrac{c}{d}} := \frac{a * d}{b * c}, \qquad b, c \neq 0.$$

There is an order relation for rational numbers. An element $a/b \in \mathbf{Q}$ is called positive if and only if $a * b > 0$. Similarly, a/b is called negative if and only if $a * b < 0$. Since, by the *Trichotomy Law*, either $a * b > 0$, $a * b < 0$ or $a * b = 0$, it follows that each element of $\mathbf{Q}$ is either positive, negative or zero. The order relations $<$ and $>$ on $\mathbf{Q}$ are defined as follows. For each $x, y \in \mathbf{Q}$,

$$x < y \text{ if and only if } x - y < 0$$

$$x > y \text{ if and only if } x - y > 0.$$

These relations are transitive but neither reflexive nor symmetric. $\mathbf{Q}$ also satisfies the Trichotomy Law. If $x, y \in \mathbf{Q}$, one and only one of

$$\text{(a) } x = y, \quad \text{(b) } x < y, \quad \text{(c) } x > y$$

holds.

Consider any arbitrary $s/m \in \mathbf{Q}$ with $m \neq 0$. Let the (positive) greatest common divisor of s and m be d and write $s = d * s_1$, $m = d * m_1$. Since $(s, m) \sim (s_1, m_1)$, it follows that $s/m = s_1/m_1$. Thus, any rational number $\neq 0$ can be written uniquely in the form a/b ($b > 0$), where a and b are relatively prime integers. Whenever s/m has been replaced by a/b, we say that s/m has been reduced to lowest terms. Hereafter, any arbitrary rational number introduced is assumed to have been reduced to lowest terms.

Theorem. If x and y are positive rationals with $x < y$, then $1/x > 1/y$.

Density Property. If x and y, with $x < y$, are two rational numbers, there exists a rational number z such that $x < z < y$.

Archimedean Property. If x and y are positive rational numbers, there exists a positive integer p such that $p * x > y$.

Consider the positive rational number a/b in which $b > 1$. Now

$$a = q_0 * b + r_0, \qquad 0 \leq r_0 < b,$$

and

$$10 * r_0 = q_1 * b + r_1, \qquad 0 \leq r_1 < b.$$

Since $r_0 < b$ and, hence, $q_1 * b + r_1 = 10 * r_0 < 10 * b$, it follows that $q_1 < 10$. If $r_1 = 0$, then

$$r_0 = \frac{q_1}{10} * b, \qquad a = q_0 * b + \frac{q_1}{10} * b, \qquad \frac{a}{b} = q_0 + \frac{q_1}{10}.$$

We write $a/b = q_0 * q_1$ and call $q_0.q_1$ the *decimal representation* of a/b. If $r_1 \neq 0$, we have

$$10 * r_1 = q_2 * b + r_2, \qquad 0 \leq r_2 \leq b$$

in which $q_2 < 10$. If $r_2 = 0$, then $r_1 = \frac{q_2}{10} * b$ so that $r_0 = \frac{q_1}{10} * b + \frac{q_2}{10^2} * b$ and the decimal representation of a/b is $q_0.q_1 q_2$. If $r_2 = r_1$, the decimal representation of a/b is the repeating decimal

$$q_0.q_1 q_2 q_2 q_2 \cdots .$$

If $r_2 \neq 0$, we repeat the process. Now the distinct remainders $r_0, r_1, r_2, \ldots$ are elements of the set $\{0, 1, 2, 3, \ldots, b-1\}$ of residues modulo b so that, in the extreme case, r_b must be identical with some one of $r_0, r_1, r_2, \ldots, r_{b-1}$, say r_c, and the decimal representation of a/b is the repeating decimal

$$q_0.q_1 q_2 q_3 \cdots q_{b-1} q_{c+1} q_{c+2} \cdots q_{b-1} q_{c+1} q_{c+2} \cdots q_{b-1} \cdots$$

Thus, every rational number can be expressed as either a terminating or a repeating decimal.

Example. For $11/6$, we find

$$\begin{aligned}
11 &= 1 * 6 + 5; & q_0 &= 1, \ r_0 = 5 \\
10 * 5 &= 8 * 6 + 2; & q_1 &= 8, \ r_1 = 2 \\
10 * 2 &= 3 * 6 + 2; & q_2 &= 3, \ r_2 = 2 = r_1
\end{aligned}$$

and $11/6 = 1.833333\ldots$

Conversely, it is clear that every terminating decimal is a rational number. For example, $0.17 = 17/100$ and $0.175 = 175/1000 = 7/40$.

Theorem. Every repeating decimal is a rational number.

The proof makes use of two preliminary theorems

(i) Every repeating decimal may be written as the sum of an infinite geometric progression.

(ii) The sum of an infinite geometric progression whose common ratio r satisfies $|r| < 1$ is a finite number.

2.5   Real Numbers

Beginning with $\mathbf{N}$, we can construct $\mathbf{Z}$ and $\mathbf{Q}$ by considering quotient sets of suitable Cartesian products. It is not possible to construct $\mathbf{R}$, the set of all real numbers, in a similar fashion and other methods must be employed.

We first note that the order relation $<$ defined on $\mathbf{Q}$ has the property that, given $a, b \in \mathbf{Q}$ such that $a < b$, there exists $c \in \mathbf{Q}$ such that $a < c$ and $c < b$. Let $\mathcal{P}(\mathbf{Q})$ be the power set of $\mathbf{Q}$, i.e. $\mathcal{P}(\mathbf{Q})$ denotes the set whose elements are the subsets of $\mathbf{Q}$. Now consider ordered pairs of elements of $\mathcal{P}(\mathbf{Q})$, (A, B) say, satisfying

(i) $A \cup B = \mathbf{Q}$, $A \cap B = \emptyset$,

(ii) A and B are both non-empty,

(iii) $a \in A$ and $b \in B$ together imply $a < b$.

Such a pair of sets (A, B) is known as a *Dedekind cut*. An equivalence relation R is defined upon the set of cuts by

$$(A, B)\,R\,(C, D)$$

if and only if there is at most one rational number which is either in both A and D or in both B and C. This ensures that the cuts $(\{x|x \le q\},\ \{x|x > q\})$ and $(\{x|x < q\}, \{x|x \ge q\})$ are equivalent for all $q \in \mathbf{Q}$. Each equivalence class under this relation is defined as a *real number*. The set of all real numbers, denoted by $\mathbf{R}$ is then the set of all such equivalence classes. If the class contains a cut (A, B) such that A contains positive rationals, then the class is a positive real number, whereas if B should contain negative rationals then the class is a negative real number. Thus, for example, $\sqrt{2}$ which contains the cut

$$(\{x \in \mathbf{Q}|x^2 < 2\} \cup \{x \in \mathbf{Q} \,|\, x < 0\},\ \{x \in \mathbf{Q}|x^2 > 2\} \cap \{x \in \mathbf{Q} \,|\, x > 0\})$$

is positive since $1 \in A = \{x | x^2 < 2\}$. To define addition of real numbers we must consider cuts (A_1, B_1) and (A_2, B_2) representing the real numbers α_1 and α_2. We define $\alpha_1 + \alpha_2$ to be the class containing the cut (A_3, B_3) where A_3 consists of all the sums $a = a_1 + a_2$ obtained by selecting a_1 from A_1 and a_2 from A_2 and similarly for B_3. Given the real number α, represented by the cut (A_1, B_1), we define $-\alpha$, negative α, to be the class containing the cut $(-B_1, -A_1)$ defined by $a \in A_1 \Leftrightarrow -a \in -A_1$ and $b \in B_1 \Leftrightarrow -b \in -B_1$. It will be observed that $\alpha + (-\alpha) = 0$, and that subtraction can now be defined by $\alpha - \beta = \alpha + (-\beta)$. Of two non-zero numbers α and $-\alpha$, one is always positive. The one which is positive is known as the *absolute value* or *modulus* of α and is denoted by $|\alpha|$. Thus $|\alpha| = \alpha$ if α is positive and $|\alpha| = -\alpha$ if α is negative. $|0|$ is defined to be 0. If α_1 and α_2 are two positive real numbers, then the product $\alpha_1 * \alpha_2$ is the class containing the cut (A_4, B_4) where A_4 consists of the negative rationals, zero, and all the products $a = a_1 * a_2$ obtained by selecting a positive a_1 from A_1 and a positive a_2 from A_2. The definition is extended to negative numbers by agreeing that if α_1 and α_2 are positive, then

$$(-\alpha_1) * \alpha_2 = \alpha_1 * (-\alpha_2) = -(\alpha_1 * \alpha_2), \qquad (-\alpha_1) * (-\alpha_2) = \alpha_1 * \alpha_2.$$

Finally, we define

$$0 * \alpha = \alpha * 0 = 0 \quad \text{for all } \alpha.$$

With these definitions it can be shown that the real numbers $\mathbf{R}$ form an *ordered field*.

By associating the element $q \in \mathbf{Q}$ with the class containing the cut

$$(\{x \mid x \le q\}, \ \{x \mid x > q\})$$

one can define a monomorphism (of fields) $\mathbf{Q} \rightarrow \mathbf{R}$. We can, therefore, consider $\mathbf{Q}$ to be a subfield of $\mathbf{R}$ (i.e. identify $\mathbf{Q}$ with a subfield of $\mathbf{R}$). Those elements of $\mathbf{R}$ which do not belong to $\mathbf{Q}$ are known as *irrational numbers*.

An important property of $\mathbf{R}$ which can now be established is that given any non-empty subset $V \subset \mathbf{R}$ for which there exists an upper bound, M, i.e. an element $M \in \mathbf{R}$ such that $v \le M$ for all $v \in V$, then there exists a *supremum (sup)* L such that if M is any upper bound of V, then $L \le M$. In a similar manner, we can define an *infimum (inf)* for any non-empty subset V or $\mathbf{R}$ which possesses a lower bound.

Not every subset in $\mathbf{R}$ possesses an upper (or lower) bound in $\mathbf{R}$, for example, $\mathbf{N}$. In order to overcome certain consequences of this, one often makes use in analysis of the *extended real number system*, $\overline{\mathbf{R}}$, consisting of $\mathbf{R}$ together with the two symbols $-\infty$ and $+\infty$ having the properties

(a) If $x \in \mathbf{R}$, then

$$-\infty < x < +\infty,$$

and

$$x + \infty = +\infty, \qquad x - \infty = -\infty, \qquad \frac{x}{+\infty} = \frac{x}{-\infty} = 0.$$

(b) If $x > 0$, then

$$x * (+\infty) = +\infty, \qquad x * (-\infty) = -\infty.$$

(c) If $x < 0$, then

$$x * (+\infty) = -\infty, \qquad x * (-\infty) = +\infty.$$

Note that $\overline{\mathbf{R}}$ does not possess all the algebraic properties of $\mathbf{R}$.

We list the basic properties of the system $\mathbf{R}$ of all real numbers.

Addition

$\mathbf{A}_1$.	Closure Law	$r + s \in \mathbf{R}$, for all $r, s \in \mathbf{R}$.
$\mathbf{A}_2$.	Commutative Law	$r + s = s + r$, for all $r, s \in \mathbf{R}$.
$\mathbf{A}_3$.	Associative Law	$r + (s + t) = (r + s) + t$, for all $r, s, t \in \mathbf{R}$.
$\mathbf{A}_4$.	Cancellation Law	If $r + t = s + t$, then $r = s$ for all $r, s, t \in \mathbf{R}$.
$\mathbf{A}_5$.	Additive Identity	There exists a unique additive identity element $0 \in \mathbf{R}$ such that $r + 0 = 0 + r = r$, for every $r \in \mathbf{R}$.
$\mathbf{A}_6$.	Additive Inverses	For each $r \in \mathbf{R}$, there exists a unique additive inverse $-r \in \mathbf{R}$ such that $r + (-r) = (-r) + r = 0$.

Multiplication

$\mathbf{M}_1$.	Closure Law	$r * s \in \mathbf{R}$, for all $r, s \in \mathbf{R}$.
$\mathbf{M}_2$.	Commutative Law	$r * s = s * r$, for all $r, s \in \mathbf{R}$.
$\mathbf{M}_3$.	Associative Law	$r * (s * t) = (r * s) * t$, for all $r, s, t \in \mathbf{R}$.
$\mathbf{M}_4$.	Cancellation Law	If $m * p = n * p$, then $m = n$ for all $m, n \in \mathbf{R}$ and $p \neq 0 \in \mathbf{R}$.
$\mathbf{M}_5$.	Multiplicative Identity	There exists a unique multiplicative identity element $1 \in \mathbf{R}$ such that $1 * r = r * 1 = r$ for every $r \in \mathbf{R}$.
$\mathbf{M}_6$.	Multiplicative Inverses	For each $r \neq 0 \in \mathbf{R}$, there exists a unique multiplicative inverse $r^{-1} \in \mathbf{R}$ such that $r * r^{-1} = r^{-1} * r = 1$.

Distributive Laws	For every $r, s, t \in \mathbf{R}$,
$\mathbf{D}_1$.	$r * (s + t) = r * s + r * t$
$\mathbf{D}_2$.	$(s + t) * r = s * r + t * r$
Density Property	For each $r, s \in \mathbf{R}$, with $r < s$, there exists $t \in \mathbf{Q}$ such that $r < t < s$.
Archimedean Property	For each $r, s \in \mathbf{R}^+$, with $r < s$, there exists $n \in \mathbf{N}$ such that $n * r > s$.
Completeness Property	Every non-empty subset of $\mathbf{R}$ having a lower bound (upper bound) has a greatest lower bound (least upper bound).

2.6 Complex Numbers

The field of *complex numbers*, $\mathbf{C}$, can be defined in several ways. Consider $\mathbf{R} \times \mathbf{R}$ and take the elements of $\mathbf{C}$ to be ordered pairs $(x, y) \in \mathbf{R} \times \mathbf{R}$. The operations of addition and multiplication of elements of $\mathbf{C}$ are defined by

$$(x_1, y_1) + (x_2, y_2) := (x_1 + x_2, y_1 + y_2)$$

and

$$(x_1, y_1) * (x_2, y_2) := (x_1 x_2 - y_1 y_2, x_1 y_2 + x_2 y_1).$$

Then we can show that $\mathbf{C}$ is a field. We can define a monomorphism (of fields),

$$r : \mathbf{R} \rightarrow \mathbf{C}, \quad \text{by} \quad r(x) = (x, 0).$$

This enables us to regard $\mathbf{R}$ as a subfield of $\mathbf{C}$. It can, moreover, be checked that

$$(x, y) = (x, 0) + (0, 1) * (y, 0).$$

Thus making use of the monomorphism defined above, one can write

$$(x, y) = x + i * y$$

where $x, y \in \mathbf{R}$ and $i = (0, 1)$ It is seen that

$$i^2 = (0, 1) * (0, 1) = (-1, 0) = -1.$$

Given a complex number

$$z = x + i * y, \qquad \text{where } x, y \in \mathbf{R},$$

we say that x is the *real part*, $\Re(z)$, and y is the *imaginary part*, $\Im(z)$, of z. The number

$$x - i * y$$

is known as the *complex conjugate* of z and is denoted by $\bar{z}$ (or z^*). An obvious geometrical representation of the complex numbers are points in the Cartesian plane $\mathbf{R} \times \mathbf{R}$. This representation is known as an *Argand diagram*. The diagram is based on a pair of perpendicular coordinate axes in the plane. The number $z = x + i * y$ is associated with the point with coordinates (x, y). With this representation, the addition of complex numbers is interpreted as the addition of vectors in the plane. The length, r, of the segment

$$(0, 0) - (x, y)$$

is known as the *absolute value* or *modulus* of z and is denoted by $|z|$. We therefore have

$$|z| = r = \sqrt{x^2 + y^2} = \sqrt{z\bar{z}}.$$

The angle which the segment Oz makes with the Ox axis is known as the *argument* (amplitude or angle) of z and is denoted by $\arg z$. We therefore have

$$\tan(\arg z) = \tan \theta = \frac{y}{x}.$$

and $\arg z$ is defined as a real number modulo 2π (provided $z \neq 0$, for $\arg 0$ is not defined). Some authors take

$$0 \leq \arg z < 2\pi$$

while others opt for

$$-\pi < \arg z \leq \pi.$$

This restricted value of the argument is often known as the *principal argument*.

The coordinates r and θ are known as *polar coordinates*. The connections between the polar coordinates of a point $x = r\cos\theta$, $y = r\sin\theta$ and the complex number $z = x + i * y$ which it represents are

$$r = |z| = \sqrt{x^2 + y^2}$$
$$\theta = \arg z$$
$$z = r(\cos\theta + i\sin\theta) = re^{i\theta}.$$

The geometry of the triangle provides the inequality

$$|z_1 + z_2| \leq |z_1| + |z_2|$$

known as the *triangle inequality*. The two formulae

$$|z_1 z_2| = |z_1||z_2|$$

and

$$\arg(z_1 z_2) \equiv \arg(z_1) + \arg(z_2) \pmod{2\pi}$$

allow one to give a geometric interpretation of the multiplication of complex numbers.

Alternative constructions of $\mathbf{C}$ are
(a) Consider the set, M, containing all matrices of $M_2(\mathbf{R})$ of the form

$$\begin{pmatrix} a & -b \\ b & a \end{pmatrix}.$$

Under matrix addition and multiplication, M forms a field with zero 0_2 (2×2 zero matrix) and identity element I_2 (2×2 identity matrix). We have

$$\begin{pmatrix} a & -b \\ b & a \end{pmatrix} \begin{pmatrix} c & -d \\ d & c \end{pmatrix} = \begin{pmatrix} ac - bd & -(ad + bc) \\ ad + bc & ac - bd \end{pmatrix}.$$

Moreover, $p : \mathbf{R} \to M$ defined by

$$p(c) = \begin{pmatrix} c & 0 \\ 0 & c \end{pmatrix}$$

is a field isomorphism between $\mathbf{R}$ and a subfield of M. Mapping the matrix

$$\begin{pmatrix} a & -b \\ b & a \end{pmatrix}$$

onto $a + i * b$, we obtain a field isomorphism between M and $\mathbf{C}$.

(b) We define $\mathbf{C}$ to be the quotient ring of $\mathbf{R}[t]$ by the principal ideal $(t^2 + 1)$, i.e.

$$\mathbf{C} = \mathbf{R}[t]/(t^2 + 1).$$

If we denote the image of the polynomial $t \in \mathbf{R}[t]$ under the canonical mapping by i, then every element of $\mathbf{C}$ can be written uniquely as $x + iy$ where $x, y \in \mathbf{R}$. The operations of addition and multiplication on $\mathbf{C}$ are the natural operations of the quotient algebra.

To facilitate the study of certain curves (e.g. the equiangular spiral) one frequently relaxes the conditions $r \geq 0$ and $0 \leq \theta < 2\pi$ on polar coordinates. One then has an extended system of polar coordinates in which r and θ can take all real values. In the extended system any pair (ρ, ω) will determine a unique point of the plane, yet every point in the plane will possess an infinite number of polar coordinates, namely

$$(\rho, \omega + 2n\pi), \quad (-\rho, \omega + (2n + 1)\pi)$$

for all $n \in \mathbf{Z}$. In the extended system the equiangular spiral is described by the single equation $r = e^{a\theta}$ (i.e. the points $(e^{a\theta}\cos\theta, e^{a\theta}\sin\theta)$), whereas in the restricted system a whole set of equations would be required to define it.

2.7 Vectors and Matrices

Let F be a field. A *vector space* (also called a *linear space*), V, over F is defined as an additive Abelian group V together with a function $F \times V \to V$, $(\lambda, \mathbf{v}) \mapsto \lambda * \mathbf{v}$, satisfying

$$\lambda * (\mathbf{a} + \mathbf{b}) = \lambda * \mathbf{a} + \lambda * \mathbf{b}$$
$$(\lambda + \mu) * \mathbf{a} = \lambda * \mathbf{a} + \mu * \mathbf{a}$$
$$(\lambda * \mu) * \mathbf{a} = \lambda * (\mu * \mathbf{a})$$
$$1 * \mathbf{a} = \mathbf{a}$$

for all $\lambda, \mu \in F$ and $\mathbf{a}, \mathbf{b} \in V$. F is called the ground field of the vector space, its elements are called *scalars* and those of V are called *vectors*. Letters representing vectors are printed in bold type. A *vector subspace* of a vector space V is defined as any subset V' of V for which

(a) V' is a subgroup of the additive group V;
(b) for all $\mathbf{a} \in V'$ and for all $\lambda \in F$, $\lambda * \mathbf{a} \in V'$.

The set of all n-tuples $(x_1, \ldots, x_n)$ where $x_i \in \mathbf{R}$, $i = 1, \ldots, n$, forms a vector space, which we denote by $\mathbf{R}^n$, over the ground field $\mathbf{R}$ when we define

$$(x_1, \ldots, x_n) + (y_1, \ldots, y_n) = (x_1 + y_1, \ldots, x_n + y_n)$$

and

$$\lambda * (x_1, \ldots, x_n) = (\lambda * x_1, \ldots, \lambda * x_n)$$

where $\lambda \in \mathbf{R}$.

A vector space V over a field F which satisfies the ring axioms in such a way that addition in the ring is addition in the vector space and such that

$$\lambda * (\mathbf{v}_1 * \mathbf{v}_2) = (\lambda * \mathbf{v}_1) * \mathbf{v}_2 = \mathbf{v}_1 * (\lambda * \mathbf{v}_2)$$

for all $\mathbf{v}_1, \mathbf{v}_2 \in V$ and $\lambda \in F$ is said to be an *algebra* over F. If, in addition, it forms a commutative ring, then we say it is a *commutative algebra*. A subset of an algebra V is termed a *sub-algebra* if it is both a vector subspace and a subring of V.

Given two vectors $\mathbf{x} = (x_1, x_2, x_3)$, $\mathbf{y} = (y_1, y_2, y_3) \in \mathbf{R}^3$ we define their *vector product* (also called *cross product*) denoted by $\mathbf{x} \times \mathbf{y}$ to be the vector

$$\mathbf{x} \times \mathbf{y} := (x_2 * y_3 - x_3 * y_2, x_3 * y_1 - x_1 * y_3, x_1 * y_2 - x_2 * y_1).$$

The vector product is not commutative. It is an example of a *Lie algebra*. We have

$$\mathbf{x} \times (\mathbf{y} + \mathbf{z}) = \mathbf{x} \times \mathbf{y} + \mathbf{x} \times \mathbf{z}$$
$$\mathbf{x} \times \mathbf{y} = -\mathbf{y} \times \mathbf{x}$$
$$\mathbf{x} \times (\lambda * \mathbf{y}) = \lambda * (\mathbf{x} \times \mathbf{y})$$

and

$$\mathbf{x} \times (\mathbf{y} \times \mathbf{z}) + \mathbf{z} \times (\mathbf{x} \times \mathbf{y}) + \mathbf{y} \times (\mathbf{z} \times \mathbf{x}) = \mathbf{0}.$$

The last equation is the *Jacobi identity*.

Given two vector spaces U and V over a field F, a *homomorphism* of U to V is a function $t : U \to V$ satisfying

$$t(\mathbf{a} + \mathbf{b}) = t(\mathbf{a}) + t(\mathbf{b})$$
$$t(\lambda * \mathbf{a}) = \lambda * t(\mathbf{a})$$

for all $\mathbf{a}, \mathbf{b} \in U$ and $\lambda \in F$. Homomorphisms of vector spaces therefore preserve linear combinations of the type

$$\lambda_1 * \mathbf{a}_1 + \lambda_2 * \mathbf{a}_2 + \cdots + \lambda_n * \mathbf{a}_n$$

where $\lambda_1, \ldots, \lambda_n \in F$, $\mathbf{a}_1, \ldots, \mathbf{a}_n \in U$. For this reason a homomorphism of vector spaces is called a *linear transformation* or *linear mapping*. The set of all linear combinations of $a_1, \ldots, a_n \in U$ forms a vector subspace of U, called the subspace generated or spanned by $\mathbf{a}_1, \ldots, \mathbf{a}_n$. A vector space, V, is said to be finitely generated if there exists a finite set of elements $\mathbf{a}_1, \ldots, \mathbf{a}_n$ which generate V.

The vectors $\mathbf{a}_1, \ldots, \mathbf{a}_n$ are said to be *linearly independent* if the only choice of $\lambda_1, \ldots, \lambda_n$ satisfying the relation

$$\lambda_1 * \mathbf{a}_1 + \lambda_2 * \mathbf{a}_2 + \cdots + \lambda_n * \mathbf{a}_n = \mathbf{0}, \qquad \mathbf{0} \in V$$

is

$$\lambda_1 = \lambda_2 = \cdots = \lambda_n = 0, \qquad 0 \in F.$$

The vectors $\mathbf{a}_1, \ldots, \mathbf{a}_n$ are linearly dependent if and only if there exist $\lambda_1, \ldots, \lambda_n$ not all zero, for which

$$\lambda_1 * \mathbf{a}_1 + \lambda_2 * \mathbf{a}_2 + \cdots + \lambda_n * \mathbf{a}_n = \mathbf{0}.$$

If $\mathbf{a}_1, \ldots, \mathbf{a}_n$ generate a vector space, V, and are linearly independent, then we say that $\mathbf{a}_1, \ldots, \mathbf{a}_n$ form a *basis* of V. If $\mathbf{a}_1, \ldots, \mathbf{a}_n$ form a basis of V and $\mathbf{b}$ is any vector of V, then there exists a unique n-tuple $(\lambda_1, \ldots, \lambda_n)$ such that

$$\mathbf{b} = \lambda_1 * \mathbf{a}_1 + \lambda_2 * \mathbf{a}_2 + \cdots + \lambda_n * \mathbf{a}_n.$$

The scalars $\lambda_1, \ldots, \lambda_n$ are then called the coordinates or components of $\mathbf{b}$ with respect to the basis $\mathbf{a}_1, \ldots, \mathbf{a}_n$. It can be shown that every finitely-generated vector space, V, has a basis and that, in particular, any two bases of V contain the same number of elements, say n. The number n is called the *dimension* of the vector space V over F and we write $\dim V = n$. The vector space V is then isomorphic to F^n, the vector space of all n-tuples $(x_1, \ldots, x_n)$ with $x_i \in F$, $i = 1, \ldots, n$. By definition, $\dim\{0\} = 0$.

If U and V are finite-dimensional vector spaces over the same field F, and if 'addition' of linear transformations and 'multiplication of linear transformations by a scalar' are defined by

$$(t_1 + t_2)(\mathbf{a}) = t_1(\mathbf{a}) + t_2(\mathbf{a})$$
$$(\lambda * t)(\mathbf{a}) = \lambda * (t(\mathbf{a})), \quad \text{for all } \mathbf{a} \in U$$

then it can easily be shown that the set of linear transformations itself forms a finite-dimensional vector space over F (having dimension $(\dim U) \times (\dim V)$). This vector space is denoted by $\mathrm{Hom}(U, V)$. Let $t \in \mathrm{Hom}(U, V)$, $\mathbf{u}_1, \ldots, \mathbf{u}_n$ be a basis for U and $\mathbf{v}_1, \ldots, \mathbf{v}_m$ be a basis for V. Then t is completely determined by the formulae which give the components $(x_1', \ldots, x_m')$ of the vector $t(\mathbf{x})$, with respect to the basis $\mathbf{v}_1, \ldots, \mathbf{v}_m$ of V in terms of $(x_1, \ldots, x_n)$, the components of the vector $\mathbf{x}$ with respect to the basis $\mathbf{u}_1, \ldots, \mathbf{u}_n$ of U. We have

$$x_1' = a_{11}x_1 + a_{12}x_2 + \cdots + a_{1n}x_n$$
$$x_2' = a_{21}x_1 + a_{22}x_2 + \cdots + a_{2n}x_n$$
$$\vdots$$
$$x_m' = a_{m1}x_1 + a_{m2}x_2 + \cdots + a_{mn}x_n, \qquad a_{ij} \in F$$

and the coefficients a_{ij} determine the homomorphism t uniquely with respect to the chosen bases. The rectangular array of coefficients

$$\begin{pmatrix} a_{11} & a_{12} & \cdots & a_{1n} \\ a_{21} & a_{22} & \cdots & a_{2n} \\ \vdots & \vdots & & \vdots \\ a_{m1} & a_{m2} & \cdots & a_{mn} \end{pmatrix}$$

is said to form a *matrix*, A, having m *rows* and n *columns* [56]. More abstractly we can think of the matrix as a function

$$A : \{1, 2, \ldots, m\} \times \{1, 2, \ldots, n\} \to F.$$

The matrix A is often abbreviated to (a_{ij}), and to denote that this represents the linear transformation t we write $(t) = (a_{ij})$. The matrix corresponding to the zero mapping, $t : \mathbf{x} \mapsto 0$ (all $\mathbf{x} \in U$), is called the *zero matrix* and is denoted by 0.

The sum of two $m \times n$ matrices, $A = (a_{ij})$ and $B = (b_{ij})$ is defined as the $m \times n$ matrix

$$A + B = (a_{ij} + b_{ij}).$$

The product of a $m \times n$ matrix $A = (a_{ij})$ with the scalar λ is defined to be the $m \times n$ matrix $\lambda * A = (\lambda * a_{ij})$. With these definitions the set of all $m \times n$ matrices with coefficients in F becomes a vector space of dimension $m * n$ and is isomorphic to $\mathrm{Hom}(U, V)$.

We can also define an entrywise product of two $m \times n$ matrices A and B as

$$A \bullet B = (a_{ij} * b_{ij})$$

which is known as the *Hadamard product* (also known as the *Schur product*). In general, the matrix product refers to the product defined below.

We define the product matrix, $C = A * B$, of two matrices A (a_{ij} an $m \times n$ matrix), and B (b_{jk} an $n \times p$ matrix), as the $m \times p$ matrix (c_{ik}) where

$$c_{ik} = \sum_{j=1}^{n} a_{ij} b_{jk}.$$

This definition is motivated by the need to form the composite linear transformation $s \circ t \in \mathrm{Hom}(U, W)$, given $t \in \mathrm{Hom}(U, V)$ corresponding to the matrix B, and $s \in \mathrm{Hom}(V, W)$ corresponding to the matrix A. The matrix product $A * B$ of the matrices A and B is not defined unless the number of columns of A is equal to the number of rows of B (is equal to the dimension of V). The existence of the product $A * B$ will not therefore imply the existence of the product $B * A$. With this definition of multiplication, the set of square matrices of order n, i.e. those matrices having n rows and n columns with coefficients in a ring K, form a non-commutative ring (provided $n > 1$ and $K \neq \{0\}$) denoted by $M_n(K)$. A matrix $A \in M_n(K)$ is said to be *non-singular* or *invertible* if there exists $B \in M_n(K)$ such that

$$A * B = B * A = I_n$$

where I_n is the $n \times n$ *identity matrix* (also called the *unit matrix*) of order n having coefficients δ_{ij} (called the *Kronecker delta*), where

$$\delta_{ij} := \begin{cases} 1 \text{ if } & i = j \\ 0 \text{ otherwise.} \end{cases}$$

The matrix B is then unique and is known as the *inverse* of A; it is denoted by A^{-1}. If there is no matrix B in $M_n(K)$ such that $A * B = B * A = I_n$, then A is said to be *singular* or *non-invertible* in $M_n(K)$. A linear mapping $U \to V$ will be an isomorphism of vector spaces if and only if it can be represented by an invertible (square) matrix. The *transpose* of $m \times n$ matrix $A = (a_{ij})$ is defined as the $n \times m$ matrix (a_{ij}) obtained from A by interchanging rows and columns. The transpose of A is denoted by A^T. When $A = A^T$ the matrix A is said to be *symmetric* with the underlying field $F = \mathbf{R}$.

For each $n > 0$ the set of non-singular $n \times n$ matrices over the field F forms a multiplicative group, called the *general linear group* $GL(n, F)$. The elements of $GL(n, F)$ having determinant 1 form a subgroup denoted by $SL(n, F)$ and are known as the *special linear group*.

An $m \times 1$ matrix having only one column is known as a column matrix or column vector, a $1 \times n$ matrix is known as a row matrix or row vector.

The equations which determine the homomorphism t with matrix A can therefore be written as

$$\mathbf{x}' = A * \mathbf{x}$$

where $\mathbf{x}'$ is a column vector whose m elements are the components of $t(\mathbf{x})$ with respect to $\mathbf{v}_1, \ldots, \mathbf{v}_m$ and $\mathbf{x}$ is a column vector having n elements, the components of $\mathbf{x}$ with respect to $\mathbf{u}_1, \ldots, \mathbf{u}_n$.

In particular, a $1 \times n$ matrix will describe a homomorphism from an n-dimensional vector space $\mathbf{u}$ with basis $\mathbf{u}_1, \ldots, \mathbf{u}_n$ to a one-dimensional vector space V with basis $\mathbf{v}_1$. The set of all $1 \times n$ matrices will form a vector space isomorphic to the vector space $\text{Hom}(U, F)$ of all homomorphisms mapping U onto its ground field F, which can be regarded as a vector space of dimension 1 over itself. In general, if V is any vector space over a field F, then the vector space $\text{Hom}(V, F)$ is called the *dual space* of V and is denoted by V^* or $\hat{V}$. Elements of $\text{Hom}(V, F)$ are known as *linear functionals*. It can be shown that every finite-dimensional vector space is isomorphic to its dual.

Example. The *trace* of a square matrix is the sum of the diagonal elements. Thus the trace of a square matrix is a linear functional. This means

$$\mathrm{tr}(A + B) = \mathrm{tr}(A) + \mathrm{tr}(B)$$
$$\mathrm{tr}(c * A) = c * \mathrm{tr}(A), \qquad c \in F.$$

Given two subspaces S and T of a vector space V, we define

$$S + T := \{\, x + y \mid x \in S,\, y \in T \,\}.$$

Then $S + T$ is a subspace of V. We say that V is the *direct sum* of S and T, written $S \oplus T$, if and only if $V = S + T$ and $S \cap T = \{0\}$. S and T are then called direct summands of V. Any subspace S of a finite-dimensional vector space V is a direct summand of V. Moreover, if $V = S \oplus T$, then

$$\dim T = \dim V - \dim S.$$

If S and T are any finite-dimensional subspaces of a vector space V, then

$$\dim S + \dim T = \dim(S \cap T) + \dim(S + T).$$

Let A be an $n \times n$ matrix over $\mathbf{C}$. Then the complex number $\lambda \in \mathbf{C}$ is called an *eigenvalue* of A if and only if the matrix $(A - \lambda I_n)$ is singular. Let $\mathbf{x}_\lambda$ be a non-zero column vector in $\mathbf{C}^n$. Then $\mathbf{x}_\lambda$ is called an *eigenvector* associated with the eigenvalue λ if and only if $(A - \lambda I_n)\mathbf{x}_\lambda = \mathbf{0}$. This equation can be written in the form

$$A\mathbf{x}_\lambda = \lambda \mathbf{x}_\lambda$$

is called the *eigenvalue equation.*

2.8   Determinants

Now, we consider some practical methods to evaluate the determinant of a matrix. The method employed usually depends on the nature of the matrix — numeric or symbolic.

- **Numeric matrix**

  Let A be an $n \times n$ matrix. The *determinant* of A is the sum taken over all permutations of the columns of the matrix, of the products of elements appearing on the principal diagonal of the permutated matrix. The sign with which each of these terms is added to the sum is positive or negative according to whether the permutation of the column is even or odd.

  An obvious case is a matrix consisting of only a single element, for which we specify that

  $$\det(A) = a_{11}, \quad \text{when } n = 1.$$

  No one computes a determinant of a matrix larger that 4×4 by generating all permutations of the columns and evaluating the products of the diagonals. A more efficient way rests on the following facts

- Adding a numerical multiple of one row (or column) of matrix A to another leaves $\det(A)$ unchanged.

- If B is obtained from A by exchanging two rows (or two columns), then $\det(B) = -\det(A)$. If A has two rows or columns proportional to each other then $\det(A) = 0$.

The idea is to manipulate the matrix A with the help of these two operations in such a way that it becomes triangular, i.e. all matrix elements below the principle diagonal are equal to zero. It follows from the definition that the determinant of a triangular matrix is the product of the elements of the principal diagonal.

- **Symbolic matrix**
  For a symbolic matrix, the determinant is best evaluated using *Leverrier's method*.

The *characteristic polynomial* of an $n \times n$ matrix A is a polynomial of degree n in terms of λ. It may be written as

$$P(\lambda) = \det(\lambda I_n - A) = \lambda^n - c_1\lambda^{n-1} - c_2\lambda^{n-2} - \cdots - c_{n-1}\lambda - c_n$$

where I_n denotes the $n \times n$ identity matrix. The *characteristic equation* is

$$P(\lambda) = \det(\lambda I_n - A) = \lambda^n - c_1\lambda^{n-1} - c_2\lambda^{n-2} - \cdots - c_{n-1}\lambda - c_n = 0.$$

The *Cayley-Hamilton theorem* states that A satisfies its characteristic equation $P(A) = 0_n$, where 0_n is the $n \times n$ zero matrix, i.e.

$$A^n - c_1 A^{n-1} - c_2 A^{n-2} - \cdots - c_{n-1}A - c_n I_n = 0.$$

Horner's rule allows us to write this equation in the form

$$A(\cdots A(\underbrace{A(\overbrace{A(A - c_1 I_n)}^{B_2} - c_2 I_n)}_{B_3} - c_3 I_n) - c_4 I_n) - \cdots - c_{n-1}I_n) - c_n I_n = 0.$$

$$\underbrace{\phantom{A(\cdots A(A(A(A - c_1 I_n) - c_2 I_n) - c_3 I_n) - c_4 I_n) - \cdots - c_{n-1}I_n)}}_{B_n}$$

Now, we present Leverrier's method to find the coefficients, c_i, of the characteristic polynomial. It is fairly insensitive to the individual peculiarities of the matrix A. The method has an added advantage that the inverse of A, if it exists, is also obtained in the process of determining the coefficients, c_i. Obviously we also obtain the determinant. The coefficients, c_i, of $P(\lambda)$ are obtained by evaluating the trace of each of the matrices, $B_1, B_2, \ldots, B_n$, generated as follows. Set $B_1 = A$ and compute $c_1 = \text{tr}(B_1)$, where tr denotes the trace. Then compute

$$B_k = A(B_{k-1} - c_{k-1}I_n), \quad c_k = \left(\frac{1}{k}\right)\text{tr}(B_k), \quad k = 2, 3, \ldots, n.$$

Since

$$B_n = A(B_{n-1} - c_{n-1}I_n) = c_nI_n$$

the inverse of a non-singular matrix A can be obtained from the relationship

$$A^{-1} = \left(\frac{1}{c_n} \right) (B_{n-1} - c_{n-1}I_n)$$

and the determinant of the matrix is

$$\begin{cases} c_n & \text{if } n \text{ is odd} \\ -c_n & \text{if } n \text{ is even.} \end{cases}$$

The Cayley-Hamilton theorem can also be used to calculate $\exp(A)$ and other entire functions for an $n \times n$ matrix A. The function $\exp(A)$ appears in the identity

$$\det(\exp(A)) \equiv \exp(\text{tr}(A)).$$

Let A be an $n \times n$ matrix over **C**. Let f be an *entire function*, i.e., an analytic function on the whole complex plane, for example $\exp(z)$, $\sin(z)$, $\cos(z)$. An infinite series expansion for $f(A)$ is not generally useful for computing $f(A)$. Using the *Cayley-Hamilton theorem* we can write

$$f(A) = a_{n-1}A^{n-1} + a_{n-2}A^{n-2} + \cdots + a_2A^2 + a_1A + a_0I_n \tag{1}$$

where the complex numbers $a_0, a_1, \ldots, a_{n-1}$ are determined as follows:
Let

$$r(\lambda) := a_{n-1}\lambda^{n-1} + a_{n-2}\lambda^{n-2} + \cdots + a_2\lambda^2 + a_1\lambda + a_0$$

which is the right hand side of (1) with A^j replaced by λ^j, where $j = 0, 1, \ldots, n-1$. For each distinct eigenvalue λ_j of the matrix A, we consider the equation

$$f(\lambda_j) = r(\lambda_j). \tag{2}$$

If λ_j is an eigenvalue of multiplicity k, for $k > 1$, then we consider also the following equations

$$f'(\lambda)|_{\lambda=\lambda_j} = r'(\lambda)|_{\lambda=\lambda_j}$$
$$f''(\lambda)|_{\lambda=\lambda_j} = r''(\lambda)|_{\lambda=\lambda_j}$$
$$\cdots = \cdots$$
$$f^{(k-1)}(\lambda)\Big|_{\lambda=\lambda_j} = r^{(k-1)}(\lambda)\Big|_{\lambda=\lambda_j}.$$

Example. We apply this technique to find $\exp(A)$ with

$$A = \begin{pmatrix} c & c \\ c & c \end{pmatrix}, \qquad c \in \mathbf{R}, \quad c \neq 0.$$

We have

$$e^A = a_1A + a_0I_2 = c \begin{pmatrix} a_1 & a_1 \\ a_1 & a_1 \end{pmatrix} + \begin{pmatrix} a_0 & 0 \\ 0 & a_0 \end{pmatrix} = \begin{pmatrix} a_0 + ca_1 & ca_1 \\ ca_1 & a_0 + ca_1 \end{pmatrix}.$$

The eigenvalues of A are 0 and $2c$. Thus we obtain the two linear equations

$$e^0 = 0a_1 + a_0 = a_0$$
$$e^{2c} = 2ca_1 + a_0.$$

Solving these two linear equations yields

$$a_0 = 1, \qquad a_1 = \frac{e^{2c} - 1}{2c}.$$

It follows that

$$e^A = \begin{pmatrix} a_0 + ca_1 & ca_1 \\ ca_1 & ca_1 + a_0 \end{pmatrix} = \begin{pmatrix} (e^{2c} + 1)/2 & (e^{2c} - 1)/2 \\ (e^{2c} - 1)/2 & (e^{2c} + 1)/2 \end{pmatrix}.$$

2.9 Quaternions

By defining multiplication suitably on $\mathbf{R} \times \mathbf{R}$, it is possible to construct a field $\mathbf{C}$ which is an extension of $\mathbf{R}$. Indeed, since $\mathbf{C}$ is a vector space of dimension 2 over $\mathbf{R}$, $\mathbf{C}$ is a commutative algebra with unity element over $\mathbf{R}$. It is natural to attempt to repeat this process and to try to embed $\mathbf{C}$ in an algebra defined upon $\mathbf{R}^n$ ($n > 2$). It is impossible to find such an extension satisfying the field axioms, but, as the following construction shows, some measure of success can be attained.

Consider an associative algebra of rank 4 with the basis elements

$$1, \quad I, \quad J, \quad K$$

where 1 is the identity element, i.e.

$$1 * I = I, \qquad 1 * J = J, \qquad 1 * K = K.$$

The compositions are

$$I * I = J * J = K * K = -1$$

and

$$I * J = K, \quad J * K = I, \quad K * I = J, \quad J * I = -K, \quad K * J = -I, \quad I * K = -J.$$

This is the so-called *quaternion algebra*. Multiplication, as thus defined, is non-commutative, so the resulting structure cannot be a field. It is a division ring and is known as the quaternion algebra. The algebra is associative.

Any quaternion q can be represented in the form

$$q := a_1 * 1 + a_I * I + a_J * J + a_K * K, \qquad \text{where} \quad a_1, a_I, a_J, a_K \in F.$$

The sum, difference, product and division of two quaternions

$$q := a_1 * 1 + a_I * I + a_J * J + a_K * K, \qquad p := b_1 * 1 + b_I * I + b_J * J + b_K * K$$

are defined as

$$q + p := (a_1 + b_1) * 1 + (a_I + b_I) * I + (a_J + b_J) * J + (a_K + b_K) * K$$
$$q - p := (a_1 - b_1) * 1 + (a_I - b_I) * I + (a_J - b_J) * J + (a_K - b_K) * K$$
$$q * p := (a_1 * 1 + a_I * I + a_J * J + a_K * K) * (b_1 * 1 + b_I * I + b_J * J + b_K * K)$$
$$= (a_1 * b_1 - a_I * b_I - a_J * b_J - a_K * b_K) * 1$$
$$+ (a_1 * b_I + a_I * b_1 + a_J * b_K - a_K * b_J) * I$$
$$+ (a_1 * b_J + a_J * b_1 + a_K * b_I - a_I * b_K) * J$$
$$+ (a_1 * b_K + a_K * b_1 + a_I * b_J - a_J * b_I) * K$$
$$q/p := q * p^{-1}$$

where p^{-1} is the inverse of p. The negative of q is

$$-q = -a_1 * 1 - a_I * I - a_J * J - a_K * K.$$

The conjugate of q, say q^*, is defined as

$$q^* := a_1 * 1 - a_I * I - a_J * J - a_K * K.$$

The inverse of q is

$$q^{-1} := \frac{q^*}{|q|^2}, \qquad q \neq 0.$$

The magnitude of q is

$$|q|^2 = a_1^2 + a_I^2 + a_J^2 + a_K^2.$$

The normalization of q is defined as

$$q/|q|.$$

A *matrix representation* of the quaternions is given by

$$1 \to \begin{pmatrix} 1 & 0 \\ 0 & 1 \end{pmatrix}$$

$$I \to -i \begin{pmatrix} 0 & 1 \\ 1 & 0 \end{pmatrix} \equiv -i\sigma_x$$

$$J \to -i \begin{pmatrix} 0 & -i \\ i & 0 \end{pmatrix} \equiv -i\sigma_y$$

$$K \to -i \begin{pmatrix} 1 & 0 \\ 0 & -1 \end{pmatrix} \equiv -i\sigma_z$$

where $i := \sqrt{-1}$ and σ_x, σ_y, σ_z are the *Pauli spin matrices*. This also shows that the quaternion algebra is associative.

The quaternion algebra can also be obtained as the subring of $M_4(\mathbf{R})$ consisting of matrices of the form

$$\begin{pmatrix} x & -y & -z & -t \\ y & x & -t & z \\ z & t & x & -y \\ t & -z & y & x \end{pmatrix}.$$

The quaternion algebra can be considered as a subcase of the octonian algebra. The *octonian algebra* $\mathcal{O}$ is an 8-dimensional non-associative algebra defined in terms of the basis elements $\{e_0, e_1, e_2, e_3, e_4, e_5, e_6, e_7\}$, where e_0 is the so called unit element. Addition of two elements of the algebra is defined in the usual way

$$(a_0 e_0 + \cdots + a_7 e_7) + (b_0 e_0 + \cdots + b_7 e_7) := (a_0 + b_0) e_0 + \cdots + (a_7 + b_7) e_7$$

where $a_0, \ldots, a_7, b_0, \ldots, b_7 \in \mathbf{R}$. Multiplication is defined in terms of the basis elements and distributivity. The multiplication rules are

$$\text{for all } a \in \mathcal{O} \quad e_0 a = a e_0 = a.$$

$$e_j e_4 = -e_4 e_j = e_{j+4}, \qquad e_4 e_{j+4} = -e_{j+4} e_4 = e_j, \qquad e_4 e_4 = -e_0$$

$$e_j e_k = -\delta_{jk} e_0 + \sum_{l=1}^{3} \epsilon_{jkl} e_l$$

$$e_{j+4} e_{k+4} - -\delta_{jk} e_0 - \sum_{l-1}^{3} \epsilon_{jkl} e_l$$

$$e_j e_{k+4} = -e_{k+4} e_j = -\delta_{jk} e_4 - \sum_{l=1}^{3} \epsilon_{jkl} e_{l+4}$$

where $j, k = 1, 2, 3$ and

$$\epsilon_{jkl} := \begin{cases} 0 & j = k \text{ or } k = l \text{ or } l = j \\ +1 & (j, k, l) \in \{(1,2,3), (2,3,1), (3,1,2)\} \\ -1 & (j, k, l) \in \{(1,3,2), (3,2,1), (2,1,3)\} \end{cases}$$

is the *permutation symbol*, also known as the *Levi-Civita symbol*. These definitions define a closed associative subalgebra over $\{e_0, e_1, e_2, e_3\}$ which is the quaternion algebra $\mathcal{Q}$. Defining $\hat{\mathcal{Q}} := e_4 \mathcal{Q} = \{e_4 e_0, e_4 e_1, e_4 e_2, e_4 e_3\}$ it follows that $\mathcal{O} = \mathcal{Q} \oplus \hat{\mathcal{Q}}$.

2.10 Polynomials

In this section we introduce polynomials. For the proof of the theorems we refer to the literature [4], [27], [33]. Functions of the form

$$1 + 2 * x + 3 * x^2, \qquad x + x^5, \qquad \frac{1}{5} - 4 * x^2 + \frac{3}{2} * x^{10}$$

are called *polynomials* in x. The coefficients in these examples are integers and rational numbers. In elementary calculus, the range of values of x (domain of definition of the function) is $\mathbf{R}$. In algebra, the range is $\mathbf{C}$. Consider, for instance, the polynomial $p(x) = x^2 + 1$. The solution of $p(x) = 0$ is given by $\pm i$. Any polynomial in x can be thought of as a mapping of a set S (range of x) onto a set T (range of values of the polynomial). Consider, for example, the polynomial $1 + \sqrt{2} * x - 3 * x^2$. If $S = \mathbf{Z}$, then $T \subset \mathbf{R}$ and the same is true if $S = \mathbf{Q}$ or $S = \mathbf{R}$; if $S = \mathbf{C}$, then

$T \subset \mathbf{C}$. Two polynomials in x are equal if they have identical form. For example, $a + b * x = c + d * x$ if and only if $a = c$ and $b = d$.

Let R be a ring and let x, called an *indeterminate*, be any symbol not found in R. By a polynomial in x over R will be meant any expression of the form

$$\alpha(x) = a_0 * x^0 + a_1 * x^1 + a_2 * x^2 + \cdots = \sum_{k=0} a_k * x^k, \qquad a_k \in R$$

in which only a finite number of the a_k's are different from z, the zero element of R. Two polynomials in x over R, $\alpha(x)$ defined above, and

$$\beta(x) = b_0 * x^0 + b_1 * x^1 + b_2 * x^2 + \cdots = \sum_{k=0} b_k * x^k, \qquad b_k \in R$$

are equal $\alpha(x) = \beta(x)$, provided $a_k = b_k$ for all values of k.

In any polynomial, as $\alpha(x)$, each of the components $a_0 * x^0$, $a_1 * x^1$, $a_2 * x^2, \ldots$ is called a *term*; in any term such as $a_i * x^i$, a_i is called the *coefficient* of the term. The terms of $\alpha(x)$ and $\beta(x)$ have been written in a prescribed (but natural) order. The i, the superscript of x, is merely an indicator of the position of the term $a_i * x^i$ in the polynomial. Likewise, juxtaposition of a_i and x^i in the term $a_i * x^i$ is not to be construed as indicating multiplication and the plus signs between terms are to be thought of as helpful connectives rather than operators. Let z be the zero element of the ring. If in a polynomial such as $\alpha(x)$, the coefficient $a_n \neq z$ while all coefficients of terms which follow are z, we say that $\alpha(x)$ is of *degree n* and call a_n its leading coefficient. In particular, the polynomial $a_0 * x^0 + z * x^1 + z * x^2 + \cdots$ is of degree zero with leading coefficient a_0 when $a_0 \neq z$ and it has no degree (and no leading coefficient) when $a_0 = z$.

Denote by $R[x]$ the set of all polynomials in x over R and, for arbitrary $\alpha(x), \beta(x) \in R[x]$, define addition $(+)$ and multiplication $*$ on $R[x]$ by

$$\alpha(x) + \beta(x) := (a_0 + b_0) * x^0 + (a_1 + b_1) * x^1 + (a_2 + b_2) * x^2 + \cdots$$
$$= \sum_{k=0} (a_k + b_k) * x^k$$

and

$$\alpha(x) * \beta(x) := a_0 * b_0 * x^0 + (a_0 * b_1 + a_1 * b_0) * x^1$$
$$+ (a_0 * b_2 + a_1 * b_1 + a_2 * b_0) * x^2 + \cdots$$
$$= \sum_{k=0} c_k * x^k$$

where

$$c_k := \sum_{i=0}^{k} a_i * b_{k-i} .$$

The sum and product of elements of $R[x]$ are elements of $R[x]$; there are only a finite number of terms with non-zero coefficients $\in R$. Addition on $R[x]$ is both associative and commutative and multiplication is associative and distributive with respect to addition. Moreover, the *zero polynomial*

$$z * x^0 + z * x^1 + z * x^2 + \cdots = \sum_{k=0} z * x^k \in R[x]$$

is the additive identity or zero element of $R[x]$ while

$$-\alpha(x) = -a_0 * x^0 + (-a_1) * x^1 + (-a_2) * x^2 + \cdots = \sum_{k=0} (-a_k) * x^k \in R[x]$$

is the additive inverse of $\alpha(x)$. Thus,

Theorem. The set of all polynomials R in x over R is a ring with respect to addition and multiplication as defined above.

Let $\alpha(x)$ and $\beta(x)$ have respective degrees m and n. If $m \neq n$, the degree of $\alpha(x) + \beta(x)$ is the larger of m, n; if $m = n$, the degree of $\alpha(x) + \beta(x)$ is at most m. The degree of $\alpha(x) * \beta(x)$ is at most $m + n$ since $a_m b_n$ may be z. However, if R is free of divisors of zero, the degree of the product is $m + n$.

Karatsuba multiplication. The *Karatsuba multiplication algorithm* [21][62] provides an efficient method for multiplying two polynomials. Suppose $\alpha(x)$ and $\beta(x)$ have degree less than 2^n, i.e.

$$\alpha(x) = a_0 + a_1 x + \cdots + a_{2^n-1} x^{2^n-1}, \qquad \beta(x) = b_0 + b_1 x + \cdots + b_{2^n-1} x^{2^n-1}.$$

We rewrite $\alpha(x)$ using

$$A_0 := a_0 + a_1 x + \cdots + a_{2^{n-1}-1} x^{2^{n-1}-1}$$
$$A_1 := a_{2^{n-1}} + a_{2^{n-1}+1} x + \cdots + a_{2^n-1} x^{2^{n-1}-1}$$

so that

$$\alpha(x) = A_0 + x^{2^{n-1}} A_1.$$

Similarly we use

$$\beta(x) = B_0 + x^{2^{n-1}} B_1.$$

Then

$$\begin{aligned}
\alpha(x) * \beta(x) &= (A_0 + x^{2^{n-1}} A_1) * (B_0 + x^{2^{n-1}} B_1) \\
&= A_0 B_0 + (A_0 B_1 + A_1 B_0) x^{2^{n-1}} + A_1 B_1 x^{2^n} \\
&= A_0 B_0 + [(A_0 + A_1)(B_0 + B_1) - A_0 B_0 - A_1 B_1] x^{2^{n-1}} + A_1 B_1 x^{2^n}.
\end{aligned}$$

Thus we recursively compute the three products $A_0 B_0$, $(A_0 + A_1)(B_0 + B_1)$ and $A_1 B_1$ using the Karatsuba multiplication algorithm. Each product is a product of polynomials of degree less than 2^{n-1}, thus the problem reduces exponentially. The base case is $n = 0$ where $\alpha(x) * \beta(x) = a_0 * b_0$.

Consider now the subset $S := \{r * x^0 \ : \ r \in R\}$ of $R[x]$ consisting of the zero polynomial and all polynomials of degree zero. The mapping

$$R \to S : r \to r * x^0$$

is an isomorphism. As a consequence, we may hereafter write a_0 for $a_0 * x^0$ in any polynomial $\alpha(x) \in R[x]$.

Let R be a ring with unity u. Then $u = u * x^0$ is the unity of $R[x]$ since $ux^0 * \alpha(x) = \alpha(x)$ for every $\alpha(x) \in R[x]$. Also, writing $x = u * x^1 = z * x^0 + u * x^1$, we have $x \in R[x]$. Now $a_k(x \cdot x \cdot x \cdot$ to k factors$) = a_k * x^k \in R[x]$ so that in $\alpha(x) = a_0 + a_1 * x + a_2 * x^2 + \cdots$ we may consider the superscript i and $a_i x^i$ as truly an exponent, juxtaposition in any term $a_i * x^i$ as (polynomial) ring multiplication, and the connective $+$ as (polynomial) ring addition. Any polynomial $\alpha(x)$ of degree m over R with leading coefficient u, the unity of R, will be called monic.

Theorem. Let R be a ring with unity u, $\alpha(x) = a_0 + a_1 * x + \cdots + a_m * x^m \in R[x]$ be either the zero polynomial or a polynomial of degree m, and $\beta(x) = b_0 + b_1 * x + \cdots + u * x^n \in R[x]$ be a monic polynomial of degree n. Then there exist unique polynomials $q_R(x)$, $r_R(x)$, $q_L(x)$, $r_L(x) \in R[x]$ with $r_R(x), r_L(x)$ either the zero polynomial or of degree $< n$ such that

$$\text{(i)} \quad \alpha(x) = q_R(x) * \beta(x) + r_R(x)$$

and

$$\text{(ii)} \quad \alpha(x) = \beta(x) * q_L(x) + r_L(x).$$

In (i) of the theorem we say that $\alpha(x)$ has been divided on the right by $\beta(x)$ to obtain the right quotient $q_R(x)$ and right remainder $r_R(x)$. Similarly, in (ii) we say that $\alpha(x)$ has been divided on the left by $\beta(x)$ to obtain the left quotient $q_L(x)$ and left remainder $r_L(x)$. When $r_R(x) = z$ ($r_L(x) = z$), we call $\beta(x)$ a right (left) divisor of $\alpha(x)$.

We consider now commutative polynomial rings with unity. Let R be a commutative ring with unity. Then $R[x]$ is a commutative ring with unity and the theorem may be restated without distinction between right and left quotients (we replace $q_R(x) = q_L(x)$ by $q(x)$), remainders (we replace $r_R(x) = r_L(x)$ by $r(x)$), and divisors. Thus (i) and (ii) of the theorem may be replaced by

$$\text{(iii)} \quad \alpha(x) = q(x) * \beta(x) + r(x)$$

and, in particular, we have

Theorem. In a commutative polynomial ring with unity, a polynomial $\alpha(x)$ of degree m has $x - b$ as divisor if and only if the remainder

$$r = a_0 + a_1 * b + a_2 * b^2 + \cdots + a_m * b^m = z.$$

When $r = z$ then b is called a zero (root) of the polynomial $\alpha(x)$.

We will use the notation $\alpha(x) \equiv r(x) \bmod \beta(x)$ when $\alpha(x) = q(x) * \beta(x) + r(x)$ and $\alpha_1(x) \equiv \alpha_2(x) \bmod \beta(x)$ whenever both $\alpha_1(x) \equiv r(x) \bmod \beta(x)$ and $\alpha_2(x) \equiv r(x) \bmod \beta(x)$ for some $r(x)$.

When R is without divisors of zero so is $R[x]$. For suppose $\alpha(x)$ and $\beta(x)$ are elements of $R[x]$, of respective degrees m and n, and that

$$\alpha(x) * \beta(x) = a_0 * b_0 + (a_0 * b_1 + a_1 * b_0)x + \cdots + a_m * b_n x^{m+n} = z \,.$$

Then each coefficient in the product and, in particular $a_m * b_n$ is z. But R is without divisors of zero; hence $a_m b_n = z$ if and only if $a_m = z$ or $b_n = z$. Since this contradicts the assumption that $\alpha(x)$ and $\beta(x)$ have degrees m and n, $R[x]$ is without divisors of zero.

Theorem. A polynomial ring $R[x]$ is an integral domain if and only if the coefficient ring R is an integral domain.

An examination of the remainder

$$r = a_0 + a_1 * b + a_2 * b^2 + \cdots + a_m * b^m$$

shows that it may be obtained mechanically by replacing x by b throughout $\alpha(x)$ and, of course, interpreting juxtaposition of elements as indicating multiplication in R. Thus, by defining $f(b)$ to mean the expression obtained by substituting b for x throughout $f(x)$, we may replace r by $\alpha(b)$. This is the familiar substitution process in elementary algebra where x is considered as a variable rather than an indeterminate. For a given $b \in R$, the mapping

$$f(x) \rightarrow f(b) \qquad \text{for all } f(x) \in R[x]$$

is a homomorphism of $R[x]$ onto R.

The most important polynomial domains arise when the coefficient ring is a field F. Every non-zero element of a field F is a unit of F. For the integral domain $F[x]$ the principal results are as follows

Division Algorithm. If $\alpha(x)$, $\beta(x) \in F[x]$ where $\beta(x) \neq z$, there exist unique polynomials $q(x)$, $r(x)$ with $r(x)$ either the zero polynomial or of degree less than that of $\beta(x)$, such that

$$\alpha(x) = q(x) * \beta(x) + r(x).$$

When $r(x)$ is the zero polynomial, $\beta(x)$ is called a divisor of $\alpha(x)$ and we write $\beta(x)|\alpha(x)$. It follows that we can find a greatest common divisor of two polynomials, in the sense of a common divisor of both polynomials with the largest degree, in using the same method that we use for integers.

Remainder Theorem. If $\alpha(x)$, $x - b \in F[x]$, the remainder when $\alpha(x)$ is divided by $x - b$ is $\alpha(b)$.

Division can be performed using the *long division algorithm.*

Algorithm for long division. Suppose $\alpha(x)$ is of degree m and $\beta(x)$ is of degree n where $m \geq n$ then

$$\frac{\alpha(x)}{\beta(x)} = \frac{a_0 + a_1 x + \cdots + a_m x^m}{b_0 + b_1 x + \cdots + b_n x^n} = \frac{a_m}{b_n} x^{m-n} + \frac{\alpha(x) - \frac{a_m}{b_n} x^{m-n} \beta(x)}{\beta(x)}$$

$$= q_1(x) + \frac{\alpha_1(x)}{\beta(x)}$$

where

$$q_1(x) := \frac{a_m}{b_n} x^{m-n}$$

and

$$\alpha_1(x) := \alpha(x) - q_1(x)\beta(x)$$

has degree less than that of $\alpha(x)$. We continue this process

$$\frac{\alpha_j(x)}{\beta(x)} = q_{j+1}(x) + \frac{\alpha_{j+1}(x)}{\beta(x)}$$

until $\alpha_{j+1}(x)$ is of degree less than n. Then we have

$$q(x) = q_1(x) + q_2(x) + \cdots + q_{j+1}(x)$$

and

$$r(x) = \alpha_{j+1}(x).$$

Alternatively we can use the *Newton iteration* for division [21][62]. The Newton iteration is a numerical technique [55] for approximating solutions to equations of the form $f(x) = 0$ using an initial value x_0. The iteration for successive approximations for a solution is given by

$$x_{j+1} = x_j - \frac{f(x_j)}{f'(x_j)}, \quad j = 0, 1, 2, \ldots$$

where $f'(x)$ is the derivative of $f(x)$. The technique may or may not converge depending on the initial value x_0. For $\alpha(x)/\beta(x)$ we have at most degree $n - m$, thus we work under mod x^{n-m+1}. Let

$$f(y) := \beta(x) - \frac{1}{y},$$

then $f(\beta^{-1}(x)) = 0$. The Newton iteration for y yields

$$y_{j+1} = y_j - \frac{f(y_j)}{f'(y_j)} = y_j - \frac{\beta(x) - \frac{1}{y_j}}{\frac{1}{y_j^2}} = 2y_j - \beta(x)y_j^2.$$

To apply the iteration we require $b_0 = 1$. If $b_0 \neq 1$ and $b_0 \neq 0$ we can use

$$\beta^{-1}(x) = \frac{1}{b_0} \left(\frac{\beta(x)}{b_0} \right)^{-1}$$

where the inverse polynomial on the right hand side has a constant term of 1. If $b_0 = 0$ we can use

$$\beta^{-1}(x) = \frac{1}{x^l} \left(\frac{\beta(x)}{x^l} \right)^{-1}$$

where l is the lowest power of x appearing in $\beta(x)$, and the inverse polynomial on the right hand side falls under the case $b_0 \neq 0$. A different approach for $b_0 = 0$ is to rewrite

$$\alpha(x) = q(x)\beta(x) + r(x)$$

as

$$x^n \alpha(1/x) = \left[x^{n-m} q(1/x) \right] \left[x^m \beta(1/x) \right] + x^{n-m+1} \left[x^{m-1} r(x) \right]$$

where $\mathrm{rev}(\alpha(x)) := x^n \alpha(x)$ reverses the order of the coefficients in $\alpha(x)$. It follows that

$$\mathrm{rev}(\alpha(x)) = \mathrm{rev}(q(x)) \, \mathrm{rev}(\beta(x)) + x^{n-m+1} \, \mathrm{rev}(r(x))$$

or equivalently

$$\mathrm{rev}(\alpha(x)) = \mathrm{rev}(q(x)) \, \mathrm{rev}(\beta(x)) \bmod x^{n-m+1}.$$

Thus we must find $\mathrm{rev}(\beta(x))^{-1} \bmod x^{n-m+1}$ which is described again by the case $b_0 \neq 0$ above. Finally we must reverse $\mathrm{rev}(\alpha(x))/\mathrm{rev}(\beta(x)) \bmod x^{n-m+1}$ to obtain $q(x)$.

Now we consider the case $b_0 = 1$. We choose the initial value $y_0 = 1$, and use the following iteration scheme.

Newton iteration for inversion. The iteration is given by

$$y_{j+1} = 2y_j - \beta(x)y_j^2 \bmod x^{2^{j+1}}, \quad j = 0, 1, 2, \ldots, \lceil \log_2(n - m + 1) \rceil.$$

This iteration yields y_j as the inverse of $\beta(x)$ under $\bmod \ x^{n-m+1}$ when $j = \lceil \log_2(n - m + 1) \rceil$. The proof follows by induction. From $y_0 = b_0 = 1$ we have

$$y_0 \beta(x) \equiv y_0 b_0 + (y_0 b_1 + y_1 b_0)x + \cdots \bmod x^{2^0} \equiv 1 \bmod x^{2^0}$$

or $1 - y_0 \beta(x) \equiv 0 \bmod x^{2^0}$. Now suppose

$$1 - y_j \beta(x) \equiv 0 \bmod x^{2^j},$$

i.e. $1 - y_j \beta(x) = cx^{2^j} + \cdots$ where c is some constant. Then it follows that

$$1 - y_{j+1}\beta(x) \equiv 1 - (2y_j - \beta(x)y_j^2)\beta(x) \equiv (1 - \beta(x)y_j)^2 \equiv 0 \bmod x^{2^{j+1}}$$

so that $y_{j+1}\beta(x) \equiv 1 \bmod x^{2^{j+1}}$.

Factor Theorem. If $\alpha(x) \in F[x]$ and $b \in F$, then $x - b$ is a *factor* of $\alpha(x)$ if and only if $\alpha(b) = z$, that is, $x - b$ is a factor of $\alpha(x)$ if and only if b is a zero of $\alpha(x)$. This leads to the following theorem.

Theorem. Let $\alpha(x) \in F[x]$ have degree $m > 0$ and leading coefficient a. If the distinct elements $b_1, b_2, \ldots, b_m$ of F are zeros of $\alpha(x)$, then

$$\alpha(x) = a * (x - b_1) * (x - b_2) * \cdots * (x - b_m).$$

Theorem. Every polynomial $\alpha(x) \in F[x]$ of degree $m > 0$ has at most m distinct zeros in F.

Theorem. Let $\alpha(x)$, $\beta(x) \in F[x]$ be such that $\alpha(s) = \beta(s)$ for every $s \in F$. Then, if the number of elements in F exceeds the degrees of both $\alpha(x)$ and $\beta(x)$, we have necessarily $\alpha(x) = \beta(x)$.

The only units of a polynomial domain $F[x]$ are the non-zero elements (i.e., the units) of the coefficient ring F. Thus the only associates of $\alpha(x) \in F[x]$ are the elements $v * \alpha(x)$ of $F[x]$ in which v is any unit of F. Since for any $v \neq z \in F$ and any $\alpha(x) \in F[x]$,

$$\alpha(x) = v^{-1} * \alpha(x) * v$$

while, whenever $\alpha(x) = q(x) * \beta(x)$,

$$\alpha(x) = \left(v^{-1} * q(x)\right) * \left(v * \beta(x)\right)$$

it follows that (a) every unit of F and every associate of $\alpha(x)$ is a divisor of $\alpha(x)$ and (b) if $\beta(x)|\alpha(x)$ so also does every associate of $\beta(x)$. The units of F and the associates of $\alpha(x)$ are called trivial divisors of $\alpha(x)$. Other divisors of $\alpha(x)$, if any, are called non-trivial divisors. A polynomial $\alpha(x) \in F[x]$ of degree $m \geq 1$ is called a prime (irreducible) polynomial over F if its divisors are all trivial.

Next we consider the polynomial domain $\mathbf{C}[x]$. Consider an arbitrary polynomial

$$\beta(x) = b_0 + b_1 * x + b_2 * x^2 + \cdots + b_m * x^m \in \mathbf{C}[x]$$

of degree $m \geq 1$. We give a number of elementary theorems related to the zeros of such polynomials and, in particular, with the subset of all polynomials of $\mathbf{C}[x]$ whose coefficients are rational numbers. Suppose $r \in \mathbf{C}$ is a zero of $\beta(x)$, i.e., $\beta(r) = 0$ and, since $b_m^{-1} \in \mathbf{C}$, also $b_m^{-1} * \beta(r) = 0$. Thus the zeros of $\beta(x)$ are precisely those of its monic associates

$$\alpha(x) = b_m^{-1} * \beta(x) = a_0 + a_1 * x + a_2 * x^2 + \cdots + a_{m-1} * x^{m-1} + x^m.$$

When $m = 1$,

$$\alpha(x) = a_0 + x$$

has $-a_0$ as zero and when $m = 2$,

$$\alpha(x) = a_0 + a_1 x + x^2$$

has

$$\frac{1}{2}\left(-a_1 - \sqrt{a_1^2 - 4a_0}\right), \qquad \frac{1}{2}\left(-a_1 + \sqrt{a_1^2 - 4a_0}\right).$$

Every polynomial $x^n - a \in \mathbf{C}[x]$ has n zeros over $\mathbf{C}$. There exist formulae which yield the zeros of all polynomials of degrees 3 and 4. It is also known that no formulae can be devised for arbitrary polynomials of degree $m \geq 5$.

Any polynomial $\alpha(x)$ of degree $m \geq 1$ can have no more than m distinct zeros. The polynomial

$$\alpha(x) = a_0 + a_1 * x + x^2$$

will have two distinct zeros if and only if the discriminant $a_1^2 - 4 * a_0 \neq 0$. We then call each a simple zero of $\alpha(x)$. However, if $a_1^2 - 4 * a_0 = 0$, each formula yields $-\frac{1}{2} * a_1$ as a zero. We then call $-\frac{1}{2} * a_1$ a zero of multiplicity two of $\alpha(x)$ and exhibit the zeros as $-\frac{1}{2} * a_1, -\frac{1}{2} * a_1$.

Fundamental Theorem of Algebra. Every polynomial $\alpha(x) \in \mathbf{C}[x]$ of degree $m > 1$ has at least one zero in $\mathbf{C}$.

Theorem. Every polynomial $\alpha(x) \in \mathbf{C}[x]$ of degree $m \geq 1$ has precisely m zeros over $\mathbf{C}$, with the understanding that any zero of multiplicity n is to be counted as n of the m zeros.

Theorem. Any $\alpha(x) \in \mathbf{C}[x]$ of degree $m \geq 1$ is either of the first degree or may be written as a product of polynomials $\in \mathbf{C}[x]$, each of the first degree.

Next we study certain subsets of $\mathbf{C}[x]$ by restricting the ring of coefficients. First, let us suppose that

$$\alpha(x) = a_0 + a_1 * x + a_2 * x^2 + \cdots + a_m * x^m \in R[x]$$

of degree $m \geq 1$ has $r = a + b * i$ as zero, i.e.,

$$\alpha(r) = a_0 + a_1 * r + a_2 * r^2 + \cdots + a_m * r^m = s + t * i = 0.$$

We have

$$\alpha(\bar{r}) = a_0 + a_1 * \bar{r} + a_2 * \bar{r}^2 + \cdots + a_m * \bar{r}^m = \overline{s + t * i} = 0$$

so that

Theorem. If $r \in \mathbf{C}$ is a zero of any polynomial $\alpha(x)$ with real coefficients, then $\bar{r}$ is also a zero of $\alpha(x)$.

Let $r = a + b * i$, with $b \neq 0$, be a zero of $\alpha(x)$. Thus $\bar{r} = a - b * i$ is also a zero and we may write

$$\alpha(x) = [x - (a + b * i)] [x - (a - b * i)] * \alpha_1(x)$$
$$= [x^2 - 2 * a * x + a^2 + b^2] * \alpha_1(x)$$

where α_1 is a polynomial of degree two less than that of $\alpha(x)$ and has real coefficients. Since a quadratic polynomial with real coefficients will have imaginary zeros if and only if its discriminant is negative, we have

Theorem. The polynomials of the first degree and the quadratic polynomials with negative discriminant are the only polynomials $\in \mathbf{R}[x]$ which are primes over $\mathbf{R}$.

Theorem. A polynomial of odd degree $\in \mathbf{R}[x]$ necessarily has a real zero.

Suppose
$$\beta(x) = b_0 + b_1 * x + b_2 * x^2 + \cdots + b_m * x^m \in \mathbf{Q}[x].$$

Let c be the greatest common divisor of the numerators of the b_i's and d be the least common multiple of the denominators of the b_i's; then

$$\alpha(x) = \frac{d}{c} * \beta(x) = a_0 + a_1 * x + a_2 * x^2 + \cdots + a_m * x^m \in \mathbf{Q}[x]$$

has integral coefficients whose only common divisors are ± 1, the units of $\mathbf{Z}$. Moreover, $\beta(x)$ and $\alpha(x)$ have precisely the same zeros. If $r \in \mathbf{Q}$ is a zero of $\alpha(x)$, i.e. if

$$\alpha(r) = a_0 + a_1 * r + a_2 * r^2 + \cdots + a_m * r^m = 0$$

it follows that

(a) if $r \in \mathbf{Z}$, then $r | a_0$;
(b) if $r = s/t$, a common fraction in lowest terms, then

$$t^m * \alpha(s/t) = a_0 * t^m + a_1 * s * t^{m-1} + a_2 * s^2 * t^{m-2} + \cdots + a_{m-1} * s^{m-1} * t + a_m * s^m = 0$$

so that $s | a_0$ and $t | a_m$. We have proved

Theorem. Let $\alpha(x) = a_0 + a_1 * x + a_2 * x^2 + \cdots + a_m * x^m$ be a polynomial of degree $m \geq 1$ having integral coefficients. If $s/t \in \mathbf{Q}$ with $(s, t) = 1$, is a zero of $\alpha(x)$, then $s | a_0$ and $t | a_m$.

Let $\alpha(x)$ and $\beta(x)$ be non-zero polynomials in $F[x]$. A polynomial $d(x) \in F[x]$ having the properties

(a) $d(x)$ is monic;
(b) $d(x) | \alpha(x)$ and $d(x) | \beta(x)$;
(c) for every $c(x) \in F[x]$ such that $c(x) | \alpha(x)$ and $c(x) | \beta(x)$, we have $c(x) | d(x)$;

is called the *greatest common divisor* of $\alpha(x)$ and $\beta(x)$.

The greatest common divisor of two polynomials in $F[x]$ can be found in the same manner as the greatest common divisor of two integers.

Theorem. Let the non-zero polynomials $\alpha(x)$ and $\beta(x)$ be in $F[x]$. The monic polynomial

$$d(x) = s(x) * \alpha(x) + t(x) * \beta(x), \qquad s(x), t(x) \in F[x]$$

of least degree is the greatest common divisor of $\alpha(x)$ and $\beta(x)$.

Theorem. Let $\alpha(x)$ of degree $m \geq 2$ and $\beta(x)$ of degree $n \geq 2$ be in $F[x]$. Then non-zero polynomials $\mu(x)$ of degree at most $n-1$ and $v(x)$ of degree at most $m-1$ exist in $F[x]$ such that

$$\mu(x) * \alpha(x) + v(x) * \beta(x) = z, \qquad \text{where } z \text{ is the zero polynomial}$$

if and only if $\alpha(x)$ and $\beta(x)$ are not relatively prime.

Theorem. If $\alpha(x)$, $\beta(x)$, $p(x) \in F[x]$ with $\alpha(x)$ and $p(x)$ relatively prime, then

$$p(x) | v(x) * \beta(x)$$

implies

$$p(x) | \beta(x).$$

Unique Factorization Theorem. Any polynomial $\alpha(x)$, of degree $m \geq 1$ and with leading coefficient a, in $F[x]$ can be written as

$$\alpha(x) = a * [p_1(x)]^{m_1} * [p_2(x)]^{m_2} * \cdots * [p_j(x)]^{m_j}$$

where the $p_i(x)$ are monic prime polynomials over F and the m_i's are positive integers. Moreover, except for the order of the factors, the factorization is unique.

We can define a formal derivative on $F[x]$ obeying the rules listed in Section 2.12.

Polynomial derivative. The derivative of a polynomial

$$\alpha(x) = a_0 + a_1 x + \cdots + a_n x^n$$

is

$$\alpha'(x) = a_1 + 2a_2 x + 3a_3 x^2 + \cdots + n a_n x^{n-1}.$$

Square-free polynomials. A polynomial $\alpha(x)$ is *square-free* if any divisor $\beta(x)$ with degree at least 1 does not divide $\alpha(x)$ twice [13][21][62], i.e.

$$\beta(x) | \alpha(x) \quad \Rightarrow \quad \left(\beta(x)\right)^2 \nmid \alpha(x).$$

It follows that a square-free polynomial $\alpha(x)$ has the property that $\alpha(x)$ and $\alpha'(x)$ have no common divisors of degree at least 1. Let $\gcd(\alpha(x), \alpha'(x))$ denote a greatest common divisor of $\alpha(x)$ and $\alpha'(x)$. Then $\alpha(x)$ is square free if $\gcd(\alpha(x), \alpha'(x)) = 1$.

Square-free decomposition. Let

$$\alpha(x) = \prod_{j=1}^{k} \left(\alpha_j(x)\right)^j$$

be a polynomial decomposition of the polynomial $\alpha(x)$, where $\alpha_j(x)$ are square-free polynomials for $j = 1, 2, \ldots, k$ and $\gcd(\alpha_i(x), \alpha_j(x)) = 1$ for $i, j = 1, 2, \ldots, k$. Such a decomposition always exists and is called the *square-free decomposition* [13][21][62]. Consequently a greatest common divisor of $\alpha(x)$ and $\alpha'(x)$ (the different greatest common divisors will differ by a constant factor) is

$$b(x) := \gcd(\alpha(x), \alpha'(x)) = \prod_{j=2}^{k} \left(\alpha_j(x)\right)^{j-1}.$$

Thus we find

$$c(x) := \frac{\alpha(x)}{b(x)} = \prod_{j=1}^{k} \alpha_j(x)$$

and

$$\gcd(c(x), \alpha'(x)) = \prod_{j=2}^{k} \alpha_j(x).$$

Finally

$$\frac{c(x)}{\gcd(c(x), \alpha'(x))} = \alpha_1(x).$$

Recursively finding the square-free decomposition of $b(x)$ yields $\alpha_2(x)$, $\alpha_3(x)$, ..., $\alpha_k(x)$. Thus we have found the square-free decomposition of $\alpha(x)$.

The *Sylvester matrix* of the polynomials

$$\alpha(x) = a_0 + a_1 x + a_2 x^2 + \cdots + a_n x^n$$

and

$$\beta(x) = b_0 + b_1 x + b_2 x^2 + \cdots + b_m x^m$$

is the $(m + n) \times (m + n)$ matrix

$$\text{Sylvester}(\alpha(x), \beta(x)) := \begin{pmatrix} A \\ B \end{pmatrix}$$

where A is the $m \times (m + n)$ matrix

$$A = \begin{pmatrix} a_n & a_{n-1} & \cdots & a_0 & & & \\ & a_n & a_{n-1} & \cdots & a_0 & & \\ & & \ddots & \ddots & \ddots & \ddots & \\ & & & \cdots & a_n & a_{n-1} & \cdots & a_0 \end{pmatrix}$$

and B is the $n \times (m+n)$ matrix

$$B = \begin{pmatrix} b_m & b_{m-1} & \cdots & b_0 & & \\ & b_m & b_{m-1} & \cdots & b_0 & \\ & & \ddots & \ddots & \ddots & \ddots \\ & & \cdots & b_m & b_{m-1} & \cdots & b_0 \end{pmatrix}$$

where all omitted entries are zero.

The *resultant* $\mathrm{res}(\alpha(x), \beta(x))$ of two polynomials $\alpha(x)$ and $\beta(x)$ is the determinant of their Sylvester matrix, i.e.

$$\mathrm{res}(\alpha(x), \beta(x)) = \det(\mathrm{Sylvester}(\alpha(x), \beta(x))).$$

Example. Let $\alpha(x) = 2x^3 - 6x - 4$ and $\beta(x) = x^2 + x - 6$ then $n = 3$, $m = 2$

$$A = \begin{pmatrix} 2 & 0 & -6 & -4 & 0 \\ 0 & 2 & 0 & -6 & -4 \end{pmatrix}, \quad B = \begin{pmatrix} 1 & 1 & -6 & 0 & 0 \\ 0 & 1 & 1 & -6 & 0 \\ 0 & 0 & 1 & 1 & -6 \end{pmatrix}$$

and therefore

$$\mathrm{Sylvester}(\alpha(x), \beta(x)) = \begin{pmatrix} 2 & 0 & -6 & -4 & 0 \\ 0 & 2 & 0 & 6 & 4 \\ 1 & 1 & -6 & 0 & 0 \\ 0 & 1 & 1 & -6 & 0 \\ 0 & 0 & 1 & 1 & -6 \end{pmatrix}.$$

The resultant is $\mathrm{res}(\alpha(x), \beta(x)) = 0$. This is due to the fact that $\alpha(x)$ and $\beta(x)$ share a common factor $x - 2$, i.e. $\gcd(\alpha(x), \beta(x)) \neq 1$.

Theorem. $\gcd(\alpha(x), \beta(x)) \neq 1$ if and only if $\mathrm{res}(\alpha(x), \beta(x)) = 0$.

2.11   Gröbner Bases

Gröbner bases techniques are useful for analyzing and solving systems of multivariate polynomial equations [21][62]. For example we want to find the set $V(f_1, f_2, f_3)$ of common zeros for the following set of polynomials

$$f_1(x_1, x_2, x_3) = x_1 - x_2 - x_3 = 0$$
$$f_2(x_1, x_2, x_3) = x_1 + x_2 - x_3^2 = 0$$
$$f_3(x_1, x_2, x_3) = x_1^2 + x_2^2 - 1 = 0.$$

A heuristic approach is as follows: From the first two equations we obtain

$$f_1 + f_2 = 2x_1 - x_3 - x_3^2 = 0.$$

Thus x_2 is eliminated and therefore

$$x_1 = \frac{1}{2}(x_3^2 + x_3).$$

Analogously
$$f_2 - f_1 = 2x_2 + x_3 - x_3^2 = 0\,.$$

Thus x_1 is eliminated and therefore
$$x_2 = \frac{1}{2}(x_3^2 - x_3)\,.$$

Substitution in f_3 gives the polynomial equation
$$x_3^4 + x_3^2 - 2 = 0\,.$$

The four solutions in $\mathbf{C}$ are $x_3 = 1$, $x_3 = -1$, $x_3 = \sqrt{2}i$, $x_3 = -\sqrt{2}i$.

In a Gauss elimination-like method we first choose x_1 in the first polynomial as the first term suitable for eliminating terms in the two other polynomials. Multiply the first polynomial by another polynomial, in this case -1, and add it to the second polynomial in order to eliminate the terms containing x_1. For the third polynomial, multiply f_1 by the polynomial $-x_1 - x_2 - x_3$, and add it to f_3. Thus

$$\begin{aligned}
V(f_1, f_2, f_3) &= V(f_1, f_2 - f_1, f_3 - (x_1 + x_2 + x_3)f_1) \\
&= V(x_1 - x_2 - x_3, 2x_2 - x_3^2 + x_3, 2x_2^2 + 2x_2x_3 + x_3^2 - 1)\,.
\end{aligned}$$

The resulting second and third polynomials have no terms that contain x_1. We call the new polynomials g_1, g_2, and g_3, respectively. Next choose the variable x_2 in g_2 as the most important variable. Then multiply g_2 by another polynomial, in this case 1, and subtract it from $2g_1$ in order to eliminate the terms containing x_2. For the third polynomial also multiply g_2 by another polynomial, in this case $-2x_2 - x_3^2 - x_3$, and add it to $2g_3$. Thus

$$\begin{aligned}
V(g_1, g_2, g_3) &= V(2g_1 + g_2, g_2, 2g_3 - (-2x_2 + x_3^2 + x_3)g_2) \\
&= V(2x_1 - x_3^2 - x_3, 2x_2 - x_3^2 + x_3, x_3^4 + x_3^2 - 2)\,.
\end{aligned}$$

The new generators are in upper triangular form: the last polynomial is only in x_3, the second one is only in x_2 and x_3, and the first one is a polynomial in x_1, x_2 and x_3.

The above methods have in common that they replace the original polynomials by simpler polynomials that have the same solution set. Here simpler means that the set of common zeros can be more easily computed from the new polynomials than from the original ones. During the elimination process, using the Gauss elimination-like method, new polynomials are formed from pairs of old ones f, g by $h = \alpha f + \beta g$, where α is a polynomial and β a scalar. h has the same common zeros as f and g and $V(f, g) = V(f, h)$. The set $I(f, g)$ of all linear combinations $\alpha f + \beta g$, where α and β are polynomials, is a called the ideal generated by f and g. The set of common zeros of the ideal $I(f, g)$ is identical to the set of common zeros of f and g. i.e. $V(f, g) = V(I(f, g))$. The Gröbner basis is a simple set of generators of an ideal.

Definition. A subset I of the polynomial ring $\mathbf{K}(x_1, x_2, \ldots, x_n)$ is an *ideal* if it satisfies:

1. 0 is an element of I,
2. if f and g are any two elements in I, then $f + g$ is an element of I,
3. if f is an element of I, then for any h in $\mathbf{K}(x_1, x_2, \ldots, x_n)$ hf is an element of I.

An example of an ideal in $\mathbf{K}(x_1, x_2, \ldots, x_n)$ is the ideal generated by a finite number of polynomials.

Lemma. Let $F := \{\, f_1, f_2, \ldots, f_s \,\}$ be a finite subset of $\mathbf{K}(x_1, x_2, \ldots, x_n)$. Then the set

$$\langle f_1, f_2, \ldots, f_s \rangle := \left\{ \sum_{i=1}^{s} h_i f_i \; : \; h_1, h_2, \ldots, h_s \text{ are in } \mathbf{K}(x_1, x_2, \ldots, x_n) \right\}$$

is an ideal.

Definition. The set $\langle f_1, f_2, \ldots, f_s \rangle$ is called the ideal generated by F. The polynomials $f_1, f_2, \ldots, f_s$ are called *generators*.

If an ideal I has finitely many generators it is said to be finitely generated and the set $\{\, f_1, f_2, \ldots, f_s \,\}$ is called a *basis* of I.

Hilbert basis theorem. Every ideal in $\mathbf{K}(x_1, x_2, \ldots, x_n)$ is finitely generated.

An important consequence of this theorem is that any ascending chain of ideals $I_1 \subset I_2 \subset I_3 \subset \cdots$ in $\mathbf{K}(x_1, x_2, \ldots, x_n)$ stabilizes with I_n for some n. This is called *ascending chain condition* and it is used to prove that the Buchberger algorithm terminates in a finite number of steps.

For the Buchberger algorithm we have to find the *leading term* in a polynomial. The rule of choosing leading terms is an example of a *term ordering*. For multivariate polynomials in $x_1, x_2, \ldots, x_n$ the *pure lexicographic ordering* is the linear ordering determined by

$$x_1^{i_1} x_2^{i_2} \cdots x_n^{i_n} \prec x_1^{j_1} x_2^{j_2} \cdots x_n^{j_n}$$

if and only if

$$\exists \ell \in \{\, 1, 2, \ldots, n-1 \,\}: \; i_\ell < j_\ell, \quad k < \ell \Rightarrow i_k = j_k.$$

For example, in the pure lexicographic ordering of three variables with $x_3 \prec x_2 \prec x_1$ we have

$$1 \prec x_3 \prec x_3^2 \prec \cdots \prec x_2 \prec x_2 x_3 \prec x_2 x_3^2 \prec \ldots$$
$$\prec x_2^2 \prec x_2^2 x_3 \prec x_2^2 x_3^2 \prec \cdots \prec x_1 \prec x_1 x_3 \prec x_1 x_3^2 \prec \ldots$$
$$\prec x_1 x_2 \prec x_1 x_2^2 \prec \cdots \prec x_1^2 \prec \cdots$$

Another ordering is the total degree inverse lexicographic ordering defined by

$$x_1^{i_1} x_2^{i_2} \cdots x_n^{i_n} \prec x_1^{j_1} x_2^{j_2} \cdots x_n^{j_n}$$

if and only if either

$$\sum_{k=1}^{n} i_k < \sum_{k=1}^{n} j_k$$

or

$$\sum_{k=1}^{n} i_k = \sum_{k=1}^{n} j_k \quad \text{and} \quad \exists \ell \in \{1, \ldots, n-1\}: \; i_\ell < j_\ell, \quad k > l \Rightarrow i_k = j_k.$$

Other term orderings are possible in Gröbner basis theory. The ordering $\prec$ only has to be admissible, i.e., satisfy

$$(i) \quad 1 \prec t \text{ for every term } t \neq 1$$

$$(ii) \quad s \prec t \Leftrightarrow s \cdot u \prec t \cdot u \text{ for all terms } s, t, u.$$

To each non-zero polynomial f we can associate the leading term

$$\mathrm{lt}(f) := \text{term that is maximal among those in } f.$$

The leading coefficient is defined as

$$\mathrm{lc}(f) := \text{the coefficient of the leading term of } f.$$

The leading term of f is the product of the leading coefficient of f and the leading monomial $\mathrm{lm}(f)$ of f

$$\mathrm{lt}(f) = \mathrm{lc}(f) \cdot \mathrm{lm}(f).$$

Example. With the pure lexicographic ordering $x_3 \prec x_2 \prec x_1$ we have

$$\mathrm{lt}(2x_2 - x_3^2 + x_3) = 2x_2, \qquad \mathrm{lc}(2x_2 - x_3^2 + x_3) = 2, \qquad \mathrm{lm}(2x_2 - x_3^2 + x_3) = x_2.$$

For non-zero polynomials f, g and polynomial $\widetilde{f}$ we say that f reduces to $\widetilde{f}$ modulo g and denote it by

$$f \to_g \widetilde{f}$$

if there exists a term t in f that is divisible by the leading term of g and

$$\widetilde{f} = f - \frac{t}{\mathrm{lt}(g)} \cdot g.$$

Admissibility of the term ordering $\prec$ guarantees that if the terms in f and $\widetilde{f}$ are ordered from high to low terms, then the first terms in which these polynomials differ are t in f and some lower term in $\widetilde{f}$.

Let $G = \{g_1, g_2, \ldots, g_m\}$ be a set of polynomials. A polynomial f reduces to $\widetilde{f}$ modulo G if there exists a polynomial g_i in G such that $f \to_{g_i} \widetilde{f}$. A *normal form* $\mathrm{normalf}(f, G)$ of f with respect to G is a polynomial obtained after a finite number of reductions which contains no terms anymore that is divisible by leading terms of polynomials of G. The normal form is in general not unique.

We define $\mathrm{lt}(I)$ as the set of the leading terms of elements of I.

Definition. For a given monomial ordering, a subset $G = \{\, g_1, g_2, \ldots, g_t \,\}$ of an ideal I is said to be a Gröbner basis if

$$\mathrm{lt}(I) = \langle\, \mathrm{lt}(g_1), \mathrm{lt}(g_2), \ldots, \mathrm{lt}(g_t) \rangle$$

This means that a subset $G = \{\, g_1, g_2, \ldots, g_t \,\}$ of an ideal I is a Gröbner basis if and only if the leading term of any element of I is divisible by one of the $\mathrm{lt}(g_t)$.

G is a *Gröbner basis* (with respect to an admissible ordering) if and only if normal forms modulo G are unique, i.e., for all f, g, h: if $g = \mathrm{normalf}(f, G)$ and $h = \mathrm{normalf}(f, G)$, then $g = h$.

Alternatively, G is a Gröbner basis if and only if $\mathrm{normalf}(g, G) = 0$ for all g in the ideal generated by G.

To compute such a basis we have to introduce the concept of an *S-polynomial*: the S-polynomial $\mathrm{spoly}(f, g)$ of polynomials f and g is defined by

$$\mathrm{spoly}(f, g) := \mathrm{lcm}\left(\mathrm{lt}(f), \mathrm{lt}(g)\right) \cdot \left(\frac{f}{\mathrm{lt}(f)} - \frac{g}{\mathrm{lt}(g)} \right)$$

where $\mathrm{lcm}(p, q)$ denotes the least common multiple of the polynomials p and q. We could also define

$$\mathrm{spoly}(f, g) = \alpha \cdot f - \beta \cdot g$$

where the polynomials α and β are chosen such that the leading terms cancel in the difference and the degree of α, β is minimal.

The *Buchberger algorithm* to find a Gröbner basis is as follows

groebnerBasis(G)

   1. $GB \leftarrow G$

   2. $B \leftarrow \{\, (f, g) : f, g \in G, f \neq g \,\}$

   3. if $B = \emptyset$ go to 11

   4.   select a pair (f, g) from B

   5. $B \leftarrow B\{\setminus (f, g)\}$

   6. $h \leftarrow \mathrm{normalf}(\mathrm{spoly}(f, g), GB)$

   7.   if $h = 0$ go to 3

   8. $GB \leftarrow GB \cup \{\, h \,\}$

   9. $B \leftarrow B \cup \{\, (f, h) : f \in GB \,\}$

 10.   go to 3

 11. return GB.

2.12   Differentiation

Let $f : I \to \mathbf{R}$ be a function, where I is an open interval. We say that f is differentiable at $a \in I$ provided there exists a linear mapping $L : \mathbf{R} \to \mathbf{R}$ such that

$$\lim_{\epsilon \to 0} \frac{f(a + \epsilon) - f(a) - L(\epsilon)}{\epsilon} = 0.$$

The linear mapping L which, when it exists, is unique and is called the *differential* of f (or *derivative* of f at a) and is denoted by $d_a f$. It is customary in traditional texts to introduce the differentials df and dx and to obtain relations such as

$$df = \frac{df}{dx} dx.$$

Using the modern notation this relation would be written as

$$d_a f = f'(a) dz \mathcal{I}$$

where $\mathcal{I}(= id)$ denotes the identity function $x \to x$. If f and g are differentiable we find that

$$\frac{d}{dx}(f + g) = \frac{df}{dx} + \frac{dg}{dx} \qquad\qquad \text{summation rule}$$

$$\frac{d}{dx}(f - g) = \frac{df}{dx} - \frac{dg}{dx} \qquad\qquad \text{difference rule}$$

$$\frac{d}{dx}(f * g) = g * \frac{df}{dx} + f * \frac{dg}{dx} \qquad\qquad \text{product rule}$$

$$\frac{d}{dx}\left(\frac{f}{g}\right) = \frac{g * \frac{df}{dx} - f * \frac{dg}{dx}}{g^2}, \quad g \neq 0 \text{ for } x \in I \quad \text{quotient rule}$$

$$\frac{d}{dx} c = 0 \qquad\qquad \text{where } c \text{ is a constant.}$$

Formally differentiation is described by differential fields. Let F be a differential field, i.e. a field with $D : F \to F$ such that

$$D(f + g) = D(f) + D(g), \qquad D(f \cdot g) = D(f) \cdot g + f \cdot D(g)$$

$\forall f, g \in F$. Here F is a field of characteristic zero, i.e. no $k \in \mathbf{Z}$ exists such that $k \cdot f = 0$ for all $f \in F$. It follows that

1. $D(0) = D(1) = 0$

2. $D(-f) = -D(f)$

3. $D(f \cdot g^{-1}) = \frac{g \cdot D(f) - f \cdot D(g)}{g^2} \quad \forall f, g \in F, \; g \neq 0$

4. $D(f^n) = n f^{n-1} D(f), \quad \forall f \in F, \; n \in \mathbf{N}.$

If we assume $F = \mathbf{Q}(x)$ the field of rational functions over x and $D(x) = 1$ then we can additionally show that

5. $\nexists r \in \mathbf{Q}(x)$ such that $D(r) = \frac{1}{x}$.

2.13   Integration

A computer algebra system should be able to integrate formally elementary functions, for example

$$\int \frac{dx}{1-x^2} = \frac{1}{2}\ln\left(\frac{1+x}{1-x}\right).$$

In general it is assumed that the underlying field is $\mathbf{R}$. Symbolic differentiation was undertaken quite early in the history of computer algebra, whereas symbolic integration (also called formal integration) was introduced much later. The reason is due to the big difference between formal integration and formal differentiation. Differentiation is an algorithmic procedure, and a knowledge of the derivatives of functions plus the sum rule, product rule, quotient rule and chain rule, enable us to differentiate any given function. The real problem in differentiation is the simplification of the result. On the other hand, integration seems to be a random collection of devices and special cases. There are only two general rules, i.e. the sum rule and the rule for integration by parts. If we integrate a sum of two functions, in general, we would integrate each summand separately, i.e.

$$\int (f_1(x) + f_2(x))dx = \int f_1(x)dx + \int f_2(x)dx.$$

This is the so-called *sum rule*. It can happen that the sum $f_1 + f_2$ could have an explicit form for the integral, but f_1 and f_2 do not have any integrals in finite form. For example

$$\int (x^x + (\ln x)x^x)dx = x^x.$$

However

$$\int x^x dx, \qquad \int (\ln x)x^x dx$$

do not have any integrals in finite form. The sum rule may only be used if it is known that two of the three integrals exist. For combinations other than addition (and subtraction) there are no general rules. For example, because we know how to integrate $\exp x$ and x^2 it does not follow that we can integrate $\exp(x^2)$. This function has no integral simpler than $\int \exp\left(x^2\right)dx$. So we learn several "methods" such as: integration by parts, integration by substitution, integration by looking up in tables of integrals, etc. In addition we do not know which method or which combination of methods will work for a given integral. In the following presentation we follow closely Davenport et al. [15], MacCallum and Wright [38], Risch [46] and Geddes [21].

Since differentiation is definitely simpler than integration, we rephrase the problem of integration as the "inverse problem" of differentiation, that is, given a function f, instead of looking for its integral g, we ask for a function g such that $dg/dx = f$.

Definition. Given two classes of functions A and B, the integration problem for A and B is to find an algorithm which, for every member f of A, either gives an element g of B such that $f = dg/dx$, or proves that there is no element g of B such that $f = dg/dx$.

For example, if $A = Q(x)$ and $B = Q(x)$, where $Q(x)$ denotes the rational functions, then the answer for $1/x^2$ must be $-1/x$, whilst for $1/x$ there is no solution in this set. On the other hand, if $B = Q(x, \ln x)$, then the answer for $1/x$ must be $\ln x$.

We consider now integration of rational functions. We deal with the case of $A = C(x)$, where C is a field of constants. Every rational function f can be written in the form $p + q/r$, where p, q and r are polynomials, q/r are relatively prime, and the degree of q is less than that of r. A polynomial p always has a finite integral, so the sum rule holds for $f_1(x) = p(x)$ and $f_2(x) = q(x)/r(x)$. Therefore the problem of integrating f reduces to the problem of the integration of p (which is very simple) and of the proper rational function q/r.

The *naive method* is as follows. If the polynomial r factorizes into linear factors, such that

$$r(x) = \prod_{i=1}^{n} (x - a_i)^{n_i}$$

we can decompose q/r into partial fractions

$$\frac{q(x)}{r(x)} = \sum_{i=1}^{n} \frac{b_i(x)}{(x - a_i)^{n_i}}$$

where the b_i are polynomials of degree less than n_i. These polynomials can be divided by $x - a_i$, so as to give the following decomposition

$$\frac{q(x)}{r(x)} = \sum_{i=1}^{n} \sum_{j=1}^{n_i} \frac{b_{i,j}}{(x - a_i)^j}$$

where the $b_{i,j}$ are constants. This decomposition can be integrated to give

$$\int \frac{q(x)}{r(x)} dx = \sum_{i=1}^{n} b_{i,1} \log(x - a_i) - \sum_{i=1}^{n} \sum_{j=2}^{n_i} \frac{b_{i,j}}{(j-1)(x - a_i)^{j-1}}.$$

Thus, we have proved that every rational function has an integral which can be expressed as a rational function plus a sum of logarithms of rational functions with constant coefficients – that is, the integral belongs to the field

$$C(x, \log(x - a_1), \ldots, \log(x - a_n)).$$

This algorithm requires us to factorize the polynomial r completely, which is not always possible without adding several algebraic quantities to C. Manipulating these algebraic extensions is often very difficult. Even if the algebraic extensions are not required, it is quite expensive to factorize a polynomial r of high degree. It also requires a complicated decomposition into partial fractions.

In the *Hermite's method* [21] we determine the rational part of the integral of a rational function without bringing in any algebraic quantity. Similarly, it finds the derivative of the sum of logarithms, which is also a rational function with coefficients in the same field. We have seen that a factor of the denominator r which appears to the power n, appears to the power $n-1$ in the denominator of the integral. This suggests square-free decomposition. A square free decomposition is a decomposition

$$a(x) = \prod_{i=1}^{n} (a_i(x))^i$$

where $a_i(x)$ has no repeated factors. Let us suppose, then, that r has a square-free decomposition of the form $\prod_{i=1}^{n} r_i^i$. The r_i are then relatively prime, and we can construct a decomposition into partial fractions

$$\frac{q(x)}{r(x)} = \frac{q(x)}{\prod_{i=1}^{n} r_i^i(x)} = \sum_{i=1}^{n} \frac{q_i(x)}{r_i^i(x)}.$$

Every element on the right hand side has an integral, and therefore the sum rule holds, and it suffices to integrate each element in turn. Integration yields

$$\int \frac{q_i(r)}{r_i^i(x)}\, dx = -\left(\frac{q_i b/(i-1)}{r_i^{i-1}}\right) + \int \frac{q_i a + d(q_i b/(i-1))/dx}{r_i^{i-1}}\, dx$$

where a and b satisfy $a r_i + b dr_i/dx = 1$. Since r_i is square-free it follows that the greatest common divisor of r_i and dr_i/dx is 1. Thus a and b are found using the division algorithm. Consequently, we have been able to reduce the exponent of r_i. We can continue in this way until the exponent becomes one, when the remaining integral is a sum of logarithms. This is the *logarithmic part*.

We can avoid the decomposition into partial fractions using properties of square-free polynomials. As above, let $r = \prod_{i=1}^{n} r_i^i$ be the square-free decomposition of r. If $n = 1$ then r is square-free, and we need to apply the technique for the logarithmic part. For $n > 1$ we define $f = r_n$ and

$$g = \frac{r}{r_n^n} = \prod_{i=1}^{n-1} r_i^i.$$

Consequently $\gcd(gf', f) = 1$. It follows that there exist polynomials s_* and t_* such that

$$s_* g f' + t_* f = 1.$$

Multiplying by q yields

$$s g f' + t f = q$$

where $s := q s_*$ and $t := q t_*$. Dividing by $r = f^n g$ then yields

$$\frac{q(x)}{r(x)} = s \frac{f'}{f^n} + \frac{t}{g f^{n-1}}.$$

Thus we can integrate using integration by parts

$$\int \frac{q(x)}{r(x)}\, dx = \int s\frac{f'}{f^n}\, dx + \int \frac{t}{gf^{n-1}}\, dx$$

$$= \frac{s}{(1-n)f^{n-1}} + \int \frac{s'}{(n-1)f^{n-1}}\, dx + \int \frac{t}{gf^{n-1}}\, dx$$

$$= \frac{s}{(1-n)f^{n-1}} + \int \frac{s'g + (n-1)t}{(n-1)gf^{n-1}}\, dx$$

where once again we have reduced the power in f and $n-1=1$ yields a logarithmic part.

Hermite's method is quite suitable for manual calculations. The disadvantage is that it needs several sub-algorithms and this involves some fairly complicated programming.

The *Horowitz method* [21] is as follows. The aim is still to be able to write

$$\int \frac{q(x)}{r(x)}\, dx = \frac{q_1}{r_1} + \int \frac{q_2}{r_2}\, dx$$

where the integral remaining gives only a sum of logarithms when it is resolved. We know that r_1 has the same factors as r, but with the exponent reduced by one, that r_2 has no multiple factors, and that its factors are all factors of r. We have $r_1 = \gcd(r, dr/dx)$, and r_2 divides $r/\gcd(r, dr/dx)$. We may suppose that q_2/r_2 is written in reduced from, and therefore $r_2 = r/\gcd(r, dr/dx)$. Then

$$\frac{q(x)}{r(x)} = \frac{d}{dx}\left(\frac{q_1}{r_1}\right) + \frac{q_2}{r_2} = \frac{1}{r_1}\frac{dq_1}{dx} - \frac{q_1}{r_1^2}\frac{dr_1}{dx} + \frac{q_2}{r_2} = \frac{r_2 dq_1/dx - q_1 s + q_2 r_1}{r}$$

where $s = (r_2 dr_1/dx)/r_1$ (the division here is without a remainder). Thus we arrive at

$$q = r_2\frac{dq_1}{dx} - q_1 s + q_2 r_1$$

where q, s, r_1 and r_2 are known, and q_1 and q_2 have to be determined. Since the degrees of q_1 and q_2 are less than the degrees m and n of r_1 and r_2 respectively we write

$$q_1(x) = \sum_{i=0}^{m-1} a_i x^i, \qquad q_2(x) = \sum_{i=0}^{n-1} b_i x^i.$$

Thus the equation for q can be rewritten as a system of $m+n$ linear equations in $n+m$ unknowns. Moreover, this system can be solved, and integration (at least this sub-problem) reduces to linear algebra.

Next we describe the *logarithmic part method* also called the *Rothstein/Trager method* [21][62]. The two methods described above can reduce the integration of any rational function to the integration of a rational function (say q/r) whose integral would be only a sum of logarithms. This integral can be resolved by completely factorizing the denominator, but this is not always necessary for an expression of the results. The real problem is to find the integral without using any algebraic numbers other than those needed in the expression of the result. Let us suppose that

$$\int \frac{q(x)}{r(x)} dx = \sum_{i=1}^{n} c_i \log v_i(x)$$

is a solution to this integral where the right hand side uses the fewest possible algebraic extensions. The c_i are constants and, in general, the v_i are rational functions. Since $\ln(a/b) = \ln a - \ln b$, we can suppose, without loss of generality, that the v_i are polynomials. Furthermore, we can perform a square-free decomposition, which does not add any algebraic extensions, and we can apply the identity

$$\ln \prod_{i=1}^{n} p_i^i \equiv \sum_{i=1}^{n} i \ln p_i.$$

From the identity

$$c * \ln(p * q) + d * \ln(p * r) = (c + d) * \ln p + c * \ln q + d * \ln r$$

we can suppose that the v_i are relatively prime, whilst still keeping the minimality of the number of algebraic extensions. Moreover, we can suppose that all the c_i are different. Differentiating the integral, we find

$$\frac{q(x)}{r(x)} = \sum_{i=1}^{n} \frac{c_i}{v_i} \frac{dv_i}{dx}.$$

The assumption that the v_i are square-free implies that no element of this summation can simplify, and the assumption that the v_i are relatively prime implies that no cancellation can take place in this summation. This implies that the v_i must be precisely the factors of r, i.e. that $r(x) = \prod_{i=1}^{n} v_i(x)$. Let us write $u_i = \prod_{j \neq i} v_j$. Then we can differentiate the product of the v_i, which shows that

$$\frac{dr(x)}{dx} = \sum_{i=1}^{n} u_i \frac{dv_i}{dx}.$$

We find that $q(x) = \sum_{i=1}^{n} c_i u_i v_i'$. These two expressions for q and r' permit the following deduction

$$\gcd(q - c_k r', r) = \gcd\left(\sum_{i=1}^{n} (c_i - c_k) u_i v_i', \ v_0 v_1 \cdots v_n \right) = v_k(x)$$

since all the other u_i are divisible by v_k, and u_k does not appear in the sum. We must still determine the values of c_k so that we can calculate $v_k = \gcd(q - c_k r', r)$. It follows that $\mathrm{res}(q - c_k r', r) = 0$. Thus the c_k are solutions for z in the polynomial equation $\mathrm{res}(q - zr', r) = 0$.

Next we consider algebraic solutions of the first order differential equation

$$\frac{dy}{dx} + f(x)y = g(x).$$

This leads to the *Risch algorithm*.

2.14   Risch Algorithm

We have introduced the problem of finding an algorithm which, given f and g belonging to a class A of functions, either finds a function y belonging to a given class B of functions, or proves that there is no element of B which satisfies the given equation. For the sake of simplicity, we consider the case when B is always the class of functions elementary over A. To solve this differential equation we substitute

$$y(x) = z(x) \exp\left(-\int^x f(s)ds\right).$$

This leads to the solution

$$y(x) = \exp\left(-\int^x f(s)ds\right)\int^x \left(g(s)\left(\exp\int^s f(t)dt\right)ds\right).$$

In general, this method is not algorithmically satisfactory for finding y, since the algorithm of integration described in the last section reformulates this integral as the differential equation we started with. Risch [46] found one method to solve these equations for the case when A is a field of rational functions, or an extension of a field over which this problem can be solved. The problem can be stated as follows: given two rational functions f and g, find the rational function y such that

$$dy/dx + f(x)y = g(x)$$

or prove that there is none. f satisfies the condition that $\exp(\int^x f(s)ds)$ is not a rational function and its integral is not a sum of logarithms with rational coefficients. The problem is solved in two stages: reducing it to a purely polynomial problem, and solving that problem. The Risch algorithm is recursive. Before applying it one has (in principle) to check that the different extension variables are not algebraically related. For rational functions the Risch algorithm is the same as for the Horowitz method. For more details of the Risch algorithm and extensions of it we refer to the literature [15], [38], [46], [21], [35].

When working with extension fields which include transcendental functions (non-algebraic) the properties of non-algebraic functions are important. A function $f(x)$ is algebraic if there exists $n \in \mathbf{N}$ and

$$a_0(x), a_1(x), \ldots, a_n(x) \in \mathbf{Q}(x)$$

such that $a_n(x) \neq 0$ and

$$\sum_{j=0}^{n} a_j(x)\big(f(x)\big)^j = 0.$$

Here $\mathbf{Q}(x)$ denotes the rational functions over x.

Example. The function $\sqrt{x}$ not algebraic.

Proof. Let $n = 2$ and $a_0(x) = -x$, $a_1(x) = 0$ and $a_2(x) = 1$. Then

$$a_2(x)(\sqrt{x})^2 + a_1(x)(\sqrt{x})^1 + a_0(x)(\sqrt{x})^0 = x + 0 - x = 0.$$

Example. The function e^x is not algebraic.

Proof. Assume e^x is algebraic. Then there exists minimal $n \in \mathbf{N}$ and $a_0, \ldots, a_n \in \mathbf{Q}(x)$ with $a_n(x) \neq 0$ such that

$$\sum_{j=0}^{n} a_j(x)(e^x)^j = 0.$$

Thus

$$(e^x)^n = -\sum_{j=0}^{n-1} \frac{a_j(x)}{a_n(x)}(e^x)^j.$$

Since $\frac{de^x}{dx} = e^x$ we find

$$\frac{d}{dx}\left(e^x\right)^n = n(e^x)^{n-1}e^x = n(e^x)^n$$

so that

$$(e^x)^n = -\frac{1}{n}\frac{d}{dx}\sum_{j=0}^{n-1}\frac{a_j}{a_n}(e^x)^j$$

$$= -\frac{1}{n}\sum_{j=0}^{n-1}\left[\left(\frac{d}{dx}\frac{a_j}{a_n}\right)(e^x)^j + j\frac{a_j}{a_n}(e^x)^j\right]$$

From $(e^x)^n - (e^x)^n = 0$ we find

$$(e^x)^n - (e^x)^n = -\sum_{j=0}^{n-1}\frac{a_j}{a_n}(e^x)^j + \frac{1}{n}\sum_{j=0}^{n-1}\left[\left(\frac{d}{dx}\frac{a_j}{a_n}\right)(e^x)^j + j\frac{a_j}{a_n}(e^x)^j\right] = 0$$

i.e.

$$\sum_{j=0}^{n-1}\left[\left(\frac{d}{dx}\frac{a_j}{a_n}\right) + j\frac{a_j}{a_n} - n\frac{a_j}{a_n}\right](e^x)^j = 0.$$

For the highest order term $j = k$, where $a_k \neq 0$, we find

$$\left(\frac{d}{dx}\frac{a_k}{a_n}\right) + (j - n)\frac{a_k}{a_n} \neq 0$$

and clearly n is not minimal, i.e. we have a contradiction.

Theorem. If $f(x)$ algebraic then $f^{-1}(x)$ (when it exists) is algebraic.

Proof. Since $f(x)$ is algebraic there exists $n \in \mathbf{N}$ and $a_0, \ldots, a_n \in \mathbf{Q}(x)$ with $a_n(x) \neq 0$ such that

$$\sum_{j=0}^{n} a_j(x) \Big(f(x) \Big)^j = 0$$

for all x in the domain, including $f^{-1}(x)$. Thus

$$\sum_{j=0}^{n} a_j(f^{-1}(x)) \Big(f(f^{-1}(x)) \Big)^j = 0$$

or

$$\sum_{j=0}^{n} a_j(f^{-1}(x)) x^j = 0.$$

Multiplying by the lowest common multiple of the denominators of $a_0(f^{-1}(x))$, $\ldots$, $a_n(f^{-1}(x))$ yields that $f^{-1}(x)$ is algebraic.

It follows that $\ln x$ is not algebraic. Since $\sin x$, $\cos x$ and many other transcendental functions are expressed in terms of e^x and $\ln x$ we confine our discussion to integrals involving only the exponential and logarithm transcendental functions.

Liouville's principle. Let F be a differential field and $f \in F$. Supposes G is an elementary extension of F over the same underlying field and that $g \in G$ satisfies $g' = f$. Then there exist $v_0, v_1, \ldots, v_m \in F$ and constants $c_1, c_2, \ldots, c_m$ such that

$$\int f \, dx = v_0 + \sum_{j=1}^{m} c_j \ln v_j.$$

For the proof we refer to [21].

Another way to describe Liouville's principle is provided in [35]:

If $f(x, y_1, y_2, \ldots, y_m)$, y_1', y_2', $\ldots$, y_m' are algebraic in $x, y_1, \ldots, y_m$ then

$$\int f(x, y_1, y_2, \ldots, y_m) \, dx$$

is elementary if and only if

$$\int f(x, y_1, y_2, \ldots, y_m) \, dx = U_0 + \sum_{j=1}^{m} C_j \ln U_j$$

where the C_j are constants and the U_j are algebraic in $x, y_1, \ldots, y_m$.

We have to apply the Hermite and Rothstein/Trager methods for rational and logarithmic parts of an integral in a given extension field. For more information we refer to [21]. For the polynomial parts we note that for $n > 1$

$$\frac{d}{dx} f(x)(\ln g(x))^n = f'(x)(\ln g(x))^n + n\frac{f(x)g'(x)}{g(x)}(\ln g(x))^{n-1}.$$

Suppose $f'(x) = 0$, then taking the integral yields

$$\int n\frac{f(x)g'(x)}{g(x)}(\ln g(x))^{n-1}dx = f(x)(\ln g(x))^n$$

i.e. integration of a polynomial of degree $n - 1$ in a logarithm in general yields a polynomial of degree n. On the other hand

$$\frac{d}{dx} f(x)(e^{g(x)})^n = f'(x)(e^{g(x)})^n + nf(x)g'(x)(e^{g(x)})^n$$

so that integration of a polynomial of degree n in an exponential in general yields a polynomial of degree n again. Consequently we apply the following two rules.

Let θ_1 denote an exponential extension of the differential field F (where $\theta_1 \notin F$ and $x \in F$), i.e. $\theta_1' = f\theta_1$ for some $f \in F$, and let $p_1(\theta_1)$ be a polynomial of degree n in θ_1 with coefficients from F. Then

$$\int p_1(\theta_1)dx = a_n\theta_n^n + a_{n-1}\theta_1^{n-1} + \cdots + a_0$$

where $a_0, a_1, \ldots, a_n \in F$.

Let θ_2 denote an logarithmic extension of the differential field F (where $\theta_2 \notin F$ and $x \in F$), i.e. $\theta_2' = f'/f$ for some $f \in F$, and let $p_2(\theta_2)$ be a polynomial of degree n in θ_2 with coefficients from F. Then

$$\int p_2(\theta_2)dx = a_{n+1}\theta_2^{n+1} + a_n\theta_2^n + \cdots + a_0$$

where $a_0, a_1, \ldots, a_{n+1} \in F$.

Given these two rules, we can differentiate these equations and use the fact that exponential and logarithmic extensions are not algebraic to solve for the coefficients a_j.

Example. We want to integrate

$$f(x) = \frac{-e^x - x + \ln(x)x + \ln(x)xe^x}{x(e^x + x)^2}.$$

The elementary field we obtain is $\mathbf{Q}(x, \theta_1, \theta_2)$ with $\theta_1 = e^x$, $\theta_2 = \ln(x)$. Thus the integrand becomes

$$\frac{-\theta_1 - x + \theta_2 x + \theta_2 x\theta_1}{x(\theta_1 + x)^2} = -\frac{1}{x(\theta_1 + x)} + \theta_2\frac{1 + \theta_1}{(\theta_1 + x)^2}.$$

Setting $A_0 = -1/(x(\theta_1 + x))$ and $A_1 = (1 + \theta_1)/(\theta_1 + x)^2$ and using (logarithmic case)

$$\int A_0 dx + \int A_1 \theta_2 dx = B_0 + B_1 \theta_2 + B_2 \theta_2^2$$

where $B_0, B_1, B_2 \in \mathbf{Q}(x, \theta_1)$ we obtain by differentiation the equation

$$A_0 + A_1 \theta = B_0' + B_1' \theta_2 + B_1 \theta_2' + B_2' \theta_2^2 + 2B_2 \theta_2' \theta_2$$

and comparing coefficients of powers of θ_2 (since $\theta_2 = \ln x$ is not algebraic)

$$0 = B_2'$$

$$A_1 = B_1' + 2B_2 \theta_2' = B_1' + 2B_2 \frac{1}{x}$$

$$A_0 = B_0' + B_1 \theta_2'$$

where $'$ denotes differentiation and $\theta_2' = 1/x$ is algebraic. Thus B_2 is a constant. Integrating the second equation, and using the fact that B_2 is constant, provides

$$\int A_1 dx = 2B_2 \theta_2 + B_1 - b_1$$

where b_1 is a constant. By recursively applying the Risch algorithm we find that

$$\int A_1 dx = \int \frac{1 + \theta_1}{(\theta_1 + x)^2} dx = \int \frac{1 + e^x}{(e^x + x)^2} dx = -\frac{1}{e^x + x} = -\frac{1}{\theta_1 + x} .$$

As no θ_2 term is involved we find that $B_2 = 0$, and we set $B_1 = \overline{B}_1 + b_1$ with $\overline{B}_1 = -1/(\theta_1 + x)$, and b_1 still an unevaluated constant. The third equation can now be written as

$$A_0 - B_1 \theta_2' = B_0'$$

where

$$A_0 - B_1 \theta_2' = A_0 - \overline{B}_1 \theta_2' - b_1 \theta_2' = -\frac{1}{x(\theta_1 + x)} - \left(-\frac{1}{\theta_1 + x}\right)\frac{1}{x} - b_1 \theta_2' = -b_1 \theta_2'.$$

Integration yields

$$\int (A_0 - B_1 \theta_2') dx = -b_1 \theta_2.$$

Returning to the third equation we find

$$\int (A_0 - B_1 \theta_2') dx = \int B_0' dx$$

which reduces to

$$-b_1 \theta_2 = B_0.$$

This shows that $b_1 = 0$ (since θ_2 is not algebraic) and also $B_0 = 0$. Thus the integral is

$$\int f(x) dx = \theta_2 B_1 = -\theta_2 \frac{1}{\theta_1 + x} = -\ln(x)\frac{1}{e^x + x} .$$

2.15 Commutativity and Non-Commutativity

In computer algebra it is usually assumed that the symbols are commutative. Many mathematical structures are non-commutative. Here we discuss some of these structures. We recall that an *associative algebra* is a vector space V over a field F which satisfies the ring axioms in such a way that addition in the ring is addition in the vector space and such that $c * (A * B) = (c * A) * B = A * (c * B)$ for all $A, B \in V$ and $c \in F$. Moreover the associative law holds, i.e. $A * (B * C) = (A * B) * C$. An example of an associative algebra is the set of the $n \times n$ matrices over the real or complex numbers with matrix multiplication as composition. There, in general, we have

$$A * B \neq B * A.$$

Another important example of a non-commutative structure is that of a *Lie algebra*. A Lie algebra is defined as follows. A vector space L over a field F, with an operation $L \times L \to L$ denoted by

$$(x, y) \to [x, y]$$

and called the commutator of x and y, is called a Lie algebra over F if the following axioms are satisfied.

(L1)   The bracket operation is bilinear.

(L2) $[x, x] = 0$ for all $x \in L$.

(L3) $[x, [y, z]] + [y, [z, x]] + [z, [x, y]] = 0, \qquad x, y, z \in L.$

A simple example of a Lie algebra is the vector product in the vector space $\mathbf{R}^3$.

Remark. The connection between an associative algebra and a Lie algebra is as follows. Let A, B be elements of the associative algebra. We define the commutator as follows

$$[A, B] := A * B - B * A.$$

It can be proved easily that the commutator defined in this way satisfies the axioms given above. Thus we have constructed a Lie algebra from an associative algebra.

Another example of a non-commutative structure are the quaternions. The quaternions have a matrix representation (Section 2.9).

2.16 Tensor and Kronecker Product

Let V, W be vector spaces over a field F. We define the tensor product [33] between elements of V and W. The value of the product should be in a vector space. If we denote $v \otimes w$ as the tensor product of elements $v \in V$ and $w \in W$, then we have the following relations. If $v_1, v_2 \in V$ and $w \in W$, then

$$(v_1 + v_2) \otimes w = v_1 \otimes w + v_2 \otimes w.$$

If $w_1, w_2 \in W$ and $v \in V$, then

$$v \otimes (w_1 + w_2) = v \otimes w_1 + v \otimes w_2.$$

If $c \in F$, then

$$(c * v) \otimes w = c * (v \otimes w) = v \otimes (c * w).$$

We now construct such a product, and prove its various properties.

Let U, V, W be vector spaces over F. By a bilinear map

$$g : V \times W \to U$$

we mean a map which to each pair of elements (v, w) with $v \in V$ and $w \in W$ associates an element $g(v, w)$ of U, having the following property.

For each $v \in V$, the map $w \mapsto g(v, w)$ of W into U is linear, and for each $w \in W$, the map $v \mapsto g(v, w)$ of V into U is linear. For the proofs of the following theorems we refer to the literature [33].

Theorem. Let V, W be finite-dimensional vector spaces over the field F. There exists a finite-dimensional space T over F, and a bilinear map $V \times W \to T$ denoted by

$$(v, w) \mapsto v \otimes w,$$

satisfying the following properties.

1. If U is a vector space over F, and $g : V \times W \to U$ is a bilinear map, then there exists a unique linear map

$$g_* : T \to U$$

such that, for all pairs (v, w) with $v \in V$ and $w \in W$ we have

$$g(v, w) = g_*(v \otimes w).$$

2. If $\{ v_1, \ldots, v_n \}$ is a basis of V, and $\{ w_1, \ldots, w_m \}$ is a basis of W, then the elements

$$v_i \otimes w_j, \qquad i = 1, \ldots, n \quad \text{and} \quad j = 1, \ldots, m$$

form a basis of T.

The space T is called the *tensor product* of V and W, and is denoted by $V \otimes W$. Its dimension is given by

$$\dim(V \otimes W) = (\dim V)(\dim W).$$

The element $v \otimes w$ associated with the pair (v, w) is also called a tensor product of v and w.

Frequently we have to take a tensor product of more than two spaces. We have associativity for this product.

Theorem. Let U, V, W be finite-dimensional vector spaces over F. Then there is a unique isomorphism

$$U \otimes (V \otimes W) \to (U \otimes V) \otimes W$$

such that

$$u \otimes (v \otimes w) \mapsto (u \otimes v) \otimes w$$

for all $u \in U$, $v \in V$ and $w \in W$.

This theorem allows us to omit the parentheses in the tensor product of several factors. Thus if $V_1, \ldots, V_r$ are vector spaces over F, we may form their tensor product

$$V_1 \otimes V_2 \otimes \cdots \otimes V_r$$

and the tensor product

$$v_1 \otimes v_2 \otimes \cdots \otimes v_r$$

of elements $v_i \in V_i$. The theorems described above give the general useful properties of the tensor product.

Next we introduce the *Kronecker product* of two matrices. It can be considered as a realization of the tensor product.

Definition. Let A be an $m \times n$ matrix and let B be a $p \times q$ matrix. Then the Kronecker product of A and B is an $(mp) \times (nq)$ matrix defined by

$$A \otimes B := \begin{pmatrix} a_{11}B & a_{12}B & \cdots & a_{1n}B \\ a_{21}B & a_{22}B & \cdots & a_{2n}B \\ \vdots & & & \\ a_{m1}B & a_{m2}B & \cdots & a_{mn}B \end{pmatrix}.$$

Sometimes the Kronecker product is also called direct product or tensor product. Obviously we have

$$(A + B) \otimes C = A \otimes C + B \otimes C$$

where A and B are matrices of the same size. Analogously

$$A \otimes (B + C) = A \otimes B + A \otimes C$$

where B and C are of the same size. Furthermore, we have

$$(A \otimes B) \otimes C = A \otimes (B \otimes C).$$

Example. Let

$$A = \begin{pmatrix} 2 & 3 \\ 0 & 1 \end{pmatrix}, \qquad B = \begin{pmatrix} 0 & -1 \\ -1 & 1 \end{pmatrix}.$$

Then

$$A \otimes B = \begin{pmatrix} 0 & -2 & 0 & -3 \\ -2 & 2 & -3 & 3 \\ 0 & 0 & 0 & -1 \\ 0 & 0 & -1 & 1 \end{pmatrix}, \quad B \otimes A = \begin{pmatrix} 0 & 0 & -2 & -3 \\ 0 & 0 & 0 & -1 \\ -2 & -3 & 2 & 3 \\ 0 & -1 & 0 & 1 \end{pmatrix}.$$

We see that $A \otimes B \neq B \otimes A$.

Example. Let

$$\mathbf{e}_1 = \begin{pmatrix} 1 \\ 0 \end{pmatrix}, \quad \mathbf{e}_2 = \begin{pmatrix} 0 \\ 1 \end{pmatrix}.$$

Then

$$\mathbf{e}_1 \otimes \mathbf{e}_1 = \begin{pmatrix} 1 \\ 0 \\ 0 \\ 0 \end{pmatrix}, \quad \mathbf{e}_1 \otimes \mathbf{e}_2 = \begin{pmatrix} 0 \\ 1 \\ 0 \\ 0 \end{pmatrix}, \quad \mathbf{e}_2 \otimes \mathbf{e}_1 = \begin{pmatrix} 0 \\ 0 \\ 1 \\ 0 \end{pmatrix}, \quad \mathbf{e}_2 \otimes \mathbf{e}_2 = \begin{pmatrix} 0 \\ 0 \\ 0 \\ 1 \end{pmatrix}.$$

Obviously, $\{\, \mathbf{e}_1, \mathbf{e}_2 \,\}$ is the standard basis in $\mathbf{R}^2$. We see that

$$\{\, \mathbf{e}_1 \otimes \mathbf{e}_1, \quad \mathbf{e}_1 \otimes \mathbf{e}_2, \quad \mathbf{e}_2 \otimes \mathbf{e}_1, \quad \mathbf{e}_2 \otimes \mathbf{e}_2 \,\}$$

is the standard basis in $\mathbf{R}^4$.

2.17   Exterior Product

Next we introduce the *exterior product* (also called *alternating product* or *Grassmann product*). Let V be a finite-dimensional vector space over $\mathbf{R}$, r be an integer ≥ 1 and $V^{(r)}$ be the set of all r-tuples of elements of V, i.e. $V^{(r)} = V \times V \times \ldots \times V$. An element of $V^{(r)}$ is therefore an r-tuple $(\mathbf{v}_1, \ldots, \mathbf{v}_r)$ with each $\mathbf{v}_i \in V$. Let U be another finite-dimensional vector space over $\mathbf{R}$. An r-multilinear map of V into U $f : V \times V \times \ldots \times V \to U$ is linear in each component. In other words, for each $i = 1, \ldots, r$ we have

$$f(\mathbf{v}_1, \ldots, \mathbf{v}_i + \mathbf{v}'_i, \ldots, \mathbf{v}_r) = f(\mathbf{v}_1, \ldots, \mathbf{v}_i, \ldots, \mathbf{v}_r) + f(\mathbf{v}_1, \ldots, \mathbf{v}'_i, \ldots, \mathbf{v}_r)$$

$$f(\mathbf{v}_1, \ldots, c * \mathbf{v}_i, \ldots, \mathbf{v}_r) = c * f(\mathbf{v}_1, \ldots, \mathbf{v}_r)$$

for all $\mathbf{v}_i, \mathbf{v}'_i \in V$ and $c \in \mathbf{R}$.

We say that a multilinear map f is alternating if it satisfies the condition

$$f(\mathbf{v}_1, \ldots, \mathbf{v}_r) = 0$$

whenever two adjacent components are equal, i.e. whenever there exists an index $j < r$ such that $\mathbf{v}_j = \mathbf{v}_{j+1}$. Note that the conditions satisfied by multilinear maps are similar to the properties of the determinants. The following theorem handles the general case of alternating products.

Theorem. Let V be a finite-dimensional vector space over F, of dimension n. Let r be an integer $1 \leq r \leq n$. There exists a finite-dimensional space over F, denoted by $\bigwedge^r V$, and an r-multilinear alternating map $V^{(r)} \to \bigwedge^r V$, denoted by

$$(\mathbf{u}_1, \ldots, \mathbf{u}_r) \mapsto \mathbf{u}_1 \wedge \cdots \wedge \mathbf{u}_r$$

satisfying the following properties.

1. If U is a vector space over F, and $g : V^{(r)} \to U$ is an r-multilinear alternating map, then there exists a unique linear map

$$g_* : \bigwedge^r V \to U$$

such that for all $\mathbf{u}_1, \ldots, \mathbf{u}_r \in V$ we have

$$g(\mathbf{u}_1, \ldots, \mathbf{u}_r) = g_*(\mathbf{u}_1 \wedge \cdots \wedge \mathbf{u}_r).$$

2. If $\{\, \mathbf{v}_1, \ldots, \mathbf{v}_n \,\}$ is a basis of V, then the set of elements

$$\{\, \mathbf{v}_{i_1} \wedge \cdots \wedge \mathbf{v}_{i_r} \,\}, \qquad 1 < i_1 < \cdots < i_r \leq n$$

is a basis of $\bigwedge^r V$.

Thus if $\{\, \mathbf{v}_1, \ldots, \mathbf{v}_n \,\}$ is a basis of V, then every element of $\bigwedge^r V$ has a unique expression as a linear combination

$$\sum_{i_1 < \cdots < i_r}^{n} c_{i_1 \ldots i_r} \mathbf{v}_{i_1} \wedge \cdots \wedge \mathbf{v}_{i_r}$$

the sum being taken over all r-tuples $(i_1, \ldots, i_r)$ of integers from 1 to n, satisfying

$$i_1 < \cdots < i_r.$$

One can shorten this notation, by writing $(i) = (i_1, \ldots, i_r)$. Thus the above sum would be written

$$\sum_{(i)}^{n} c_{(i)} \mathbf{v}_{i_1} \wedge \cdots \wedge \mathbf{v}_{i_r}.$$

The structure introduced above is also called a *Grassmann algebra*.

The dimension of the vector space $\bigwedge^r V$ is given by

$$\binom{n}{r}.$$

Example. We consider the vector space $\mathbf{R}^4$ with the standard basis

$$\mathbf{e}_1, \quad \mathbf{e}_2, \quad \mathbf{e}_3, \quad \mathbf{e}_4.$$

Then a basis for the two forms is given by

$$\mathbf{e}_1 \wedge \mathbf{e}_2, \quad \mathbf{e}_1 \wedge \mathbf{e}_3, \quad \mathbf{e}_1 \wedge \mathbf{e}_4, \quad \mathbf{e}_2 \wedge \mathbf{e}_3, \quad \mathbf{e}_2 \wedge \mathbf{e}_4, \quad \mathbf{e}_3 \wedge \mathbf{e}_4.$$

A basis for the three forms is given by

$$\mathbf{e}_1 \wedge \mathbf{e}_2 \wedge \mathbf{e}_3, \quad \mathbf{e}_1 \wedge \mathbf{e}_2 \wedge \mathbf{e}_4, \quad \mathbf{e}_1 \wedge \mathbf{e}_3 \wedge \mathbf{e}_4, \quad \mathbf{e}_2 \wedge \mathbf{e}_3 \wedge \mathbf{e}_4.$$

The basis for the four forms consists of only one element, namely

$$\mathbf{e}_1 \wedge \mathbf{e}_2 \wedge \mathbf{e}_3 \wedge \mathbf{e}_4.$$

The *determinant* of a square matrix can be calculated using the exterior product. Consider the 4×4 matrix

$$\begin{pmatrix} 1 & 2 & 5 & 2 \\ 0 & 1 & 2 & 3 \\ 1 & 0 & 1 & 0 \\ 0 & 3 & 0 & 7 \end{pmatrix}$$

then

$$\begin{pmatrix} 1 \\ 0 \\ 1 \\ 0 \end{pmatrix} \wedge \begin{pmatrix} 2 \\ 1 \\ 0 \\ 3 \end{pmatrix} \wedge \begin{pmatrix} 5 \\ 2 \\ 1 \\ 0 \end{pmatrix} \wedge \begin{pmatrix} 2 \\ 3 \\ 0 \\ 7 \end{pmatrix} = 24\, \mathbf{e}_1 \wedge \mathbf{e}_2 \wedge \mathbf{e}_3 \wedge \mathbf{e}_4$$

where $\mathbf{e}_1, \mathbf{e}_2, \mathbf{e}_3, \mathbf{e}_4$ is the standard basis of $\mathbf{R}^4$. Thus 24 is the determinant of the matrix. In Chapter 9, we give an implementation of this (obviously slow) technique to find the determinant.

Chapter 3

Computer Algebra Systems

3.1   Introduction

We survey some computer algebra systems and give some applications. There are a large number of computer algebra systems available, so we concentrate on a few of them. Reduce, Axiom and Maxima are based on LISP. On the other hand Mathematica, Maple and MuPAD are based on C. GiNaC is a symbolic computation framework for C++. Besides symbolic manipulations, all systems can also do numerical manipulations. The systems are not only an interactive environment operating in response to on-line commands, but all of them (except GiNaC) are also programming languages.

Reduce is one of the oldest computer algebra systems around. It is based on Portable Standard Lisp. The system was designed in the late 1960s by Anthony C. Hearn [24][48][50]. It allows one to mix Reduce and Lisp code. Reduce version 3.8 is available at

http://www.zib.de/Symbolik/reduce, http://www.reduce-algebra.com.

Maple [12] is a product of Waterloo Maple Software. It is a system for mathematical computation — symbolic, numerical and graphical.

Axiom is a symbolic, numerical and graphical system developed at the IBM Thomas J. Watson Research Center. It gives the user all foundation and algebra instruments necessary to develop a computer realization of sophisticated mathematical objects in exactly the way a mathematician would do it. Axiom is distributed by The Numerical Algorithms Group Limited [29]. NAG agreed to release the Axiom source code after withdrawing the product in October 2001. Axiom source and information can be found at

http://savannah.nongnu.org/projects/axiom,
http://axiom-wiki.newsynthesis.org/

The FriCAS system, which is derived from Axiom, is available from http://fricas.sourceforge.net.

Mathematica [64] is a product of Wolfram Research, Inc. It is a general computer software system and language which handles symbolic, numerical and graphical computations.

MuPAD is a symbolic-numerical computer algebra system developed by the Department of Mathematics of the University of Paderborn [41]. MuPAD lets us define our own data types.

Macsyma is a product of Macsyma Inc. It is based on LISP. It implements the Aslaksen test for complex functions [3]. A detailed description of Macsyma is given by Davenport et al. [15]. Maxima is derived from the Macsyma system. It is implemented in common LISP. Macsyma was developed at MIT for the Department of Energy. In 1998, Professor W. F. Schelter obtained permission from the Department of Energy to release the DOE Macsyma source code under the GNU Public License as the Maxima project at http://maxima.sourceforge.net.

GiNaC [8] stands for GiNaC is not a CAS, (where CAS is short for computer algebra system). It is written entirely in C++, providing both a limited interactive shell as well as a framework for symbolic computation in C++. It was originally designed to provide efficient handling of multivariate polynomials, algebras and special functions which are needed for loop calculations in theoretical quantum field theory. GiNaC is available under the GNU Public License from http://www.ginac.de.

Besides the computer algebra systems described above there are a number of other excellent computer algebra systems available

- Derive is a small but powerful computer algebra system of the Software Warehouse, Hawaii. It is also based on LISP. Derive applies the rules of algebra, trigonometry, calculus, and matrix algebra to solve a wide range of mathematical problems. It also has graphical capabilities. Derive is available from mathware

  http://www.mathware.com/derive.html.

- Magma (Computational Algebra Group, School for Mathematics and Statistics, University of Sydney) is a system for computation in algebraic, geometric and combinatorical structures such as groups, rings, fields, algebras, modules, graphs and codes. It can be accessed at

  http://magma.maths.usyd.edu.au/magma.

- MathCad is a product of MathSoft, Cambridge, Massachusetts. It provides a platform for engineers, scientists and academics to perform, share and document symbolic and numerical calculations.

- Yacas [43], available at `http://yacas.sourceforge.net`, provides a high level weakly typed functional programming language for quick prototyping of computer algebra algorithms and for general symbolic manipulation.

Further packages exist for special tasks such as number theory, graph theory and algebra.

3.2 Reduce

3.2.1 Basic Operations

We give a summary of the most commonly used commands in Reduce. Reduce does not distinguish between capital and small letters. Thus the commands `sin(x)`, `Sin(x)` or `SIN(X)` are the same.

The commands we use most in this book are differentiation and integration. To differentiate the polynomial $x^3 + 2x$ with respect to x we write

```
df(x**3+2*x,x);
```

The output is `3*x^2 + 2`. To integrate $x^2 + 1$ we write

```
int(x**2+1,x);
```

The output is `x*(x^2 + 3)/3`. Both `**` and `^` denote the power operator.

The command `solve()` solves a number of algebraic equations, systems of algebraic equations and transcendental equations. For example, the command

```
solve(x**2 + (a+1)*x + a=0,x);
```

gives the solution `x = -1` and `x = -a`.

Another important command is the substitution command `sub()`. For example, the command

```
sub(x=2,x*y+x**2);
```

yields `2*y + 4`.

Amongst others, Reduce includes the following mathematical functions: `sqrt(x)` (square root, $\sqrt{x}$), `exp(x)` (exponential function, $\exp(x)$), `log(x)` (natural logarithm, $\ln(x)$), and the trigonometric functions `sin(x)`, `cos(x)`, `tan(x)` with arguments in radians.

Reduce reserves `i` (or `I`) to represent $\sqrt{-1}$ and `pi` (or `PI`, `pI`, `Pi`) for the number π. Thus the input

```
i*i;
```

gives `-1` and

```
sin(pi);
```

results in 0. Other predefined constants are T (or t), nil, E, Infinity where T
stands for true and nil stands for false.

For differentiation Reduce offers two options for implementation. In the first option
we declare the function to be differentiated as operator. The following example
shows how to use this option. The 2 in df() indicates that we differentiate twice.

```
operator f;
f(x) := x*x + sin(x);
result := df(f(x),x,2);
```

The output is

```
result := -sin(x) + 2
```

On the other hand we can also declare that f depends on x and then differentiate
f with respect to x.

```
depend f, x;
f := x*x + sin(x);
result := df(f,x,2);
```

In Reduce the default data type for numbers is rational numbers. The command

```
2 + 0.1 + 1/3;
```

gives the output

```
73/30
```

The switch **on rounded** allows the calculation with real numbers. The commands

```
on rounded;
2 + 0.1 + 1/3;
```

gives the output

```
2.43333333333
```

Reduce only knows the most elementary identities such as

```
cos(-x)    = cos(x)
sin(pi)    = 0
log(e)     = 1
e^(i*pi/2) = i
```

The user can add further rules for the reduction of expressions by using the LET
command, such as the trigonometry identity

```
for all x let sin(x)^2 + cos(x)^2 = 1;
```

then

```
sin(y)^2 + cos(y)^2 - 5;
```

gives the output

```
-4
```

For other commands we refer to the user's manual for Reduce [24].

3.2.2 Example

We show how soliton equations can be derived from pseudospherical surfaces, using the sine-Gordon equation can be derived. Extensions to other soliton equations are straightforward.

Soliton equations can be described by pseudospherical surfaces, i.e. surfaces of constant negative Gaussian curvature. An example is the *sine-Gordon equation*

$$\frac{\partial^2 u}{\partial x_1 \partial x_2} = \sin(u).$$

Here we show how Reduce can be used to find the sine-Gordon equation from the line element of the surface. The *metric tensor field* is given by

$$g = dx_1 \otimes dx_1 + \cos(u(x_1, x_2))dx_1 \otimes dx_2 + \cos(u(x_1, x_2))dx_2 \otimes dx_1 + dx_2 \otimes dx_2$$

i.e. the *line element* is

$$\left(\frac{ds}{d\lambda}\right)^2 = \left(\frac{dx_1}{d\lambda}\right)^2 + 2\cos(u(x_1, x_2))\frac{dx_1}{d\lambda}\frac{dx_2}{d\lambda} + \left(\frac{dx_2}{d\lambda}\right)^2.$$

Here u is a smooth function of x_1 and x_2. First we have to calculate the Riemann curvature scalar R from g. Then the sine-Gordon equation follows when we impose the condition

$$R = -2.$$

The calculation of the curvature scalar is well described in many textbooks (see for example [53]). For the sake of completeness we give the equations. We have

$$g_{11} = g_{22} = 1, \qquad g_{12} = g_{21} = \cos(u(x_1, x_2)).$$

The quantity g can be written in matrix form

$$g = \begin{pmatrix} g_{11} & g_{12} \\ g_{21} & g_{22} \end{pmatrix}.$$

Then the inverse of g is given by

$$g^{-1} = \begin{pmatrix} g^{11} & g^{12} \\ g^{21} & g^{22} \end{pmatrix}$$

where

$$\cdot g^{11} = g^{22} = \frac{1}{\sin^2 u}, \qquad g^{12} = g^{21} = -\frac{\cos u}{\sin^2 u}.$$

Next we have to calculate the *Christoffel symbols*. They are defined as

$$\Gamma^a_{mn} := \frac{1}{2} g^{ab}(g_{bm,n} + g_{bn,m} - g_{mn,b})$$

where the *sum convention* (i.e. we sum over b from 1 to 2 in the present case) is used and

$$g_{bm,1} := \frac{\partial g_{bm}}{\partial x_1}, \qquad g_{bm,2} := \frac{\partial g_{bm}}{\partial x_2}.$$

Next we have to calculate the *Riemann curvature tensor* which is given by

$$R^r_{msq} := \Gamma^r_{mq,s} - \Gamma^r_{ms,q} + \Gamma^r_{ns}\Gamma^n_{mq} - \Gamma^r_{nq}\Gamma^n_{ms}.$$

The *Ricci tensor* follows as

$$R_{mq} := R^a_{maq} = -R^a_{mqa}$$

i.e. the Ricci tensor is constructed by contraction. From R_{nq} we obtain R^m_q via

$$R^m_q = g^{mn} R_{nq}.$$

Finally the *curvature scalar* R is given by

$$R := R^m_m.$$

With the metric tensor field given above we find that

$$R = -\frac{2}{\sin u} \frac{\partial^2 u}{\partial x_1 \partial x_2}.$$

If $R = -2$ then we obtain the sine-Gordon equation.

We apply the concept of operators. Operators are the most general objects available in Reduce. They are usually parametrized, and can be parametrized in a completely general way. Only the operator identifier is declared in an operator declaration. The number of parameters is not declared. Operators represent mathematical operators or functions. We declare u as an operator and it depends on x. x itself is also declared as an operator and depends on 1 and 2. Since terms of the form $\cos^2(x)$ and $\sin^2(x)$ result from our calculation we have to include the identity

$$\sin^2(u) + \cos^2(u) \equiv 1$$

in order to simplify expressions.

```
% tensor.red

matrix g(2,2);
matrix g1(2,2);  % inverse of g;
array gamma(2,2,2); array R(2,2,2,2); array Ricci(2,2);

operator u, x;
```

```
g(1,1) := 1;   g(2,2) := 1;
g(1,2) := cos(u(x(1),x(2)));  g(2,1) := cos(u(x(1),x(2)));

g1 := g^(-1);     % calculating the inverse
for a := 1:2 do
   for m := 1:2 do
      for n := 1:2 do
      gamma(a,m,n) := (1/2)*
                         (for b := 1:2 sum g1(a,b)*(df(g(b,m),x(n))
                             + df(g(b,n),x(m)) - df(g(m,n),x(b))));

for a := 1:2 do
   for m := 1:2 do
      for n := 1:2 do
      write "gamma(",a,",",m,",",n,") = ", gamma(a,m,n);

for b := 1:2 do
   for m := 1:2 do
      for s := 1:2 do
         for q := 1:2 do
         R(b,m,s,q) := df(gamma(b,m,q),x(s))-df(gamma(b,m,s),x(q))
                         + (for n := 1:2 sum gamma(b,n,s)*gamma(n,m,q))
                         - (for n := 1:2 sum gamma(b,n,q)*gamma(n,m,s));

cos(u(x(1),x(2)))**2 := 1 - sin(u(x(1),x(2)))**2;

for m := 1:2 do
   for q := 1:2 do
   Ricci(m,q) := for s := 1:2 sum R(s,m,s,q);

for m := 1:2 do
   for q := 1:2 do
   write "Ricci(",m,",",q,") = ", Ricci(m,q);

array Ricci1(2,2);
for m := 1:2 do
   for q := 1:2 do
   Ricci1(m,q) := (for b := 1:2 sum g1(m,b)*Ricci(q,b));

CS := for m := 1:2 sum Ricci1(m,m);
```

The calculation of the curvature scalar from a metric tensor field (for example Gödel metric, Schwarzschild metric, Kerr metric, anti-de Sitter metric) is one of the oldest applications of computer algebra. Here we showed that it can be extended to find soliton equations. By modifying the metric tensor field we can obtain other soliton equations. For example

```
g(1,2) := cos(u(x(1),x(2)));   g(2,1) := cos(u(x(1),x(2)));
```

could be replaced by

```
g(1,2) := cosh(u(x(1),x(2)));   g(2,1) := cosh(u(x(1),x(2)));
```

Additionally, we have to include the identity rule

```
sinh(u(x(1),x(2)))**2 := cosh(u(x(1),x(2)))**2 - 1;
```

in order to simplify expressions.

A large number of application programs in Reduce can be found in [14], [25], [50], [52], [53]. In [50] applications in quantum mechanics are described. In [52] applications for differential equations are given. In [51] applications for nonlinear dynamical systems are provided.

3.3 Maple

3.3.1 Basic Operations

Maple distinguishes between small and capital letters. The command `Sin(0.1)` gives `Sin(.1)`, whereas `sin(0.1)` gives the desired result `0.09983341665`.

In Maple the differentiation command is `diff()`. The input

```
diff(x^3 + 2*x,x);
```

yields as output `3x^2 + 2`. The integration command is `int()`. The input

```
int(x^2 + 1,x);
```

yields `x^3/3 + x`.

Maple has two different commands for solving equations. The command

```
solve(x^2 + (1+a)*x + a=0,x);
```

solves the equation `x^2+(1+a)x+a=0` and gives the result `x = -1` and `x = -a`. The command

```
fsolve(x^2 - x - 1=0,x)
```

solves `x^2-x-1.0=0` and gives the output `-0.6180339887` and `1.618033989`.

The substitution command is given by `subs()`. For example, the command

```
subs(x=2,x*y + x^2);
```

gives `2y + 4`.

Amongst others, Maple includes the following mathematical functions: `sqrt(x)` (square root, $\sqrt{x}$), `exp(x)` (exponential function, e^x), `log(x)` (natural logarithm, $\ln(x)$), and the trigonometric functions `sin(x)`, `cos(x)`, `tan(x)` with arguments in radians.

Predefined constants are

```
Catalan, E, Pi, false, gamma, infinity, true.
```

For other commands we refer to the user manual for Maple [12].

3.3.2 Example

We consider a quantum mechanical problem. Given a trial function for a one-dimensional potential, we find an approximation for the ground state energy. The *eigenvalue equation* in one-space dimension is given by

$$-\frac{\hbar^2}{2m}\frac{d^2u}{dx^2} + V(x)u(x) = Eu(x).$$

We use the variational principle to estimate the ground state energy of a particle in the potential

$$V(x) := \begin{cases} cx & \text{for } x > 0 \\ \infty & \text{for } x < 0 \end{cases}$$

where $c > 0$. Owing to this potential the spectrum is discrete and bounded from below. We use

$$u(x) = \begin{cases} x\exp(-ax) & \text{for } x > 0 \\ 0 & \text{for } x < 0 \end{cases}$$

as a *trial function*, where $a > 0$. Note that the trial function is not yet normalized. From the eigenvalue equation we find that the *expectation value* for the energy is given by

$$\langle E \rangle := \frac{\langle u|\hat{H}|u\rangle}{\langle u|u\rangle} = \frac{\int_0^\infty xe^{-ax}\left(-\frac{\hbar^2}{2m}\frac{d^2}{dx^2} + cx\right)xe^{-ax}dx}{\int_0^\infty x^2\exp(-2ax)dx} = \frac{3c}{2a} + \frac{\hbar^2a^2}{2m}$$

where $\langle \,|\, \rangle$ denotes the scalar product in the Hilbert space $L_2(0,\infty)$. The expectation value for the energy depends on the parameter a. The expectation value has a minimum for

$$a = \left(\frac{3mc}{2\hbar^2}\right)^{1/3}.$$

In the program we evaluate $\langle E \rangle$ and then determine the minimum of $\langle E \rangle$ with respect to the parameter a. Thus the ground-state energy is greater than or equal to

$$\frac{9}{4}\left(\frac{2\hbar^2c^2}{3m}\right)^{1/3}.$$

```
# energy.map

# potential
V := c*x;
# trial ansatz
u := x*exp(-a*x);
# eigenvalue equation
Hu := -hb^2/(2*m)*diff(u,x,x) + V*u;

# integrating for finding expectation value
# not normalized yet
res1 := int(u*Hu,x);

# collect the exponential functions
```

```
res2 := collect(res1,exp);

# substitution of the boundary
res3 := 0 - subs(x=0,res2);

# finding the norm of u to normalize
res4 := int(u*u,x);
res5 := -subs(x=0,res4);

# normalized expectation value
expe := res3/res5;

# finding the minimum with respect to a
minim := diff(expe,a);
res6 := solve(minim=0,a);
a := res6[1];

# approximate ground state energy
appgse = subs(a=0,expe);
```

Remark. Only the real solution of `res6`, namely `res6[1]` is valid in our case.

3.4　Axiom

3.4.1　Basic Operations

Axiom emphasizes strict type checking. Unlike other computer algebra systems, types in Axiom are dynamic objects. They are created at run-time in response to user commands. Types in Axiom range from algebraic types (e.g. polynomials, matrices and power series) to data structures (e.g. lists, dictionaries and input files). Types may be combined in meaningful ways. We may build polynomials of matrices, matrices of polynomials of power series, hash tables with symbolic keys and rational function entries and so on.

Categories in Axiom define the algebraic properties which ensure mathematical correctness. Through categories, programs may discover that polynomials of continued fractions have commutative multiplication whereas polynomials of matrices do not. Likewise, a greatest common divisor algorithm can compute the "gcd" of two elements for any Euclidean domain, but foil the attempts to compute meaningless "gcds" of two hash tables. Categories also enable algorithms to be compiled into machine code that can be ruled with arbitrary types.

Type declarations in Axiom can generally be omitted for common types in the interactive language. Basic types are called *domains of computation*, or simply, *domains*. Domains are defined in the form

```
Name(...): Exports == Implementation
```

Each domain has a capitalized `Name` that is used to refer to the class of its members. For example, `Integer` denotes "the class of integers", whereas `Float` denotes "the class of floating point numbers" and so on. The "..." part following `Name` lists the parameter(s) for the constructor. Some basic types like `Integer` take no parameters. Others, like `Matrix`, `Polynomial` and `List`, take a single parameter that again must be a domain. For example,

> `Matrix(Integer)`   denotes "matrices over the integers"
> `Polynomial(Float)` denotes "polynomial with floating point coefficients"

There is no restriction on the number of types of parameters of a domain constructor.

The `Exports` part in Axiom specifies the operations for creating and manipulating objects of the domain. For example, the `Integer` type exports constants 0 and 1, and operations `+`, `-` and `*`. The `Implementation` part defines functions that implement the exported operations of the domain. These functions are frequently described in terms of another lower-level domain used to represent the objects of the domain.

Every Axiom object belongs to a unique domain. The domain of an object is also called its type. Thus the integer 7 has type `Integer` and the string `"willi"` has type `String`. The type of an object, however, is not unique. The type of the integer 7 is not only an `Integer` but also a `NonNegativeInteger`, a `PositiveInteger` and possibly any other "subdomain" of the domain `Integer`.

A subdomain is a domain with a "membership predicate". `PositiveInteger` is a subdomain of `Integer` with the predicate "is the integer > 0?". Subdomains with names are defined by abstract data type programs similar to those for domains. The `Exports` part of a subdomain, however, must list a subset of the exports of the domain. The `Implementation` part optionally gives special definitions for subdomain objects. The following gives some examples in Axiom.

Axiom uses `D` to differentiate an expression

```
f := exp exp x

            x
         %e
   (1)  %e

D(f,x)

                x
         x  %e
   (2)  %e %e
```

An optional third argument `n` in `D` instructs Axiom to find the `n`-th derivative of `f`, e.g. `D(f,x,3)`.

Axiom has extensive library facilities for integration. For example

```
integrate((x**2+2*x+1)/((x+1)**6+1),x)
```

yields

$$\frac{\arctan(x^3 + 3x^2 + 3x + 1)}{3}.$$

Axiom uses the `rule` command to describe the transformation rules one needs. For example

```
sinCosExpandRules := rule
   sin(x+y) == sin(x)*cos(y) + sin(y)*cos(x)
   cos(x+y) == cos(x)*cos(y) - sin(x)*sin(y)
   sin(2*x) == 2*sin(x)*cos(x)
   cos(2*x) == cos(x)**2 - sin(x)**2
```

Thus the command

```
sinCosExpandRules(sin(a+2*b+c))
```

applies the rules implemented above. For more commands, we refer to the literature [29].

3.4.2   Example

In solving systems of polynomial equations, *Gröbner basis* theory plays a central role [15], [29], [38]. If the polynomials

$$\{\, p_j \ : \ j = 1, \ldots, n \,\}$$

vanish totally, so does the combination

$$\sum_j c_j p_j$$

where the coefficients c_j are in general also polynomials. All possible such combinations generate a space called an *ideal*. The ideal generated by a family of polynomials consists of the set of all linear combinations of those polynomials with polynomial coefficients. A system of generators (or a basis) G for an ideal I is a Gröbner basis (with respect to an ordering $<$) if every complete reduction of an $f \in I$ (with respect to G) gives zero. The ordering could be lexicographic (meaning lexicographic variable ordering followed by variable degree ordering), inverse lexicographic (meaning inverse lexicographic then variable degree), total degree (meaning total degree then lexicographic) and inverse total degree (meaning total degree then inverse lexicographic). Most computer algebra systems can find the Gröbner basis for a given system of polynomials. Here we consider Axiom.

DMP stands for DistributedMultivariatePolynomial. Consider the commands

```
(d1, d2, d3) : DMP([z,y,x],FRAC INT)
d1 := -4*z + 4*y**2*x + 16*x**2 + 1
d2 := 2*z*y**2 + 4*x + 1
d3 := 2*z*x**2 - 2*y**2 - x
groebner [d1,d2,d3]
```

This gives the output

$$z - \frac{1568}{2745}x^6 - \frac{1264}{305}x^5 + \frac{6}{305}x^4 + \frac{182}{549}x^3 - \frac{2047}{610}x^2 + \frac{103}{2745}x - \frac{2857}{10980}$$

$$y^2 + \frac{112}{2745}x^6 - \frac{84}{305}x^5 - \frac{1264}{305}x^4 - \frac{13}{549}x^3 + \frac{84}{305}x^2 + \frac{1772}{2745}x + \frac{2}{2745}$$

$$x^7 + \frac{29}{4}x^6 - \frac{17}{16}x^4 - \frac{11}{8}x^3 + \frac{1}{32}x^2 + \frac{15}{16}x + \frac{1}{4}.$$

3.5 Mathematica

3.5.1 Basic Operations

Mathematica distinguishes between capital letters and small letters. The command `sin[0.1]` for the evaluation of the sine of 0.1 gives the error message, `possible spelling error`, whereas `Sin[0.1]` gives the right answer 0.0998334.

In Mathematica the differentiation command

```
D[x^3 + 2*x,x]
```

gives the output `2 + 3 x^2`. The command

```
Integrate[x^2 + 1,x]
```

gives `x + x^3/3`.

The command `Solve[ ]` can solve a number of algebraic equations, systems of algebraic equations and transcendental equations. For example the command

```
Solve[x^2 + (a+1)*x + a==0,x]
```

gives the solution `x = -1` and `x = -a`.

The replacement operator `/.` applies rules to expressions. Consider the expression

```
x*y + x*x
```

Then

```
x*y + x*x /. x -> 2
```

yields as output `4 + 2*y`.

Amongst others, Mathematica includes the following mathematical functions: `Sqrt[x]` (square root, $\sqrt{x}$), `Exp[x]` (exponential function, e^x), `Log[x]` (natural logarithm, $\ln(x)$), and the trigonometric functions `Sin[x]`, `Cos[x]`, `Tan[x]` with arguments in radians.

Predefined constants are

```
 I, Infinity, Pi, Degree, GoldenRatio, E, EulerGamma, Catalan
```

For other commands we refer to the user's manual for Mathematica [64].

3.5.2   Example

We consider the *spin-1 matrices*

$$s_+ := \begin{pmatrix} 0 & \sqrt{2}\hbar & 0 \\ 0 & 0 & \sqrt{2}\hbar \\ 0 & 0 & 0 \end{pmatrix}, \qquad s_- := \begin{pmatrix} 0 & 0 & 0 \\ \sqrt{2}\hbar & 0 & 0 \\ 0 & \sqrt{2}\hbar & 0 \end{pmatrix}.$$

We calculate the *commutator* of the two matrices

$$[s_+, s_-] := s_+ s_- - s_- s_+$$

and then determine the eigenvalues of the commutator. The Mathematica program is as follows

```
(* spin.m *)

sp = {{ 0, Sqrt[2]*hb, 0}, {0, 0, Sqrt[2]*hb}, {0, 0, 0}}
sm = {{0, 0, 0}, {Sqrt[2]*hb, 0, 0}, {0, Sqrt[2]*hb, 0}}
comm = sp . sm - sm . sp
Eigenvalues[comm]
```

The output is

```
            2       2
{0, -2 hb , 2 hb }
```

3.6   MuPAD

3.6.1   Basic Operations

MuPAD is a computer algebra system which has been developed mainly at the University of Paderborn. It is a symbolic, numeric and graphical system. MuPAD provides native parallel instructions to the user. MuPAD syntax is close to that of MAPLE and has object-oriented capabilities close to that of AXIOM. MuPAD distinguishes between small and capital letters. The command Sin(0.1) gives Sin(0.1), whereas sin(0.1) gives the desired result 0.09983341664. In MuPAD the differentiation command is diff(). The input

```
diff(x^3 + 2*x,x);
```

yields as output

```
         2
    3 x  + 2
```

The integration command is int(). The input

```
int(x^2 + 1,x);
```

yields

```
             3
            x
       x + --
            3
```

The command

```
solve(x^2 + (1+a)*x + a=0,x);
```

solves the equation `x^2 + (1+a)*x + a=0` and gives the result `{-a, -1}`. The substitution command is given by `subs()`. For example, the command

```
subs(x*y + x^2,x=2);
```

gives `2 y + 4`. Amongst others, MuPAD includes the following mathematical functions: `sqrt(x)` (square root, $\sqrt{x}$), `exp(x)` (exponential function, e^x), `ln(x)` (natural logarithm), and the trigonometric functions `sin(x)`, `cos(x)`, `tan(x)` with arguments in radians.

In MuPAD we use `:=` for assignment and `=` for equality (equations).

Predefined constants are

```
I, PI, E, EULER, TRUE, FALSE, gamma, infinity
```

3.6.2 Example

We consider *Picard's method* to approximate a solution to the differential equation $dy/dx = f(x, y)$ with initial condition $y(x_0) = y_0$, where f is an analytic function of x and y. Integrating both sides yields

$$y(x) = y_0 + \int_{x_0}^{x} f(s, y(s))ds.$$

Now starting with y_0 this formula can be used to approach the exact solution iteratively if the procedure converges. The next approximation is given by

$$y_{n+1}(x) = y_0 + \int_{x_0}^{x} f(s, y_n(s))ds.$$

The example approximates the solution of $dy/dx = x + y$ using five steps of Picard's method with initial value $x_0 = 0$ and $y(x_0) = 1$. To input a file the command `read(filename)` is used. In this case the command `read("picard");` gives the output

```
              3    4    5    6
         2   x    x    x    x
    x + x  + -- + -- + -- + --- + 1
              3   12   60   720
```

```
/* picard.mpd */

x0:=0:          /*initial x*/
y0:=1:          /*initial y*/
y:=y0:
y1:=subs(y,x=s):
f:=(x,y)->(x+y):            /*declare function f(x,y)=x+y*/

for i from 1 to 5 do
   y:=(y0+subs(int(f(s,y1),s),s=x)-subs(int(f(s,y1),s),s=x0)):
   y1:=subs(y,x=s):
end_for:

print(y);
```

3.7   Maxima

3.7.1   Basic Operations

Maxima distinguishes between small and capital letters. Thus for `sin(0.1)` Maxima outputs `0.0998334166468282` whereas for `Sin(0.1)` Maxima outputs `Sin(0.1)`.

The command for differentiation is `diff()`. For example,

```
diff(x^3 + 2*x,x);
```

yields the result

$$3\,x^2 + 2$$

and the command for integration is `integrate()`,

```
integrate(x^2 + 1,x);
```

yields

$$\frac{x^3}{3} + x$$

Equations are solved using `solve()`

```
solve(x^2 + (a+1)*x + a, x);
                [x = - a,  x = - 1]
```

Substitution is performed using the `substitute()` command, or the shorter form `subst`

```
substitute(x=2, x*y +x^2);
                2 y + 4
subst(x=2, x*y +x^2);
                2 y + 4
```

Maxima includes, amongst others, the functions `sqrt(x)` (square root, $\sqrt{x}$), `exp(x)` or equivalently `%e^x` (exponential function, e^x), `log(x)` (natural logarithm, $\ln x$), and the trigonometric function `sin(x)`, `cos(x)`, `tan(x)` with arguments in radians.

Maxima has positive infinity when working with real values (`inf`) negative infinity for real values (`minf`) and infinity when working with complex numbers (`infinity`) as well as the numerical constants e and π written as `%e` and `%pi`. The Boolean constants `true` and `false` are also available.

3.7.2 Example

The Risch algorithm (Section 2.14) is implemented in Maxima. We integrate the function

```
f:(-exp(x)-x+log(x)*x+log(x)*x*exp(x))/(x*(exp(x)+x)^2);
                    x                      x
           x %e  log(x) + x log(x) - %e  - x
           --------------------------------
                        x    2
                  x (%e  + x)
integrate(f,x);
                       log(x)
                       ------
                         x
                       %e  + x
```

3.8 GiNaC

3.8.1 Basic Operations

GiNaC is implemented in C++ and relies on C++ to provide the standard programming constructs such as data types, assignment, loops etc. The class `symbol` provides symbolic variables while `ex` is for arbitrary symbolic expressions. GiNaC provides overloaded operators for the standard arithmetic operations.

Since GiNaC is embedded in C++ it distinguishes between small and capital letters.

In GiNaC the differentiation member function is `diff()`. The input

```
(pow(x,3) + 2*x).diff(x);
```

yields $3x^2 + 2$.

The substitution member function is given by `subs()`. For example, the command

```
(x*y + pow(x,2)).subs(x==2);
```

gives $2y + 4$.

Amongst others, GiNaC includes the following overloaded mathematical functions: sqrt(x) (square root, $\sqrt{x}$), exp(x) (exponential function, e^x), log(x) (natural logarithm, $\ln(x)$), and the trigonometric functions sin(x), cos(x), tan(x) with arguments in radians.

Predefined constants are

```
Catalan,  Pi,  Euler
```

For other methods we refer to the user manual and tutorial for GiNaC available from http://www.ginac.de.

3.8.2 Example

The following example is taken from [8]. A similar example in SymbolicC++ is given in Section 9.6.1.

The *Hermite polynomials* are defined by

$$H_n(x) := (-1)^n e^{x^2} \frac{d^n}{dx^n} e^{-x^2}$$

where $n = 0, 1, 2, \ldots$. The following C++ program uses GiNaC to generate the Hermite polynomials.

```cpp
#include <ginac/ginac.h>
using namespace GiNaC;

ex HermitePoly(const symbol &x,int n)
{
 const ex HGen = exp(-pow(x,2));
 return normal(pow(-1,n)*HGen.diff(x,n)/HGen);
}


int main(void)
{
 symbol z("z");
 ex H = HermitePoly(z,11);
 // output the 11th Hermite polynomial
 cout << H << endl;
 // output the numerical value at z = 0.8
 cout << H.subs(z == 0.8) << endl;
 // output the numerical value at z = 0.8 (exact)
 cout << H.subs(z == numeric("4/5")) << endl;

 return 0;
}
```

Chapter 4

Tools in C++

C++ not only corrects most of the deficiencies found in C, it also introduces many completely new features that were designed for the language to provide data abstraction and object-oriented programming. Here are some of the features:

- *Classes*, the basic language construct that consists of data structure and operations applicable to the class.

- *Member variables*, which describe the attributes of the class.

- *Member functions*, which define the permissible operations of the class.

- *Operator overloading*, which gives additional meaning to operators so that they can be used with user-defined data types.

- *Function overloading*, which is similar to operator overloading. It allows the same function to have several definitions whereby reducing the need for unusual Tnames, making code easier to read.

- *Programmer-controlled automatic type conversion*, which allows us to blend user-defined types with the fundamental data types provided by C++.

- *Derived classes*, also known as subclasses, inherit member variables and member functions from their base classes (also known as superclasses). They can be differentiated from their base classes by adding new member variables, member functions or overriding existing functions.

- *Virtual functions*, which allow a derived class to redefine member functions inherited from a base class. Through *dynamic binding*, the run-time system will choose an appropriate function for the particular class.

- *Templates*, which address many of the problems encountered with C macros.

In the next few sections, we demonstrate how to apply these new features to create data types for particular applications, and combine these abstract data types into object-oriented programs.

4.1   Pointers and References

A *pointer* is a reference to data or code in a program. It is literally the address in memory of the item pointed at. Pointers enables us to write more flexible programs, especially for object-oriented programs. We use pointers when the following situations occur:

- *If the program uses data of unknown size at compile time.*

- *If the program uses temporary data buffers.*

- *If the program handles multiple data types.*

- *If the program uses linked lists and trees.*

- *If the program uses polymorphism.*

Another common use of pointers is to tie together *linked lists*. In many simple database-type applications, we can hold data records in arrays or typed files. However sometimes we need something more flexible than a fixed-size array. By allocating dynamic records so that each record has a pointer that points to the next record, we can construct a list that contains as many elements as we need.

A pointer is a type of variable which is used to store a computer memory address. It could be the address of a variable, a data record, a function or a pointer. Normally, we do not care where the variable resides in memory. All we need is to refer to it by name. The compiler knows where to look for it. That is exactly what happens when we declare a variable. For example, if the program includes the following code,

```
int number;
```

The compiler will set aside an area in memory which is referred to as **number**. We can find out the memory address of **number** by using the & operator. This memory address can be assigned to a *pointer variable*, which holds the address of the data or code in memory. So far we have seen how values are assigned to pointers, but that is not much use if we cannot get the values back. We could treat a typed pointer as if it were a variable of the type by *dereferencing* it. To dereference a pointer, we use the *(prefix) operator. Suppose p contains the memory address of **number**, *p gets the value of **number**. Consider the following example

```
// pointer.cpp

#include <iostream>
using namespace std;

int main(void)
{
  double *p1, *p2;        // declares two pointers p1 and p2 to double
  double a = 2.5, b = 5.1;
  p1 = &a;                // p1 holds the address of a
  p2 = p1;
  cout << "*p1 = " << *p1 << " and *p2 = " << *p2 << endl;
```

```
  *p2 = b;                // *p2 is a variable of type double
  cout << "*p1 = " << *p1 << " and *p2 = " << *p2 << endl;
  return 0;
}
/*
Result
======
*p1 = 2.5 and *p2 = 2.5
*p1 = 5.1 and *p2 = 5.1
*/
```

In the first `cout` statement, both `p1` and `p2` pointed at `a`. Therefore, `*p1` and `*p2` have values of `2.5`. In the second `cout` statement, the content of `p2` is assigned the value of `b` using the dereferencing operator (`*`). Thus both `p1` and `p2` point at a value of `5.1`.

Now we describe how pointers can be used dynamically:

- *Allocating dynamic variables*: C++ uses the keywords `new` and `delete` for allocating and deallocating dynamic variables respectively. The `new` operator allocates dynamics variables on the heap. For example,

```
new int;
```

allocates one object of type `int` whereas

```
new int[100];
```

allocates an array of 100 objects of type `int`. The `new` operator returns a pointer to the type specified. It can be assigned to a pointer variable, such as

```
int *p = new int[500];
```

The dynamically allocated `int` array here is uninitialized. To initialize class objects allocated with `new`, we use their constructor. It has the following form

```
Thing *p = new Thing(argument list);
```

The `new` operator allocates an object of the class `Thing`, which is initialized by its constructor that matches the argument list.

- *Deallocating dynamic variables*: Variables allocated with `new` must be deallocated when we are finished with them to make the heap space available for other dynamic variables. To achieve this purpose, we use the `delete` operator. For example,

```
delete p;
```

It returns the memory, previously allocated by the `new` operator, back to free store. To release an array of objects, we use

```
delete[] p;
```

The brackets [] are used here to free an array of objects.

Pointers are powerful in C++. However a couple of common problems need to be avoided:

- *Dereferencing uninitialized pointers*: One common source of errors with pointers is to dereference a pointer that has not been initialized. Like all other C++ variables, a pointer's value remains undefined until we assign it. In principle, it could point at any location in the memory. If we dereference such a pointer, we will get some random bits. This becomes disastrous if we assign some values to the item pointed to, because we may overwrite some important data segment, such as the program code or even the operating system. This sounds a little ominous, but with a little discipline it is easy to manage.

  To avoid dereferencing uninitialized pointers, which is potentially dangerous, C++ provides the keyword NULL that can be used for pointers that point at *nothing*. A NULL pointer is valid, but unattached. One of the common uses of a NULL pointer is to terminate a linked list. Suppose L_1, L_2, L_3 are elements of a linked list. To indicate that the linked list has ended, we usually make the last element point at the NULL pointer, as shown below:

$$L_1 \rightarrow L_2 \rightarrow L_3 \rightarrow \text{NULL}$$

- *Losing heap memory ("heap leaks")*: Another common problem when using dynamic variables is known as the *heap leak*. A heap leak is a situation where space is allocated on the heap and then lost – for some reason the pointer no longer points at the allocated memory area, so that it cannot be referred to or deallocated. A common cause of heap leaks is by reassigning dynamic variables without disposing of the previous ones. For instance,

```
#include <iostream>
using namespace std;

int main(void)
{
  int* ptr;
  ptr = new int[500];
  ptr = NULL;
  return 0;
}
```

The pointer variable ptr is first allocated to an int array of 500 elements. It is then reassigned to NULL straight away. This action has made the memory area (the int array) initially pointed to by ptr lost, and it is not recoverable. In such a situation, memory has leaked. Serious memory leakage during run-time may cause a program to halt abnormally due to memory exhaustion.

C++ introduces a concept called *reference*. Basically, it defines an alias or alternative name for any instance of data. The syntax is to append an ampersand (&) to the name of the data type. For example,

```
int   i = 5;      // An automatic variable i
int *pi = &i;     // A pointer to int which points at i
int &ri = i;      // A reference to the variable i
```

Now, we can use `ri` anywhere just as we use `i` or `*pi`. Suppose we write

```
ri *= 10;   // multiply ri by 10
```

The values of `i`, `*pi` and `ri` become 50.

By far the most important use for references is in the passing of arguments to functions. Some applications require pointers and references together. An example is given below

```cpp
// PointerReference.cpp

#include <iostream>
#include <fstream>
#include <cstdlib>
using namespace std;

void readFile(double* array,int& length)
{
    ifstream inFile("input.dat");
    inFile >> length;
    if(length <= 0)
    exit(0);
    cout << "length of array = " << length << endl;
    array = new double[length];
    for(int n=0;n<length;n++)
    { inFile >> array[n]; cout << array[n] << " "; }
    cout << endl;
}

double arithMean(double*& array,const int length)
{
  double sum = 0.0;
  for(int n=0;n<length;n++)
  {
  sum += array[n];
  }
  return sum/length;
}

int main(void)
{
  double* array = NULL;
  int length = 0;
  readFile(array,length);
  cout << "arithmetic mean = " << arithMean(array,length);
  delete [] array;
  return 0;
}
```

The input file with the name `input.dat` contains the data

```
5 1.1 1.3 1.5 2.3 0.7
```

Here `double*&` array specifies that array is an alias or reference to a pointer to `double`. Thus, if necessary, we could change the memory location referred to by array and the change would be reflected in `main`.

4.2   `this` Pointer

There is a special pointer in C++ called the `this` pointer. It is a constant pointer to an object of the class containing the member function, i.e. it denotes an implicitly declared *self-referential pointer* of the object. Let us illustrate the idea with a simple program:

(1) The member function increment uses the implicitly declared pointer `this` to return the newly incremented value of both c1 and c2.

(2) The member function `where_am_I` displays the address of the given object.

(3) The `this` keyword provides a built-in self-referential pointer.

```cpp
// this.cpp

#include <iostream>
using namespace std;

// declaration of class Cpair
class Cpair
{
  private:
    char c1, c2;
  public:
    Cpair(char);           // constructor
    Cpair increment();
    void print();
    void where_am_I();
};

// Definition of class Cpair
Cpair::Cpair(char c) : c2(c), c1(1+c) { }

Cpair& Cpair::increment() { c1++; c2++; return *this; }
void Cpair::print() { cout << c1 << " and " << c2 << endl; }
void Cpair::where_am_I() { cout << this << endl; }

int main(void)
{
  Cpair x('A');
  x.print();                 // output : B and A
  x.where_am_I();            // output : 0x7fffaee0
  Cpair z('X');
  z.where_am_I();            // output : 0x7fffaedc
  z.increment();
  z.print();                 // output : Z and Y
  z.where_am_I();            // output : 0x7fffaedc
```

```
  Cpair n('1');
  n.increment().print();  // output : 3 and 2
  n.where_am_I();         // output : 0x7fffaed8
  return 0;
}
```

4.3   Classes

Class is a primary C++ construct used to create abstract data types (ADTs). It describes the behaviors of an object, such as attributes, operations, argument types of the operations and so on. The general syntax for declaring a class is:

```
class class-name
{
    private:
        <private data members>
        <private member functions>
    protected:
        <protected data members>
        <protected member functions>
    public:
        <public data members>
        <public member functions>
};
```

Classes in C++ offer three levels of visibility for the data members and member functions — `public`, `protected` and `private`. They are called the *access specifier*. Each access has its own merits, as described below:

- `private`: Only member functions of the class have access to the private members. Class instances are denied from accessing them.

- `protected`: Only member functions of the class and its descendant classes have access to the protected members. Class instances are denied from accessing them.

- `public`: Members are visible to the member functions of the class, class instances, member functions of descendant classes and their instances.

In the following, we list some simple guide-lines for the proper use of classes:

- The access specifier may appear in any order.

- The same access specifier may appear more than once.

- If an access specifier is omitted, the compiler treats the members as `protected`.

- Avoid placing data members in the `public` region, unless such a declaration significantly simplifies the program design.

- Data members are usually placed in the **protected** region so that the member functions of descendant classes can access them.

- Use member functions to alter or query the values of data members.

After a class has been declared, the class name can be used to declare new class instances. The syntax resembles declaring variables.

In the rest of the discussion, we consider an *expression tree*. Along with the construction, we explain the concepts and the C++ language constructs.

The listing shows a typical class description and implementation of a class for an expression tree. The class **ExprNode** has four private data fields. The first field **_op** is used for operations **+**, **-** and **#** (integer value). The field **_value** provides the integer value. The two pointers **_left** and **_right** are used to construct the tree. They are used in the constructor

```
ExprNode(char oper,ExprNode& left,ExprNode& right)
```

A copy constructor is also implemented and the operator **=** is overloaded. Since **new** is used we also implement a destructor.

```
// ExpressionTree.cpp

#include <iostream>
#include <cstdlib>
using namespace std;

class ExprNode
{
  public:
    ExprNode();                         // default constructor
    ExprNode(const ExprNode&);          // copy constructor
    ExprNode(char,const ExprNode&,const ExprNode&);
    ExprNode(int);
    ~ExprNode();                        // destructor
    const ExprNode& ExprNode::operator = (const ExprNode&);
    int eval() const; //

  private:
    char _op;
    int  _value;
    ExprNode* _left;
    ExprNode* _right;
};

ExprNode::ExprNode() : _op('#'), _left(NULL), _right(NULL) { }

ExprNode::ExprNode(const ExprNode& expr)
{
  _op = expr._op;
  _value = expr._value;
  if(expr._left != NULL) _left = new ExprNode(*expr._left);
  else _left = NULL;
  if(expr._right != NULL) _right = new ExprNode(*expr._right);
```

```
    else _right = NULL;
}

ExprNode::ExprNode(char oper,const ExprNode &left,const ExprNode &right)
{
  _op = oper;
  _left = new ExprNode(left);
  _right = new ExprNode(right);
}

ExprNode::ExprNode(int v)
{
  _op = '#';
  _value = v;
  _left = NULL;
  _right = NULL;
}

ExprNode::~ExprNode()
{
  if(_left != NULL) delete _left;
  if(_right != NULL) delete _right;
}

const ExprNode& ExprNode::operator = (const ExprNode& expr)
{
  if(&expr != this)
  {
  _op = expr._op;
  _value = expr._value;
  if(_left != NULL) delete _left;
  if(_right != NULL) delete _right;
  if(expr._left != NULL) _left = new ExprNode(*expr._left);
  else _left = NULL;
  if(expr._right != NULL) _right = new ExprNode(*expr._right);
  else _right = NULL;
  }
  return *this;
}

int ExprNode::eval() const
{
  int result;
  switch(_op)
  {
  case '+':
    result = _left -> eval() + _right -> eval();
    break;
  case '-':
    result = _left -> eval() - _right -> eval();
    break;
  case '#':
    result = _value;
    break;
  }
  return result;
}
```

```
int main(void)
{
  ExprNode add1 = ExprNode('+',ExprNode(3),ExprNode(-5));
  ExprNode sub1 = ExprNode('-',add1,ExprNode(7));
  ExprNode add2 = ExprNode('+',sub1,ExprNode(21));
  cout << "result1 = " << add2.eval() << endl;
  ExprNode expr1 = sub1;
  cout << "result2 = " << expr1.eval() << endl;
  ExprNode expr2;
  expr2 = add1;
  cout << "result3 = " << expr2.eval() << endl;
  return 0;
}
```

4.4   Constructors and Destructor

The two most important class member functions are the *constructor* and *destructor*.
There could be many constructors in a class but it can have only one destructor.
A constructor is responsible for the creation of class instances or objects. It can be
used to handle initialization and allocation of dynamic memory for an object. Note
that the constructors always have the same name as the class. A destructor, on the
other hand, is used to clean-up after an object is destroyed. The clean up process
usually releases the unused memory back to the system. It has the same name as
the class except for a tilde ~ prefix.

For our `ExpressionTree` class there are three constructors. Each of them is responsible for a different type of construction:

1) The first is the default constructor.

```
ExprNode::ExprNode() : _op('#'), _left(NULL), _right(NULL) { }
```

2) The constructor

```
ExprNode::ExprNode(char oper,ExprNode& left,ExprNode& right)
{
  _op = oper;
  _left = left; _right = right;
}
```

sets up a tree with a left branch specified by `left` and a right branch `right`.

3) The following constructor is used to provide the value.

```
ExprNode::ExprNode(int v)
{
  _op = '#';
  _value = v;
  _left = NULL; _right = NULL;
}
```

The destructor deallocates the memory allocated by the constructors.

```
ExprNode::~ExprNode()
{
 if(_left != NULL) delete _left;
 if(_right != NULL) delete _right;
}
```

4.5   Copy Constructor and Assignment Operator

As the name suggested, the *copy constructor* is used to duplicate the content of an instance during a declaration statement. On top of that, it will also be invoked to generate temporary values when class arguments are passed as value parameters. If the data fields do not include any pointers that have to be initialized dynamically, the copy constructor generated by the compiler which performs member wise copy, works correctly. However, a proper copy constructor is needed when the data fields involve pointer variables.

Let us take our `ExprNode` class for example. The copy constructor is given by

```
ExprNode::ExprNode(const ExprNode& expr)
{
  _op = expr._op;
    _value = expr._value;
    if(expr._left != NULL) _left = new ExprNode(*expr._left);
    else _left = NULL;
    if(expr._right != NULL) _right = new ExprNode(*expr._right);
  else _right = NULL;
}
```

It is good practice to provide a copy constructor for every user-defined class.

An *assignment operator* is needed whenever the data fields require dynamic memory allocation, or whenever the copy constructor of the class needs to be rewritten. For our `ExprNode` class, its definition is as follows

```
const ExprNode& ExprNode::operator = (const ExprNode& expr)
{
 if(&expr != this)
 {
 _op = expr._op;
 _value = expr._value;
 if(_left != NULL) delete _left;
 if(_right != NULL) delete _right;
 if(expr._left != NULL) _left = new ExprNode(*expr._left);
 else _left = NULL;
 if(expr._right != NULL) _right = new ExprNode(*expr._right);
 else _right = NULL;
 }
 return *this;
}
```

From the above listing, we notice that this function is similar to the copy constructor but there are two major differences. The assignment operator needs to handle the special case when an object is assigned to itself. This explains why the function started with an `if` statement. The conditional statement ensures that the operator works properly when an `ExprNode` is assigned to itself. The return reference type has made the left hand side assignment possible.

4.6   Type Conversion

C++ has many type conversion rules which the compiler obeys when converting the value of an object from one fundamental type to another. These rules make mixed type operations possible. They are convenient but potentially dangerous if they are not used properly. Note that the ASCII table is used to convert the characters such as 'A' and 'P' into integers and vice versa. Furthermore, implicit conversion may induce subtle run-time bugs which are hard to detect. Therefore, it is useful to know how the conversion rules work in C++. The general rules for implicit type conversion involve two steps.

1. First, a `char`, `short`, `enum` is promoted to `int` whenever an `int` is expected. It is converted to `unsigned int` if it is not representable as `int`.

2. If the expression still consists of mixed data types, then lower types are promoted to higher types, according to the following hierarchy:

$$\text{int} \rightarrow \text{unsigned int} \rightarrow \text{long} \rightarrow \text{unsigned long}$$
$$\rightarrow \text{float} \rightarrow \text{double} \rightarrow \text{long double}$$

The resultant value of the mixed expression has the higher type.
Suppose `double a = 2;` and `long b = 3;` then `a+b` does the following:

* `b` is promoted to `double`.

* Perform `double` addition `a+b`, and the type of the result is also `double`.

Explicit conversion uses *cast*. In traditional C, cast has the following form:

   (*type*) expression

whereas in C++, there is another alternative (functional notation)

   type (expression)

For example, the following two equivalent expressions convert the `int` n to a `double`

```
int n = 10;
double x, y;
x = (double) n;
y = double(n);
```

However, the functional notation is preferred.

Example. Consider the built-in `string` class. It can convert `char*` to `string` as follows.

```
// mystring.cpp

#include <iostream>
#include <string>
using namespace std;

int main(void)
{
    char* cs = new char[4];
    cs[0] = 'a'; cs[1] = 'b'; cs[2] = 'c'; cs[3] = '\0';
    string s(cs);
    cout << s;
    delete[] cs;
    return 0;
}
```

Type conversions also apply to pointers. Any pointer type can be converted to a generic pointer of type `void*`. However, a generic pointer needs to be cast to an explicit type when it is assigned to a non-generic pointer variable.

```
// generic.cpp

#include <iostream>
using namespace std;

int main(void)
{
    int *n = NULL;
    void *generic_ptr;
    generic_ptr = n;         // OK
    n = (int*) generic_ptr;  // OK
    n = generic_ptr;         // Error !!!
    return 0;
}
```

The code generates a compilation error message:

```
a value of type "void*" cannot be assigned to an entity of type "int*"
  n = generic_ptr;
```

The null pointer can be converted to any pointer type

```
char *ptr1 = 0;
int  *ptr2 = ptr1;  // Not OK: need (int*) ptr1;
int  *y = 0;        // OK
```

Here we see that the same memory location can be interpreted as both `int` and `char`. This is referred to as weakly typed. A pointer to a class can be converted to a pointer to a publicly derived base class. This also applies to references.

4.7 Operator Overloading

In C++, a built-in operator may be given more than one meaning, depending on its arguments. Consider the binary addition operator `a + b`. The data type of `a` and b could be `int`, `float`, `double` or even a user-defined type such as `Matrix`. The operation for matrix addition is certainly very different from the floating point addition. We say that the operator + possesses more than one meaning. In fact, there are many more C++ operators that can be overloaded, as listed in the following:

Type	Operator Notation
Unary	++(prefix/postfix) --(prefix/postfix) &(address of) *(dereference) + - ~ ! (*type*)
Arithmetic	* / % + -
Shift	<< >>
Relational	> >= < <= == !=
Bitwise	& ^ |
Logical	&& ||
Assignment	= *= /= %= += -= <<= >>= &= ^= |=
Data Access	[] -> ->*
Others	() , new delete

Almost all C++ operators can be overloaded except

Member access operator	a.b
Dereferencing pointer to member	a.*b
Scope resolution operator	a::b
Conditional operator	a?b:c

Only predefined operators can be overloaded in C++. We cannot introduce any new operator notations, such as the exponentiation `**` used in Fortran. Although the operators can be given extra meaning, the order of precedence cannot be changed. Suppose we wish to overload the bitwise XOR operator ^ as the exponentiation operator. We have to be very careful about the order of precedence. The expression `a+b*c^d` will be evaluated as `(a+(b*c))^d`, not as `a+(b*(c^d))`. Extra parentheses are needed for the expression to be correct.

In fact, there are two kinds of overloading in C++: *member overloading* and *global overloading*. Consider again the addition operator `a + b` where a and b are objects of class C. The general form for member overloading is as follows:

```
C C::operator + (C) { ... }
```

In the program, the expression `a + b` means `a.operator+(b)`. We could also overload the operator using global overloading:

```
friend C operator + (C,C) { ... }
```

where a and b are the first and second parameters of the function, respectively. A global overloaded operator usually needs a `friend` declaration in the class. The friendship between functions and classes will be discussed in Section 4.10.

Example. In the built-in `string` class the operator + is overloaded to perform concatenation.

```
// concat.cpp

#include <iostream>
#include <string>
using namespace std;

int main(void)
{
   string s1 = "AGTCA";
   string s2 = "AATTCAC";
   string s3 = s1 + s2;
   cout << "The concatenated string is " << s3 << endl;
   return 0;
}
```

Example. Consider a family of linear operators $\{\, b, b^\dagger \,\}$ on an inner product space V. The commutation relations for these operators (so-called *Bose operators*) are given by

$$[b, b^\dagger] = I$$

where I is the identity operator and

$$[b, b] = [b^\dagger, b^\dagger] = 0$$

where 0 is the zero operator and $[\,,\,]$ denotes the commutator. The operator b is called an *annihilation operator* and the operator $b^\dagger$ is called a *creation operator*. The vector space V must be infinite-dimensional for $[b, b^\dagger] = I$ to hold. Let $|0\rangle$ be the vacuum state, i.e.

$$b|0\rangle = 0, \quad \langle 0|0\rangle = 1.$$

A state is given by

$$(b^\dagger)^n |0\rangle$$

where $n = 0, 1, 2, \ldots$. As an example, consider the operator $\hat{H} = b^\dagger b b^\dagger b$ and the state $|\phi\rangle = b^\dagger |0\rangle$, then

$$
\begin{aligned}
\hat{H}|\phi\rangle &= b^\dagger b b^\dagger b b^\dagger |0\rangle \\
&= b^\dagger b b^\dagger (I + b^\dagger b)|0\rangle \\
&= b^\dagger b b^\dagger |0\rangle \\
&= b^\dagger (I + b^\dagger b)|0\rangle \\
&= b^\dagger |0\rangle.
\end{aligned}
$$

We overload the addition operator as the creation operator and the subtraction operator as the annihilation operator. These operators apply on the vacuum state $|0\rangle$ and the outcome is another state. Several operators have been used for demonstrations. In order to simplify the program we use the fact that

$$\left[b, \left(b^\dagger\right)^m \right] = m \left(b^\dagger\right)^{m-1}$$

for $m = 0, 1, 2, \ldots$.

```cpp
// bose1.cpp

#include <iostream>
using namespace std;

// Declaration of class State
class State
{
  private:
    int m;
    int factor;
  public:
    State();                                // constructor
    State(const State&);                    // copy constructor
    const State& operator = (const State&); // overloading =
    void bose_state(int);                   // member function
    void display();                         // member function
};

// Implementation of class State
State::State()
{
  m = 0;
  factor = 1;
}

State::State(const State& arg)
{
  m = arg.m;
  factor = arg.factor;
}

const State& State::operator = (const State& arg)
{
  cout << "overloaded = invoked" << endl;
  m = arg.m;
  factor = arg.factor;
  return *this;
}

void State::bose_state(int cd)
{
  if(cd == -1) factor *= m--;
  else m++;
}

void State::display()
{
  if(factor == 0) cout << "0";
  else
  {
    cout << factor << "*";
    if(m != 0)
    {
      cout << "(";
      for(int i=0;i<m;i++) cout << "b+";
      cout << ")";
```

```
    }
  cout << "|0>";
  }
}


// Declaration of class Bose
class Bose
{
  public:
    Bose();                              // constructor
    State operator + (const State&);  // overloading +
    State operator - (const State&);  // overloading -
};

// Implementation of class Bose
Bose::Bose() { }  // default constructor

State Bose::operator + (const State& st2)
{
  State st1 = st2;
  st1.bose_state(1);
  return st1;
}

State Bose::operator - (const State& st2)
{
  State st1 = st2;
  st1.bose_state(-1);
  return st1;
}

int main(void)
{
  State g1;
  Bose b;
  g1 = b- (b- (b+ (b+ (b+ (b+ (b+ g1)))))));
  cout << "g1 = ";
  g1.display(); // output 30*(b+b+b+b+)|0>
  cout << endl;
  State g2;
  g2 = b+ (b- g2);
  g2.display();          // output 0
  return 0;
}
```

Example. Consider a family of linear operators (*Fermi operators*) $c_j, c_j^\dagger$ for $j = 1, \ldots, n$ defined on a finite-dimensional vector space V satisfying the *anticommutation relations*

$$[c_j, c_k]_+ = [c_j^\dagger, c_k^\dagger]_+ = 0$$
$$[c_j, c_k^\dagger]_+ = \delta_{jk} I$$

where 0 is the zero operator and I is the unit operator with $j, k = 1, 2, \ldots, n$. Operators satisfying these relations are called the annihilation and creation operators for fermions. We define a state $|0\rangle$ (the so-called *vacuum state*) with the properties

$$c_j|0\rangle = 0 \quad \text{for } j = 1, 2, \ldots, n$$

and
$$\langle 0|0 \rangle = 1 \, .$$

This means that the state $|0\rangle$ is normalized. Other states can now be constructed from $|0\rangle$ and the creation operators $c_j^\dagger$.

Applying the rules given above, we find

$$
\begin{aligned}
c_1 c_4 c_1^\dagger c_4^\dagger |0\rangle &= -c_1 c_1^\dagger c_4 c_4^\dagger |0\rangle \\
&= -c_1 c_1^\dagger (I - c_4^\dagger c_4)|0\rangle \\
&= -c_1 c_1^\dagger |0\rangle \\
&= -(I - c_1^\dagger c_1)|0\rangle \\
&= -|0\rangle.
\end{aligned}
$$

In the following program, we implement Fermi operators and several examples are given.

```cpp
// fermi1.cpp

#include <iostream>
#include <cassert>
#include <string>
using namespace std;

class Power
{
  public:
    string c; // the operator name, e.g. c1, c4+
    int n;    // the degree of the operator, e.g. (c4+)^2
};

class State
{
  private:
    int factor; // the multiplication of the state
    int m;      // number of distinctive operators
    Power *p;   // a pointer to Power

  public:
    State();                      // default constructor
    void operator = (const State&); // assignment operator
    void Fermi_creator(string,int); // operation on the state
    void display() const;         // display the state
    void reset();                 // reset the state
};

    // Fermi operators
class Fermi
{
  private:
    string f;  // store the name of the operator, e.g. c1, c4

  public:
    Fermi(string);                    // constructor
```

```
      State operator + (const State&);  // creation operator
      State operator - (const State&);  // annihilation operator
};

State::State() : factor(1), m(0), p(NULL) {}

void State::operator = (const State& s2)
{
  m = s2.m;
  p = new Power[m]; assert(p);
  for(int i=0;i<m;i++) p[i] = s2.p[i];
  factor = s2.factor;
}

void State::Fermi_creator(string ch,int s)
{
  int i;
  if(factor)
  {
  // [c1,c2]_+ = [c1+,c2+]_+ = [c1,c2+]_+ = 0
  for(i=0;i<m && (ch != (p+i)->c);i++)
   if((p+i)->n % 2) factor *= (-1);
  // if there is a new operator
  if(i==m)
  {
    if(s==1) // creation operator
    {
     Power *p2;
     m++;
     p2 = new Power[m]; assert(p2);
     for(int j=0;j<m-1;j++) p2[j] = p[j];
     (p2+m-1)->c = ch; (p2+m-1)->n = 1;
     delete[] p;
     p = p2;
    }
    else  // annihilation operator
    { m=0; factor=0; delete[] p;}
  }
  else // operator appears before
  {
   // creation operator, c1+ (c1+)^n = (c1+)^(n+1)
   if(s==1) (p+i)->n++;
   else // annihilation operator, [c1,c2+]_+ = I
   {
    if((p+i)->n%2) // if power of operator is odd
    {
     (p+i)->n--;
     if(!(p+i)->n)
     {
      for(int j=i+1;j<m;j++) p[j-1] = p[j];
      m--;
     }
    }
    else // if power of operator is even
    { m=0; factor=0; delete[] p; }
   }
  }
  }
}
```

```
}
void State::display() const
{
  if(!factor) cout << "0";
  else
  {
   if(factor != 1) cout << "(" << factor << ")*";
   for(int i=0;i<m;i++)
   {
    cout << " " << (p+i)->c << "+";
    if((p+i)->n != 1) cout << "^" << (p+i)->n;
   }
   cout << "|0>";
  }
} // end display()

void State::reset()
{
  m = 0; factor = 1;
  delete[] p; p = NULL;
}

Fermi::Fermi(string st) : f(st) { }

State Fermi::operator + (const State &s2)
{
  State s1(s2);
  s1.Fermi_creator(f,1);
  return s1;
}

State Fermi::operator - (const State &s2)
{
  State s1(s2);
  s1.Fermi_creator(f,-1);
  return s1;
}

int main(void)
{
  State g;
  Fermi c1("c1"), c4("c4");
  g = c1- g;
  cout << "g = "; g.display(); cout << endl;
  g.reset();
  g = c1+ g;
  cout << "g = "; g.display(); cout << endl;
  g.reset();
  g = c1- (c1+ g);
  cout << "g = "; g.display(); cout << endl;
  g.reset();
  g = c4- (c4+ (c1+ (c4+ g)));
  cout << "g = "; g.display(); cout << endl;
  g.reset();
  g = (c4- (c4+ (c4+ (c1+ (c4+ g)))));
  cout << "g = "; g.display(); cout << endl;
  g.reset();
```

```
    g = c4- (c4+ (c4- (c4+ (c4- (c4+ g)))));
    cout << "g = "; g.display(); cout << endl;
    return 0;
}
```

4.8   Class Templates

Templates are usually used to implement data structures and algorithms that are
independent of the data types of the objects they operate on. It allows the same
code to be used with respect to different types, where the type acts as a parameter
of the data structures or algorithms. They are also known as *parametrized types*.
One important use for templates is the implementation of *container classes*, such
as stacks, queues, lists, etc. A stack template may contain objects of type `int`,
`double` or even user-defined types. The compiler will automatically generate the
implementations of the `Stack` classes. The syntax of the class declaration is

```
template <class T>
class Stack
{
 public: Stack(int);
   ...
};
```

where the symbol `T` serves as the argument for the template which stands for arbitrary type. To declare a `Stack`, we use

```
Stack<char> s1(100); // 100 elements Stack of char
Stack<int>  s2(50);  //  50 elements Stack of int
```

The keyword `template` always precedes every forward declaration and definition
of a template class. It is followed by a formal template list surrounded by angle
brackets (`<` `>`). The formal parameter list cannot be empty. Multiple parameters are
separated by commas and declarations with specific types such as `int` are allowed,
e.g.

```
template <class T1,class T2,class T3,class T4> class Myclass1
template <class T,int size> class Myclass2
```

When the compiler encounters the object definitions, it substitutes the template
type names by the actual data types. If the class template uses more than one
template, the compiler will perform the type substitutions one at a time, beginning
with the first class referred to.

Example. Consider a simplified version of a `Vector` template class

```
// svector.h

#include <iostream>
#include <string>
#include <cassert>
using namespace std;
```

```cpp
// definition of class Vector
template <class T> class Vector
{
  private:
   // Data Fields
   int size;
   T *data;
  public:
   // Constructors
    Vector();
    Vector(int);
    Vector(const Vector<T>&);
    ~Vector();
    // Member Functions
    T& operator[] (int) const;
    void resize(int);
    // Assignment Operator
    const Vector<T> &operator = (const Vector<T>&);
};

// implementation of class Vector
template <class T> Vector<T>::Vector() : size(0), data(NULL) {}

template <class T> Vector<T>::Vector(int n) : size(n), data(new T[n])
   { assert(data != NULL); }

template <class T> Vector<T>::Vector(const Vector<T>& v)
     : size(v.size), data(new T[v.size])
{
   assert(data != NULL);
   for(int i=0;i<v.size;i++) data[i] = v.data[i];
}

template <class T> Vector<T>::~Vector()
   { delete[] data; }

template <class T> T &Vector<T>::operator[] (int i) const
{
   assert(i >= 0 && i < size);
   return data[i];
}

template <class T> void Vector<T>::resize(int length)
{
   int i;
   T *newData = new T[length]; assert(newData != NULL);
   if(length <= size) for(i=0;i<length;i++) newData[i] = data[i];
   else for(i=0;i<size;i++) newData[i] = data[i];
   delete[] data;
   size = length;
   data = newData;
}

template <class T>
   const Vector<T> &Vector<T>::operator = (const Vector<T> &v)
{
  if(this == &v) return *this;
  if(size != v.size)
```

```
  {
  delete[] data;
  data = new T[v.size]; assert(data != NULL);
  size = v.size;
  }
  for(int i=0;i<v.size;i++) data[i] = v.data[i];
  return *this;
}

int main(void)
{
  // The symbol Vector must always be accompanied by a
  // data type in angle brackets
  Vector<int> x(5);        // generates a vector of integers
  int i, j;
  for(i=0;i<5;++i) x[i] = i*i;
  for(i=0;i<5;++i) cout << x[i] << " ";
  cout << endl; cout << endl;

  Vector<char> ch(3);      // generates a vector of characters
  ch[0] = 'a'; ch[1] = 'b'; ch[2] = 'c';
  cout << ch[1] << endl;
  cout << ch[1]-ch[2] << endl;
  cout << endl;

  Vector<string> c(2);     // generates a vector of strings
  c[0] = "aba"; c[1] = "bab";

  for(i=0;i<2;i++) cout << c[i] << endl;

  Vector<Vector<int> > vec(3);
  for(i=0;i<3;i++) vec[i].resize(4);
  vec[0][0] = 7;  vec[0][1] = 4;  vec[0][2] = -3;  vec[0][3] = -17;
  vec[1][0] = 4;  vec[1][1] = 9;  vec[1][2] = 0;   vec[1][3] = 5;
  vec[2][0] = 7;  vec[2][1] = 11; vec[2][2] = 8;   vec[2][3] = 77;

  for(i=0;i<3;i++)
    for(j=0;j<4;j++) cout << vec[i][j] << "  ";
  cout << endl << endl;

  cout << "vec[0][0] = " << vec[0][0] << endl;
  cout << "vec[0][0]*vec[2][3] = " << vec[0][0]*vec[2][3] << endl;
  return 0;
}
/*
Result
======
0 1 4 9 16

b
-1

aba
bab
bab

7  4  -3  -17
4  9  0  5
```

```
7   11   8   77

vec[0][0] = 7
vec[0][0]*vec[2][3] = 539
*/
```

4.9   Function Templates

In the last section, we discussed the class templates. Functions can also be parametrized. In function template, all the type arguments must be mentioned in the arguments of the function. Therefore, we do not need to specify any template argument when calling a template function. The compiler will figure out the template type arguments from the actual arguments.

By using function templates, we can define a family of related functions by making the data type itself a parameter. It enhances the compactness of the code without forfeiting any of the benefits of a statically-typed language. Traditionally, a function has to be defined for a specific data type to make it work. If we wish to write a mathematical library containing Bessel functions, then we have to define the Bessel functions for `int`, `float`, `double`, `long double` and `complex`, etc. If we want to improve the algorithm at a later stage, all the Bessel functions with different types have to be changed. This is cumbersome and prone to errors. By defining the functions on templates, the compiler will automatically generate the code for each data type.

The keyword `template` always precedes the definition and forward declaration of a template function. It is followed by a comma-separated list of parameters, which are enclosed by a pair of angle brackets (`< >`). This list is known as the formal parameter list of the template. It cannot be empty. Each formal parameter consists of the keyword `class` followed by an identifier, which could be a built-in or user-defined type. The name of a formal parameter can occur only once within the parameter list. The function definition follows the formal parameter list. An example is

```
template <class T> T min(T x,T y) { ... }
```

Example. We illustrate the function templates using a function that swaps two pointers. In the first version of the program we use pointers, whereas in the second version we use references.

```
// First version
// swap1.cpp

#include <iostream>
using namespace std;

template <class T> void swap(T *a,T *b)
{
   T temp;
   temp = *a; *a = *b; *b = temp;
```

```
}

int main(void)
{
  int i, *ip;
  ip = &i;
  cout << "&i = " << &i   << endl;   // 0x0003bf8c
  cout << "ip = " << ip   << endl;   // 0x0003bf8c
  cout << "&ip = " << &ip  << endl;  // 0x0003bf88

  int j, *jp;
  jp = &j;
  cout << "&j  = " << &j  << endl;   // 0x0003bf84
  cout << "jp = " << jp  << endl;    // 0x0003bf84
  cout << "&jp = " << &jp << endl;   // 0x0003bf80

  swap(&ip,&jp);

  cout << "&i  = " << &i  << endl;   // 0x0003bf8c
  cout << "ip  = " << ip  << endl;   // 0x0003bf84
  cout << "&ip = " << &ip << endl;   // 0x0003bf88
  cout << "&j  = " << &j  << endl;   // 0x0003bf84
  cout << "jp  = " << jp  << endl;   // 0x0003bf8c
  cout << "&jp = " << &jp << endl;   // 0x0003bf80
  return 0;
}
```

```
// Second version
// swap2.cpp

#include <iostream>
using namespace std;

template <class T> void swap(T& a,T& b)
{
  T temp;
  temp = a; a = b; b = temp;
}

int main(void)
{
  int i, *ip;

  ip = &i;
  cout << "&i =  " << &i  << endl; // 0x0003bf8c
  cout << "ip =  " << ip  << endl; // 0x0003bf8c
  cout << "&ip = " << &ip << endl; // 0x0003bf88

  int j, *jp;
  jp = &j;
  cout << "&j =  " << &j  << endl; // 0x0003bf84
  cout << "jp =  " << jp  << endl; // 0x0003bf84
  cout << "&jp = " << &jp << endl; // 0x0003bf80

  swap(ip,jp);

  cout << "&i  = " << &i  << endl; // 0x0003bf8c
  cout << "ip  = " << ip  << endl; // 0x0003bf84
```

```
cout << "&ip = " << &ip << endl; // 0x0003bf88
cout << "&j  = " << &j  << endl; // 0x0003bf84
cout << "jp  = " << jp  << endl; // 0x0003bf8c
cout << "&jp = " << &jp << endl; // 0x0003bf80
return 0;
}
```

A *template specialization* is a definition which overrides the default template definition for a specific value of the template parameter. For example, we can use the XOR to swap two integers.

```
template <class T> void swap(T& a,T& b)
{
   T temp;
   temp = a; a = b; b = temp;
}
```

The specialization for `int` is

```
template <> void swap<int>(int& a,int& b)
{
   a ^= b; b ^= a; a ^= b;
}
```

If only some of the template parameters are specified, then it is called a *partial template specialization*.

4.10 Friendship

From the previous discussions, we know that only member functions can access the private data of a class. Sometimes this rule is too restrictive and inefficient. In such cases, we may want to allow a non member function to access the private data directly. This can be achieved by declaring that non member function as a `friend` of the class concerned.

A friend **F** of class **X** is a function (or class) that, although not a member function of **X**, has full access rights to the private and protected members of **X**. In all other aspects, **X** is a normal function (or class) with respect to scope, declarations and definitions. Class friendship is not *transitive*: **X** is a friend of **Y** and **Y** is a friend of **Z** does not imply **X** is a friend of **Z**. However, friendship can be inherited. Like a member function, a friend function is explicitly specified in the declaration part of a class. The keyword `friend` is the function declaration modifier. It takes the general form

> `friend` *return-type function-name* (*parameter list*)

for functions and

> `friend class` *class-name*

for classes.

Example. Suppose we want to construct a simplified `Complex` number class. Instead of defining the `add()` and `display()` functions as members, we declare them as `friend` functions to the `Complex` class.

```cpp
// myComplex.cpp

#include <iostream>
using namespace std;

class Complex
{
  private:
    double real, imag;
  public:
    Complex();
    Complex(double);
    Complex(double,double);
    friend Complex add(Complex,Complex);
    friend void display(Complex);
};

Complex::Complex() : real(0.0), imag(0.0) { }
Complex::Complex(double r) : real(r), imag(0.0) { }
Complex::Complex(double r,double i) : real(r), imag(i) { }

Complex add(Complex c1,Complex c2)
{
  Complex result;
  result.real = c1.real + c2.real;
  result.imag = c1.imag + c2.imag;
  return result;
}

void display(Complex c)
{ cout << "(" << c.real << " + i*" << c.imag << ")"; }

int main(void)
{
  Complex a(1.2,3.5), b(3.1,2.7), c;
  c = add(a, b);
  display(a); cout << " + "; display(b); cout << " = "; display(c);
  return 0;
}
/*
Result
======
(1.2 + i*3.5) + (3.1 + i*2.7) = (4.3 + i*6.2)
*/
```

4.11 Inheritance

Inheritance is a powerful code reuse mechanism. It enables us to categorize related classes for sharing common properties. It is a mechanism for deriving new classes from old ones. For example `Circle`, `Triangle` and `Square` can be grouped under a more general abstraction — `Shape`. Instances of `Shape` possess some common properties such as area, perimeter, etc. `Circle`, `Triangle` and `Square` are subclasses (or derived classes) of `Shape`, whereas `Shape` is the superclass (or base class). Derived classes inherit properties and functionalities from the base class. Through inheritance, a class hierarchy can be created for sharing code and interface. The general form of a subclass that is derived from an existing base class is

```
class class-name : [ public/protected/private ] base-class-name
{
    ...
};
```

The keywords `public`, `protected` and `private` describe how the derived class inherits information from the base class. They allow us to control how the members of the superclass appear in the subclass.

- `public` inheritance: If the superclass is inherited publicly, the access of all the members of the superclass remain unchanged in the subclass.

- `protected` inheritance: In this type of inheritance, all public members of the superclass become protected in the subclass.

- `private` inheritance: Whenever the superclass is inherited privately, all the public and protected members of the superclass become private in the subclass. Therefore, the data members and member functions of the superclass are not accessible to the subclass.

The following table summarizes the behaviors of the three different kinds of inheritance:

Type of inheritance	Access in superclass $\Longrightarrow$ Access in subclass
Public inheritance	public members $\Longrightarrow$ public members protected members $\Longrightarrow$ protected members private members $\Longrightarrow$ private members
Protected inheritance	public members $\Longrightarrow$ protected members protected members $\Longrightarrow$ protected members private members $\Longrightarrow$ private members
Private inheritance	public members $\Longrightarrow$ private members protected members $\Longrightarrow$ private members private members $\Longrightarrow$ private members

Public inheritance corresponds to the *is-a* relationship, i.e. **X** *is-a* **Y** (**X** is derived from **Y**). It is used when the inheritance is part of the interface. For example, the inheritance described above pertaining to the geometric shape is a public inheritance because `Circle` is a `Shape`, `Triangle` is also a `Shape` and so on.

Protected and private inheritance have other meanings. Protected inheritance is used when the inheritance is part of the interface to the derived classes, but is not part of the interface to the users. On the other hand, private inheritance is used when the inheritance is not part of the interface but is an implementation detail.

In the following program, the base class `Shape` has a protected data member `area` and a public member function `print_Area()`. The classes `Triangle` and `Circle` are subclasses of `Shape` (Figure 4.1). Since `Triangle` is publicly inherited, the public member function `print_Area()` remain public. Therefore, it can be invoked from the main program. However, `Circle` is privately inherited, so the function `print_Area()` becomes private in the class. Hence, an attempt to call the function from the main program generates a compilation error.

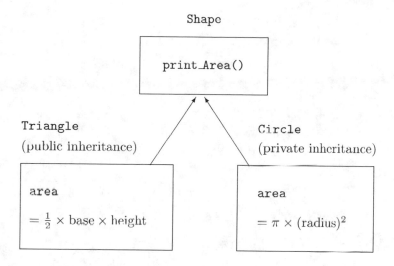

Figure 4.1: *Schematic diagram of the relationships among* Shape, Triangle *and* Circle.

```
// shape1.cpp

#include <iostream>
using namespace std;

const double PI = 3.14159265;

class Shape
{
  protected:
    double area;
  public:
    void print_Area() const;
};

void Shape::print_Area() const {   cout << area << endl; }

class Triangle : public Shape
{
```

```
  private:
    double base, height;
  public:
    Triangle(double,double);
};

Triangle::Triangle(double b,double h) : base(b), height(h)
{   area = 0.5*b*h; }

class Circle : private Shape
{
  private:
    double radius;
  public:
    Circle(double);
};

Circle::Circle(double r) : radius(r) { area = PI*radius*radius; }

int main(void)
{
  Triangle t(2.5,6.0);
  t.print_Area();   // 7.5
  Circle r(3.5);
  r.print_Area(); // function Shape::print_Area is inaccessible
                  // r.print_Area();

  return 0;
}
```

4.12   Virtual Functions

Dynamic binding in C++ is implemented by the *virtual function*. It allows the
system to select the right method for a particular class during run-time. Virtual
functions are defined in the base class using the keyword `virtual`. The definition
of these functions can be deferred or overridden in any subclass. If a derived class
does not supply its own implementation, the base class version will be used.

Let us consider the same example as in Section 4.11. An inheritance hierarchy
is constructed with `Shape` as the abstract base class and `Circle`, `Triangle` as
the derived classes (Figure 4.2). A virtual function named `Calculate_Area()` is
declared in the abstract base class `Shape`. However, its implementation is deferred
and redefined in its derived classes, where well-defined formulae can be applied to
calculate the area of the geometric shapes.

```
// shape2.cpp

#include <iostream>
using namespace std;

const double PI = 3.14159265;

class Shape
{
```

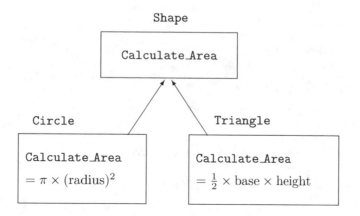

Figure 4.2: *Schematic diagram of the inheritance hierarchy.*

```
  public:
     virtual double Calculate_Area() const = 0;
};

class Circle : public Shape
{
  private:
    double radius;
  public:
    Circle(double);
    double Calculate_Area() const;
};

class Triangle : public Shape
{
  private:
    double base, height;
  public:
    Triangle(double,double);
    double Calculate_Area() const;
};

Circle::Circle(double r) : radius(r) {}

double Circle::Calculate_Area() const
{  return PI*radius*radius; }

Triangle::Triangle(double b,double h) : base(b), height(h) {}

double Triangle::Calculate_Area() const
{  return 0.5*base*height; }

int main(void)
{
  Shape *s;
  Circle c(5);
  Triangle t(3,8);
  s = &c;     // s points at Circle c
```

```
  cout << "s->Calculate_Area() = " << s->Calculate_Area() << endl;
  s = &t;    // s points at Triangle t
  cout << "s->Calculate_Area() = " << s->Calculate_Area() << endl;
  return 0;
}
Result
======
s -> Calculate_Area() = 78.5398
s -> Calculate_Area() = 12
```

From the result, we observe that the appropriate `Calculate_Area()` function is invoked depending on the object type pointed to by `s` during run-time. This form of polymorphism is called inclusion polymorphism.

4.13   Wrapper Class

In C++, data types are classified as basic and abstract data types. In some applications it would be useful to have the basic data type as an abstract data type. This leads to a concept called *wrapper class*. For example, Java has wrapper classes for all basic data types. In the following program, we show how the basic data type `double` is converted into an abstract data type `Double`.

```cpp
// WDouble.cpp

#include <iostream>
#include <cstddef>
using namespace std;

class Double
{
  private:
    double value;
  public:
    Double(double=0.0); // constructor
    void* operator new(size_t);
    void operator delete(void*,size_t);
    operator double() const; // type conversion operator
};

Double::Double(double f) : value(f) {  }

void* Double::operator new(size_t s)
{ return (void*) new char[s]; }

void Double::operator delete(void *p,size_t)
{ delete[] p; }

Double::operator double() const { return value; }

int main(void)
{
  Double *s1, s2(5.23);
  s1 = new Double(3.14);
  cout << "content of s1 = " << *s1 << endl;
  cout << "content of s2 = " << s2 << endl;
```

```
// conversion operator to double is used
cout << "*s1 + s2 = " << *s1 + s2 << endl;
delete s1;
return 0;
}
```

4.14 Recursion

Recursion plays a central role in computer science. For example a string and a linear linked list are recursive structures. A recursive function is one whose definition includes a call to itself. A recursive definition needs a condition which halts the recursion. Of course, it is not allowed to use the `main` function in a recursive call.

Example 1. The program determines the length of a string recursively. A string consists of contiguous characters in memory, ending with the nul character '\0'. Conceptually, we can think of a string as either the empty string, consisting of just the nul character, or as a character followed by a string. This definition of a string describes it as a recursive data structure. Thus we can use this to code basic string-handling functions recursively. The function `length()` determines the length of a string using recursion.

```cpp
// recursion.cpp

#include <iostream>
#include <cstring>     // for strcpy
using namespace std;

int length(char *s)
{
  if(*s == '\0')
    return 0;
  else
    return (1 + length(s+1));
}

int main(void)
{
  char* st = "willi hans";
  int res1 = length(st);
  cout << "The length of the string st is: " << res1; // => 10
  cout << endl;

  char* empty = "\0"; // empty string
  int res2 = length(empty);
  cout << "The length of the string empty is: " << res2; // => 0
  cout << endl;

  return 0;
}
```

Example 2. The following program shows how to implement a linked list recursively.

```cpp
// rlist.h

#ifndef RLIST_HEADER
#define RLIST_HEADER

#include <cassert>

template <class T>
class RList
{
 public:
   RList();                // default constructor
   RList(const RList&);    // copy constructor
   ~RList();               // destructor

   RList &operator=(const RList&);
   int operator==(const RList&);

   void Insert(const T&);
   int Search(const T&);
   int Delete(const T&);

   T Head(void);
   RList* Tail(void);
   int Empty(void);
   RList* Reverse(RList*);
 private:
   T head;
   RList* tail;
   int empty;
};

template <class T>
RList<T>::RList()
{ empty = 1; }

template <class T>
RList<T>::RList(const RList<T>& RL)
{
  empty = RL.empty;
  head = RL.head;
  if(!empty) tail = new RList<T>(*RL.tail);
}

template <class T>
RList<T>::~RList()
{ if(!empty) delete tail; }

template <class T>
RList<T> &RList<T>::operator = (const RList<T> &RL)
{
 if(this == &RL) return;
 if(!empty) delete tail;
 empty = RL.empty;
 head = RL.head;
```

```
   if(!empty) tail = new RList<T>(*RL.tail);
}

template <class T>
int RList<T>::operator==(const RList<T> &RL)
{
   if(empty != RL.empty) return 0;
   if(empty && RL.empty) return 1;
   if(this == &Rl) return 1;
   if(head != RL.head) return 0;
   return (*tail == *RL.tail);
}

template <class T>
void RList<T>::Insert(const T &toInsert)
{
 if(empty)
 {
  head = toInsert;
  tail = new RList<T>;
  empty = 0;
 }
 else tail->Insert(toInsert);
}

template <class T>
int RList<T>::Search(const T &toSearch)
{
   if(empty)
     return 0;
   else
    if(head == toSearch) return 1;
   else
     return tail -> Search(toSearch);
}

template <class T>
int RList<T>::Delete(const T &toDelete)
{
   if(empty) return 0;
   else if(head==toDelete)
   {
   head = tail -> head;
   empty = tail -> empty;
   tail -> Delete(tail -> head);
   if(empty) delete tail;
   return 1;
   }
   else return tail -> Delete(toDelete);
}

template <class T>
T RList<T>::Head(void)
{
   assert(!empty);
   return head;
}
```

```
template <class T>
RList<T> *RList<T>::Tail(void)
{
  assert(!empty);
  return tail;
}

template <class T>
int RList<T>::Empty(void)
{ return empty; }

template <class T>
RList<T> *RList<T>::Reverse(RList<T> *RL)
{
 if(RL->Empty())
  {
   RList<T> *temp;
   temp = new RList<T>;
   return temp;
  }
 else
  {
   RList<T> *R;
   R = Reverse(RL->Tail());
   (*R).Insert(RL->Head());
   return R;
  }
}
#endif
```

An application is as follows:

```
// rlisteg.cpp

#include <iostream>
#include "rlist.h"
using namespace std;

int main(void)
{
  RList<int> L;
  for(int i=1;i<=8;i++) L.Insert(i);

  RList<int>* LX = &L;
  cout << "The initial RList is: " << endl;
  while(!LX -> Empty())
  {
  cout << LX -> Head() << ' ';
  LX = LX -> Tail();
  }
  cout << endl << endl;

  RList<int>* R = L.Reverse(&L);
  RList<int>* LP = R;

  while(!LP -> Empty())
  {
  cout << LP -> Head() << ' ';
```

```
LP = LP -> Tail();
}
cout << endl << endl;

cout << "what happened to the initial list: "<< endl;
LP = &L;
while(!LP -> Empty())
{
cout << LP -> Head() << ' ';
LP = LP -> Tail();
}
cout << endl;

cout << "remove some items: "<< endl;
L.Delete(1);
L.Delete(4);
L.Delete(8);
LP = &L;
while(!LP -> Empty())
{
cout << LP -> Head() << ' ';
LP = LP -> Tail();
}
cout << endl;
cout << "is 3 in the list: " << L.Search(3) << endl;
cout << "is 4 in the list: " << L.Search(4) << endl;

return 0;
}
```

The output is

```
The initial RList is:
1 2 3 4 5 6 7 8

8 7 6 5 4 3 2 1

what happened to the initial list:
1 2 3 4 5 6 7 8
remove some items:
2 3 5 6 7
is 3 in the list: 1
is 4 in the list: 0
```

Example 3. In *mutual recursion* we have two functions which call each other. As an example we consider an implementation of sine and cosine. The identities

$$\sin(x) \equiv 2\sin(x/2)\cos(x/2)$$

$$\cos(x) \equiv \cos^2(x/2) - \sin^2(x/2) \equiv 2\cos^2(x/2) - 1$$

are used. Both sine and cosine call themselves and are therefore recursive. The sine function calls the cosine function. For the cosine function we have two options. For efficiency we select the second option where cosine calls only itself (once).

```cpp
// sincos.cpp

#include <iostream>
#include <cmath>
using namespace std;

double sine(double,double);    // forward declaration
double cosine(double,double); // forward declaration

int main(void)
{
   double x = 3.14159;
   double eps = 0.001;
   double res1 = sine(x,eps);
   cout << "res1 = " << res1 << endl;
   double res2 = cosine(x,eps);
   cout << "res2 = " << res2 << endl;
}

double sine(double x,double eps)
{
   double s;
   if(fabs(x) < eps) { s = x*(1.0 - x*x/6.0); }
   else s = 2.0*sine(x/2.0,eps)*cosine(x/2.0,eps);
   return s;
}

double cosine(double x,double eps)
{
   double c;
   if(fabs(x) < eps) { c = 1.0 - x*x/2.0; }
   else { c = cosine(x/2,eps); c = 2.0*c*c - 1.0; }
   return c;
}
```

Example 4. The *Jacobi elliptic functions* can be defined as inverses of the elliptic integral of first kind. Thus, if we write

$$x(\phi, k) = \int_0^\phi \frac{ds}{\sqrt{1 - k^2 \sin^2 s}}$$

where $k \in [0, 1]$ we can define the following functions

$$\mathrm{sn}(x, k) := \sin(\phi), \qquad \mathrm{cn}(x, k) := \cos(\phi), \qquad \mathrm{dn}(x, k) := \sqrt{1 - k^2 \sin^2 \phi}.$$

For $k = 0$ we obtain

$$\mathrm{sn}(x, 0) \equiv \sin(x), \qquad \mathrm{cn}(x, 0) \equiv \cos(x), \qquad \mathrm{dn}(x, 0) \equiv 1$$

and for $k = 1$ we have

$$\mathrm{sn}(x, 1) \equiv \tanh(x), \qquad \mathrm{cn}(x, 1) \equiv \mathrm{dn}(x, 1) \equiv \frac{2}{e^x + e^{-x}}.$$

We have the following identities

$$\mathrm{sn}(x,k) \equiv \frac{2\mathrm{sn}(x/2,k)\mathrm{cn}(x/2,k)\mathrm{dn}(x/2,k)}{1-k^2\mathrm{sn}^4(x/2,k)}$$

$$\mathrm{cn}(x,k) \equiv \frac{1-2\mathrm{sn}^2(x/2,k)+k^2\mathrm{sn}^4(x/2,k)}{1-k^2\mathrm{sn}^4(x/2,k)}$$

$$\mathrm{dn}(x,k) \equiv \frac{1-2k^2\mathrm{sn}^2(x/2,k)+k^2\mathrm{sn}^4(x/2,k)}{1-k^2\mathrm{sn}^4(x/2,k)}.$$

The expansions of the Jacobi elliptic functions in powers of x up to order 3 are given by

$$\mathrm{sn}(x,k) = x - (1+k^2)\frac{x^3}{3!} + \cdots$$

$$\mathrm{cn}(x,k) = 1 - \frac{x^2}{2!} + \cdots$$

$$\mathrm{dn}(x,k) = 1 - k^2\frac{x^2}{2!} + \cdots$$

For x sufficiently small this will be a good approximation.

We can now use these identities and the expansions to implement the Jacobi elliptic functions using one recursive call. The recursive call in scdn uses half of the provided parameter x, in other words the absolute value of the parameter passed in the recursive call is always smaller (by $\frac{1}{2}$). This guarantees that for fixed $\epsilon > 0$ the parameter x will satisfy $x < \epsilon$ after a finite number of recursive calls. At this point a result is returned immediately using the polynomial approximation given by the expansions above. This ensures that the algorithm will complete successfully. The recursive call is possible due to the identities for the sn, cn and dn functions. Since the identities depend on all three functions sn, cn and dn we can calculate all three at each step instead of repeating calculations for each of sn, cn and dn. Lastly some optimization was done to reduce the number of multiplications used in the double angle formulas. We also use the fact that the denominator of all three identities is the same.

The advantage of this approach is that all three Jacobi elliptic functions are found with one function call. Furthermore the cases $k = 0$ and $k = 1$ include the sine, cosine, tanh and sech functions. Obviously, for these special cases faster routines are available. Elliptic functions belong to the class of doubly periodic functions in which $2K$ plays a similar role to π in the theory of circular functions, where $K = F(1,k)$ is the complete elliptic integral of first kind. We have the identities

$$\mathrm{sn}(x \pm 2K, k) \equiv -\mathrm{sn}(x,k)$$

$$\mathrm{cn}(x \pm 2K, k) \equiv -\mathrm{cn}(x,k)$$

$$\mathrm{dn}(x \pm 2K, k) \equiv \mathrm{dn}(x,k).$$

To reduce the argument of the Jacobi elliptic functions we can also apply these identities.

The recursion method described above can be implemented using C++ as follows.

```cpp
// jacobi.cpp

#include <cmath>
#include <iostream>
using namespace std;

// forward declaration
void scdn(double,double,double,double&,double&,double&);

int main(void)
{
 double x = 3.14159, k, k2, eps = 0.01, res1, res2, res3;

 //sin,cos,1 of x
 k = 0.0; k2 = k*k;
 scdn(x,k2,eps,res1,res2,res3);
 cout << "sin(x) = " << res1 << endl;
 cout << "cos(x) = " << res2 << endl;
 cout << "1(x) = " << res3 << endl;

 //tanh,sech,sech of x
 k = 1.0; k2 = k*k;
 scdn(x,k2,eps,res1,res2,res3);
 cout << "tanh(x) = " << res1 << endl;
 cout << "sech(x) = " << res2 << endl;
 cout << "sech(x) = " << res3 << endl;
}

void scdn(double x,double k2,double eps,double &s,double &c,double &d)
{
  if(fabs(x) < eps)
  {
   double x2 = x*x/2.0;
   s = x*(1.0 - (1.0 + k2)*x2/3.0);
   c = 1.0 - x2;
   d = 1.0 - k2*x2;
  }
  else
  {
   double sh,ch,dh;

   scdn(x/2.0,k2,eps,sh,ch,dh);  // recursive call

   double sh2 = sh*sh;
   double sh4 = k2*sh2*sh2;
   double denom = 1.0 - sh4;

   s = 2.0*sh*ch*dh/denom;
   c = (1.0 - 2.0*sh2+sh4)/denom;
   d = (1.0 - 2.0*k2*sh2+sh4)/denom;
  }
}
```

4.15   Error Handling Techniques

An exception is an error that occurs at run-time. A typical example is division by zero. Thus it may be a good idea to classify a program into distinct subsystems that either execute successfully or fail. Thus, local error checking should be implemented throughout the system for a sound program. It is, therefore, important to identify the possible source of errors. They could be caused by the programmer himself. This case includes the detection of an internal logic error, such as an assertion failure. Sometimes, special attention has to be paid to the preconditions of calling a function. For example, an attempt to get the seventh character of a three-character `String` would cause an error. However, some problems are not logic errors, but a failure to get some resource during run-time. These errors might include running out of memory, a write failure due to disk full, etc.

A fault-tolerant system is usually designed hierarchically, with each level coping with as many errors as possible. However, some errors cannot be handled locally. Thus, we need a global error communication mechanism, which would be one of the following:

- Error state, i.e. the different return values of a function or class method represent the different status of the function execution.

- Exception handling.

In C we can use the function `void assert(int test)` to test a condition and possibly abort. This function is a macro that expands to an `if` statement; if `test` evaluates to zero, `assert` prints a message on `stderr` and aborts the program by calling `abort`. Using C++'s exception handling subsystem we can handle run-time errors in a structured and controlled manner. C++ exception handling is built upon three keywords: `try`, `catch` and `throw`.

One can implement *error states* via class variables. Typically, each object has an error flag (error state variable). If an error occurs, the error state variable is set to a value indicating the error type. For example, the `<iostream>` library of C++ has an integer state variable called `state`. The different bits of this variable indicate different error states. The error state can be queried via the class methods `bad()`, `eof()`, `fail()`. The user can then implement his own error-handling mechanism.

4.16   Exception Handling

A more sophisticated method that controls error detection and error handling is called *exception handling*. It is a non-local mechanism which provides a means of communicating errors between the classes/functions and programs that make use of these classes/functions. It also lets the programmers write their code without worrying about errors at every function call.

In the exception handling mechanism in C++, the function that detects an error raises an exception using the keyword `throw`. The handler function then catches and handles the exception. The function that raises the exception must come within a `try`-block. The handlers, which are declared using the keyword `catch`, are placed at the end of the `try`-block.

As an example, consider the following `Rational` class, together with the `main()` program:

```cpp
// except.cpp

#include <iostream>
#include <string>
using namespace std;

class Rational
{
  private:
    long num, den;
  public:
    Rational(const long=0,const long=1);
};

class Check_Error
{
  private:
    string reason;
  public:
    Check_Error(const string &s) : reason(s) { }
    string diagnostic() const { return reason; }
};

Rational::Rational(const long int N,const long int D) : num(N), den(D)
{ if(D==0) throw Check_Error("Division by zero !"); }

int main(void)
{
  try
  {
  long a, b;
  a = 5; b = 0;
  Rational R1(a,b);
  }
  catch(Check_Error diag)
  {
  cerr << "Internal error : " << diag.diagnostic() << endl;
  return 1;
  }
  return 0;
}
/*
Result
======
Internal error : Division by zero !
*/
```

The constructor of the `Rational` class checks if the denominator of the number is zero. It throws an exception when this error occurs. The thrown object contains the diagnostic message, which is

```
Division by zero !
```

in this case.

The class `Check_Error` handles all the exceptions that might happen within the program. It contains a `private` data member which stores a string containing the reason for the exception. The member function `diagnostic()` prints the diagnostic message of the failure. It is used in the `catch` function of the program.

In the `main()` function the whole program is placed in the `try`-block. Immediately after the block is the `catch` function, which is the actual exception handler. The `catch` function specifies the type of exception it handles.

4.17 Run-Time Type Identification

In polymorphic languages such as C++, there can be situations in which the type of an object is unknown at compile time, since the precise nature of that object is not determined until the program is executed. C++ implements polymorphism through the use of class hierarchies, virtual functions, and the base class pointers. A base class pointer may be used to point to objects of the base class or any object derived from that base. Thus, if is not always possible to know in advance what type of object will be pointed to by a base pointer at any given moment in time. This determination must be made at run time by using run-time type identification.

To obtain an object's type, use the command `typeid`. We include the header file `<typeinfo>` in order to use `typeid`. An example is

```cpp
// rtti.cpp

#include <iostream>
#include <typeinfo>
#include <string>
using namespace std;

int main(void)
{
    int i = 7;
    double d = 3.14159;
    char c = 'X';
    string s("egoli");
    cout << typeid(i).name() << endl;
    cout << typeid(d).name() << endl;
    cout << typeid(c).name() << endl;
    cout << typeid(s).name() << endl;
    if(typeid(i) == typeid(c)) cout << "The types are the same";
    return 0;
}
```

The type information for pointers to instances of derived classes depends on whether
the classes are virtual. This is illustrated below.

```cpp
// rtti2.cpp

#include <iostream>
#include <typeinfo>
#include <string>
using namespace std;

class base1
{
 public: string whatami(void) { return typeid(*this).name(); }
};
class derived11: public base1 {};
class derived12: public base1 {};

class base2
{
 public: virtual string whatami(void) { return typeid(*this).name(); }
         virtual ~base2() {}
};
class derived21: public base2 {};
class derived22: public base2 {};

int main(void)
{
   derived11 d11;
   derived12 d12;
   base1*    b1;
   derived21 d21;
   derived22 d22;
   base2*    b2;
   b1 = &d11;
   cout << b1->whatami() << endl;
   b1 = &d12;
   cout << b1->whatami() << endl;
   b2 = &d21;
   cout << b2->whatami() << endl;
   b2 = &d22;
   cout << b2->whatami() << endl;
   return 0;
}
```

Chapter 5

String Class

5.1 Introduction

A string is a sequence of characters for example `"xyz"`. We also include the empty string with no characters (i.e. `""`). In fact, nearly all the operations associated with arrays or vectors can be applied equally well to strings. However, a string is also a much more abstract entity, in addition to simple vector operators, the string data type provides a number of useful and powerful high-level operations. One confusing aspect of the `string` facility is the presence of two different include files. The statement

```
#include <cstring>
```

includes a file that declares a number of functions useful in conjunction with older, C-style character values, for example `strlen`, `strcpy`, `strcmp` etc..

A string type is lacking in traditional C. Strings in C are represented as a pointer to a `char` and are manipulated accordingly. This means strings can be considered as arrays of type `char`. For example

```
#include <iostream>
using namespace std;

int main(void)
{
  char* s = NULL;
  s = new char[4];
  s[0] = 'o'; s[1] = 'l'; s[2] = 'a'; s[3] = '\0';
  cout << "s = " << s;
  delete[] s;
  return 0;
}
```

In this representation, the end-of-string is denoted by the nul character `'\0'`. The nul character `'\0'` requires one byte to store and it is part of the string. This means `"ola"` requires 4 bytes of memory to hold. This convention has an important drawback, in that many basic string manipulations are proportional to string length. When the string length is known, the efficiency of operations on strings can be improved. Therefore, to avoid possible errors and improve efficiency of the string operations, it is best to implement the string as an abstract data type. We give a string class in the next section, where the strings are considered as abstract data types. C++ has a built-in string class. We consider this string class in Section 5.3. Applications are given in Section 5.4.

5.2   A String Class

The program `string1.cpp` gives a comprehensive implementation of a string class. The operators

```cpp
// string1.cpp

#include <iostream>
#include <cstring>  // for strlen, strcpy, strcat, strncpy
#include <cassert>  // for assert
using namespace std;

            // Declaration of class String
class String
{
  private:
     char* s;
     int length;
  public:
    String(const char*);
    String(unsigned);
    String(unsigned,char);
    String(const String&);     // copy constructor
    ~String();                 // destructor
    String& operator = (const String&);
    String operator () (unsigned,unsigned) const;
    String operator () (unsigned) const;
    char& operator [] (unsigned);
    operator const char* () const;
    friend String operator + (const String&,const String&);
    friend int operator == (const String&,const String&);
    friend int operator != (const String&,const String&);
    friend int operator < (const String&,const String&);
    friend int operator <= (const String&,String&);
    friend ostream& operator << (ostream&,const String&);
    friend istream& operator >> (istream&,String&);
    void display();
    void read(unsigned);
    String swap_char(unsigned,unsigned);
    int replace(String,String);
    String reverse() const;
    void fill(char);
};

String::String(const char* p)
{
  length = strlen(p);
```

```
  s = new char[length+1];
  strcpy(s,p);
}

String::String(unsigned len)
{
  length = len;
  s = new char[length+1];
  s[length] = '\0';
}

String::String(unsigned len,char tofill)
{
  length = len;
  s = new char[length+1];
  fill(tofill);
  s[length] = '\0';
}

String::String(const String& toCopy)
{
  length = toCopy.length;
  s = new char[length+1];
  strcpy(s,toCopy.s);
}

String::~String()
{ delete[] s; }

String& String::operator = (const String& toSet)
{
  if(this == &toSet) return *this;
  if(length != toSet.length)
  {
    delete[] s;
    length = toSet.length;
    s = new char[length+1];
  }
  strcpy(s,toSet.s);
  return *this;
}

String String::operator () (unsigned start,unsigned end) const
{
  assert(start > 0 && end <= length && start <= end);
  int newlength = end-start+1;
  String S(newlength);
  strncpy(S.s,s+start-1,newlength);
  return S;
}

String String::operator() (unsigned end) const
{
  assert(end > 0 && end <= length);
  return (*this)(1,end);
}

char& String::operator[] (unsigned i)
```

```cpp
{
  assert(i > 0 && i <= length);
  return s[i-1];
}

String::operator const char* () const
{ return s; }

String operator + (const String& s1,const String& s2)
{
  String S(s1.length + s2.length);
  strcpy(S.s,s1.s);
  strcat(S.s,s2.s);
  return S;
}

int operator == (const String& s1,const String& s2)
{ return !strcmp(s1.s,s2.s); }

int operator != (const String& s1,const String& s2)
{ return !(s1 == s2); }

int operator < (const String& s1,const String& s2)
{ return strcmp(s1.s,s2.s) < 0; }

int operator <= (const String& s1,const String& s2)
{ return (s1 < s2 || s1 == s2); }

ostream& operator << (ostream& out,const String& S)
{
  out << S.s;
  return out;
}

istream& operator >> (istream& in,String& S)
{
  const int max = 1024;
  delete[] S.s;
  S.s = new char[max];
  in.width(max);
  in >> S.s;
  S.length = strlen(S.s);
  return in;
}

void String::display()
{ cout << s; }

void String::read(unsigned i)
{
  length = i;
  delete[] s;
  s = new char[length+1];
  cin.width(i+1);
  cin >> s;
}

String String::swap_char(unsigned int n,unsigned int m)
```

```
{
  String S(length);
  char temp;
  strcpy(S.s,s);
  temp = S[n]; S[n] = S[m]; S[m] = temp;
  return S;
}

int String::replace(String sub,String new_sub)
{
  char *temp, *ptr, *base;
  int len, diff;
  if(length == 0) return 0;
  else
  {
    diff = new_sub.length-sub.length;
    if(diff > 0) // string must grow
    {
      // after substitution, at most length/sub.length
      // substrings will have been replaced causing the
      // string length to grow by length*diff/sub.length
      temp = new char[length+length*diff/sub.length];
      strcpy(temp,s);
      delete[] s;
      s = temp;
    }
    temp = new char[length];
    len = sub.length;
    base = ptr = s;
    while((base = strstr(base,sub.s)) != NULL)
    {
      ptr = base+len;
      strncpy(temp,ptr,strlen(ptr)+1);
      strcpy(base,new_sub.s);
      strcpy(base + new_sub.length,temp);
      // the string length changed after substitution
      length += diff;
      // the substituted string is not subject to substitution again
      base = base + new_sub.length;
    }
    if(ptr == s)
      cout << "sorry substring cannot be replaced.\n";
    else
      cout << "The new string is " << s << endl;
    delete[] temp;
    return 1;
  }
}

String String::reverse() const
{
  String S(length);
  for(int i=0;i<length;i++) S.s[i] = s[length-i-1];
  return S;
}

void String::fill(char tofill)
{ for(int i=0;i<length;s[i++]=tofill); }
```

```
int main(void)
{
  String S1("Johannesburg");
  cout << "S1 = " << S1 << endl;   // Johannesburg
  S1.display();                    // Johannesburg
  cout << endl;

  String S2(3);
  cout << "S2 = " << S2 << endl;

  String S3(6,'$');
  cout << "S3 = " << S3 << endl;   // $$$$$$

  String S4(S1);
  cout << "S4 = " << S4 << endl;

  S2 = S4;
  cout << "S2 = " << S2 << endl;   // Johannesburg

  S4.fill('+');
  cout << "S4 = " << S4 << endl;   //  ++++++++++++
  cout << "S2 = " << S2 << endl;   //  Johannesburg

  cout << "reverse of S1: " << S1.reverse() << endl; // grubsennahoJ
  S1.reverse().display();                            // grubsennahoJ
  cout << endl;

  String S5(30);
  S5 = S1 + S1.reverse();
  cout << "S5 = " << S5 << endl;      // JohannesburggrubsennahoJ

  cout << "S1 == S2 --> " << (S1 == S2) << endl;   // 1
  cout << "S1 == S4 --> " << (S1 == S4) << endl;   // 0

  cout << "S1.swap_char(2,6): " << S1.swap_char(2,6); // Jnhanoesburg
  cout << endl;

  String S6(S1);
  cout << "S6 = " << S6 << endl;
  S6.replace("hannes", "");
  cout << "S6.replace(\"hannes\", \"\") --> " << S6 << endl; //Joburg

  String S7(S1);
  cout << "S7 = " << S7 << endl;
  S7.replace("n", "nn");
  cout << "S7.replace(\"n\", \"nn\") --> " << S7 << endl; //Johannnnesburg

  return 0;
}
```

5.3   C++ String Class

The C++ Standard provides a library **string** class, to make string manipulation easier and less error-prone. The string class does not belong to STL but it is supported by all newer C++ compilers.

To define string variables, we have to use the following #include line:

```
#include <string>
```

Now we can do the following assignments

```
string s, t;
s = "ABC";
t = "DEFGH";
s = t;        // s = "DEFGH"
s = "A";      // s = "A"
s = "";       // empty string
```

The following table summarizes the constructors and member functions (methods) of the string class.

Table. Operations for the string data type

```
Constructors
string s;                 default constructor
string s("text");         initialized with literal string
string s(aString);        copy constructor


Character Access
s[i]                      subscript access
s.substr(pos,len)         return substring starting at position
                          of given length
s.c_str()                 return a C-style character pointer


Length
s.length()                number of characters in string
s.resize(int,char)        change size of string, padding with char
s.empty()                 true if string has no characters


Assignment and Catenation
s = s2;   assignment of string
s += s2;  append second string to end of first
s + s2    new string containing s followed by s2


Iterators
string::iterator t   declaration of new iterator
s.begin()            starting iterator
s.end()              past-the-end iterator


Insertion, Removal, Replacement
s.insert(pos,str)             insert string after given position
s.erase(start,length)         remove length characters after start
```

```
s.replace(start,length,str) insert string, replacing
                            indicated characters
```

Comparisons
```
s == s2   comparison for equality
s != s2   comparison for inequality
s < s2    comparisons for ordering
s > s2    comparisons for ordering
s <= s2   comparisons for ordering
s >= s2   comparisons for ordering
```

Searching Operations
```
s.find(str)                  find start of argument string
                             in receiver string
s.find(str,pos)              find with explicit starting position
s.find_first_of(str,pos)     first position of first character
                             from argument
s.find_first_not_of(str,pos) first character not from argument
```

Input / Output Operations
```
stream << str                output string on stream
stream >> str                read word from stream
getline(stream,str,char)     read line of input from stream
```

Notice that the comparison operators

```
==   !=
```

as well as the relational operators

```
<   <=   >   >=
```

compare the contents of the two string arguments, not their addresses in memory, as is the case with character pointers.

The following program shows an application of the string class.

```cpp
// mystring.cpp

#include <iostream>
#include <string>
using namespace std;

int main(void)
{
  string str1("The string class provides C++");
  string str2(" with string handling.");
  string str3;

  // assignment
  str3 = str1;
```

```
cout << str3 << endl; // => The string class provides C++

// comparison
int result = (str1 != str2);
cout << "result = " << result << endl; // => 1 (true)

// concatenate two strings using overloaded +
str3 = str1 + str2;
cout << str3 << endl;

// length of string
unsigned long a = str3.length();
cout << "length of string str3 = " << a << endl;           // => 51
cout << "length of string str3 = " << str3.size() << endl; // => 51

bool b = str1.empty();
cout << "b = " << b << endl; // => 0 (false)

string str4("Good Morning Egoli");
string str5("XYZ1234");

cout << "Initial string:\n";
cout << "str4: " << str4 << endl;
cout << "str5: " << str5 << endl;

// method insert
cout << "insert str5 into str4:\n";
str4.insert(5,str5);
cout << str4 << "\n"; // => Good XYZ1234Morning Egoli

// method erase
cout << "remove 7 characters from str4:\n";
str4.erase(5,7);
cout << str4 << "\n";  // => Good Morning Egoli

// method replace
cout << "replace 2 characters in str4 with str5:\n";
str4.replace(5,2,str5);
cout << str4 << endl;  // => Good XYZ1234rning Egoli

// substring
cout << "substring = " << str4.substr(2,5); // => od XY

return 0;
}
```

5.4 Applications

Here we give a collection of examples where the **string** class is applied.

Example 1. Consider the mathematical expression

```
a*b*cos(0)+sin(0)*c-d+cos(0)*sin(0)+2*d*1
```

This expression will be stored as a string. We know that $\sin(0)=0$ and $\cos(0)=1$. Thus we want to use these rules to simplify the expression. Furthermore if we find the substrings "1*" or "*1" it should be replaced by the empty string, since 1*x = x and x*1 = x. The string "+0+" should be replaced by "+". We have a map between strings and what they should be replaced by. Thus we use the STL map class with the strings that should be replaced as indices and the replacement strings as values. For example, to simplify "1*x" to "x" we use the rule

```
simplify["1*"] = "";
```

We apply the method `find()` in the `string` class to find the substring and then the method `replace()` to replace the substring.

```
// stringexpr.cpp

#include <iostream>
#include <map>
#include <string>
using namespace std;

int main(void)
{
  map<string,string> simplify;
  map<string,string>::iterator i;
  int count = 1;

  simplify["sin(0)"] = "0";
  simplify["cos(0)"] = "1";
  simplify["1*"]     = "";
  simplify["*1"]     = "";
  simplify["+0+"]    = "+";

  string exp = "a*b*cos(0)+sin(0)*c-d+cos(0)*sin(0)+2*d*1";
  cout << "exp = " << exp << endl;

  while(count)
   for(count=0,i=simplify.begin();i!=simplify.end();i++)
   {
    string::size_type pos = exp.find(i->first,0);
    while(pos < exp.npos)
    {
     count++;
     exp.replace(pos,i->first.length(),i->second);
     pos = exp.find(i->first,0);
     cout.setf(ios_base::left,ios_base::adjustfield);
     cout.width(50);
     cout << "exp = " + exp;
     if(i->second == "")
      cout << "(delete " + i->first + ")" << endl;
     else
      cout << "(" + i->first + " -> " + i->second + ")" << endl;
    }
   }
  return 0;
}
```

Example 2. We implement a string splitting. It divides a string into smaller strings based on a separator character. These smaller strings are also known as tokens. We store the smaller strings in the vector container class of STL.

```cpp
// tokenizer.cpp

#include <iostream>
#include <string>
#include <vector>
using namespace std;

vector<string>* split(string s,char sep)
{
  // dynamically allocate the result vector
  // so that the returning pointer will be valid
  vector<string>* results = new vector<string>;
  string t = "";
  int c, slength;
  slength = s.size();
  // iterate through string s character by character
  for(c=0;c<slength;c++)
   if(s[c] == sep)
    {
     // end of token found, add to result vector
     results -> push_back(t);
     t = "";  // start new token
    }
   else { t += s[c]; }  // add char to token
  if(t != "") results -> push_back(t); // get last token
  return results;              // return pointer to result vector
}  // end split function

int main(void)
{
  // pointer to a vector of strings
  vector<string>* sp;
  sp = split("a+b;c*d;d/f",';');
  int vsize = sp -> size();
  cout << "number of terms: " << vsize << endl;
  // display each substring token one by one
  for(int i=0;i<vsize;i++) cout << sp -> at(i) << endl;
  delete sp;
  return 0;
}
```

Example 3. The following program enumerates the binary representations of integers. The function N converts the binary representation in string s to an integer, while the function Ninv converts an integer to the binary representation. The function mapinv(i) finds the *i*-th binary representation, where the binary representations are ordered first by length and second by integer value.

```cpp
// convertbinary.cpp

#include <iostream>
#include <string>
#include <cmath>
using namespace std;
```

```
unsigned long N(string s)
{
  unsigned long sum = 0, p2=1;
  for(int i=s.length()-1;i>=0;i--,p2*=2) sum += (s[i]-'0')*p2;
  return sum;
}

string Ninv(unsigned long x)
{
  string s = "";  // empty string
  for(;x>0;x/=2) s = char((x%2)+'0')+s;
  return s;
}

unsigned long map(string s)
{
  unsigned long n = N(s);
  if(s.length()>1) n += (1 << s.length())-2;
  return n;
}

string mapinv(unsigned long x)
{
  unsigned long n = (unsigned long)floor(log(x+2.0)/log(2.0));
  string b;
  if(n == 1) b = Ninv(x);
  else b = Ninv(x-(1<<n)+2);
  while(b.length()<n) b = '0' + b;
  return b;
}

int main(void)
{
  string s;
  for(unsigned long i=0;i<20;i++)
  {
   s = mapinv(i);
    cout.width(3); cout << map(s) << " -> " << s << endl;
  }
  return 0;
}
```

Example 4. The following program provides a simple implementation of string tokenizer class using the **string** class.

```
// tokenizer2.cpp

#include <iostream>
#include <string>
using namespace std;

class StringTokenizer
{
  public:
    StringTokenizer(string s,string d=string(" \t"))
    { toTokenize = s; delimiters = d; first = 0; }
```

```cpp
    string nextToken()
    {
     if(first == string::npos) return "";
     string::size_type next = toTokenize.find_first_of(delimiters,first);
     string token = toTokenize.substr(first,next-first);
     first = toTokenize.find_first_not_of(delimiters,next);
     return token;
    }

  private:
    string toTokenize, delimiters;
    string::size_type first;
};

int main(void)
{
  string s = "The quick brown fox";
  StringTokenizer st0(s);

  cout << st0.nextToken() << endl; cout << st0.nextToken() << endl;
  cout << st0.nextToken() << endl; cout << st0.nextToken() << endl;

  // try two StringTokenizers non-interleaved
  StringTokenizer st1(string("This is a test"));
  StringTokenizer st2(string("Now is the time"));

  cout << "non-interleaved:" << endl;
  cout << st1.nextToken() << endl; cout << st1.nextToken() << endl;
  cout << st1.nextToken() << endl; cout << st1.nextToken() << endl;
  cout << st2.nextToken() << endl; cout << st2.nextToken() << endl;
  cout << st2.nextToken() << endl; cout << st2.nextToken() << endl;

  // now try two StringTokenizers interleaved
  StringTokenizer st3(string("This is a test"));
  StringTokenizer st4(string("Now is the time"));

  cout << "interleaved:" << endl;
  cout << st3.nextToken() << endl; cout << st4.nextToken() << endl;
  cout << st3.nextToken() << endl; cout << st4.nextToken() << endl;
  cout << st3.nextToken() << endl; cout << st4.nextToken() << endl;
  cout << st3.nextToken() << endl; cout << st4.nextToken() << endl;
  return 0;
}
```

Chapter 6

Standard Template Library

6.1 Introduction

The *Standard Template Library*, or STL, is a C++ library of container classes, algorithms and iterators [1]. It provides many of the basic algorithms and data structures of computer science. The intent of the STL is to provide a set of container classes that are both efficient and functional. The presence of such a library will simplify the creation of complex programs, and because the library is standard the resulting programs ultimately will have a high degree of portability. This section is about the new standard version of STL. The use of STL is likely to make software more reliable, more portable and more general and to reduce the cost of producing it. One of the most interesting aspects of the STL is the way it radically departs in structure from almost all earlier libraries. Since C++ is an object-oriented language, these earlier libraries have tended to rely heavily on object-oriented techniques, such as inheritance. The STL uses almost no inheritance. To see this non-object-oriented perspective, consider that object-oriented programming holds encapsulation as a primary ideal. A well-designed object will try to encapsulate all the state and behavior necessary to perform the task for which it is designed, and at the same time, hide as many of the internal implementation details as possible. In almost all previous object-oriented container class libraries this philosophical approach was manifested by collection classes with exceedingly rich functionality, and consequently with large interfaces and complex implementations. The designers of STL moved in an entirely different direction. Each component is designed to operate in conjunction with a rich collection of generic algorithms. These generic algorithms are independent of the containers and can therefore operate with many different containers types. By separating the functionality of the generic algorithms from the container classes themselves, the STL realizes a great saving in size, in both the library and the generated code. Instead of duplication of algorithms in each of the dozen or so different container classes, a single definition of a library function can be used with any container. Furthermore, the definition of these functions is so general that they can be used with ordinary C-style arrays and pointers as well as with other data types.

The boolean data type `bool` is also implemented in the STL. At the core of the STL are three foundational items:

1. containers

2. algorithms

3. iterators.

These items work in conjunction with one another. Containers are objects that hold other objects. Algorithms act on the contents of containers. Iterators are objects that behave like pointers. They provide us with the ability to cycle through the contents of a container in the same way that we apply pointers to cycle through an array. The classes provided by the STL are:

`algorithm` `deque` `functional` `iterator` `list` `map`

`memory` `numeric` `queue` `set` `stack` `utility` `vector`

In principle, STL assumes that all object containers have at least the following

An assignment operator (=)

An equality comparison operator (==)

A less than comparison operator (<)

A copy constructor

Both the `<utility>` and `<functional>` templates are viewed as support libraries. We use the `<functional>` STL for defining several templates that help construct predicates for the templates defined in `<algorithm>` and `<numeric>`.

The STL `<iterator>` template is an important component in STL. The ANSI/ISO Committee defines an *iterator* as a generalized pointer that allows a programmer to work with different data structures (or containers) in a uniform manner. Structurally, an iterator is a pointer data type. Like all pointers, for example

```
int *pi;   float *pf;   MyClass *pmyclass;
```

iterators are defined in terms of what they point to. Their syntax is

```
container<Type>::iterator iterator_instance;
```

where the type name begins with the type of container they are associated with. Sometimes we have to add the keyword `typename` at the beginning of the line to instruct the compiler that `::iterator` is a type and not a member.

6.2   Namespace Concept

Namespaces are a recent addition to C++. A namespace creates a declarative region in which various program elements can be placed. Elements declared in one namespace are separated from elements declared in another. The line `using namespace std`; tells the compiler to use the standard STL namespace.

```cpp
// namesp.cpp

#include <iostream>
using namespace std;

namespace A { int i = 10; }

namespace B { int i = 5; }

void fA()
{
    using namespace A;
    cout << "In fA: " << A::i << " " << B::i << " " << i << endl;
    // => 10 5 10
}

void fB()
{
    using namespace B;
    cout << "In fB: " << A::i << " " << B::i << " " << i << endl;
    // => 10 5 5
}

int main(void)
{
    fA();  fB();
    cout << A::i << " " << B::i << endl; // => 10 5
    using A::i;
    cout << i << endl; // => 10
    return 0;
}
```

6.3   Vector Class

The vector class is a container class. The vector class supports a dynamic array. This is an array that can grow as needed. We can use the standard array subscript notation [] to access its elements.

The constructors and methods (member functions) in the class `vector` are summarized in the following.

```
Constructors
vector<T> v;                default constructor
vector<T> v(int);           initialized with explicit size
vector<T> v(int,T);         size and initial value
```

```
vector<T> v(aVector);        copy constructor
```

Element Access
```
v[i]                         subscript access, can be assignment target
v.front()                    first value in collection
v.back()                     last value in collection
```

Insertion
```
v.push_back(T)               push element on to back of vector
v.insert(iterator,value)     insert new element after iterator
v.swap(vector<T>)            swap values with another vector
```

Removal
```
v.pop_back()                 pop element from back of vector
v.erase(iterator)            remove single element
v.erase(iterator,iterator)   remove range of values
```

Size
```
v.capacity()                 maximum number of elements buffer can hold
v.size()                     number of elements currently held
v.resize(unsigned,T)         change to size, padding with value
v.reserve(unsigned)          set physical buffer size
v.empty()                    true if vector is empty
```

Iterators
```
vector<T>::iterator itr      declare a new iterator
v.begin ()                   starting iterator
v.end ()                     ending iterator
```

```cpp
// mvector.cpp

#include <iostream>
#include <vector>
using namespace std;

int main(void)
{
  vector<char> w(4);
  w[0] = 'X';    w[1] = 'Y';    w[2] = '+';    w[3] = '-';
  cout << w[2] << endl;  // => +
  int j;
  for(j=0;j<4;j++)
  cout << "w[" << j << "] = " << w[j] << endl;

  vector<double> v;
  double x;
  cout << "enter double number, followed by 0.0:\n";
  while(cin >> x,x != 0.0)
  v.push_back(x);    // adds the double value x at the end
                     // of the vector v
  vector<double>::iterator i;
```

```
    for(i=v.begin();i!= v.end();++i)
    cout << *i << endl; // dereferencing

    bool b = v.empty();
    cout << "b = " << b << endl;

    // insert element in the vector
    double y;
    cout << "enter value to be inserted: ";
    cin >> y;
    vector<double>::iterator p = v.begin();
    p += 2;
    v.insert(p,4,y);  // point to 3rd element, insert 4 elements

    for(i=v.begin();i!=v.end();++i)
    cout << *i << endl;

    int length = v.size();
    cout << "length = " << length << endl;
    return 0;
}
```

We read a text file line by line into a vector. We read the file, line by line, until `eof` is encountered. The only catch is that with the istream's operator `>>` string cannot be used. The function only reads a word, and not a whole line. The global function

`getline(istream&,string&)`

is the solution. Let the text file (ASCII file) be `william.txt` with the contents

```
to be or
not to be
```

After compiling and linking to obtain the executable file at the command line we enter:

`readin1 william.txt`

```
// readin1.cpp

#include <iostream>
#include <fstream>
#include <string>
#include <vector>
using namespace std;

int main(int argc,char *argv[])
{
    ifstream ifs(argv[1]);
    vector<string> v;
    string s;

    while(getline(ifs,s))
    {
```

```
  v.push_back(s);
  cout << s <<endl;
  }
  cout << "v[1] = " << v[1] << endl;   // => not to be
  return 0;
}
```

The `vector` class can also be used to construct two and higher dimensional arrays.
For example, a two-dimensional array can be constructed as follows.

```cpp
// vecvec.cpp

#include <iostream>
#include <vector>
using namespace std;

int main(void)
{
  vector<vector<int> > v(3);
  for(int i=0;i<=2;i++)
  v[i].resize(4);
  v[0][0] = 7; v[0][1] = 4; v[0][2] = 3; v[0][3] = -7;
  v[1][0] = 4; v[1][1] = 9; v[1][2] = 6; v[1][3] = -9;
  v[2][0] = 7; v[2][1] = 11; v[2][2] = 1; v[2][3] = 3;

  double sum = 0.0;
  for(int k=0;k<=2;k++)
  for(int l=0;l<=3;l++)
  {
  sum += v[k][l];
  }
  cout << "sum = " << sum;
  return 0;
}
```

A three-dimensional array can be constructed as follows

```cpp
// vecvecvec.cpp

#include <iostream>
#include <vector>
using namespace std;

int main(void)
{
  vector<vector<vector<int> > > v(2);
  int i, j, k;
  for(i=0;i<2;i++)
  v[i].resize(3);
  for(i=0;i<2;i++)
  for(j=0;j<3;j++)
  v[i][j].resize(4);

  for(i=0;i<2;i++)
  for(j=0;j<3;j++)
  for(k=0;k<4;k++)
  v[i][j][k] = 1;
```

```
    int sum = 0;
    for(i=0;i<2;i++)
    for(j=0;j<3;j++)
    for(k=0;k<4;k++)
    sum += v[i][j][k];
    cout << "sum = " << sum;
    return 0;
}
```

The **vector** class is a container class, i.e. not a vector in the mathematical sense. No arithmetic operations are implemented so that we could, for example, add two vectors. However, we can add the overloaded operator + so that we can add two vectors (of the same size)

```
// overloadvector.cpp

#include <iostream>
#include <vector>
using namespace std;

template <class T>
vector<T> operator + (vector<T>& v1,vector<T>& v2)
{
    int length1 = v1.size();
    int length2 = v2.size();
    if(length1 != length2)
    { cerr << "length of v1 and v2 must be the same"; exit(0); }
    vector<T> result(length1);
    for(int i=0;i<length1;i++) result[i] = v1[i] + v2[i];
    return result;
}

int main(void)
{
    vector<int> v1(2);
    v1[0] = 3; v1[1] = 5;
    vector<int> v2(2);
    v2[0] = 6; v2[1] = -3;

    vector<int> w(2);
    w = v1 + v2;
    for(int i=0;i<2;i++) cout << w[i] << endl;

    return 0;
}
```

6.4 List Class

The list class is also a container class. A linked list is the data structure we choose when the number of elements in a collection cannot be bounded, or varies widely during the course of execution. Like the vector class, the linked list class maintains values of uniform type. Lists are not indexed. The following table gives the constructors and the member functions of the class.

Constructors and Assignment
```
list<T> l;                      default constructor
list<T> l(aList);               copy constructor
l = aList                       assignment
l.swap (aList)                  swap values with another list
```

Element Access
```
l.front()                       first element in list
l.back()                        last element in list
```

Insertion and Removal
```
l.push_front(value)             add value to front of list
l.push_back(value)              add value to end of list
l.insert(iterator,value)        insert value at specified location
l.pop_front()                   remove value from front of list
l.pop_back()                    remove value from end of list
l.erase(iterator)               remove referenced element
l.erase(iterator,iterator)      remove range of elements
l.remove(value)                 remove all occurrences of value
l.remove_if(predicate)          removal all values that match condition
```

Size
```
l.empty()                       true if collection is empty
l.size()                        return number of elements in collection
```

Iterators
```
list<T>::iterator itr           declare a new iterator
l.begin()                       starting iterator
l.end()                         ending iterator
l.rbegin()                      starting backwards moving iterator
l.rend()                        ending backwards moving iterator
```

Miscellaneous
```
l.reverse()                     reverse order of elements
l.sort()                        place elements into ascending order
l.sort(comparison)              order elements using comparison function
l.merge(list)                   merge with another ordered list
```

The following program gives an application of the list class.

```
// mlist.cpp

#include <iostream>
#include <list>
#include <string>
using namespace std;

int main(void)
{
```

```
    list<char> lst;
    for(int i=0;i<10;i++)
    lst.push_back('X'+i);
    cout << "size of list = " << lst.size() << endl;
    cout << "contents of list: ";
    list<char>::iterator p = lst.begin();

    while(p != lst.end())  {  cout << *p << " ";  p++;  }
    lst.push_back('A');
    cout << endl;

    lst.sort();
    cout << "sorted contents:\n";
    p = lst.begin();
    while(p != lst.end())  {  cout << *p << " ";  p++;  }
    cout << endl;

    list<string> str;
    str.push_back("Good");
    str.push_back(" ");
    str.push_back("Night");
    str.push_back(" ");
    str.push_back("Egoli");
    list<string>::iterator q = str.begin();
    while(q != str.end())  {  cout << *q << " ";  q++;  }
    return 0;
}
```

6.5 Stack Class

The *stack* class provides a restricted subset of container functionality. By default this underlying container type is `vector`. The underlying container could also be a `list`. It is a last-in-first-out (LIFO) data structure. The stack class does not allow iteration through its elements. It is a collection of data items organized in a linear sequence, together with the following five operations:

1. *CreateStack* brings a stack into existence.

2. *MakeStackEmpty* deletes all items, if any, from the stack.

3. *Push* adds an item at one end, called the top, of the stack.

4. *Pop* removes the item at the top of a stack and makes it available for use.

5. *StackIsEmpty* tests whether a stack contains any items.

The items in a stack might be integers, real numbers, characters, or abstract data types. The stack template has two arguments

```
stack<T,vector<T> > s;
```

The following list gives the operations provided by the STL for the stack class.

```
Insertion and Removal
s.push(value)   push value on front of stack
s.top()         access value at front of stack
                no change is made to the stack
s.pop()         remove value from front of stack

Size
s.size()        number of elements in collection
s.empty()       true if collection is empty

Assignment
=

Comparisons
==              equality
!=              inequality
<               lexicographical less than for stacks
<=
>               lexicographical greater than for stacks
>=
```

Example 1. The following program shows a simple application of the stack class. We push characters to the stack. The method **pop()** pops an item from the stack but does not give this item to the programmer to use. The **top()** method gives the programmer a reference to the top of stack item; no change is made to the stack.

```cpp
// stack1.cpp

#include <iostream>
#include <stack>
#include <vector>
using namespace std;

int main(void)
{
  stack<char,vector<char> > S, T, U;
  cout << "&S = " << &S << endl; // address

  S.push('X');
  S.push('Y');
  S.push('Z');
  cout << "size of S = " << S.size() << endl; // => 3
  cout << "top of stack S = " << S.top() << endl; // => Z
  S.pop();
  cout << "top of stack S = " << S.top() << endl; // => Y

  T = S;
  cout << "after T = S; we have ";
  cout << (S==T ? "S==T" : "S!=T") << endl; // => S==T
```

```
  S.pop();
  cout << "top of stack S = " << S.top() << endl; // => X

  bool b = (S == T);
  cout << "b = " << b << endl; // => 0
  return 0;
}
```

Example 2. The following program uses the `stack` class to convert an infix expression to a prefix expression.

```cpp
// infix.cpp

#include <iostream>
#include <string>
#include <stack>
#include <list>
using namespace std;

enum operators { leftP, plusO, minusO, multiplyO, divideO };

string opString(operators theOp)
{
  switch(theOp)
  {
  case plusO:     return " + ";
  case minusO:    return " - ";
  case multiplyO: return " * ";
  case divideO:   return " / ";
  }
}

void process(operators theOp,stack<operators,list<operators> >& opStack,
             string& result)
{
  while((!opStack.empty()) && (theOp < opStack.top()))
  {
  result += opString(opStack.top());
  opStack.pop();
  }
  opStack.push(theOp);
}

string infixToPrefix(string infixStr)
{
  stack<operators,list<operators> > opStack;
  string result = "";
  int i = 0;

  while(infixStr[i] != '\0')
  {
  if(isdigit(infixStr[i]))
  {
  while(isdigit(infixStr[i]))
  result += infixStr[i++];
  result += " ";
  }
  else
```

```
switch(infixStr[i++])
{
case '(': opStack.push(leftP);
          break;
case ')': while(opStack.top() != leftP)
          {
          result += opString(opStack.top());
          opStack.pop();
          }
          opStack.pop();
          break;
case '+': process(plus0,opStack,result);
          break;
case '-': process(minus0,opStack,result);
          break;
case '*': process(multiply0,opStack,result);
          break;
case '/': process(divide0,opStack,result);
          break;
}
}
while(!opStack.empty())
{
result += opString(opStack.top());
opStack.pop();
}
return result;
}

int main(void)
{
  string input = "5*(27+3*7)+22";
  cout << infixToPrefix(input) << endl;
  return 0;
}
```

For the given input "5*(27+3*7)+22" the output is

```
5  27  3  7  *  +  *  22  +
```

6.6   Queue Class

A *queue* is a first-in-first-out (FIFO) data structure. This means that elements are added to the back of the queue and may be removed from the front. Queue is a container adapter, meaning that it is implemented on top of some underlying container type. Queue does not allow iteration through its elements. A queue is a collection of data items organized in a linear sequence, together with the following five operations:

1. *CreateQueue* brings a queue into existence.

2. *MakeQueueEmpty* deletes all items, if any, from the queue.

3. *EnQueue* adds an item at one end, called the rear, of the queue.

4. *DeQueue* removes the item from the other end, called the front, of the queue, and makes it available for use.

5. *QueueIsEmpty* tests whether a queue is empty.

The queue template has two arguments

```
queue<T,list<T> > q;
```

The following list gives the operations provided by the STL for the queue class.

```
Insertion and Removal
q.push(value)       push value on back of queue
q.front()           access value at front of queue
q.back()            access value at back of queue
q.pop()             remove value from front of queue


Size
q.size()            number of elements in collection
q.empty()           true if collection is empty
```

The following program shows a simple application of the queue class.

```
// queue1.cpp

#include <iostream>
#include <list>
#include <queue>
using namespace std;

int main(void)
{
  queue<double,list<double> > Q;
  cout << "&Q = " << &Q << endl;    // => address
  Q.push(3.14);
  Q.push(4.5);
  Q.push(6.9);
  cout << "after pushing 3.14, 4.5, 6.9:\n";
  cout << "Q.front() = " << Q.front() << endl; // => 3.14
  cout << "Q.back() = " << Q.back() << endl;    // => 6.9

  Q.pop();
  cout << "after Q.pop():\n";
  cout << "Q.front() = " << Q.front() << endl; // => 4.5
  cout << "Q.size() = " << Q.size() << endl; // => 2

  queue<double,list<double> > P;
  P = Q;
  cout << "P.front() " << P.front() << endl; // => 4.5
  cout << "P.back() " << P.back() << endl;   // => 6.9
  return 0;
}
```

6.7   Deque Class

A deque (double-ended queue) is an abstract data type in which insertions and deletions can be made at either the front or the rear. Thus it combines features of both stacks and queues. In practice, the deque has two variations. The first is the input-restricted deque, where insertions are restricted to one end only, and the second is the output-restricted deque, where deletions are restricted to a single end.

The following list gives the operations provided by the STL for the `deque` class.

```
Constructors and Assignment
deque<T> d;                    default constructor
deque<T> d(int);               construct with initial size
deque<T> d(int,value);         construct with initial size and initial
                               value
deque<T> d(aDeque);            copy constructor
d = aDeque;                    assignment of deque from another deque
d.swap(aDeque);                swap contents with another deque

Element Access and Insertion
d[i]                           subscript access, can be assignment target
d.front()                      first value in collection
d.back()                       final value in collection .
d.insert(iterator,value)       insert value before iterator
d.push_front(value)            insert value at front of container
d.push_back(value)             insert value at back of container

Removal
d.pop_front()                  remove element from front of vector
d.pop_back()                   remove element from back of vector
d.erase(iterator)              remove single element
d.erase(iterator,iterator)     remove range of elements

Size
d.size()                       number of elements currently held
d.empty()                      true if vector is empty

Iterators
deque<T>::iterator itr         declare a new iterator
d.begin()                      starting iterator
d.end()                        stopping iterator
d.rbegin()                     starting iterator for reverse access
d.rend()                       stopping iterator for reverse access
```

The following small program shows an application of the deque class.

```
// deque1.cpp

#include <iostream>
#include <deque>
using namespace std;

int main(void)
{
  deque<char> D(10,'A');
  deque<char>::iterator i;
  for(i=D.begin();i!=D.end();++i)
    cout << *i << " ";  // => A A A A A A A A A A
  cout << endl;

  deque<char> E(D);  // copy constructor
  E.push_front('Z'); E.push_back('Y');

  for(i=E.begin();i!=E.end();++i)
  cout << *i << " ";  // => Z A A A A A A A A A A Y
  cout << endl;

  cout << E.empty() << endl; // => 0 (false)
  cout << E.size() << endl;  // => 12
  char c = E[0];
  cout << "c = " << c << endl;  // => Z
  c = E[5];
  cout << "c = " << c << endl;  // => A
  return 0;
}
```

6.8 Bitset Class

The bit set class in the STL implements the bitwise operations. This class is very similar to `vector<bool>` (also known as bit vector). It contains a collection of bits and provides constant-time access to each bit. There are two main differences between `bitset` and `vector<bool>`. First the size of a `bitset` cannot be changed: `bitset`'s template parameter N, which specifies the number of bits in the bitset must be an integer constant. Second, `bitset` is not a sequence; in fact it is not an STL container at all. In general bit 0 is the least significant bit (on the right hand side) and bit N-1 is the most significant bit.

```
Constructors
bitset<N> s              construct bitset for N bits
bitset<N> s(aBitSet)     copy constructor

Bit level operations
s.flip()                 flip all bits
s.flip(i)                flip position i
s.reset(0)               set all bits to false
s.reset(i)               set bit position i to false
s.set()                  set all bits to true
```

```
s.set(i)                set bit position i to true
s.test(i)               test if bit position i is true
```

Operations on entire collection
```
s.any()                 return true if any bit is true
s.none()                return true if all bits are false
s.count()               return number of true bits
```

Assignment
```
=
```

Combination with other bitsets
```
s1 & s2                 bitwise AND
s1 | s2                 bitwise inclusive OR
s1 ^ s2                 bitwise exclusive OR
s == s2                 return true if two sets are the same
```

Other operations
```
s << n                  shift set left by one
s >> n                  shift set right by one
s.to_string()           return string representation of set
```

The following program shows an application of the bitset class.

```cpp
// bitset1.cpp

#include <iostream>
#include <bitset>
#include <string>
using namespace std;

int main(void)
{
  const unsigned long n = 32;
  bitset<n> s;
  cout << s.set() << endl;  // set all bits to 1
  cout << s.flip() << endl; // flip at position 12

  bitset<n> t;
  cout << t.reset() << endl;
  t.set(23);
  t.set(27);
  cout << "t.count() = " << t.count() << endl; // number of bits
                                               // that are set
  bool b1 = t.any(); // returns true if any bits are set
  cout << "b1 = " << b1 << endl;
  bool b2 = t.none(); // returns true if no bits are set
  cout << "b2 = " << b2 << endl;

  bitset<n> u;
  u = s & t;  // bitwise AND operation
  cout << "u = " << u << endl;
```

```
bitset<n> v;
v = s | t;  // bitwise OR operation
cout << "v = " << v << endl;

bitset<n> w;
w = s ^ t;  // bitwise XOR operation
cout << "w = " << w << endl;

bitset<n> z;
z = w ^ w;
cout << "z = " << z << endl;

cout << "w.to_string() = " << w.to_string();

return 0;
}
```

6.9   Set Class

The abstract data type set is defined in terms of objects and operations. The objects
are just sets in the mathematical sense with the representation unspecified. The
set and multiset data types in the STL are both template data structures, where
the template argument represents the type of the elements the collection contains.
In the set class each element of a set is identical to its key, and keys are unique.
Because of this, two distinct elements of a set cannot be equal. A multiset differs
from a set only in that it can contain equal elements. The operations in the set
class are given below.

```
Constructors
set<T> s;                      default constructor
multiset<T> m;                 default constructor
set<T> s(aSet);                copy constructor
multiset<T> m(aMultiset)       copy constructor
s = aSet                       assignment
s.swap(aSet)                   swap elements with argument set

Insertion and Removal
s.insert(value_type)           insert new element
s.erase(value_type)            remove all matching elements
s.erase(iterator)              remove element specified by iterator
s.erase(iterator,iterator)     remove range of values

Testing for Inclusion
s.empty()                      true if collection is empty
s.size()                       number of elements in collection
s.count(value_type)            count number of occurrences
s.find(value_type)             locate value
s.lower_bound(value_type)      first occurrence of value
```

```
s.upper_bound(value_type)    next element after value
s.equal_range(value_type)    lower and upper bound pair


Iterators
set<T>::iterator itr         declare a new iterator
s.begin()                    starting iterator
s.end()                      stopping iterator
s.rbegin()                   starting iterator for reverse access
s.rend()                     stopping iterator for reverse access
```

The following program shows an application of the set class.

```cpp
// set1.cpp

#include <iostream>
#include <set>
using namespace std;

int main(void)
{
  set<int> s;
  s.insert(23);
  s.insert(45);
  s.insert(-1);
  s.insert(-2);
  s.insert(23);
  s.insert(51);

  cout << s.empty() << endl;    // => 0 (false)
  cout << s.size() << endl;     // => 5
  cout << s.count(23) << endl;  // => 1

  set<int>::iterator i = s.begin();
  for(i=s.begin();i!=s.end();++i)
  cout << *i << " ";
  cout << endl;
  s.erase(45);

  for(i=s.begin();i!=s.end();++i)
   cout << *i << " ";
  cout << endl;
  cout << s.size() << endl; // => 4

  return 0;
}
```

Using the set class we can also do set manipulations such as union, intersection, difference, size (cardinality), includes and empty.

```cpp
// setmanipulation.cpp

#include <iostream>
#include <string>
#include <set>
#include <algorithm>
using namespace std;
```

```
int main(void)
{
  const int M = 3;
  const int N = 4;
  const string a[M] = { "Steeb", "C++", "80.00" };
  const string b[N] = { "Hardy", "Java", "23456", "80.00" };

  set<string> S1(a,a+M);
  set<string> S2(b,b+N);
  set<string> S3;

  // display union of sets
  cout << "union of sets S1 and S2: " << endl;
  set_union(S1.begin(),S1.end(),S2.begin(),S2.end(),
            ostream_iterator<string>(cout," "));
  cout << endl;

  // display intersection of sets
  cout << "intersection of sets S1 and S2: " << endl;
  set_intersection(S1.begin(),S1.end(),S2.begin(),S2.end(),
                   ostream_iterator<string>(cout," "));
  cout << endl;

  // display difference
  cout << "set S3 difference of S1 and S2: " << endl;
  set_difference(S1.begin(),S1.end(),S2.begin(),S2.end(),
                 inserter(S3,S3.begin()));
  copy(S3.begin(),S3.end(),ostream_iterator<string>(cout," "));
  cout << endl;

  // displays include
  bool b1 = includes(S1.begin(),S1.end(),S2.begin(),S2.end());
  cout << "b1 = " << b1 << endl;

  const string c[1] = { "80.00" };
  set<string> S4(c,c+1);
  bool b2 = includes(S2.begin(),S2.end(),S4.begin(),S4.end());
  cout << "b2 = " << b2 << endl;

  // test whether set is empty
  bool b3 = S4.empty();
  cout << "b3 = " << b3 << endl;

  return 0;
}
```

6.10 Pair Class

The pair template class is defined in <utility> to keep pairs of values. Thus we use the pair template class to store pairs of objects. The program illustrates the declaration and use of pair with its two public fields first and second. The header file <utility> defines global versions of the <=, >, >=, and ! = operators, which are all defined in terms of the operators < and ==.

```
// pair.cpp

#include <iostream>
#include <utility>
#include <string>
using namespace std;

int main(void)
{
  pair<string,int> p1;
  pair<string,int> p2("ola",1);
  pair<string,int> p3("ili",2);
  cout << p2.first << "  " << p2.second << endl;

  p1 = p3;
  cout << p1.first << "  " << p1.second << endl;

  pair<char,char> p4;
  p4 = make_pair('O','A');
  cout << p4.first << "  " << p4.second << endl;

  // pointers
  pair<string,double>* pp = new pair<string,double>("xxx",3.14159);
  cout << pp -> first << "  " << pp -> second << endl;
  delete pp;

  return 0;
}
```

The `pair` template class is mostly used in connection with the `map` and `multimap` class.

6.11   Map Class

The `map` class (also called dictionary or table) is an indexed collection. The index values need not be integers, but can be any ordered data values. Therefore a map is a collection of associations of *key-value pairs*. For maps no two keys can be equal. A multimap differs from a map in that duplicated keys are allowed. The following table gives the operations.

```
Constructors
map<T1,T2> m;                   default constructor
multimap<T1,T2> m;              default constructor
map<T1,T2> m(aMap)              copy constructor
multimap<T1,T2> m(aMultiMap)    copy constructor
m = aMap                        assignment

Insertion and Removal
m[key]                          return reference to value with key
m.insert(value_type)            insert given key value pair
m.erase(key)                    erase value with given key
```

```
m.erase(iterator)            erase value at given iterator
```

Testing for Inclusion
```
m.empty()                    true if collection is empty
m.size()                     return size of collection
m.count(key)                 count number of elements with given key
m.find(key)                  locate element with given key
m.lower_bound(key)           first occurrence of key
m.upper_bound(key)           next element after key
m.equal_range(key)           lower and upper bound pair
```

Iterators
```
map<T>::iterator itr;        declare new iterator
m.begin()                    starting iterator
m.end()                      ending iterator
m.rbegin()                   backwards moving iterator start
m.rend()                     backwards moving iterator end
```

The following two programs show applications of the **map** class.

```cpp
// mmap.cpp

#include <iostream>
#include <map>
#include <string>
using namespace std;

int main(void)
{
  map<char,int> m1;
  for(int i=0;i<26;i++)
   m1.insert(pair<char,int>('A'+i,65+i)); // ASCII table
  char ch;
  cout << "enter key: ";
  cin >> ch;

  map<char,int>::iterator p1;
  p1 = m1.find(ch); // find value given key
  if(p1 != m1.end()) cout << "ASCII table number: " << p1 -> second << endl;
  else cout << "key not in map.\n";

  map<string,double> m2;
  m2.insert(pair<string, double>("Willi",0.0));
  m2.insert(pair<string, double>("Fritz",1.0));
  m2.insert(pair<string, double>("Charles",2.0));

  map<string,double>::iterator p2;
  p2 = m2.find("Fritz");
  cout << p2 -> second << endl;   // => 1
  cout << p2 -> first << endl;    // => Fritz

  return 0;
}
```

The next programs show how to use map with a map. We also use the pair class.

```cpp
// mapmap1.cpp

#include <iostream>
#include <string>
#include <map>
using namespace std;

int main(void)
{
  map<string,pair<string,string> > m;
  pair<string,string>
  pa[] = { pair<string,string>("one","1"),
           pair<string,string>("two","2"),
           pair<string,string>("three","3") };
  m.insert(pair<string,pair<string,string> >("willi",pa[0]));
  map<string,pair<string,string> >::iterator it;
  it = m.find("willi");
  cout << it -> first << endl;
  cout << it -> second << endl;

  return 0;
}
```

```cpp
// MapPair1.cpp

#include <iostream>
#include <map>
using namespace std;

int main(void)
{
  pair<int,int> pi(4,5);
  pair<pair<int,int>,double> pid(pi,6.7);

  map<pair<int,int>,double> M;
  M.insert(pid);
  cout << M[pi] << endl; // => 6.7

  pair<int,int> i00(0,0);
  pair<pair<int,int>,double> v00(i00,2.2);
  map<pair<int,int>,double> M;
  M.insert(v00);
  cout << M[i00] << endl;
  pair<int,int> i02(0,2);
  pair<pair<int,int>,double> v02(i02,1.7);
  M.insert(v02);
  cout << M[i02] << endl;
  pair<int,int> i21(2,1);
  pair<pair<int,int>,double> v21(i21,2.3);
  M.insert(v21);
  cout << M[i21] << endl;
  double r = M[i00]*M[i21];
  cout << "r = " << r << endl;

  return 0;
}
```

The following program implements a finite *group* with two elements a and b. Thus, for example, when we enter a*b it outputs b. Here we use string concatenation to form a product. In general we would use make_pair.

```cpp
// group.cpp

#include <iostream>
#include <map>
#include <string>
using namespace std;

int main(void)
{
  map<string,string> group;
  string a = "a", b = "b", c;
  group[ a + "*" + a ] = a;
  group[ a + "*" + b ] = b;
  group[ b + "*" + a ] = b;
  group[ b + "*" + b ] = a;
  cin >> c;
  cout << group[c] << endl;

  return 0;
}
```

6.12   Algorithm Class

Algorithms act on the contents of containers. They include functions for initializing, sorting, searching, and transforming the contents of containers. The functions can also be applied to arrays of basic data types. Functions which modify their arguments are modifying, while functions which do not modify their arguments are nonmodifying. Some of the nonmodifying functions are

find, for_each, count, equal, search

Some of the modifying functions are:

sort, copy, merge, reverse, replace, swap, transform

The following code shows an application of some of the functions

```cpp
// algorith1.cpp

#include <iostream>
#include <algorithm>
#include <vector>
using namespace std;

int main(void)
{
  const int n = 5;
  double x[n];
  x[0] = 3.13; x[1] = 4.5; x[2] = 1.2; x[3] = 0.8; x[4] = 0.1;
```

```
sort(x,x+n);
cout << "array after sorting: \n";
double* p;
for(p=x;p!=x+n;p++)
cout << *p << " " << endl;

double y;
cout << "double value to be searched for: ";
cin >> y;

double* q = find(x,x+n,y);
if(q == x+n)
 cout << "not found\n";
else
{
 cout << "found";
 if(q == x)
  cout << " as the first element";
 else cout << " after " << *--q;
}
cout << endl;

reverse(x,x+n);
double* r;
cout << "reversed array: " << endl;
for(r=x;r!=x+n;r++)
cout << *r << " " << endl;

int a[6] = { 4, 5, 10, 20, 30, 84 };
int b[5] = { 7, 9, 14, 35, 101 };
int c[11];
merge(a,a+6,b,b+5,c);
int i;
for(i=0;i<11;i++)
 cout << "c[" << i << "] = " << c[i] << endl;

char str[] = "otto";
int length = strlen(str);
replace(str,str+length,'t','l');
cout << str << endl;  // => ollo

vector<int> v(4);
v[0] = 1; v[1] = 6; v[2] = 3; v[3] = 2;
vector<int> w(5);
w[0] = 8; w[1] = 10; w[2] = 3; w[3] = 12; w[4] = 5;

sort(v.begin(),v.end());
sort(w.begin(),w.end());

vector<int> z(9);
merge(v.begin(),v.end(),w.begin(),w.end(),inserter(z,z.begin()));

for(i=0;i<9;i++)
 cout << "z[" << i << "] = " << z[i] << endl;

return 0;
}
```

An example of a modifying operation is `transform`. It applies a function to each element in an input sequence and stores the result in an output sequence (possibly the same input sequence). The following example makes use of `transform`.

```cpp
// transformation.cpp

#include <algorithm>
#include <cctype>
#include <iostream>
#include <string>
#include <vector>
using namespace std;

double cube(double x) { return x*x*x; }

string concat(string s) { return s + "x"; }

char lowercase(char c) { return char(tolower(int(c))); }

int main(void)
{
 vector<double> v;
 v.push_back(0.5);
 v.push_back(1.0);
 v.push_back(1.5);
 transform(v.begin(),v.end(),v.begin(),cube);
 for(int i=0;i<v.size();i++)
   cout << "v[" << i << "] = " << v[i] << endl;

 vector<string> s;
 s.push_back("a");
 s.push_back("ab");
 s.push_back("aba");
 transform(s.begin(),s.end(),s.begin(),concat);
 for(int j=0;j<s.size();j++)
   cout << "s[" << j << "] = " << s[j] << endl;

 string str = "lower CASE";
 transform(str.begin(),str.end(),str.begin(),lowercase);
 cout << "str = " << str << endl;
 return 0;
}
```

6.13 Complex Class

The `complex` class implements complex numbers where each component is of some specified type, for example the basic data types `float`, `double` and `long double`. We can also use user defined data types, for example `Verylong` from SymbolicC++. The operators `+`, `-`, `*`, `/` are overloaded for the operations on complex numbers. The following functions are implemented

```
exp(), sin(), cos(), sinh(), sqrt(), tan(), tanh(), cosh(),
log(), log10()
```

An example is as follows

```cpp
// mycomplex.cpp

#include <iostream>
#include <complex>
#include <string>
#include "verylong.h"
using namespace std;

int main(void)
{
  complex<double> z1(1.0,-2.0);
  complex<double> z2(3.0,4.0);
  complex<double> v = z1 + z2;
  complex<double> w = z1*z2;
  cout << "v = " << v << endl;
  cout << "w = " << w << endl;

  complex<double> pi(3.141592653589793235360287);
  complex<double> i(0,1);
  cout << exp(pi*i) - 1.0 << endl;

  Verylong two("2");
  Verylong three("3");
  Verylong four("4");
  Verylong five("5");
  complex<Verylong> u1(two,three);
  complex<Verylong> u2(four,five);
  complex<Verylong> u3;
  u3 = u1 + u2;
  cout << "u3 = " << u3 << endl;

  complex<double> a(2.0,3.0);
  complex<double> b(1.0,-4.0);
  complex<complex<double> > c(a,b);
  cout << "c = " << c << endl;
  complex<double> d(-3.5,7.0);
  complex<double> e(2.5,-5.5);
  complex<complex<double> > f(d,e);
  complex<complex<double> > h;
  h = c + f;
  cout << "h = " << h << endl;

  double rparta = a.real();
  cout << "rparta = " << rparta << endl;
  double iparta = a.imag();
  cout << "iparta = " << iparta << endl;

  complex<double> sz = sin(a);
  cout << "sz = " << sz << endl;

  // principle branch
  cout << "log(i) = " << log(i) << endl;
  cout << "i^i = " << pow(i,i) << endl;

  return 0;
}
```

Chapter 7

Classes for Computer Algebra

In this chapter we introduce the basic building classes of our symbolic system. Some of the classes appeared in earlier versions of SymbolicC++ [60], the classes listed below feature substantial improvements, new implementations and some new classes. The chapter deals with many structures in mathematics as well as some common data structures in computer science. The description of the classes are arranged in such a way that primitive structures like the very long integer and rational classes are placed earlier in the chapter than the more sophisticated structures like the vector and matrix classes. There are nine classes presented in this chapter:

(1) `Verylong` provides the integer numbers abstract data type without upper and lower bound.
(2) `Rational` provides the rational numbers abstract data type.
(3) `Quaternion` provides the quaternion abstract data type.
(4) `Derive` provides the exact differentiation class.
(5) `Vector` provides the vector data structure.
(6) `Matrix` provides the matrix data structure.
(7) `Array` provides the array data structure.
(8) `Polynomial` provides the polynomial abstract data type.
(9) `Multinomial` provides the multi-variable polynomial abstract data type.

In each class, the basic ideas and the theory of the class are explained, followed by the abstraction. Different parts of the class, like data fields, constructors, operators and member functions, etc. are also described. Only short examples are given in each section. More advanced applications will be presented in Chapter 9.

7.1 Identity Elements

Some of the classes below are template classes. We assume numeric data types with standard arithmetic operations such as + and *. Thus the template parameter

171

T data type should have an additive identity 0 and a multiplicative identity 1. It is unreasonable to expect every class we could use to provide a constructor from `int` to obtain `T(0)` and `T(1)`. For example, the `Vector` class uses the constructor from `int` to denote the dimension of the vector space. Thus `Vector(1)` creates an instance of a vector in a one-dimensional vector space.

Consequently we need another means for specifying 0 and 1, which is provided in the header file `identity.h`. Of course, some assumptions must be made. Here we assume the default constructor `T()` always creates an instance with the same value.

```
template <class T> T zero(T)
{
 return T() - T();
}


template <class T> T one(T)
{
 cerr << "one() not implemented" << endl;
 abort();
}
...
template <> int zero(int) { return (int) 0; }
template <> int one(int) { return (int) 1; }
...
template <> double zero(double) { return (double) 0.0; }
template <> double one(double) { return (double) 1.0; }
```

The definitions of `zero<T>` and `one<T>` are defined for most of the standard data types such as T=char, T=short, T=int, T=long, T=float, T=double and the template class T=complex<C>. Other classes, such as `Verylong`, provide their own definitions for `zero` and `one`.

Thus in a template class with data type T we invoke the default constructor `T()` in `zero(T())` to obtain 0 and `one(T())` to obtain 1.

7.2   Verylong Integer Class

7.2.1   Abstraction

The integer data type is implemented internally in most programming languages. However, the effective range is limited due to the nature of the registers of the CPU. For a 4-byte integer, the effective range is between -2^{31} to $2^{31}-1$ or $-2,147,483,648$ to $2,147,483,647$. This range is usually not sufficient for an elaborate computation which requires very large positive integer numbers or negative integer numbers. The purpose of this section is to break the limitation of the built-in integer data type by introducing the concept of a *verylong integer* class.

For a data type to be able to store an arbitrary long integer number, we have to figure out a storage method in memory that imposes no limitation on representing an integer number. Using a string of characters to represent a very long integer is one possibility. Unlike the built-in integer type which could only be stored in 4 bytes of memory, we go beyond this limit. By using a string to represent an integer number, we could in principle make the string as long as possible (subject to availability of memory). This has the implication that an integer number could be represented to any number of digits.

The string that stores the very long number contains only character digits '0', '1', '2', ... , '9'. Each digit in the string represents a decimal digit of the number. For example, the string "123" represents the value 123 in decimal. We store an additional integer representing the sign, i.e. 0 for positive and 1 for negative. With this representation, the arithmetic operations could be implemented using the usual manipulation algorithms. Although this is not the only possible representation and it may not be the best way, some other representations such as binary representation may need some less straightforward algorithm for implementing the arithmetic operations. Since simplicity is one of our primary goals, we choose the decimal representation.

An *abstract data type* (ADT) defines not only the representation of the data (for example, the string of characters of integers in the case of the `Verylong` class) but also the operations which may be performed on the class. However, both the data representation and the implementation details of the operations should be hidden. The user only needs to know the behaviors of the ADT and the public interfaces. It is generally a good idea to strive for complete but minimal class interfaces. This applies to the `Verylong` class as well. In the following we summarize the behaviors of the `Verylong` class ADT:

- We have to create new instances of `Verylong` number abstraction easily.

- We have to use the arithmetic operators such as +, -, *, /, %, ++, -- to manipulate the instances of the `Verylong` number.

- We have to assign a `Verylong` number value to a `Verylong` variable using the operator =.

- The modification forms of assignment, such as +=, -=, *=, /=, %=, have to be supported.

- The relational operators >, >=, <, <=, ==, != should be available.

- We have to convert instances of `Verylong` number to other standard data types like `int`, `double`, `char*` and `string` when necessary.

- Some common functions like absolute value functions `abs()`, integer power function `pow()`, integer square root function `sqrt()` and double division operator `div()` have to be included as well.

- We have to perform input and output operations using the `Verylong` numbers.

7.2.2   Data Fields

The class `Verylong` is based on the STL class `string`. There are two data fields in the `Verylong` class, namely `vlstr` and `vlsign`. Following the philosophy of *information hiding* the data fields are declared as `private` which makes them inaccessible from outside the class. To access or manipulate the private data of the class, one has to use the member functions or operators available.

Below is a description of each data field in the class:

- The variable `vlstr` stores a `string` of characters consisting of integers which represents the very long integer number. The string could in principle be stored as the usual ordering or stored in the reversed order. Since we use the STL `string` class we can use both of the iterators `string::iterator` and `string::reverse_iterator`, thus we have the advantages of both orderings.

- The variable `vlsign` stores 0 or 1 which indicates a positive or negative number, respectively.

7.2.3   Constructors

This section shows how a `Verylong` number is created. As with all data types, the simplest way to create a `Verylong` number is through the declaration statement. For example,

```
Verylong x;
```

It creates a new variable named `x`. This simple statement actually invokes the default constructor provided in the class. During the construction, the `Verylong` number is initialized to zero internally. However, one may think about initializing the `Verylong` number to a specific value other than zero. C++ allows the constructors to be overloaded with multiple definitions. Different constructors could be differentiated by different argument lists in the declaration statement. The following statement, for example, creates a new variable named `y`, and the variable `y` is initialized to the integer value 3

```
Verylong y(3);
```

Again, there exists some problem with this specification. What happens if the user wants to initialize a value that exceeds the built-in integer type? One solution to this problem is to provide a constructor that reads in a `string` as its argument. In addition, automatic type casting allows us to use the traditional notation `"12345678"` which has data type `const char*`. With this implementation, one could declare a `Verylong` number yet initialize it to any possible value. For example,

```
Verylong u("123")
```
initializes the variable `u` to the value of 123.

```
Verylong v("1234567890123")
```
initializes the variable `v` to a value that exceeds the bound of the built-in integer type.

```
Verylong w("-567890")
```
initializes a negative integer number.

Below is a brief description of the constructors available in the class:

- `Verylong(const string&)` takes in a `string` as argument. It checks and assigns the sign of the integers, and stores them internally. If there is no argument during the construction, it is initialized to zero.

- `Verylong(int)` takes an integer as argument, converts it to a `string` and stores it internally.

- The copy constructor `Verylong(const Verylong&)` is crucial for the class. Although the C++ compiler automatically generates one if it is omitted, the generated copy constructor may not be correct whenever it involves dynamic allocation of data fields as in this case. The assignment operator would simply copy the pointers. This is not what we want. Therefore the programmer must define a copy constructor to ensure that the duplication of instances are correct.

- The destructor simply releases memory that is no longer in use. This is handled automatically for the `string` class member `vlstr`.

7.2.4 Operators

Various operators have been overloaded for the `Verylong` class. They are

```
++, --, -(unary), +, -, *, /, %, =,
+=, -=, *=, /=, %=, ==, !=, <, <=, >, >=.
```

In the following, we describe the functions and the algorithms used for each operator:

- The assignment operator `=` is used to assign a `Verylong` number to another. Its implementation is similar to the copy constructor but their functions are different.

- The increment and decrement operators are overloaded in the class. They can be used in two different ways — prefix and postfix. To overload these operators, we need to know how to distinguish between them. The operator with no parameter is for prefix usage and the operator with an `int` parameter is for postfix usage:

```
class Verylong
{
  ...
  Verylong operator ++ ();     // prefix:  ++Verylong
  Verylong operator ++ (int);  // postfix: Verylong++
  ...
}
```

When implementing the two functions, we should remember that:
(1) For prefix use, change the value and then use it.
(2) For postfix usage, use the value and then change it.

This explains the structure of these functions:

```
// Prefix increment operator
Verylong Verylong::operator ++ ()
{ return *this = *this + one; }

// Postfix increment operator
Verylong Verylong::operator ++ (int)
{
 Verylong result(*this);
 *this = *this + one;
 return result;
}
```

- The addition operator + adds two **Verylong** integers.

  The exclusive OR (XOR) operator ^ is used to determine the signs of each argument which in turn determine how the operations are to be carried out. Suppose we are evaluating u+v, where u, v are instances of the **Verylong** class, we perform the following:

  1. If u and v are of different sign then
     - if u is positive then return u-|v| else return v-|u|
     - in both cases, the result is evaluated using the subtraction operator.

     else get digit by digit from each operand and add them together using the usual addition arithmetic.

  2. Finally determine the correct sign and return the result.

- The subtraction operator − subtracts one **Verylong** integer from another. Similar to addition, the exclusive or operator ^ is used to determine the signs of each argument. Suppose we are evaluating u-v, where u, v are instances of the **Verylong** class, we perform the following:

  1. If u and v are of different sign then
     - if u is positive then return u+|v| else return -(v+|u|)
     - in both cases, the result is evaluated using the addition operator.

     else get digit by digit from each operand and subtract them using the usual subtraction arithmetic.

  2. Finally determine the correct sign and return the result.

  The addition and subtraction operators invoke each other during the manipulation.

- Suppose we are evaluating u*v, where u, v are instances of the **Verylong** class. The multiplication is carried out using the usual method:

  For each digit in v, we multiply it by u using a private member function **multdigit()**. The summation of these results gives the product of u and v.

- For the *division operator* /, the algorithm used is the usual long division. Consider the expression u/v. First we make sure that the divisor v is non-zero. A zero value is return if u<v. The rest of the operations involves finding the quotient digit by digit. Finally, we assign the correct sign to the value and return it.

- The *modulo operator* %, to find the remainder after the division u/v, is calculated using u - v*(u/v).

- There are 6 relational operators defined in the class. For ==, we compare the signs and the contents of the two numbers. If they are both equal to each other, then the two numbers must be equal. The != could be calculated using !(u==v).

To check if u<v is true, we do the following:

   - Compare the signs of the two numbers u and v; a positive number is always greater than a negative number.
   - If both numbers have the same sign, compare the lengths of each number taking their signs into consideration.
   - If both numbers have the same length, compare the string values using the operator == in the **string** class.
   - Return the boolean value 1 or 0 (True/False).

The rest of the relational operators could be constructed based on the less than operator <:

   - <= could be constructed by (u<v or u==v)
   - > could be constructed by (!(u<v) and u!=v)
   - >= could be constructed by (u>v or u==v)

7.2.5 Type Conversion Operators

In C++, a floating point number could be assigned to an integer number and vice versa. The type conversion is done implicitly. In the case of the **Verylong** class, we include these properties as well. There are two types of conversion:

1. Convert a **Verylong** number to a built-in data type.
   This type of conversion is accomplished by the conversion operator provided in C++. Three conversion operators are used in the class:

   - operator int() converts the **Verylong** number into a built-in integer type if it is within the valid range, otherwise an error message is reported.
   - operator double() converts the **Verylong** number to a double precision floating point number where applicable.

- operator char*() converts the Verylong number to a pointer to character type. This conversion is useful when we apply the routines in the library <cstring>.

- operator string() converts the Verylong number to a string.

2. Convert a built-in data type to a Verylong number.
   This type of conversion is carried out by the constructors of the class. The constructors read the data type of the arguments and perform the appropriate transformation which converts a built-in data type to a Verylong number.

7.2.6   Private Member Functions

Some operations on the data fields should not be visible to the user. These member functions are declared as private. There are 3 private member functions in the class. Their behaviors are described as follows:

- multdigit(int num) multiplies this Verylong number by num where num is an integer ranged between 0 and 9. This function is invoked during the multiplication of two Verylong numbers.

- mult10(int num) multiplies this Verylong number by 10^{num}.

  e.g. v.mult10(5) is equivalent to v*100000.

7.2.7   Other Functions

For a class to be useful, one must include a complete set of public interfaces. A Verylong class could not be considered as complete if the following functions were omitted:

- abs(const Verylong&) returns the absolute value of a Verylong number.

- Integer square root function sqrt(const Verylong&)

  Given a positive integer b, the integer square root of b is given by a provided

  $$a^2 \leq b < (a+1)^2.$$

  For example, the integer square root of 105 is 10 since $10^2 \leq 105 < 11^2$.

  There are many ways to find an integer square root of a positive integer. The method we use here is based on the identity $(a+b)^2 \equiv a^2 + 2ab + b^2$. The algorithm is as follows:

  1. Start from the rightmost digit towards the left and split the number into 2 digits each. The number of segments is equal to the number of digits of the integer square root.
  2. Get the first digit a of the result by taking the integer square root of the first segment. Record the result immediately.
  3. Subtract a^2 from the first segment and get the remainder.

4. Divide the remainder by $2a$ to obtain the second digit of the root b. Record the result.

5. Subtract $2ab$ and b^2 from the remainder in the appropriate position.

6. If all the digits required have been obtained, return the final result.

7. Go to Step 4 to find the next few digits of the integer square root.

For the purpose of illustration, consider the following example: $\sqrt{394384} = ?$

$$
\begin{array}{rll}
3\,9\,4\,3\,8\,4 & & \\
-\,3\,6 \quad\cdots\cdots & a \to 6 & \cdots\cdots \text{First digit} \\
\hline
3\,4\,3 \quad\cdots\cdots & 34 \div 2*6 \to 2 & \cdots\cdots \text{Second digit} \\
-\quad 2\,4 \quad\cdots\cdots & 2a*b = 2*6*2 = 24 & \\
-\quad\quad 4 \quad\cdots\cdots & b^2 = 4 & \\
\hline
9\,9\,8\,4 \cdots\cdots & 998 \div 2*62 \to 8 & \cdots\cdots \text{Third digit} \\
-\quad 9\,9\,2 \quad\cdots\cdots & 2\,ab*c = 2*62*8 = 992 & \\
-\quad\quad 6\,4 \cdots\cdots & c^2 = 64 & \\
\hline
0 & &
\end{array}
$$

Therefore, `sqrt(394384) = 628`.

- `pow(const Verylong&,const Verylong&)`

Suppose we want to compute x^{29}, we could start with x and multiply by x twenty-eight times. However we can obtain the same answer with only seven multiplications: start with x, square, multiply by x, square, multiply by x, square, square, multiply by x, forming the sequence

$$ x \to x^2 \to x^3 \to x^6 \to x^7 \to x^{14} \to x^{28} \to x^{29}. $$

This sequence could be obtained by its binary representation 11101: replace each "1" by the pair of letters SX, replace each "0" by S, we get SX SX SX S SX and remove the leading SX to obtain the rule SXSXSSX, where "S" is interpreted as squaring and "X" is interpreted as multiplying by x. This method can readily be programmed. However it is more convenient to do so from right to left. Here, we present an algorithm based on a right-to-left scan of the number:

Consider for positive N:

Step 1. Set $N \leftarrow n$, $Y \leftarrow 1$, $Z \leftarrow x$

Step 2. If N is odd, $Y \leftarrow Y \times Z$, $N \leftarrow \lfloor N/2 \rfloor$
        If $N = 0$ return the answer Y
        else $N \leftarrow \lfloor N/2 \rfloor$

Step 3. $Z \leftarrow Z \times Z$ and goto Step 2

where $\lfloor x \rfloor$ denotes the *floor function*, i.e. we round down (for example $\lfloor 3.7 \rfloor = 3$). It is defined as the largest integer value smaller than x. Let us consider the steps in the evaluation of x^{29}:

	N	Y	Z
After Step 1.	29	1	x
After Step 3.	14	x	x^2
After Step 3.	7	x	x^4
After Step 3.	3	x^5	x^8
After Step 3.	1	x^{13}	x^{16}
After Step 3.	0	x^{29}	–

- `div(const Verylong&,const Verylong&)`

  The usual division operator / performs integer division. In many cases, however, we need the floating point value of the quotient of `Verylong` numbers. The function `div(u,v)` is used to perform floating point division and the return type of the function is `double`. In this function, we perform the following:

  - First, ensure that the denominator is not equal to 0.

  - Next, find the appropriate scale factor to start with.

  - Then, find the quotient digit by digit.

  - The division is completed if there is no remainder or 16 significant digits have been obtained.

7.2.8 Streams

The `>>` and `<<` operators are used for input and output, respectively. These operators work with all the built-in data types. For the `Verylong` class we overload the stream operators. For example, one could read a value from an input stream into a `Verylong` variable named `x` and then display the value.

```cpp
#include <iostream>
#include "Verylong.h"
using namespace std;

int main(void)
{
  Verylong x;
  cin >> x;
  cout << "The value entered is " << x << endl;
  return 0;
}
```

The implementation for these functions are straightforward:

- `istream& operator >> (istream&,Verylong&)`
  It reads in a sequence of characters of integers from the stream (e.g. keyboard), assuming a maximum string length of 1000 characters.

- `ostream& operator << (ostream&,const Verylong&)`
  It displays the value of the `Verylong` number.

7.2.9 BigInteger Class in Java

Java has the built-in classes `BigInteger` and `BigDecimal` to deal with large integers and large decimal numbers. The following program uses the `BigInteger` class.

```java
// Verylong.java

import java.math.*;
public class Verylong
{
  public static void main(String[] argv)
  {
   BigInteger b1 = new BigInteger("12345678901234567");
   BigInteger b2 = new BigInteger("56789");
   BigInteger b3 = new BigInteger("-56712");
   BigInteger b4 = b1.add(b2);
   System.out.println("b4 = " + b4);
   BigInteger b5 = b1.multiply(b3);
   System.out.println("b5 = " + b5);
   BigInteger b6 = b3.abs();
   System.out.println("b6 = " + b6);
   BigInteger b7 = b1.and(b2);
   System.out.println("b7 = " + b7);
   double x = b1.doubleValue();
   System.out.println("x = " + x);
  }
}
```

7.3 Rational Number Class

7.3.1 Abstraction

In mathematics a number system is built up level by level. In this section, we construct a `Rational` number class, which is the natural extension of integer **Z**. The mathematics of *rational numbers* has been described in Section 2.4.

Here, we encounter again the problem that the integral data type in C++ has limited range. To solve this problem, instead of using the data type `int`, we use the data type `Verylong` to represent the numerator and denominator of a `Rational` number. However, we extend it further; we make use of the class template feature provided in C++ to implement the class. This allows the users to select the data type that suits their purposes.

We have to specify the behavior of the `Rational` number ADT:

- It is a `template` class for which the data type of the numerator and denominator are to be specified by the users.

- Creation of a new instance of `Rational` number is simple.

- The `Rational` number is reduced and stored in its simplest form, using the greatest common divisor algorithm.

- Arithmetic operators such as +, -, *, / are available.

- Assignment and modification forms of assignment =, +=, -=, *=, /= are available.

- The relational operators >, >=, <, <=, ==, != are available.

- Conversion to the type `double` is supported.

- Functions that return the numerator, denominator, and fractional part of `Rational` numbers are available.

- Input and output operations with `Rational` numbers are supported.

7.3.2   Template Class

`Rational` numbers with numerator and denominator of type `int` could not represent the whole class of rational numbers due to the limitation on the data type `int`. This is possible only if the numerator and denominator can represent the whole range of the integer. The `Verylong` class developed in the previous section in principle allows an arbitrarily long integer number. Therefore, we need to incorporate the `Verylong` number into the `Rational` class.

The template in C++ provides *parametrized types*. Template classes give us the ability to reuse code in a simple type safe manner that allows the compiler to automate the process of type instantiation.

We have developed the `Rational` number as a template class. The reason is that the data items, numerator and denominator, could be of type `int` or `Verylong` as specified by the user. With this desirable feature, one can perform extensive computation without the concern that a number could possibly run out of range.

7.3.3   Data Fields

The `Rational` class declares two fields of data type `T`. This means that the user has to specify the actual data type represented in order to use the class. The first field `p` maintains the numerator, while the second field `q` maintains the denominator of the `Rational` number. To enforce the concept of *data hiding*, both fields are declared as `private`. Thus they are accessible only within the class.

7.3.4   Constructors

To declare a `Rational` number, we proceed as follows:

```
// To declare a Rational number u of int-type which is initialized to 0
Rational<int> u;

// To declare v and initialize it to 5
Rational<int> v(5);
```

```
// To declare w and initialize it to 2/3
Rational<int> w(2,3);

// To declare w1 and initialize it to 2/3
Rational<int> w("2/3");

// To declare x of Verylong-type and initialize it to 127762/2384623
Rational<Verylong> x(Verylong("127762"),Verylong("2384623"));

// To declare y and initialize it to a value
// that exceeds the int-bound
Rational<Verylong> y(Verylong("32872134727762"),
                     Verylong("2348972938479822384623"));

// To declare y1 and initialize it to a value
// that exceeds the int-bound
Rational<Verylong> y1("32872134727762/2348972938479822384623");

// To declare z and initialize it to a value from data type double
Rational<Verylong> z(1e-16);
```

There are five constructors in the class which cater for different ways of construction of a `Rational` number:

1. The default constructor declares a `Rational` variable and it is initialized to zero.

2. `Rational(T N)` declares a `Rational` variable and initializes it to N.

3. `Rational(T N,T D)` declares a `Rational` variable and initializes it to N/D.

4. `Rational(const string&)` constructs a rational number from a string of the form "a/b" or the decimal notation"a.b".

5. `Rational(const double&)` constructs a rational number from data type `double`.

During the construction, a private member function `gcd()` is invoked to reduce the `Rational` number into its standard form. The copy constructor and destructor are also included in the class.

7.3.5 Operators

Since the `Rational` class is a mathematical object, the common operators used in the class are the arithmetic operators. We have included many operators in the class, namely

`-(unary), +, -, *, /, =, +=, -=, *=, /=, ==, !=, >, >=, <, <=.`

The definitions of some of these operators are as follows:

$$-\frac{a}{b} = \frac{-a}{b},$$

$$\frac{a}{b} + \frac{c}{d} = \frac{a*d+b*c}{b*d}, \qquad \frac{a}{b} - \frac{c}{d} = \frac{a*d-b*c}{b*d},$$

$$\frac{a}{b} * \frac{c}{d} = \frac{a*c}{b*d}, \qquad \frac{a}{b} \div \frac{c}{d} = \frac{a*d}{b*c},$$

$$(a = c) \text{ and } (b = d) \quad \Rightarrow \quad \frac{a}{b} = \frac{c}{d}, \qquad (a*d) < (b*c) \quad \Rightarrow \quad \frac{a}{b} < \frac{c}{d}.$$

The implementations of these operators are straightforward.

7.3.6   Type Conversion Operators

A *type conversion operator* allows a data type to be cast into another when needed. A floating point representation of a **Rational** number is always useful. Here, we have included a conversion operator to the data type **double**. The conversion operator in the class exists in two forms. One is the general form which allows any data type to be cast into **double**. The other is only specific to the conversion from **Rational<Verylong>** to **double**. One may ask why we need an extra conversion operator for the **Rational<Verylong>** when the other one could do the job as well. In the following example we show that there exists a better and more accurate method for the **Rational<Verylong>** number. The general method which works for any data type simply returns the **double**-cast value of the **Rational** number:

```
return double(p)/double(q);
```

This method works well for the **Rational<Verylong>** number in general, but it fails sometimes due to the limited accuracy in **double**. There is a function named **div()** which could do the double division for **Verylong** numbers. With this function we could do a better job:

```
return div(p,q);
```

For the purpose of illustration, consider the example.

```
// division.cpp

#include <iostream>
#include "verylong.h"
using namespace std;

int main(void)
{
  Verylong P("999"), Q("111"), D("105");
  P = pow(P,D);        // 999^105 = 9.00277e+314 (exceeded double limit)
  Q = pow(Q,D);        // 111^105 = 5.74001e+214
  cout << div(P,Q) << endl;            // 1.56842e+100 - OK
  cout << div(Q,P) << endl;            // 6.37583e-101 - OK
  cout << double(P)/double(Q) << endl; // NaN
  cout << double(Q)/double(P) << endl; // NaN
  return 0;
}
```

The output is

```
1.56842e+100
6.37583e-101
nan
nan
```

The statement `double(P)/double(Q)` does not work when `P` or/and `Q` exceed the limit of the data type `double` which is approximately $\pm 1.7977 \times 10^{308}$ as shown above. The word `NaN` stands for Not A Number. From this example, we conclude that the definition for the `Rational<Verylong>` is necessary.

7.3.7 Private Member Functions

Private member functions are internal member functions that are only known to the class itself. They are inaccessible from outside the class. There is only one private member function in our class. The private member function

```
gcd(T a,T b)
```

returns the *greatest common divisor* of `a` and `b` using the following algorithm:

1. While $b > 0$, do the following:

   - $m \leftarrow a \bmod b$
   - $a \leftarrow b$ and $b \leftarrow m$

2. Return the answer a.

For example, gcd(4,8) returns 4.

7.3.8 Other Functions

There are only three member functions, other than the arithmetic operators, defined in the class: `num()`, `den()` and `frac()`. They return the numerator, denominator and the fractional part of the `Rational` number, respectively.

```
template <class T> T Rational<T>::num() const
{ ... }
template <class T> T Rational<T>::den() const
{ ... }
template <class T> Rational<T> Rational<T>::frac() const
{ ... }
```

They are declared as `const` functions. This indicates that the functions do not alter the value of the instance. Any value can be declared as constant in C++. A constant variable is bound to a value and its value can never be changed. Therefore, only constant operations can be performed on the value. An application of these methods is given in the following program

```
// testfrac.cpp

#include <iostream>
#include "Rational.h"
using namespace std;

int main(void)
{
  Rational<int> r(7,3);
  int n = r.num();
```

```
  int d = r.den();
  Rational<int> f;
  f = r.frac();
  cout << "n = " << n << endl; // => 7
  cout << "d = " << d << endl; // => 3
  cout << "f = " << f << endl; // => 1/3
  return 0;
}
```

7.3.9   Streams

We describe how the >> and << operators are overloaded to perform the input and output for a Rational number:

- The input stream operator >> is implemented in the way that one could input a fraction like a/b from the keyboard and the class could recognize that a is the numerator whereas b is the denominator. If a non-fraction is entered, the class should be able to recognize that it is an integer (i.e. a rational number with denominator equal to one).

  In order to fulfill the requirement, the function make use of some functions from <iostream> library. The manipulator ws clears any leading white space from the input. The function peek() is used to have a sneak preview of the next character and the function get() reads a character from the input stream. No precaution against erroneous input is taken. The users take the full responsibility for handling the function in a proper manner.

- The implementation for the output stream operator << is much simpler. If the denominator of the Rational number is equal to one, then output only the numerator. Otherwise, output the value of the fraction.

7.3.10   Rational Class for Java

In the following program we show how rational numbers can be implemented using Java. The class Rational.java implements rational numbers and their arithmetic functions: addition (add), subtraction (subtract), multiplication (multiply), division (divide). This class extends java.lang.Number, implementing that class's abstract methods. The methods equals, toString, clone from the Object class are overridden.

```
// Rational.java

import java.lang.*;

class Rational extends Number
{
  private long num;
  private long den;

public Rational(long num,long den) { this.num = num; this.den = den; }
```

```
private void normalize()
{
   long num = this.num;
   long den = this.den;
   if(den < 0) { num = (-1)*num; den = (-1)*den; }
}

private void reduce()
{
   this.normalize();
   long g = gcd(this.num,this.den);
   this.num /= g;
   this.den /= g;
}

private long gcd(long a,long b)
{
   long g;
   if(b == 0) { return a; }
   else
   {
   g = gcd(b,(a%b));
   if(g < 0) return -g;
   else return g;
   }
}

public long num() { return this.num; }

public long den() { return this.den; }

public void add(long num,long den)
{
   this.num = (this.num*den)+(num*this.den);
   this.den = this.den*den;
   this.normalize();
}

public void add(Rational r)
{
   this.num = (this.num*r.den())+(r.num()*this.den);
   this.den = this.den*r.den();
   this.normalize();
}

public void subtract(long num,long den)
{
   this.num = (this.num*den)-(num*this.den);
   this.den = this.den*den;
   this.normalize();
}

public void subtract(Rational r)
{
   this.num = (this.num*r.den())-(r.num()*this.den);
   this.den = this.den*r.den();
   this.normalize();
```

```java
}

public void multiply(long num,long den)
{
   this.num = (this.num*num);
   this.den = (this.den*den);
   this.normalize();
}

public void multiply(Rational r)
{
   this.num = (this.num*r.num());
   this.den = (this.den*r.den());
   this.normalize();
}

public void divide(long num,long den)
{
   this.num = (this.num*den);
   this.den = (this.den*num);
   this.normalize();
}

public void divide(Rational r)
{
   this.num = (this.num*r.den());
   this.den = (this.den*r.num());
   this.normalize();
}

public static boolean equals(Rational a,Rational b)
{
   if((a.num()*b.den()) == (b.num()*a.den())) { return true; }
   else { return false; }
}

public boolean equals(Object a)
{
   if(!(a instanceof Rational)) { return false; }
   return equals(this,(Rational) a);
}

public Object clone() { return new Rational(num,den); }

public String toString()
{
   StringBuffer buf = new StringBuffer(32);
   long num, den, rem;
   this.reduce();
   num = this.num;
   den = this.den;
   if(num == 0) return "0";
   if(num == den) return "1";
   if(num < 0) { buf.append("-"); num = -num; }
   rem = num%den;
   if(num > den)
   {
    buf.append(String.valueOf(num/den));
```

```
    if(rem == 0) { return buf.toString(); }
    else { buf.append(" "); }
  }
  buf.append(String.valueOf(rem));
  buf.append("/");
  buf.append(String.valueOf(den));
  return buf.toString();
}

public float floatValue()
{ return (float) ((float)this.num/(float)this.den); }

public double doubleValue()
{ return (double) ((double)this.num/(double)this.den); }

public int intValue()
{ return (int) ((int)this.num/(int)this.den); }

public long longValue()
{ return (long) ((long)this.num/(long)this.den); }

public void print()
{ System.out.print(this.toString()); }

public void println()
{ System.out.println(this.toString()); }

// main() method used for testing other methods.
public static void main(String args[])
{
    Rational r1 = new Rational(-4,6);
    Rational r2 = new Rational(13,6);
    r1.add(r2);
    System.out.println(r1.toString());

    Rational r3 = new Rational(123,236);
    Rational r4 = new Rational(-2345,123);
    r3.multiply(r4);
    System.out.println(r3.toString());

    Rational r5 = new Rational(3,6);
    Rational r6 = new Rational(1,2);
    boolean b1 = equals(r5,r6);
    System.out.println("b1 = " +b1);
    boolean b2 = equals(r4,r5);
    System.out.println("b2 = " +b2);

    Rational r7 = new Rational(3,17);
    Rational r8 = (Rational) r7.clone();
    System.out.println("r8 = " + r8.toString());
} // end main
} // end class Rational
```

7.4   Quaternion Class

7.4.1   Abstraction

The *quaternion* described in Section 2.9 is a higher level mathematical structure compared with the basic numeric structures like rational and complex numbers. It is therefore based on the basic structures. The behaviors of the `Quaternion` ADT can be summarized as follows:

- It is a template class for which the underlying data type for the coefficients of each component is to be specified.

- The construction of instances of `Quaternion` is simple.

- Arithmetic operators for `Quaternion` are overloaded. For example, (unary)-, +, -, *, /.

- The assignment operator = is available.

- Operations like finding the magnitude, conjugate and inverse of a `Quaternion` are available.

- Input and output stream operations are supported.

7.4.2   Template Class

The coefficients of a `Quaternion` could be of type `int`, `double` or `Verylong`, etc. Thus a class like this is best implemented in template form. We list some possible ways to use the `Quaternion` class:

- `Quaternion<int>`, `Quaternion<double>` declare `Quaternion` with coefficients of built-in type `int` and `double`, respectively.

- `Quaternion<Verylong>` declares a `Quaternion` with coefficients of user-defined type `Verylong`.

- `Quaternion<Rational<int> >`, `Quaternion<Rational<Verylong> >` declare `Quaternion` with coefficients of type `Rational`.

- `Quaternion<Complex<int> >`, `Quaternion<Complex<Verylong> >`, `Quaternion<Complex<Rational<int> > >`, `Quaternion<Complex<Rational<Verylong> > >` are different ways to declare `Quaternion` with coefficients of type `Complex`.

It is interesting to couple four user-defined types successfully to form a new data type.

7.4.3   Data Fields

To define a quaternion uniquely, we have to specify the coefficients of the four components. This is exactly what we need to maintain in the data fields of the `Quaternion` class. The entries `r`, `i`, `j`, `k` represent the coefficients of $1, I, J, K$ respectively.

7.4.4   Constructors

The construction of a `Quaternion` is simple. The users have to provide the four coefficients and a `Quaternion` is created. The users could also opt for not providing any coefficient whereby the default constructor would be invoked and the coefficients would be initialized to zero. Below are some examples of how instances of `Quaternion` could be constructed

```
// To declare u of int-type that is initialized to 0
Quaternion<int> u;

// To declare v of double-type and initialize to 1 + 2I - 3J + 4K
Quaternion<double> v(1,2,-3,4);

// To declare w and initialize to 1/2 - 2/3 I + 3/4 J - 4/5 K
Quaternion<Rational<int> > w(Rational<int>(1,2),Rational<int>(-2,3),
                            Rational<int>(3,4),Rational<int>(-4,5));
```

The copy constructor and destructor are trivial.

7.4.5   Operators

Suppose q and p are two arbitrary quaternions,

$$q + p, \qquad q - p, \qquad q * p, \qquad q/p$$

are overloaded as the addition, subtraction, multiplication, division of q and p respectively; whereas $-q$ is the negative of q. The mathematics has been described in Section 2.9.

7.4.6   Other Functions

In this section, we would like to demonstrate the usage of the functions `conjugate()`, `inverse()`, `magnitude()` and the normalization operator (`~`):

```
// squater.cpp

#include <iostream>
#include "quatern.h"
using namespace std;

int main(void)
{
   Quaternion<double> Q1(3,4,5,6),
                      Q2 = Q1.conjugate(),
                      Q3 = Q1.inverse();
   double Mag  = Q1.magnitude(), Magz = (~Q1).magnitude();

   cout << "Q1 = "                    << Q1 << endl;
   cout << "Q2 = Conjugate of Q1 = "  << Q2 << endl;
   cout << "Q3 = Inverse of Q1 = "    << Q3 << endl;
   cout << "Mag  = Magnitude of Q1 = " << Mag << endl;
   cout << "Magz = Magnitude of normalized Q1 = " << Magz << endl << endl;

   cout << "Q1 * Q2 = " << Q1 * Q2 << endl;
```

```
    cout << "Q2 * Q1 = " << Q2 * Q1 << endl;
    cout << "Mag^2 = Square of magnitude = " << Mag * Mag << endl << endl;

    cout << "Q1 * Q3 = " << Q1 * Q3 << endl;
    cout << "Q3 * Q1 = " << Q3 * Q1 << endl;

    return 0;
}
```

The output is

```
Q1 = (3,4,5,6)
Q2 = Conjugate of Q1 = (3,-4,-5,-6)
Q3 = Inverse of Q1 = (0.0348837,-0.0465116,-0.0581395,-0.0697674)
Mag  = Magnitude of Q1 = 9.27362
Magz = Magnitude of normalized Q1 = 1

Q1 * Q2 = (86,0,0,0)
Q2 * Q1 = (86,0,0,0)
Mag^2 = Square of magnitude = 86

Q1 * Q3 = (1,0,0,0)
Q3 * Q1 = (1,0,0,0)
```

7.4.7   Streams

The input and output stream functions in the class are straightforward. The input stream function simply reads in the four coefficients of each component. For the output stream function, the quaternion

$$q = a_1 * 1 + a_I * I + a_J * J + a_K * K$$

is formatted and exported as (a_1, a_I, a_J, a_K).

7.5   Derive Class

7.5.1   Abstraction

So far we have been dealing with numeric types; here we are going to specify a somehow quite different abstraction. The `Derive` class provides an operator which applies to a numeric type, and the result of the operation is again the numeric type. The behaviors of the ADT are as follows:

- The data type of the result is the template `T` which is to be specified by the users.

- The construction of an expression is simple, using arithmetic operators such as (unary)-, +, -, *, /.

- The member function `set()` is used to specify the point (a number) where the derivative is taken.

- The value of the derivative can be obtained using the function df().

- Output operation with the derivative is supported.

7.5.2 Data Fields

There are only two data fields in the class: one being the variable u, which stores the value of the point where the derivative is evaluated, whereas the other variable du stores the value of the derivative of u.

7.5.3 Constructors

The constructors of the class are as follows:

- The default constructor declares an independent Derive variable. The dependent variable is declared using the assignment operator or copy constructor.

- Derive(const T num) declares a constant number num.

- The private constructor Derive(const T,const T) is used to define the values of the point and its derivative.

- The copy constructor and assignment operator are trivial in this case, and perform no more than member-wise copying.

- The destructor is trivial.

As an example, the declaration of $y = 2x + 1$ requires the following statements:

```
Derive<int> x;          // This declares the independent variable x
Derive<int> y = 2*x + 1; // and the dependent variable.
```

7.5.4 Operators

The arithmetic operators

```
(unary)-, +, -, *, /
```

are overloaded to perform operations that obey the *derivative rules* as described in Section 2.12.

7.5.5 Member Functions

The followings member functions are available in the class:

- The function set(const T) operates on the independent variable. It is used to specify the value of u which is the point where the derivative of f is evaluated.

- The function df(const Derive &x) returns the value of the derivative evaluated at x.

- The output stream operator << returns the value of u.

Example 1. Consider the function $f(x) = 2x^3 + 5x + 1$. Suppose we intend to evaluate the value of $df(x = 2)/dx$, we do the following:

```
// sderive1.cpp

#include <iostream>
#include "derive.h"
using namespace std;

int main(void)
{
   Derive<int> x;
   x.set(2);
   Derive<int> y = 2*x*x*x + 5*x + 1;
   cout << "The derivative of y at x = " << x << " is " << df(y)<< endl;
   return 0;
}
```

The output is

```
The derivative of y at x = 2 is 29
```

Example 2. Consider the function $g(x) = x^2 + 3/x$. Suppose we want to evaluate the value of $dg(x = 37/29)/dx$, then

```
// sderive2.cpp

#include <iostream>
#include "derive.h"
#include "rational.h"
using namespace std;

int main(void)
{
   Derive<Rational<int> > x;
   x.set(Rational<int>(37,29));
   Derive<Rational<int> > c(3);
   Derive<Rational<int> > y = x*x + c/x;
   cout << "The derivative of y at x = " << x << " is " << df(y) << endl;
   return 0;
}
```

The output is

```
The derivative of y at x = 37/29 is 28139/39701
```

After the declaration of the independent variable x, we always set the value where the derivative is taken. This has to be done before the declaration of the dependent variable y. Otherwise, unpredictable results will be obtained.

7.5.6   Possible Improvements

The major drawback of this class is the inflexibility in specifying the function f. For example, to specify the expression $y = x^5 + 2x^3 - 3$, it requires a long statement like y = x*x*x*x*x + 2*x*x*x - 3. What happens if the function required is of the order of x^{100}? Other drawbacks include the fact that the derived function f' is not known. Only the value of $f'(a)$ can be evaluated. This imposes a strict restriction on the usefulness and applications of the class. These shortcomings can be overcome by implementing a more elaborate class which is shown in the next chapter – the Symbolic Class. The class solves all the problems mentioned above, it also includes many more features.

7.6 Vector Class

7.6.1 Abstraction

A *vector* is a common mathematical structure in linear algebra and vector analysis (see Section 2.7). This structure could be constructed using arrays. However, C and C++ arrays have some weaknesses. They are effectively treated as pointers. It is therefore useful to introduce a `Vector` class as an abstract data type. With bound checking and mathematical operators overloaded (for example vector addition), we built a comprehensive and type-safe `Vector` class. Many of the features we require can, to a large degree, be obtained using the STL `vector` class. Thus we have chosen to derive the mathematical `Vector` class from the STL `vector` container class, i.e. we use a `Vector` as a container with additional mathematical properties.

The `Vector` class is a structure that possesses many interesting properties for manipulation. The behavior of the `Vector` ADT is summarized as follows.

- It is implemented as a template class, because it is a *container class* whereby the data items could be of any type.

- The construction of a `Vector` is simple.

- Arithmetic operators such as +, -, *, / with `Vector` and numeric constants are available.

- The assignment and modification forms of assignment =, +=, -=, *=, /= are overloaded. We could also copy one `Vector` to another by using the assignment operator.

- The subscript operator [] is overloaded. This is useful for accessing individual data items in the `Vector`.

- The equality (==) and inequality (!=) operators check if the two vectors contain the same elements in the same order.

- Operations such as dot product and cross product for `Vector` are available.

- The member function `length()` returns the size of a `Vector` while `resize()` reallocates the `Vector` to the size specified.

- The `Matrix` class is declared as a `friend` of the class. This indicates that the `Matrix` class is allowed to access the `private` region of the class.

- Input and output stream operations with `Vector` is supported.

- The auxiliary file `VecNorm.h` contains different types of norm operators: $\|\mathbf{v}\|_1$, $\|\mathbf{v}\|_2$, $\|\mathbf{v}\|_\infty$ and the normalization function for the `Vector`.

7.6.2 Templates

A *container class* implements some data structures that "contain" other objects. Examples of containers include arrays, lists, sets and vectors, etc. Templates work especially well for containers since the logic to manage a container is often largely independent of its contents. We see how templates can be used to build one of the fundamental data structures in mathematics — the `Vector` class.

The container we implement here is *homogeneous*, i.e. it contains objects of just one type as opposed to a container that contains objects of a variety of types. It also has *value semantics*. Therefore it contains the *object* itself rather than the *reference* to the object.

7.6.3 Data Fields

The `Vector` class has no data members of its own, since the STL `vector` class provides all the container class facilities we require. There is no item in the data fields that records the lower or upper index bound of the vector; this means that the index will run from `0, 1, ..., size-1`. To make a vector that starts from an index other than zero, we may introduce a derived class that inherits all the properties and behaviors of the `Vector` class and adds an extra data field that indicates the lower index bound of the `Vector`. The implementation of such a bound vector is left as an exercise for the readers.

7.6.4 Constructors

Whenever an array of n vectors is declared, the compiler automatically invokes the default constructor. Therefore, in order to ensure the proper execution of the class, we need to initialize the data items properly in the default constructor. There are two more overloaded constructors

```
Vector(int n)
Vector(int n,const T &value)
```

The first constructor allocates `n` memory locations for the `Vector`, whereas the other additionally initializes the data items to `value`.

The copy constructor `Vector(const Vector& source)` constructs a new `Vector` identical to `source`. It will be invoked automatically to make temporary copies when needed, for example for passing function parameters and return values. It could also be used explicitly during the construction of a `Vector`.

In the following, we list some common ways to construct a `Vector`:

```
// declare a vector of 10 numbers of type int
Vector<int> u(10);

// declare a vector of 10 numbers and initialize them to 0
Vector<int> v(10,0);

// use the copy constructor to create and duplicate a vector
Vector<int> w(v);
```

7.6.5 Operators

Most of the operators applicable to `Vector` are implemented in the class, namely

```
(unary)+, (unary)-, +, -, *, /,
=, +=, -=, *=, /=, ==, !=, |, %, [].
```

Suppose u, v, w are vectors and c is a numeric constant, then the available operations are defined as follows:

- u+v, u-v, u*v, u/v adds, subtracts, multiplies or divides corresponding elements of u and v.

- u+=v, u-=v, u*=v, u/=v adds, subtracts, multiplies or divides corresponding elements of v into u.

- u+=c, u-=c, u*=c, u/=c adds, subtracts, multiplies or divides each element of u with the scalar.

- The assignment operator = should be overloaded in the class. Should one omit to define an assignment operator, the C++ compiler will write one. The code produced by the compiler simply makes a byte-for-byte copy of the data members. In the case where the class allocates memory dynamically, we usually have to write our own assignment operator. This is because the byte-for-byte operation copies only the memory address of the pointer not the memory content.

  Two forms of the assignment operator have been overloaded:

  - u=v makes a duplication of v into u and is provided by the STL `vector` class.

  - u=c assigns the constant c to every entry of u.

  The assignment operator is defined such that it returns a reference to the object itself, thereby allowing constructs like u = v = w.

- u==v returns *true* if u and v contain the same elements in the same order and returns *false* otherwise.

- u!=v is just the converse of u==v.

- We use the symbol | as the *dot product* operator (also called the *scalar product* or *inner product*). It is defined as $u|v = u \cdot v = \sum_{j=1}^{n} u_j v_j$.

- The *vector product* (also called the *cross product*) is operated by % in the class.

- The [] operator allows u[i] to access the i-th entry of the Vector u. It must return a reference to, not the value of, the entry because it must be usable on the left hand side of an assignment. In C++ terminology, it must be an lvalue.

The following shows some examples of the usage of the dot product and cross product of Vector. Suppose u, v, w, t are four vectors in $\mathbf{R}^3$, then

$$u \times (v \times w) + v \times (w \times u) + w \times (u \times v) = 0$$
$$(u \times v) \times (w \times t) = v(u \cdot w \times t) - u(v \cdot w \times t)$$
$$= w(u \cdot v \times t) - t(u \cdot v \times w)$$
$$u \cdot (v \times w) = (u \times v) \cdot w.$$

The following program demonstrates that the identities are obeyed for some randomly selected vectors:

```
// vproduct.cpp

#include <iostream>
#include "vector.h"
using namespace std;

int main(void)
{
    Vector<double> A(3), B(3), C(3), D(3);
    A[0] = 1.2; A[1] = 1.3; A[2] = 3.4;
    B[0] = 4.3; B[1] = 4.3; B[2] = 5.5;
    C[0] = 6.5; C[1] = 2.6; C[2] = 9.3;
    D[0] = 1.1; D[1] = 7.6; D[2] = 1.8;
    cout << A%(B%C) + B%(C%A) + C%(A%B) << endl;
    cout << (A%B)%(C%D) << endl;
    cout << B*(A|C%D)-A*(B|C%D) << endl;
    cout << C*(A|B%D)-D*(A|B%C) << endl;
    // precedence of | is lower than <<
    cout << (A|B%C) - (A%B|C) << endl;
    return 0;
}
```

The output is

```
[1.42109e-14]
[0]
[0]

[372.619]
[376.034]
[540.301]

[372.619]
```

```
[376.034]
[540.301]

[372.619]
[376.034]
[540.301]

0
```

The small, non-zero value `1.42109e-14` is due to rounding errors. Thus to obtain the correct result, namely the zero vector, we use the data type `Vector<Rational<T> >`, where `T` could be `int` or `Verylong` etc.

7.6.6 Member Functions and Norms

Other than the arithmetic operators, there exist some useful operations for the `Vector` class. Their definitions and properties are listed as follows:

- The function `length()` returns the size of the `Vector`.

- `resize(int n)` sets the `Vector`'s length to `n`. All elements are unchanged, except that if the new size is smaller than the original, than trailing elements are deleted, and if greater, trailing elements are uninitialized.

- `resize(int n,T value)` behaves similar to the previous function except when the new size is greater than the original, trailing elements are initialized to `value`.

In the auxiliary file `VecNorm.h`, we implement three different vector norms and the normalization function:

- `norm1(x)` is defined as $\|\mathbf{x}\|_1 := |x_0| + |x_1| + \cdots + |x_{n-1}|$.

- `norm2(x)` is defined as $\|\mathbf{x}\|_2 := \sqrt{x_0^2 + x_1^2 + \cdots + x_{n-1}^2}$, the return type of `norm2()` is `double`.

- `normI(x)` is defined as $\|\mathbf{x}\|_\infty := \max\{|x_0|, |x_1|, \ldots, |x_{n-1}|\}$.

- The function `normalize(x)` is used to *normalize* a vector `x`. The normalized form of the vector `x` is defined as `x/|x|` where `|x|` is the 2-norm of `x`.

An example is:

```
// vnorm.cpp

#include <iostream>
#include "vector.h"
#include "vecnorm.h"
using namespace std;

int main(void)
{
   Vector<int> v;
   v.resize(5,2);
```

```
cout << "The size of vector v is " << v.size() << endl;
cout << endl;

Vector<double> a(4,-3.1), b;
b.resize(4);
b[0] = 2.3; b[1] = -3.6; b[2] = -1.2; b[3] = -5.5;

// Different vector norms
cout << "norm1() of a = " << norm1(a) << endl;
cout << "norm2() of a = " << norm2(a) << endl;
cout << "normI() of a = " << normI(a) << endl;
cout << endl;

cout << "norm1() of b = " << norm1(b) << endl;
cout << "norm2() of b = " << norm2(b) << endl;
cout << "normI() of b = " << normI(b) << endl;
cout << endl;

// The norm2() of normalized vectors a and b is 1
cout << "norm2() of normalized a = " << norm2(normalize(a)) << endl;
cout << "norm2() of normalized b = " << norm2(normalize(b)) << endl;

return 0;
}
```

The output is

```
The size of vector v is 5

norm1() of a = 12.4
norm2() of a = 6.2
normI() of a = 3.1

norm1() of b = 12.6
norm2() of b = 7.06682
normI() of b = 5.5

norm2() of normalized a = 1
norm2() of normalized b = 1
```

7.6.7 Streams

The overloaded output stream operator `<<` exports all the entries in the vector v and puts them in between a pair of square brackets $[v_0, v_1, v_2, \ldots, v_{n-1}]$.

The input stream operator `>>` first reads in the size of the **Vector** followed by the data entries.

7.7 Matrix Class

7.7.1 Abstraction

Matrices are two-dimensional arrays with a certain number of rows and columns. They are important structures in *linear algebra* (Section 2.7). To build a matrix class we make use of the advantages (reusability and extensibility) of object-oriented programming to build a new class based on the `Vector` class. A vector is a special case of a matrix with the number of columns being equal to one. Thus we are able to define a matrix as a vector of vectors.

```
template <class T> class Matrix
{
  private:
    // Data Fields
    int rowNum, colNum;
    Vector<T>* mat;
    ...
}
```

We have declared the `Matrix` as a template class. Defining the matrix as a vector of vectors has the advantage that the matrix class methods can use many of the vector operations defined for the vector class. For example, we could perform vector additions on each row of the matrix to obtain the result of a matrix addition. This reduces the amount of code duplication.

Below, we summarize the properties of the `Matrix` ADT:

- It is implemented as a template class. This indicates that the data type of the entries could be of any type including built-in types and user-defined types.

- There are several simple ways to construct a `Matrix`.

- Arithmetic operators such as +, -, * with `Matrix` and +, -, *, / with numeric constants are available.

- The assignment and modification forms of assignment =, +=, -=, *= and /= are overloaded.

- The *vectorize operator* is available.

- The *Kronecker product* of two matrices is supported.

- The *Hadamard product* of two matrices is supported.

- The subscript operator [] is overloaded to access the row vector of the matrix while the parenthesis operator () is overloaded to access the column vector of the matrix.

- The equality (==) and inequality (!=) operators check if two matrices are identical.

- The transpose, trace and determinant of a matrix are implemented.

- The inverse of a square matrix is implemented.

- The member function `resize()` reallocates the memory for row and column vectors according to the new specification provided in the arguments of the function.

- The member functions `rows()` and `cols()` return the number of rows or columns of the matrix, respectively.

- Input (`>>`) and output (`<<`) stream operators are supported.

- The auxiliary file `MatNorm.h` contains the three different matrix norms: $||A||_1$, $||A||_\infty$ and $||A||_H$.

7.7.2   Data Fields

The data fields `rowNum` and `colNum` specify the number of rows and columns of the matrix respectively.

The data member `Vector<Vector<T> >` `mat` stores the data items of the matrix. It is declared as a `Vector` of `Vectors`, `mat` is a `Vector` of size m and each component vector is a `Vector` of size n. This results in a total of $m \times n$ memory space for the matrix.

7.7.3   Constructors

There are a number of ways to construct a `Matrix` in the class. One prime criterion for a matrix to exist is the specification of the number of rows and columns:

- `Matrix()` declares a matrix with no size specified. Such a matrix is not usable. To activate the matrix, we make use of a member function called `resize()`, which reallocates the matrix with the number of rows and columns specified in the arguments of the function.

- `Matrix(int nr,int nc)` declares an `nr` $\times$ `nc` matrix with undefined entry values.

- `Matrix(int nr,int nc,T value)` declares an `nr` $\times$ `nc` matrix with all the entries initialized to `value`.

- `Matrix(const Vector<T>& v)` constructs a matrix from a vector `v`. It is understood that the resultant matrix will contain only one column.

- The copy constructor duplicates a matrix. It is invoked automatically by the compiler when needed and it can be invoked by the users explicitly as well.

- The destructor releases the unused memory back to the free memory pool.

Below are some examples on how to declare a `Matrix`

```
// declare a 2-by-3 matrix of type int
Matrix<int> m(2,3)

// declare a 3-by-4 matrix and initialize the entries to 5
Matrix<int> n(3,4,5)

// duplicate a matrix using the copy constructor
Matrix<int> p(n);
```

```
// construct a matrix q from a vector v
Vector<double> v(3,0);
Matrix<double> q(v);
```

7.7.4 Operators

There are many matrix operators implemented in the class, namely

(unary)+, (unary)-, +, -, *, /,
=, +=, -=, *=, /=, [], (), ==, !=.

Some of the operators are overloaded with more than one meaning! The users are advised to read the documentation carefully before using the class.

In the following, we discuss the behavior and usage of the operators. Suppose A, B are matrices, v is a vector and c is a numeric constant,

- The operations A+B, A-B and A*B add, subtract and multiply two matrices according to their normal definitions.

- The operations A+c, A-c, A*c and A/c are defined as A+cI, A-cI, A*cI and A/cI respectively where I is the identity matrix.

- The operations c+A, c-A, c*A and c/A have similar definitions as above.

- A=B makes a duplication of B into A whereas A=c assigns the value c to all the entries of the matrix A.

- A+=B, A-=B and A*=B are just the modification forms of assignments which perform two operations in one shot. For example, A+=B is equivalent to A = A+B.

- A+=c, A-=c, A*=c and A/=c are just the modification forms of assignments. For example, A+=c is equivalent to A = A+cI.

- The function vec(A) (*vec operator*) is used to create a vector that contains elements of the matrix A, one column after the other. Suppose

$$A = \begin{pmatrix} 2 & x & a \\ 0 & 3 & -3 \end{pmatrix}$$

then

$$\text{vec(A)} = \begin{pmatrix} 2 \\ 0 \\ x \\ 3 \\ a \\ -3 \end{pmatrix}.$$

- The *Kronecker product* of two matrices is described in Section 2.16. The function `kron(A,B)` is used for calculating the Kronecker product of the matrices A and B. Note that $A \otimes B \neq B \otimes A$ in general (if $A \otimes B$ and $B \otimes A$ are of the same size) and $(A \otimes B)(C \otimes D) = (AC) \otimes (BD)$ (if A is compatible with C and B is compatible with D).

- The function `hadamard(A,B)` is used for calculating the *Hadamard product* (also known as the *Schur product*) of the matrices A and B.

- The function `dsum(A,B)` is used for calculating the *direct sum* of the matrices A and B.

- The subscript operator `[]` is overloaded to access a specific row vector of a matrix. For example, `A[i]` returns the i-th row vector of matrix `A`.

- The parenthesis operator `()` is overloaded to access a specific column vector of a matrix. For example, `B(j)` returns the j-th column vector of matrix `B`.

- The equality (`==`) and inequality (`!=`) operators compare whether the individual entries of two matrices match each other in the right order.

The precedence of `==` and `!=` are lower than the output stream operator `<<`. This means that a pair of brackets is required when the users write statements that resemble the following:

```
cout << (u != v) << endl;
cout << (u == v) << endl;
```

otherwise, the compiler may complain about it.

7.7.5   Member Functions and Norms

Many useful operations have been included in the class. Their properties are described as follows:

- The `transpose()` of an $m \times n$ matrix A is the $n \times m$ matrix A^T such that the ij entry of A^T is the ji entry of A.

- The `trace()` of an $n \times n$ matrix A is the sum of all the diagonal entries of A, $\mathrm{tr}A := a_{00} + a_{11} + \cdots + a_{(n-1)(n-1)}$.

- `determinant()`: The method for evaluating the determinant of a matrix has been described in Section 2.8. The method employed depends on whether the matrix is symbolic or numeric. Since our system is meant to solve symbolic expressions, we use the Leverrier's method for solving the determinant.

- `inverse()` is used to obtain the inverse of an invertible matrix. The methods used for finding the inverse of a matrix are different for numeric and symbolic matrices. For a numeric matrix, we can use the LU-decomposition [44] and backward substitution routines, whereas for a symbolic matrix we use *Leverrier's method*. This method can also be used for a numeric matrix.

- `resize()` reallocate the number of rows and columns according to the new specification provided in the arguments of the function.

- `rows()` and `cols()` return the number of rows and columns of the matrix, respectively.

In the auxiliary file `MatNorm.h`, we implement three different matrix norms:

- `norm1(A)` is defined as the maximum value of the sum of the entries in column vectors,

$$||A||_1 := \max_{0 \le j \le n-1} \left\{ \sum_{i=0}^{n-1} |a_{ij}| \right\}$$

- `normI(A)` is defined as the maximum value of the sum of the entries in row vectors,

$$||A||_\infty := \max_{0 \le i \le n-1} \left\{ \sum_{j=0}^{n-1} |a_{ij}| \right\}$$

- `normH(A)` is the *Hilbert–Schmidt norm*, defined as

$$||A||_H := [\text{tr} A^* A]^{1/2} - [\text{tr} \Lambda \Lambda^*]^{1/2} = \sqrt{\sum_{i=0}^{n-1} \sum_{j=0}^{m-1} |a_{ij}|^2}.$$

Example 1. We demonstrate the usage of the Kronecker product of matrices. We declare and define four matrices and then calculate the Kronecker product.

```
// kronecker.cpp

#include <iostream>
#include "matrix.h"
using namespace std;

int main(void)
{
   Matrix<int> A(2,3), B(3,2), C(3,1), D(2,2);

   A[0][0] =  2; A[0][1] = -4; A[0][2] = -3;
   A[1][0] =  4; A[1][1] = -1; A[1][2] = -2;
   B[0][0] =  2; B[0][1] = -4;
   B[1][0] =  2; B[1][1] = -3;
   B[2][0] =  3; B[2][1] = -1;
   C[0][0] =  2;
   C[1][0] =  1;
   C[2][0] = -2;
   D[0][0] =  2; D[0][1] =  1;
   D[1][0] =  3; D[1][1] = -1;

   cout << kron(A,B) << endl;
   cout << kron(B,A) << endl;
   cout << kron(A,B)*kron(C,D) - kron(A*C,B*D) << endl;
   cout << dsum(A,B) << endl;
   return 0;
}
```

The output is

```
[ 4    -8   -8   16   -6   12]
[ 4    -6   -8   12   -6   9 ]
[ 6    -2  -12    4   -9   3 ]
[ 8   -16   -2    4   -4   8 ]
[ 8   -12   -2    3   -4   6 ]
[ 12   -4   -3    1   -6   2 ]

[ 4    -8   -6   -8   16   12]
[ 8    -2   -4  -16    4   8 ]
[ 4    -8   -6   -6   12   9 ]
[ 8    -2   -4  -12    3   6 ]
[ 6   -12   -9   -2    4   3 ]
[ 12   -3   -6   -4    1   2 ]

[0 0]
[0 0]
[0 0]
[0 0]
[0 0]
[0 0]

[ 2 -4 -3  0  0]
[ 4 -1 -2  0  0]
[ 0  0  0  2 -4]
[ 0  0  0  2 -3]
[ 0  0  0  3 -1]
```

Example 2. We demonstrate that

$$\mathrm{tr}(AB) = \mathrm{tr}(BA) \qquad \text{and} \qquad \mathrm{tr}(AB) \neq \mathrm{tr}(A)\mathrm{tr}(B)$$

in general.

```cpp
// trace.cpp

#include <iostream>
#include "matrix.h"
using namespace std;

int main(void)
{
   Matrix<int> A(3,3), B(3,3,-1);

   A[0][0] = 2; A[0][1] = -1; A[0][2] =  1;
   A[1][0] = 1; A[1][1] = -2; A[1][2] = -1;
   A[2][0] = 3; A[2][1] =  2; A[2][2] = 2;

   cout << "A =\n" << A << endl;
   cout << "B =\n" << B << endl;
   cout << "tr(A) = " << A.trace() << endl;
   cout << "tr(B) = " << B.trace() << endl;
   cout << "tr(AB) = " << (A*B).trace() << endl;
   cout << "tr(BA) = " << (B*A).trace() << endl;
   cout << "tr(A)tr(B) = " << A.trace() * B.trace() << endl;

   return 0;
}
```

The output is

```
A =
[ 2 -1  1]
[ 1 -2 -1]
[ 3  2  2]

B =
[-1 -1 -1]
[-1 -1 -1]
[-1 -1 -1]

tr(A) = 2
tr(B) = -3
tr(AB) = -7
tr(BA) = -7
tr(A)tr(B) = -6
```

Example 3. We demonstrate the usage of the determinant function.   .

```cpp
// deter.cpp

#include <iostream>
#include "matrix.h"
using namespace std;

int main(void)
{
   Matrix<double> A(2,2);
   A[0][0] = 1; A[0][1] = 2;
   A[1][0] = 3; A[1][1] = 4;
   cout << A;
   cout << "Determinant of the matrix is " << A.determinant() << endl << endl;

   for(int i=3;i<5;i++)
   {
      A.resize(i,i,i);
      cout << A;
      cout << "Determinant of the matrix is " << A.determinant() << endl << endl;
   }
   return 0;
}
```

The output is

```
[1 2]
[3 4]
Determinant of the matrix is -2

[1 2 3]
[3 4 3]
[3 3 3]
Determinant of the matrix is -6

[1 2 3 4]
[3 4 3 4]
[3 3 3 4]
[4 4 4 4]
Determinant of the matrix is 8
```

7.7.6   Matrix Class for Java

The following programs shows how a Matrix class could be implemented. Two constructors are implemented and the methods equals and toString from the Object class are overridden.

```java
// Matrix.java

class Matrix
{
   private int rows, columns;
   public double entries[][];

   Matrix(int m,int n)    // constructor
   {
   rows = m;
   columns = n;
   entries = new double[m][n];
   for(int i=0;i<rows;i++)
   for(int j=0;j<columns;j++)
   entries[i][j] = 0.0;
   }

   Matrix(int m,int n,double[][] A) // constructor
   { rows = m; columns = n; entries = A; }

   public void add(Matrix M)
   {
   if((this.rows != M.rows) || (this.columns != M.columns))
   {
   System.out.println("matrices cannot be added");
   System.exit(0);
   }
   for(int i=0;i<columns;i++)
   for(int j=0;j<rows;j++)
   this.entries[i][j] = this.entries[i][j] + M.entries[i][j];
   }

   public Matrix multiply(Matrix M)
   {
   int i, j, t;
   if(columns != M.rows)
   {
   System.out.println("matrices cannot be multiplied");
   System.exit(0);
   }
   Matrix product = new Matrix(rows,M.columns);
   for(i=0;i<rows;i++)
   {
   for(j=0;j<M.columns;j++)
   {
   double tmp = 0.0;
   for(t =0;t<columns;t++)
   tmp = tmp + entries[i][t]*M.entries[t][j];
   product.entries[i][j] = tmp;
   }
   }
   return product;
```

```
}

public void randomize()
{
for(int i=0;i<rows;i++)
for(int j=0;j<columns;j++)
entries[i][j] = Math.random();
}

public boolean equals(Matrix A,Matrix B)
{
for(int i=0;i<rows;i++)
{
for(int j=0;j<columns;j++)
{
if(A.entries[i][j] != B.entries[i][j])
return false;
}
}
return true;
}


public boolean equals(Object ob)
{
if(!(ob instanceof Matrix)) { return false; }
return equals(this,(Matrix) ob);
}

public Object clone() { return new Matrix(rows,columns,entries); }

public String toString()
{
String result = new String();
for(int i=0;i<rows;i++)
{
for(int j=0;j<columns;j++)
{
result = result + String.valueOf(entries[i][j])+"   ";
}
result = result + "\n";
}
result = result + "\n";
return result;
}

public void onStdout()
{
System.out.println(toString());
}

public static void main(String args[])
{
Matrix M = new Matrix(2,2);
M.entries[0][0] = 3.4; M.entries[0][1] = 1.2;
M.entries[1][0] = 4.5; M.entries[1][1] = 5.8;
Matrix N = new Matrix(2,2);
N.entries[0][0] = 6.4; N.entries[0][1] = -1.2;
N.entries[1][0] = 8.5; N.entries[1][1] = 6.8;
```

```
    M.add(N);
    System.out.println("M = \n" + M.toString());
    Matrix X = new Matrix(2,2);
    X = M.multiply(N);
    System.out.println("X = \n" + X.toString());
    Matrix Y = new Matrix(2,2);
    Matrix Z = new Matrix(2,2);
    boolean b1 = Y.equals(Z);
    System.out.println("b1 = " +b1);
    Z.randomize();
    System.out.println("Z = \n" +Z);
    boolean b2 = Y.equals(Z);
    System.out.println("b2 = " +b2);
    double d[][] = new double[2][2];
    d[0][0] = 2.1; d[0][1] = -3.4;
    d[1][0] = 0.9; d[1][1] = 5.6;
    Matrix B = new Matrix(2,2,d);
    B.add(B);
    System.out.println("B = \n" + B.toString());
    Matrix U = (Matrix) X.clone();
    System.out.println("U = \n" + U.toString());
    }
}
```

7.8   Array Class

7.8.1   Abstraction

Array is a common data structure in computer programming. Although C and C++ provide built-in array data structures, there are some weaknesses. The bounds of the array are not checked to prevent possible run-time errors. It is therefore useful to introduce an `Array` class as an abstract data type. With bound checking and some simple mathematical operators overloaded (for example array addition), we built a comprehensive and type-safe `Array` class. On the other hand, it could replace the array supported by C and C++ as a collection of data objects. We use the STL `vector` class to implement one-dimensional arrays. The `vector` class provides many of the facilities we require in an array class. However, higher dimensional arrays are achieved via the template parameter. For example we would use `vector<vector<vector<double> > >` for a three-dimensional array of type `double`. This is obviously cumbersome. Instead we provide an implementation where the dimension is stored as a template parameter, i.e. for a three-dimensional array of type `double` we would use `Array<double,3>`.

We implement arbitrary dimensional arrays using templates. Every higher-dimensional array makes use of the member functions and operators of the lower-dimensional one. Their close relationship shows up transparently in the template parameters:

```
template <class T,int d>
class Array : public vector<Array<T,d-1> >
{
```

```
  ...
};

//Override the template definition for the 1-dimensional case
template <class T>
class Array<T,1> : public vector<T>
{
  ...
};
```

Thus we have greatly simplified the job of construction and the code is more concise and informative.

The applications for one- and two-dimensional `Array` are common. The reasons we extended the `Array` class to three- and four-dimensional is because some applications for tensor fields need it. For example, the *Christoffel symbols* require a three-dimensional `Array` and for the *curvature tensor*, we need a four-dimensional `Array`.

The behavior and functionality of the `Array` class is as follows.

- It is implemented as a template class, because templates work very well for container classes.

- The construction of an `Array` is simple.

- Simple arithmetic operators such as +, -, * are available.

- The assignment and modification forms of assignment =, +=, -=, *= are overloaded.

- We copy one `Array` to another by using the assignment operator =.

- The subscript operator [] is overloaded. It provides individual element access and returns a reference to the indexed element.

- The equality (==) and inequality (!=) operators check if the two `Arrays` contain the same elements in the same order.

- For `Array` with dimensionality D, the member function `size()` returns the number of elements in the 1st dimension of the `Array`.

- `resize(int)` and `resize(vector<int>)` reallocates the `Array` to the size specified. The helper function `dimensions<n>(int d1,int d2,...,int dn)` is provided to simplify resizing.

- Output stream operations with `Array` is supported.

7.8.2 Data Fields

The `Array` class is derived from **vector** and no additional data fields are required.

7.8.3 Constructors

In the following we give some examples of the usage of the constructors:

```
// declare u as a one-dimensional array of size 20,
// the entries are uninitialized
Array<int,1> u(20);

// declare v as a two-dimensional array of size 2 x 3,
// the entries are initialized to zero
Array<double,2> v(2,3,0.0);

// declare w as a three-dimensional array of double,
// with size 2 x 3 x 2, all its entries are initialized to 2.0
Array<double,3> w(2,3,2,2.0);

// declare x as a four-dimensional array of size 1 x 2 x 3 x 4,
// all the entries in the array are uninitialized
Array<double,4> x(1,2,3,4);

// copy constructor is invoked to duplicate the array.
Array<double,3> y(w);
```

7.8.4 Operators

Some common operators are overloaded in the class, namely (unary)+, (unary)-, +, -, *, =, +=, -=, *=, ==, !=, [], << and >>. Suppose u, v, w are arrays and c is a constant, then the operations are defined as follows:

- +u has no effect on u while -u negates each element of u.

- u+v, u-v adds or subtracts corresponding elements of u and v.

- u*c multiplies each element of u by the scalar c.

- u+=v, u-=v adds or subtracts corresponding elements of v into u.

- u*=c multiplies the scalar c into each element of u.

- u=v makes a duplicate copy of v and stores into u. The assignment operator returns a reference to the object itself. Thus multiple duplication like u = v = w is allowed.

- u=c overwrites all the entries in u with the constant c.

- u==v returns *true* if u and v are identical, otherwise returns *false*.

- u!=v is just the converse of u==v.

- The [] operator is overloaded to perform individual element accessing. The construct u[i] will return a reference to the i-th entry of the array. Since a reference value is returned, it can be used on either side of an assignment expression.

- The output stream operator `<<` simply exports all the entries and encloses them with a pair of square brackets. An instance `u` of `Array<T,1>` is formatted as

$$[\, u_0 \, u_1 \ldots u_{n-1} \,]$$

An instance `v` of `Array<T,2>` is formatted as

$$\begin{array}{llll}
[& v_{00} & v_{01} & \cdots & v_{0\ \text{cols-1}} &] \\
[& v_{10} & v_{11} & \cdots & v_{1\ \text{cols-1}} &] \\
[& \vdots & & \cdots & &] \\
[& v_{\text{rows-1 }0} & v_{\text{rows-1 }1} & \cdots & v_{\text{rows-1 cols-1}} &]
\end{array}$$

The output for `Array<T,3>` is less obvious; it can be considered as a cube of data entries to be output. The result is formatted in such a way that every layer of the cube is output accordingly, separated by an empty line. An instance of `Array<T,3>` `w(2,3,2)` will be printed as

$$[w_{000}\ w_{001}]$$
$$[w_{010}\ w_{011}]$$
$$[w_{020}\ w_{021}]$$

$$[w_{100}\ w_{101}]$$
$$[w_{110}\ w_{111}]$$
$$[w_{120}\ w_{121}]$$

The idea for exporting an instance of `Array<T,3>` is extended to `Array<T,4>`. A four-dimensional array can be visualized as a collection of three-dimensional blocks of `Array<T,3>`. Therefore, in printing an instance of `Array<T,4>`, we print a block of `Array<T,3>` followed by another until all the fourth dimensional entries are printed. For example we print an instance of the class `Array<T,4>` `x(2,2,2,3)` as follows

$$[x_{0000}\ x_{0001}\ x_{0002}]$$
$$[x_{0010}\ x_{0011}\ x_{0012}]$$

$$[x_{0100}\ x_{0101}\ x_{0102}]$$
$$[x_{0110}\ x_{0111}\ x_{0112}]$$

$$[x_{1000}\ x_{1001}\ x_{1002}]$$
$$[x_{1010}\ x_{1011}\ x_{1012}]$$

$$[x_{1100}\ x_{1101}\ x_{1102}]$$
$$[x_{1110}\ x_{1111}\ x_{1112}]$$

Each layer in `Array<T,3>` is separated by an empty line whereas each block in `Array<T,4>` is separated by two empty lines.

7.8.5   Member Functions

The member functions available for each array class include:

- `int size() const;`
  Returns the number of data entries in the first dimension.

- `void resize(int,...);`
  and
  `void resize(vector<int>);`
  It is an important function for the array classes. The following statement

```
Array<int,1> u;
```

declares a variable u but no memory space is allocated. Such a variable is not usable. When the statement

```
u.resize(100);
```

is executed, the compiler allocates 100 memory spaces for u. After this the variable becomes operational. To fill the array with some initial values upon memory allocation, another form of the function can be used

```
void resize(vector<int> n,T value);
```

In addition to the memory allocation, all the entries are initialized to `value`.

The function `resize()` can actually be used to change the sizes of arrays in use. An expansion of array size would result in the old storage being copied into the new array with the expanded region left undefined or initialized to `value` depending on which form of `resize()` is used. When `resize()` shrinks an array only the part of the old storage that will fit is copied into the new array and the extra entries are "truncated".

The following program demonstrates the usage of the function `resize()`:

```cpp
// aresize.cpp

#include <iostream>
#include "array.h"
using namespace std;

int main(void)
{
   Array<double,1> M;
   M.resize(3);
   M[0] = 5.0;  M[1] = 8.0; M[2] = 4.0;
   cout << "M = \n" << M << endl; cout << endl;

   Array<double,2> A;
   A.resize(2,3);
   A[0][0] = 1.0;   A[0][1] = 3.0;   A[0][2] = 2.0;
   A[1][0] = 4.0;   A[1][1] = 5.0;   A[1][2] = 6.0;
```

```
A.resize(dimensions<2>(3,4),2);
cout << "A = \n" << A << endl;
A.resize(2,2);
cout << "A = \n" << A << endl;
A.resize(dimensions<2>(4,4),9);
cout << "A = \n" << A << endl;

return 0;
}
```

The output is

```
M =
[5 8 4]

A =
[1 3 2 2]
[4 5 6 2]
[2 2 2 2]

A =
[1 3]
[4 5]

A =
[1 3 9 9]
[4 5 9 9]
[9 9 9 9]
[9 9 9 9]
```

7.9 Polynomial Class

7.9.1 Abstraction

The behavior of the class is specified as follows:

- It is a `template` class for which the data type of the polynomial coefficients are to be specified by the user.

- Creation of a new instance of `Polynomial` is simple.

- Arithmetic operators such as +, -, *, / are available.

- Assignment and modification forms of assignment =, +=, -=, *=, /=, %= are available.

- The relational operators == and != are available.

- `Polynomial` variables can be raised to an integer power with the ^ operator.

- A `Polynomial` can be used as a function and can be evaluated for a given value for the polynomial's variable.

- Symbolic differentiation and integration of a `Polynomial` is possible.

- Output operations with `Polynomial` are supported.

Since polynomials consist of a sum of polynomial terms it makes sense to implement a polynomial using a linked list.

7.9.2   Template Class

The `Polynomial` class is a `template` class so that the user can select a data type for the polynomial coefficients and evaluation. For example

```
Polynomial<Rational<int> >
```

can be used to create a polynomial with rational coefficients and

```
Polynomial<Polynomial<double> >
```

can be used to create a two-variable polynomial with coefficients of type `double`. Thus the `template` class provides multi-variable polynomials.

7.9.3   Data Fields

The polynomial's terms are represented as pairs of coefficients and powers. The class uses a linked list of pairs to form the polynomial.

7.9.4   Constructors

There are three ways to construct a `Polynomial`:

1. Default constructor for the zero polynomial.

2. Construct a `Polynomial` by specifying the symbol string of the polynomial's variable (the only argument).

3. Construct a constant polynomial from a number.

There is also a copy constructor. Some examples are:

```
// Declare a Polynomial p1 with integer coefficients and
// symbolic variable "x"
Polynomial<int> p1("x");

// Declare a Polynomial q with double coefficients and
// symbolic variable "x" and initialize to x^2+2x+1
Polynomial<double> x("x");
Polynomial<double> q=(x^2)+2.0*x+1;

// Declare a Polynomial r with polynomial coefficients
// and symbolic variable "y" and initialize to q(x)*(y^2+y);
Polynomial<Polynomial<double> > y("y");
Polynomial<Polynomial<double> > r=q*((y^2)+y);
```

7.9.5 Operators

The `Polynomial` class overrides the following operators:

$$+(\text{unary}),$$
$$-(\text{unary}), \ +, \ -, \ *, \ /, \ \%, \ \hat{} , \ =, \ +=, \ -=, \ *=, \ /=, \ \%=, \ ==, \ !=, \ ().$$

Care must be taken with the `^` operator since the precedence is lower than `+`,`-` and `*`. Most operators have their usual meaning except for `^` which raises a polynomial to a positive integer power and `()` which is used to evaluate the polynomial at a point.

7.9.6 Type Conversion Operators

A constant polynomial can be created from a number using `Polynomial<T>(T)` constructor (type cast).

7.9.7 Private Member Functions

The `Polynomial` class has a private member function `remove_zeros(void)`. Its only purpose is to remove terms with zero coefficients from a polynomial. This is to save memory resources but also makes output of `Polynomials` easier to read.

7.9.8 Other Functions

The function `Diff(Polynomial,string)` differentiates a polynomial with respect to its variable. The function `Int(Polynomial,string)` integrates a polynomial with respect to its variable. The constant of integration is assumed to be zero. The function

```
gcd(Polynomial,Polynomial)
```

determines the greatest common divisor of two polynomials. The member function

```
karatsuba(Polynomial)
```

uses the *Karatsuba multiplication algorithm* to multiply two polynomials. The member function `newton(Polynomial)` uses the *Newton iteration* to divide two polynomials. The function `squarefree()` returns a list of factors from the *square-free* decomposition.

7.9.9 Streams

The output stream operator `<<` first checks if the linked list for the polynomial is empty and if so outputs zero. Otherwise the operator outputs each term in the linked list, i.e. the coefficient, variable and exponent of each pair.

7.9.10   Example

Example 1. We show how we can add, multiply and divide polynomials.

```cpp
// pexample.cpp

#include <iostream>
#include "polynomial.h"
#include "rational.h"
using namespace std;

int main(void)
{
  Polynomial<double> x("x");
  Polynomial<double> p1 = (x^3)+2.0*(x^2)+7.0;
  cout << "p1(x) = " << p1 << endl;
  Polynomial<double> p2 = (x^2) - x - 1.0;
  cout << "p2(x) = " << p2 << endl;
  Polynomial<double> p3 = p1 + 2.0*p2;
  cout << "p3(x) = " << p3 << endl;
  Polynomial<double> p4 = 3.0*p1*p2 + 2.0;
  cout << "p4(x) = " << p4 << endl;
  Polynomial<double> p5 = (x^2)-1.0;
  Polynomial<double> p6 = x-1.0;
  Polynomial<double> p7 = p5/p6;
  cout << "p7(x) = " << p7 << endl;
  return 0;
}
```

The output is

```
p1(x) = x^3+2x^2+7
p2(x) = x^2-1x-1
p3(x) = x^3+4x^2-2x+5
p4(x) = 3x^5+3x^4-9x^3+15x^2-21x-19
p7(x) = x+1
```

Example 2. Here we show how we can differentiate and integrate a polynomial.

```cpp
// pexample2.cpp

#include <iostream>
#include "polynomial.h"
#include "rational.h"
using namespace std;

int main(void)
{
 Polynomial<double> x("x");
 Polynomial<Polynomial<double> > y("y"); //multivariable term

 Polynomial<double> p=(x^3)+2.0*(x^2)+7.0;

 cout << "p(x) = " << p <<endl;
 cout << "Diff(p) = " << Diff(p,"x") << endl;
 cout << "Int(p) = " << Int(p,"x") << endl;
 cout << "p(x)^2 = " << (p^2) << endl << endl;

 // multivariable polynomial
```

```
// differentiation and integration are with respect to y
Polynomial<Polynomial<double> > q=p+(4.0*p)*(y^2);
cout << "q(y) = " << q << endl;
cout << "Diff(q,y) = " << Diff(q,"y") << endl;
cout << "Diff(q,x) = " << Diff(q,"x") << endl;
cout << "Int(q,y) = " << Int(q,"y") << endl;
cout << "Int(q,x) = " << Int(q,"x") << endl;
cout << "q(y)^2 = " << (q^2) << endl;
cout << "gcd((x^2)-1.0,(x^2)-2.0*x+1.0) = "
     << x.gcd((x^2)-1.0,(x^2)-2.0*x+1.0) << endl;
Polynomial<Rational<int> > t("t");
cout << "gcd(..) = "
     << t.gcd(Rational<int>(48)*(t^3)-Rational<int>(84)*(t^2)
             +Rational<int>(42)*t-Rational<int>(36),
             Rational<int>(-4)*(t^3)-Rational<int>(10)*(t^2)
             +Rational<int>(44)*t-Rational<int>(30)) << endl;
list<Polynomial<Rational<int> > > l =
  (Rational<int>(5)*(t^8)-Rational<int>(10)*(t^6)+Rational<int>(10)*(t^2)
                     -Rational<int>(5)).squarefree();
cout << "squarefree(...) = " << l.front(); l.pop_front();
for(int i=1;!l.empty();i++)
{
  cout << " * (" << l.front() << ")^" << i;
  l.pop_front();
}
cout << endl;
return 0;
}
```

The output is

```
p(x) = x^3+2x^2+7
Diff(p) = 3x^2+4x
Int(p) = 0.25x^4+0.666667x^3+7x
p(x)^2 = x^6+4x^5+4x^4+14x^3+28x^2+49

q(y) = (4x^3+8x^2+28)y^2+(x^3+2x^2+7)
Diff(q,y) = (8x^3+16x^2+56)y
Diff(q,x) = (12x^2+16x)y^2+(3x^2+4x)
Int(q,y) = (1.33333x^3+2.66667x^2+9.33333)y^3+(x^3+2x^2+7)y
Int(q,x) = (x^4+2.66667x^3+28x)y^2+(0.25x^4+0.666667x^3+7x)
q(y)^2 = (16x^6+64x^5+64x^4+224x^3+448x^2+784)y^4
        +(8x^6+32x^5+32x^4+112x^3+224x^2+392)y^2
        +(x^6+4x^5+4x^4+14x^3+28x^2+49)
gcd((x^2)-1.0,(x^2)-2.0*x+1.0) = x-1
gcd(..) = t-3/2
squarefree(...) = 5 * (t^2+1)^1 * (1)^2 * (t^2-1)^3
```

7.10 Multinomial Class

Supporting multinomials using the template parameter of the Polynomial class is cumbersome. It would be useful to have a simpler mechanism to construct polynomials. Here we achieve this by explicitly constructing the coefficients of the polynomial in the same way as in the template Polynomial class. However, in the Polynomial class the template parameter and the types of variables determines which variable should be considered a coefficient of another. In the `Multinomial` class we assume the inverse lexicographical order, i.e. *a* is considered a coefficient of *b* and *b* is considered a coefficient of *c*.

7.10.1 Abstraction

The behavior of the class is specified as follows:

- It is a `template` class for which the data type of the numeric coefficients are to be specified by the user.

- Creation of a new instance of `Multinomial` is simple.

- Arithmetic operators such as `+`, `-`, `*` are available.

- Assignment and modification forms of assignment `=`, `+=`, `-=`, `*=` are available.

- The relational operators `==` and `!=` are available.

- `Multinomial` variables can be raised to an integer power with the `^` operator.

- Symbolic differentiation of a `Multinomial` is possible.

- Output operations with `Multinomial` are supported.

Since multinomials consist of a sum of multinomial (product) terms it makes sense to implement a multinomial using a linked list.

7.10.2 Template Class

The `Multinomial` class is a `template` class so that the user can select a data type for the multinomial term coefficients. For example

```
Multinomial<Rational<int> >
```

can be used to create a multinomial with rational coefficients.

7.10.3 Data Fields

The `Multinomial` has 5 data fields of which 2 are used at any one time. The enumeration `type` can take on the value `number`, `univariate` or `multivariate`. In the case of `number` the data member `n` stores the numeric value. In the case of `univariate` the data member `u` is a list of pairs of coefficients and powers for in the variable. In the case of `multivariate` the data member `m` is a list of pairs of multinomial coefficients and powers. The last data member if `variable` with data type `string` which is used as the name of the variable for univariate and multivariate polynomials.

7.10.4 Constructors

There are three ways to construct a `Multinomial`:

1. Default constructor for the zero multinomial.

2. Construct a `Multinomial` by specifying the symbol string of the polynomial's variable (the only argument).

3. Construct a constant multinomial from a number.

There is also a copy constructor. Some examples are:

```
// Declare a Multinomial m1 with integer coefficients and
// symbolic variable "x"
Multinomial<int> m1("x");

// Declare a Multinomial m2 with double coefficients and
// symbolic variable "x" and initialize to x^2+2x+1
Multinomial<double> x("x");
Multinomial<double> m2=(x^2)+2.0*x+1;

// Declare a Multinomial r with double coefficients
// and symbolic variable "y" and initialize to m2(x)*(y^2+y);
Multinomial<double> y("y");
Multinomial<double> r=m2*((y^2)+y);
```

7.10.5 Operators

The `Multinomial` class overrides the following operators:

$$+(\text{unary}), \ -(\text{unary}), \ +, \ -, \ *, \ \char`\^, \ =, \ +=, \ -=, \ *=, \ ==, \ !=.$$

Care must be taken with the `^` operator since the precedence is lower than `+`, `-` and `*`. Most operators have their usual meaning except for `^` which raises a polynomial to a positive integer power.

7.10.6 Private Member Functions

The `Multinomial` class has three private member function. The first is `remove_zeros`. Its only purpose is to remove terms with zero coefficients from a polynomial. This is to save memory resources but also makes output of a `Multinomial` easier to read. The second is `toarray` which returns a vector of strings, where each string is a string representation of a term in the multinomial. The last is `reconcile` which takes to multinomials and recreates them such that they have the same variables in the same order.

7.10.7 Other Functions

The function `Diff(Multinomial,string)` differentiates a multinomial with respect to the given variable.

7.10.8   Streams

The stream operator $<<$ calls the private member `output` to output the multinomial. In the case of a number the value of `n` is output. Otherwise `output` first checks if the linked list for the multinomial is empty and if so outputs zero. Otherwise the operator outputs each term in the linked list, i.e. the coefficient, variable and exponent of each pair. In the case of a true multinomial however, the multinomial is converted to a `vector` of strings using `toarray` and then each string is output in an appropriate way.

7.10.9   Example

```
// pexample3.cpp

#include <iostream>
#include "multinomial.h"
using namespace std;

int main(void)
{
 Multinomial<double> x("x");
 Multinomial<double> y("y");

 Multinomial<double> p=(x^3)+2.0*(x^2)+7.0;

 cout << "p(x)       = " << p <<endl;
 cout << "Diff(p,x) = " << Diff(p,"x") << endl;
 cout << "p(x)^2    = " << (p^2) << endl << endl;

 // multivariable polynomial differentiation
 Multinomial<double> q=p+(4.0*p)*(y^2);
 cout << "q(x,y)    = " << q << endl;
 cout << "Diff(q,y) = " << Diff(q,"y") << endl;
 cout << "Diff(q,x) = " << Diff(q,"x") << endl;
 cout << "q(y)^2    = " << (q^2) << endl;
 return 0;
}
```

The output is

```
p(x)       = x^3 + (2)x^2 + (7)
Diff(p,x) = (3)x^2 + (4)x
p(x)^2    = x^6 + (4)x^5 + (4)x^4 + (14)x^3 + (28)x^2 + (49)

q(x,y)     = (4)x^3y^2 + (8)x^2y^2 + (28)y^2 + x^3 + (2)x^2 + (7)
Diff(q,y) = (8)x^3y + (16)x^2y + (56)y
Diff(q,x) = (12)x^2y^2 + (16)xy^2 + (3)x^2 + (4)x
q(y)^2     = (16)x^6y^4 + (64)x^5y^4 + (64)x^4y^4 + (224)x^3y^4 + (448)x^2y^4
            + (784)y^4 + (8)x^6y^2 + (32)x^5y^2 + (32)x^4y^2 + (112)x^3y^2
            + (224)x^2y^2 + (392)y^2 + x^6 + (4)x^5 + (4)x^4 + (14)x^3 + (28)x^2
            + (49)
```

Chapter 8

Symbolic Class

Computer algebraic systems which perform symbolic manipulations have proved useful in many respects and they have become indispensable tools for research and scientific calculation. However, most of the software systems available are independent systems and the transfer of mathematical expressions from them to other programming environments such as C is rather tedious, time consuming and error prone. It is therefore useful to use a high level language that provides all the necessary tools to perform the task elegantly. This is the aim of this chapter.

In the next few sections we construct, step by step using object-oriented techniques, a computer algebra system — **SymbolicC++**. The system can be used in many areas of study. We describe the structures, functions and special features of the system. Examples are also included to demonstrate the usage of the functions.

The computer algebra system is built upon the concept of classes in object-oriented programming. Therefore, it inherits all the advantages and flexibilities of object-oriented programming. Some major advantages of object-oriented programming are modularity, reusability and extensibility. These features are very important especially for computer algebra systems because mathematics is built up level by level. For example, complex numbers are built upon real numbers, rational numbers are built upon integers and so on. From this point of view, the mathematical hierarchy is parallel with the philosophy of object-oriented programming. It seems to indicate that object-oriented programming is a natural choice for developing a computer algebra system. On the other hand, reusability and extensibility play a crucial role in clean coding which produce a more robust system.

The programs in this chapter use `symbolicc++.h`. This header file makes use of

`cloning.h, constants.h, equation.h, functions.h, number.h, product.h,`

`sum.h, symbol.h, symbolic.h, symerror.h` and `symmatrix.h`.

The header also uses the classes `pair`, `list`, `vector` and `string` from the Standard Template Library.

8.1   Main Header File

The main header file is `symbolicc++.h` which then includes all the other header files required for symbolic computation. Since many classes are inter-related, i.e. `Sum` must know about `Product` and vice-versa, header files are included in three phases.

We identify the three phases as the forward declaration of classes, the declaration phase (of classes and their members, which may now refer to forward declared classes) and finally the definition phase which provides the implementations. The phases are achieved using C macros. The header file `symbolicc++.h` includes itself recursively for each of the three phases.

This header file also provides many convenience definitions before the definition phase. For example all the operator overloading is implemented in this header file as well as overloading of functions such as `sin`.

Below we give a brief outline of how the three phases are achieved.

```
// symbolicc++.h

// phased include headers
#include "product.h"   //   Product  : CloningSymbolicInterface
#include "sum.h"       //   Sum      : CloningSymbolicInterface
...


#ifndef SYMBOLIC_CPLUSPLUS
#define SYMBOLIC_CPLUSPLUS

// forward declarations of all classes first
#define SYMBOLIC_FORWARD
#include "symbolicc++.h"
#undef  SYMBOLIC_FORWARD

// declarations of classes without definitions
#define SYMBOLIC_DECLARE
#include "symbolicc++.h"
#undef  SYMBOLIC_DECLARE

// definitions for non-member functions

Symbolic operator+(const Symbolic &s1,const Symbolic &s2)
{ return Sum(s1,s2); }
...


Symbolic sin(const Symbolic &s)
{ return Sin(s); }

// definitions for classes, member functions
#define SYMBOLIC_DEFINE
#include "symbolicc++.h"
#undef  SYMBOLIC_DEFINE

#endif
```

Below we give a brief example of the header file for the Sum class which checks the current include phase and then provides the appropriate forward declarations, declarations or definitions.

```
// sum.h

#ifndef SYMBOLIC_CPLUSPLUS_SUM

using namespace std;

#ifdef   SYMBOLIC_FORWARD
#ifndef SYMBOLIC_CPLUSPLUS_SUM_FORWARD
#define SYMBOLIC_CPLUSPLUS_SUM_FORWARD

class Sum;

#endif
#endif

#ifdef   SYMBOLIC_DECLARE
#ifndef SYMBOLIC_CPLUSPLUS_SUM_DECLARE
#define SYMBOLIC_CPLUSPLUS_SUM_DECLARE

class Sum
{
 public: Sum();
 ...
};

#endif
#endif

#ifdef   SYMBOLIC_DEFINE
#ifndef SYMBOLIC_CPLUSPLUS_SUM_DEFINE
#define SYMBOLIC_CPLUSPLUS_SUM_DEFINE
#define SYMBOLIC_CPLUSPLUS_SUM

Sum::Sum() {}

...

#endif
#endif

#endif
```

8.2 Memory Management

Computer algebra systems written in LISP have the advantage of automatic memory management. The C++ standard does not currently require or recommend any form of automatic memory management beyond that for local variables. The C++ programmer is required to perform all memory management. However, C++ allows the programmer to overload the member indirection -> and unary * dereference operators. The constraint is that -> must return a pointer or an object that

can act as a pointer. We use these features to isolate memory management from
the symbolic computation system, thus simplifying the implementation (i.e. classes
implement the mathematical rules and not memory management).

In SymbolicC++ all memory management is provided in a single class `CloningPtr`.
Any class that wishes to be used with the memory management facilities of
`CloningPtr` should be derived from `Cloning`. A class derived from `Cloning` can
clone (or copy) itself when necessary and must provide a `clone()` member function.
`Cloning` provides a template member function `clone` which can be used for this
purpose whenever a copy constructor is available (which may be the default copy
constructor). For example

```
class CloningClass: public Cloning
{
 public: ...
         Cloning *clone() const { return Cloning::clone(*this); }
};
```

allows `CloningClass` to be used with the memory management system.

`Cloning` provides two more methods, `reference` and `unreference` which are used
by `CloningPtr`. The method `reference` is called whenever a pointer to the underlying data is needed (which may or may not make a clone of the data, depending
on whether the object memory was allocated automatically, for example on the
stack, or is already managed by the `Cloning` and `CloningPtr` classes). The method
`unreference` is called whenever a pointer will no longer point to the data so that
any allocated memory may be freed if necessary. The method returns an integer
which indicates whether the underlying data should be freed. In addition to these
members, `Cloning` has a default constructor, a copy constructor, and a member
`refcount` of type `int` which is used to count references to the underlying data.
`Cloning` is an abstract base class.

`CloningPtr` provides pointer operations and has a member `value` pointer to an
instance of `Cloning`. The class provides a default constructor, a copy constructor
and a constructor from class `Cloning` which clones the instance of `Cloning` and sets
`value` to point to it. All the constructors, except for the default constructor, can
take an additional integer argument specifying that a clone is required (argument
1) or that `CloningPtr` should clone as necessary (argument 0, the default). Assignment operators are also provided for assignment from `Cloning` and `CloningPtr`
which has a similar implementation to the constructors. Finally, `CloningPtr` provides the member indirection operator `->` which returns a `Cloning*` and unary
dereferencing operator `*` which returns a `Cloning&` for the underlying data.

The template class `CastPtr<T>` is derived from `CloningPtr` and allows us to cast
the underlying pointer to the appropriate data type when the underlying type is
known. For example `CastPtr<CloningClass>` is a `CloningPtr` that points to a
`CloningClass`.

`Cloning` and `CloningPtr` allow us to choose between a pure copying scheme (which is generally) slow or a reference counting garbage collection scheme or any other other memory management scheme. This choice is isolated from the symbolic computation system.

8.3 Object-Oriented Design

We first define what is meant by an *expression*. There are a number of ways to define an expression. We consider it as a recursive data structure — that is, expressions are defined in terms of themselves. Suppose an expression can only use +, -, *, / and parentheses, then it can be defined with the following rules:

expression $\rightarrow$ (sum) or (product) or (symbol) or (number) or ...
sum $\rightarrow$ (expression) [+ (expression)]
product $\rightarrow$ (expression) [* (expression)]
symbol $\rightarrow$ $a, b, c, \ldots$ or sin(expression), cos(expression), ...
number $\rightarrow$ (bounded integer) or (unbounded integer)
                       or (rational number) or (real number).

The square brackets designate an optional element, and $\rightarrow$ means produces. These rules are usually called the *production rules* of the expression. Based on these rules, we could construct classes that work closely among themselves to represent mathematical expressions in a simple and clear form.

8.3.1 The Expression Tree

For the symbolic system an expression is organized in a tree-like structure. Every node in the tree may represent a variable, a number, a function or a root of a subtree. A subtree may represent a term or an expression. The class `SymbolicInterface` defines all the operations for a node in the expression tree. The class `Symbolic` acts as a pointer to `SymbolicInterface` and provides methods for convenient access to the underlying `SymbolicInterface`. Classes indirectly derived from `SymbolicInterface` include `Symbol` (for symbolic variables), `Sum`, `Product`, `Numeric` (and its derived classes `Number<int>`, `Number<double>` and `Number<void>` amongst others), `Equation` and functions such as `Cos` and `Sin`. This representation conforms to the production rules mentioned above.

The `Symbolic` class is the most used class as it can be used to define a symbolic variable as well as a numeric number or symbolic expression. It can be added or multiplied by another instance of `Symbolic` to form an expression, which is also represented by the `Symbolic` class. All the operators, functions and interfaces are defined in the class. The `Symbolic` class also defines special functions like `sin(x)` and `cos(x)` which will be described in detail in later sections.

Another important class is `Equation`. `Equation` is used primarily for substitution.

On the other hand, the other classes (Sum, Product etc.) describe are facilitator classes and their existence is generally not known by the users.

Example. Consider the expression

$$y = (a+b)*(a+c).$$

The object y has a structure as given in Figure 8.1. The memory management system allows subtrees to be shared, so in reality the symbol a is shared as shown in Figure 8.2. However, note that the two objects labeled a of type Symbolic are copies of each other. The symbolic variables a, b and c are represented by Symbol nodes. They are the leaf nodes of the expression tree. The expression a+b is composed of a Sum node which points at the variables a and b. a+c is constructed similarly. A Symbolic node y points at the final expression.

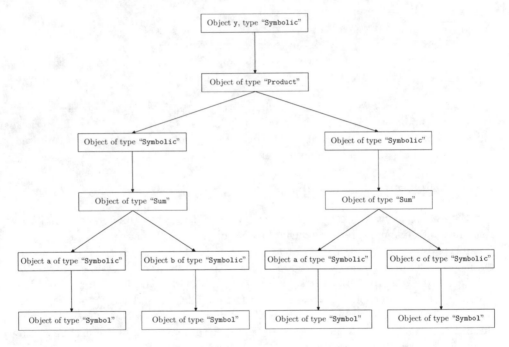

Figure 8.1: *Schematic diagram of the expression* y = (a+b)*(a+c).

8.3.2 Polymorphism of the Expression Tree

The Symbolic node could either hold a variable, a number or a special function. This is achieved by using *inheritance* and *polymorphism* of object-oriented techniques.

Inheritance possesses the ability to create new types by importing or reusing the description of existing types. Polymorphism with dynamic dispatching enables functions available for the base type to work on the derived types. Such a function can

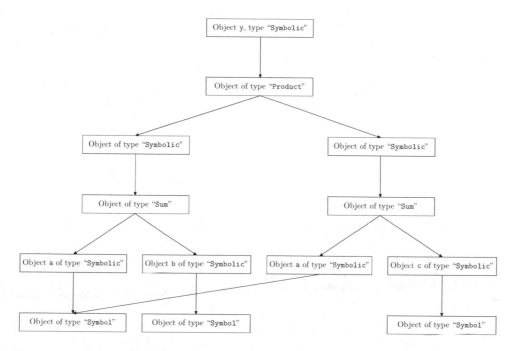

Figure 8.2: *Schematic diagram of* `y = (a+b)*(a+c)` *with shared subexpressions.*

have different implementations which are invoked by a run-time determination of the derived types. The class hierarchy for SymbolicC++ is given in Figure 8.3.

In the figure we see the central class `SymbolicInterface` with a tree structure to the left which provide memory management and convenience classes, and a tree structure to the right which implements the features necessary for symbolic computation.

The class `SymbolicInterface` is a an abstract base class with members (without implementation) that describe all the different operations that any derived class should implement. Thus `SymbolicInterface` is used to describe any symbolic expression, i.e. we use a pointer or reference to an instance of `SymbolicInterface` which could actually be a `Sum` or `Product`. The class `CloningSymbolicInterface` is derived from `SymbolicInterface` and `Cloning` and is consequently a class providing the `clone` member necessary for automatic memory management and all the features necessary for a symbolic expression. All the classes used in symbolic expressions (on the right side of the tree) are derived from `CloningSymbolicInterface`. The class `SymbolicInterface` requires that the following members are implemented in derived classes:

```
class SymbolicInterface
{
 public:
         ...
         virtual void print(ostream&) const = 0;
```

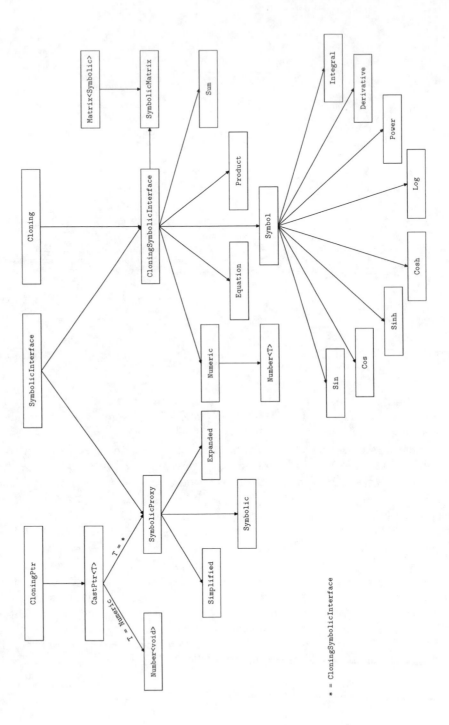

Figure 8.3: *Class Hierarchy for SymbolicC++.*

```
        virtual const type_info &type() const;
        virtual Symbolic subst(const Symbolic&,
                          const Symbolic&,int &n) const = 0;
        virtual Simplified simplify() const = 0;
        virtual int compare(const Symbolic&) const = 0;
        virtual Symbolic df(const Symbolic&) const = 0;
        virtual Symbolic integrate(const Symbolic&) const = 0;
        virtual Symbolic coeff(const Symbolic&) const = 0;
        virtual Expanded expand() const = 0;
        virtual int commute(const Symbolic&) const = 0;
};
```

SymbolicInterface provides a default definition for type that returns the derived class type. This definition does not usually have to be overloaded except for Numeric which returns the type_info corresponding to Numeric and not to one of the underlying Number<T> implementations.

The Numeric class describes all operations we can do when the underlying data is numeric (i.e. int, double etc). This class is an abstract base class, and so derived classes provide the implementations. The implementations are provided in the derived template classes Number<T> which is a numeric type based on the type T. Thus we implement Number<int>, Number<double>, Number<Verylong> etc. A special case is Number<void>, which could be read as a Number<T> implementation with no (specific) underlying supporting data type. Number<void> uses the different Number<int> Number<double> etc. implementations to create a single Number type that can mix the underlying representations. Number<void> acts as a pointer to an underlying Numeric.

```
class Numeric: public CloningSymbolicInterface
{
 public: Numeric();
        Numeric(const Numeric&);
        virtual const type_info &numerictype() const = 0;
        virtual Number<void> add(const Numeric&) const = 0;
        virtual Number<void> mul(const Numeric&) const = 0;
        virtual Number<void> div(const Numeric&) const = 0;
        virtual Number<void> mod(const Numeric&) const = 0;
        virtual int isZero() const = 0;
        virtual int isOne() const = 0;
        virtual int isNegative() const = 0;
        virtual int cmp(const Numeric&) const = 0;
        pair<Number<void>,Number<void> >
            match(const Numeric&,const Numeric&) const;
        ...
};
```

The method numerictype provides derived classes the opportunity to identify the underlying implementation. Thus, when necessary, Number<int> can be treated differently to Number<void>. The method match is used to find the common Number<T> that can represent both instances of Numeric. For example Number<int> and Number<double> should both be represented as Number<double> since double can represent all instances of int and double. The Number<T> classes support automatic promotion and demotion of the implementation type:

```
// demoting to type int
template <> Simplified Number<Verylong>::simplify() const
{
 if(n <= Verylong(numeric_limits<int>::max())
    && n > Verylong(numeric_limits<int>::min()))
  return Number<int>(n);
 return *this;
}

// promoting to type Verylong
template <> Number<void> Number<int>::mul(const Numeric &x) const
{
 assert(numerictype() == x.numerictype());
 CastPtr<const Number<int> > p = x;
 int product = n*p->n;
 if(n != 0 && product/n != p->n)
  return Number<Verylong>(Verylong(n)*Verylong(p->n));
 return Number<int>(product);
}
```

The class **Symbol** is used to represent an unknown quantity, i.e. independent variables x and dependent variables (functions) $y(x)$. Consequently functions such as $\sin(x)$ are just **Symbols** with additional information and can be implemented as derived classes. Extending the system with new functions is usually as simple as providing a class derived from **Symbol** with the appropriate definitions and providing an overloaded function. **Derivatives** and **Integrals** are treated in the same way.

```
class Sin: public Symbol
{
 public: Sin(const Sin&);
         Sin(const Symbolic&);

         Simplified simplify() const;
         Symbolic df(const Symbolic&) const;
         Symbolic integrate(const Symbolic&) const;

         Cloning *clone() const { return Cloning::clone(*this); }
};

Sin::Sin(const Symbolic &s) : Symbol(Symbol("sin")[s]) {}

Simplified Sin::simplify() const
{
 const Symbolic &s = parameters.front().simplify();
 if(s == 0) return Number<int>(0);
 if(s.type() == typeid(Product))
 {
   CastPtr<const Product> p(s);
   if(p->factors.front() == -1) return -Sin(-s);
 }
 if((s.type() == typeid(Numeric)) &&
    (Number<void>(s).numerictype() == typeid(double)))
  return Number<double>(sin(CastPtr<const Number<double> >(s)->n));
 return *this;
}
```

```
Symbolic Sin::df(const Symbolic &s) const
{ return cos(parameters.front()) * parameters.front().df(s); }

Symbolic sin(const Symbolic &s)
{ return Sin(s); }
```

The convenience classes include the memory management classes described in Section 8.2 and Number<void>, SymbolicProxy, Symbolic, Simplified and Expanded. SymbolicProxy is derived from SymbolicInterface and thus implements the methods for symbolic computation. Furthermore, SymbolicProxy is derived from

CastPtr<CloningSymbolicInterface>

so that it acts as a pointer to CloningSymbolicInterface. Therefore Symbolic-Proxy acts as a proxy for the CloningSymbolicInterface that it points to by forwarding any method calls to the object which is pointed to. This provides the polymorphism we require. Two important operations in symbolic computation is the expansion (application of the distributive law) and simplification of expressions. It is important to simplify an expression once since simplification routines can be expensive and allowing simplification to be repeated can lead to implementation problems such as infinite recursion if not managed correctly. The classes Simplified and Expanded are convenience classes derived from SymbolicProxy that mark the data that is pointed to as expanded or simplified so that these operations only happen once. Thus we have the simple member declarations in class SymbolicInterface

```
class SymbolicInterface
{
 public: int simplified, expanded;
        ...
        virtual Simplified simplify() const = 0;
        ...
        virtual Expanded expand() const = 0;
        ...
};
```

The classes Simplified and Expanded have constructors from

CloningSymbolicInterface,

SymbolicProxy and Number<void> which makes it possible to return any symbolic expression in a simplify or expand member function which will then automatically be marked as Simplified or Expanded.

Consider the expression

z = 3*(a^2)*(b^4)-5+6.2*cos(x+c)

given in Figure 8.4. The smaller boxes attached to the class nodes are instance specific data. The arrows represent class Symbolic which acts as a pointer, which may be labelled with a variable name if they were constructed implicitly. Here we assume the Symbols have their printed representation. This is usually achieved, using a as an example, as follows

```
Symbolic a("a");
```

We can construct the above expression as follows

```
Symbolic a("a"), b("b"), c("c"), x("x");
Symbolic z = 3*(a^2)*(b^4)-5+6.2*cos(x+c);
```

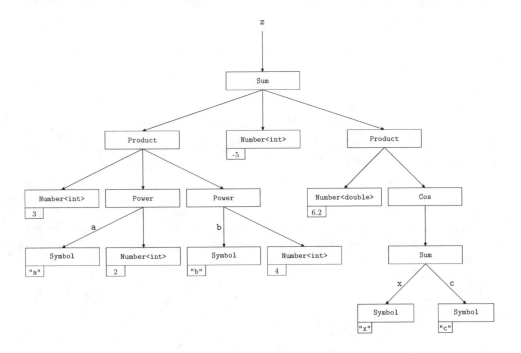

Figure 8.4: *Schematic diagram of* `z = 3*(a^2)*(b^4)-5+6.2*cos(x+c)`.

8.4   Constructors

In the following we omit discussing the default and copy constructors. In many cases the default constructor is automatically generated by the C++ compiler.

Most of the constructors listed below are used internally by SymbolicC++ and are rarely used by the user. We use constructors explicitly when we need to avoid infinite recursion, for example to return a formal derivative instead of attempting to recursively differentiate an expression.

- `Product(const Symbolic&,const Symbolic&)` creates a product of the two arguments, if one or both of the arguments are products they will be incorporated into the resulting product (flattening the expression tree). `Product` has a single data member `factors` which is of type `list<Symbolic>`.

- `Sum(const Symbolic&,const Symbolic&)` creates a sum of the two arguments, if one or both of the arguments are sum they will be incorporated into the resulting product (flattening the expression tree). `Sum` has a single data member `summands` which is of type `list<Symbolic>`.

- `Symbol(const string&,int=1)` creates a variable with the given name. The integer argument specifies whether the variable is a member of a commutative or non-commutative algebra. The default is a commutative algebra. `Symbol` has three data members. The `name` of the symbol which is a `string`, an `int commutes` which specifies whether the symbol belongs to a commutative algebra or not and a `list<Symbolic> parameters` for parameter lists when the symbol represents a function.

- `Symbol(const char*,int=1)` creates a variable with the given name. The integer argument specifies whether the variable is a member of a commutative or non-commutative algebra. The default is a commutative algebra.

The `Symbol` class forms the basis for a number of derived classes which represent functions, formal derivatives and formal integrals. These classes do not introduce any additional data members.

- `Sin(const Symbolic&)` constructs a sin function of the first argument.

- `Cos(const Symbolic&)` constructs a cos function of the first argument.

- `Sinh(const Symbolic&)` constructs a sinh function of the first argument.

- `Cosh(const Symbolic&)` constructs a cosh function of the first argument.

- `Log(const Symbolic &a,const Symbolic &b)` constructs a log function $\log_a b$ of the second argument to the base of the first argument.

- `Power(const Symbolic &a,const Symbolic &b)` constructs a power a^b of the first argument to the second argument.

- `Derivative(const Symbolic&,const Symbolic&)` constructs a formal derivative of the first argument with respect to the second argument.

- `Integral(const Symbolic&,const Symbolic&)` constructs a formal integral of the first argument with respect to the second argument.

The class `Equation` represents an equality or substitution. It has a two data members of type `Symbolic` for the left hand side (`lhs`) and right hand side (`rhs`) on an equation respectively.

- `Equation(const Symbolic &lhs,const Symbolic &rhs)` constructs an equation where the left hand side `lhs` is considered equivalent to the right hand side `rhs`.

The class `SymbolicMatrix` is derived from `Matrix<Symbolic>` so that it can easily be integrated into the symbolic computation system.

- `SymbolicMatrix(int m,int n)` creates a $m \times n$ `Matrix<Symbolic>`.

- `SymbolicMatrix(const Matrix<Symbolic>&)` invokes the copy constructor of the underlying `Matrix<Symbolic>` to create an instance of `Matrix<Symbolic>`.

- `SymbolicMatrix(const list<list<Symbolic> >&)` creates an instance of `Matrix<Symbolic>` with dimensions given by the lengths of the lists and with entries given by the contents of the lists.

- `SymbolicMatrix(const string &A,int m,int n)` constructs a matrix of variables

$$
\begin{pmatrix}
A_{0,0} & A_{0,1} & \cdots & A_{0,n-1} \\
A_{1,0} & A_{1,1} & \cdots & A_{1,n-1} \\
\vdots & \vdots & \ddots & \vdots \\
A_{m-1,0} & A_{m-1,1} & \cdots & A_{m-1,n-1}
\end{pmatrix}
$$

  or a vector if m or n are 1.

- `SymbolicMatrix(const char *A,int m,int n)` constructs a matrix of variables

$$
\begin{pmatrix}
A_{0,0} & A_{0,1} & \cdots & A_{0,n-1} \\
A_{1,0} & A_{1,1} & \cdots & A_{1,n-1} \\
\vdots & \vdots & \ddots & \vdots \\
A_{m-1,0} & A_{m-1,1} & \cdots & A_{m-1,n-1}
\end{pmatrix}
$$

  or a vector if m or n are 1.

- `SymbolicMatrix(const Symbolic &s,int m,int n)` constructs an $m \times n$ matrix with every entry set to s.

The `Numeric` class provides the functionality that the symbolic system requires to identify and manipulate numbers. `Numeric` is an abstract base class. The template class `Number<T>` is derived from `Numeric` and provides the specific functionality using the implementation of `T`, for example `T=int` and `T=Verylong`. `Number<T>` has a single data member `n` of type `T` which stores the numerical value.

- `Number<T>(const T&)` constructs a number with underlying implementation `T` and initializes an instance of `T`

The class `Number<void>` is a convenience class which acts as a pointer to a `Numeric`. Thus we can use a `Number<T>` without knowing what `T` is by setting a `Number<void>` to point to it. `Number<void>` also enables us to add two `Number<T>` of different underlying data types.

- `Number<void>(const Numeric&)` creates a `Number<void>` which acts as a pointer to the instance of `Numeric`

- `Number<void>(const Symbolic&)` creates a `Number<void>` which acts as a pointer to the instance of `Numeric`. The instance of `Symbolic` is assumed to point to an instance of `Numeric`, if this is not the case the exception

```
SymbolicError(SymbolicError::NotNumeric)
```

is thrown.

The class `SymbolicProxy` acts as a pointer to `CloningSymbolicInterface` and forwards all method calls to the underlying `SymbolicInterface` instance. Besides the default and copy constructors it has two additional constructors.

- `SymbolicProxy(const CloningSymbolicInterface&)` creates a `Symbolic-Proxy` that forwards all `SymbolicInterface` method calls to a copy of the first argument, i.e. to its underlying interface `CloningSymbolicInterface`.

- `SymbolicProxy(const Number<void>&)` creates a `SymbolicProxy` from the underlying `Numeric` (which is a `CloningSymbolicProxy`) which the `Number<void>` points to.

The classes `Simplified` and `Expanded` have the same constructors as `SymbolicProxy` and respectively set the `simplified` and `expanded` members of `SymbolicInterface` to 1.

The class `Symbolic` has a variety of constructors.

- `Symbolic(const CloningSymbolicInterface&)` is used most often internally to cast a symbolic result to the general type `Symbolic`.

- `Symbolic(const Number<void>&)` provides the same functionality as the above constructor, but for a generic number.

- `Symbolic(const int&)` is the type conversion from `int` to `Symbolic` by pointing to a `Number<int>`.

- `Symbolic(const double&)` is the type conversion from `double` to `Symbolic` by pointing to a `Number<double>`.

- `Symbolic(const string &name)` constructs a variable with the given `name`.

- `Symbolic(const char *name)` constructs a variable with the given `name`.

- `Symbolic(const string &v,int n)` constructs a vector of variables

$$(v_0, v_1, \ldots, v_{n-1})$$

by pointing to an underlying `SymbolicMatrix`.

- `Symbolic(const char *v,int n)` constructs a vector of variables

$$(v_0, v_1, \ldots, v_{n-1})$$

by pointing to an underlying `SymbolicMatrix`.

- `Symbolic(const string &A,int m,int n)` constructs a matrix by pointing to the instantiated class `SymbolicMatrix(A,m,n)`.

- Symbolic(const char *A,int m,int n) constructs a matrix by pointing to the instantiated class SymbolicMatrix(A,m,n).

- Symbolic(const list<Symbolic>&) constructs a vector with dimension given by the length of the list and with components given by the contents of the list. Thus to construct the vector

$$(a, b, c)^T$$

we use

```
Symbolic v = (Symbolic("a"),Symbolic("b"),Symbolic("c"));
```

- Symbolic(const list<list<Symbolic> >&) constructs a matrix with dimensions given by the lengths of the lists and with entries given by the contents of the lists. Thus to construct the 2×2 matrix

$$\begin{pmatrix} a & b \\ c & d \end{pmatrix}$$

we use

```
Symbolic M = ( (Symbolic("a"),Symbolic("b")),
               (Symbolic("c"),Symbolic("d")) );
```

8.5   Operators

In this section, we describe the arithmetic and mathematical operators available in the system:

```
+(unary), -(unary), ++, --, +, -, *, /, ^, =, +=, -=, *=, /=, ==,
!=, ~, |, %, [], (), (comma),
```

Suppose x, y are symbolic variables of type Sum and c is a numerical constant of type int or double, the operators do the following:

- x++ and x-- are the increment and decrement operators.

- x+y, x+c, c+x add symbolic variables, expressions or numerical constants.

- x-y, x-c, c-x subtract symbolic variables, expressions or numerical constants.

- x*y, x*c, c*x multiply symbolic variables, expressions or numerical constants.

- x/y, x/c, c/x divide symbolic variables, expressions or numerical constants.

- x=y assigns a symbolic variable/expression y to x.

- x=c assigns a numerical constant c to x.

- +=, -=, *= and /= are just the modification forms of the assignment operator.

- == creates an equation of two expressions. For example, x+y==y+x is the equation stating that $x + y$ is equivalent to $y + x$. In an if statement the equation is automatically cast to bool and returns true. The result actually depends on the commutativity of the * operator as well. For example, a*b==b*a will return *true* if it is commutative, otherwise it is false.

- != compares the equality of two expressions. For example, the statement x!=x returns false.

- s^n raises expression s to the power n. An alternative way to express the term making use of the exponential function is given by

$$x^x \equiv \exp(x \ln(x)).$$

- When x is a variable belonging to a commutative algebra (or matrix of variables belonging to a commutative algebra) ~x returns the variable x belonging to a non-commutative algebra and vice-versa.

- x|y denotes the scalar product between vectors x and y.

- x%y denotes the cross product between vectors x and y.

- x(i) denotes the i-th component of the vector x.

- x(i,j) denotes the $x_{i,j}$ entry in the matrix x.

- x[y] denotes $x(y)$ i.e. x as a function of y.

- x[y,z] denotes $x(y, z)$.

- x[y==c] denotes $x_{y=c}$.

- x[y==c,y==c] denotes $x_{y=c,...}$.

- x,y creates a list of symbolic expressions, i.e. list<Symbolic>. This is also true of equations. For example

```
Symbolic x("x"), y("y"), z("z");
Equations e = (x==3, y==2.1, z==2);
```

creates a list of equations which can be used in substitution.

8.6 Functions and Member Functions

- df(const Symbolic &y,const Symbolic &x) evaluates the partial derivative of y with respect to x. The evaluation of df(y,x) proceeds as follows:

    1. Suppose y is expressed in terms of x, then each term of y which contains the variable x is differentiated according to the standard rules.

2. If z is another variable then its derivative with respect to x is zero, unless z has been declared as dependent on x, in which case the derivative is `df(z,x)`.

Consider a polynomial P of degree n,

$$P(x) = a_n x^n + a_{n-1} x^{n-1} + \cdots + a_k x^k + \cdots + a_2 x^2 + a_1 x + a_0.$$

Since the derivative of the monomial $a_k x^k$ is $k a_k x^{k-1}$, we obtain by using `diff()` in the function `df()`:

$$\frac{dP}{dx} = n a_n x^{n-1} + (n-1) a_{n-1} x^{n-2} + \cdots + k a_k x^{k-1} + \cdots + 2 a_2 x + a_1.$$

In the process, we have made use of several derivative rules.

– The function `SymbolicInterface::df(const Symbolic&)` is overloaded to manipulate expressions following the properties listed below:

1. `Number::df(const Symbolic&)` implements $\dfrac{dc}{dx} = 0$ where c is a numerical constant.

2. `Symbol::df(const Symbolic&)` implements

$$\frac{dy}{dx} = \begin{cases} \dfrac{dy}{dx} & \text{if } y \text{ is dependent on } x \\ 1 & \text{if } y = x \\ 0 & \text{otherwise} \end{cases}$$

3. The differentiation operator is a *linear operator*, i.e.

$$\frac{d}{dx} [a * u(x) + b * v(x)] = a * \frac{du(x)}{dx} + b * \frac{dv(x)}{dx}$$

where a, b are numerical constants, x is an independent variable and $u(x)$, $v(x)$ are functions of x. This is implemented in part by **Sum**'s member function `df(const Symbolic&)` and by the `df(const Symbolic&)` member function of **Product** via the product rule.

4. *Chain-rule*:

$$\frac{d}{dx} f(u(x)) = \frac{df}{du} * \frac{du}{dx}$$

where f is a function of u which is dependent on x. This is implemented by `Symbol::df(const Symbolic&)`.

5. Integration is the inverse of differentiation as implemented in the `df()` member function of **Integral**

$$\frac{d}{dx} \int y(x) dx = y(x).$$

- The `Product::df(const Symbolic&)` method implements the *product rule* for differentiation

$$\frac{d}{dx}(u_1(x) * u_2(x)) = \frac{du_1}{dx} * u_2 + u_1 * \frac{du_2}{dx}.$$

- The `Power::df(const Symbolic&)` method implements the derivative rule for monomials

$$\frac{d}{dx}(a * (u(x))^n) = n * a * u^{n-1} * \left(\frac{du}{dx}\right)$$

where a is a constant. This is achieved by the more general formula

$$\frac{d}{ds}a^b = \left(\ln a \frac{db}{ds} + \frac{b}{a}\frac{da}{ds}\right)a^b.$$

- The `Log::df(const Symbolic&)` method implements the general formula for derivatives of logarithms

$$\frac{d}{ds}\log_a b = \frac{\frac{1}{b}\frac{db}{ds} - \frac{\log_a b}{a}\frac{da}{ds}}{\ln a} = \frac{a\frac{db}{ds} - b\log_a b\frac{da}{ds}}{ab\ln a}.$$

- The remaining classes implement the standard rules for integration of various functions etc., for example `Sin::df(const Symbolic&)` implements the rule

$$\frac{d}{dx}\sin x = \cos x.$$

The differentiation operator needs knowledge of the dependency between various variables. Such dependency may be declared by the indexing operator. For example, `y[x,z]` is the symbol `y` that is explicitly dependent on `x` and `z`. Thus `df(y[x,z],x)` would be evaluated to itself `df(y[x,z],x)`, instead of 0 as is the case for `df(y,x)`.

- `SymbolicInterface::integrate(const Symbolic &x)` evaluates simple integrals with respect to `x`, neglecting constants of integration.

Consider a polynomial P of degree n,

$$P(x) = a_n x^n + a_{n-1}x^{n-1} + \cdots + a_k x^k + \cdots + a_2 x^2 + a_1 x + a_0.$$

Since the integral of the monomial $a_k x^k$ is $a_k x^{k+1}/(k+1)$, we obtain by using `Power::integrate(const Symbolic&)` and `Sum::integrate(const Symbolic&)`

$$\int P(x)dx = \frac{a_n}{n+1}x^{n+1} + \frac{a_{n-1}}{n}x^n + \cdots + \frac{a_k}{k+1}x^{k+1} + \cdots + \frac{a_2}{3}x^3 + \frac{a_1}{2}x^2 + a_0 x.$$

In the process, we have made use of several integration rules.

– The function `integrate(const Symbolic&,const Symbolic&)` manipulates expressions following the properties listed below:

1. For c a numerical constant and x an independent variable the `integrate()` member function of **Product** and the `integrate()` member function of **Symbol** provide

$$\int c\, dx = c * x, \qquad \int x\, dx = \frac{x^2}{2}.$$

2. $$\int y\, dx = \begin{cases} \int y\, dx & \text{if } y \text{ is dependent on } x \\ y * x & \text{otherwise.} \end{cases}$$

3. The integration operator is a *linear operator*, i.e.

$$\int [a * u(x) + b * v(x)]\, dx = a * \int u(x)\, dx + b * \int v(x)\, dx$$

where a, b are numerical constants, x is an independent variable and $u(x)$, $v(x)$ are functions of x. This is implemented in the `integrate()` member functions of **Sum** and **Product**.

4. Integration is the inverse operation of differentiation:

$$\int \frac{d}{dx} y(x)\, dx = y(x)$$

as implemented in `Derivative::integrate(const Symbolic&)`.

5. The member function `integrate()` of **Product** and **Power** implement the integration rules for monomials

$$\int a * x^n dx = \begin{cases} \frac{a}{n+1} x^{n+1} & n \neq -1 \\ a * \ln(x) & n = -1 \end{cases}$$

where a is a constant. Thus, for example, we have

$$\int \frac{1}{x}\, dx = \int 1 * x^{-1}\, dx = \ln(x).$$

The integration operator sometimes needs knowledge of the dependency between various variables. Such dependency may be declared using the indexing operator. For example, `y[x]` declares that `y` is dependent on `x`. Thus `integrate(y[x],x)` would be evaluated to itself `Integral(y[x],x)`, instead of `y*x` as is the case for `integrate(y,x)`. The description of `df()` expands on this discussion.

- `integrate(const Symbolic &y,const Symbolic &x,unsigned int n)` evaluates the n-th integral of `y` with respect to `x`.

- expand() expands the product of expressions that may be raised to a power n and applies the distributivity law. The integer variable Symbolic::auto_ expand can be used to control whether all expressions are automatically expanded. By default Symbolic::auto_expand is 1, setting it to 0 suppresses automatic expansion. Power::expand() effectively implements the *binomial theorem* to perform the binomial expansion

$$(a + b)^n = {}^nC_0 a^n + {}^nC_1 a^{n-1}b + {}^nC_2 a^{n-2}b^2 + \cdots + {}^nC_{n-1}ab^{n-1} + {}^nC_n b^n \,.$$

For a multinomial expansion, where the expression is a sum of more than two components, we further extend the idea of the binomial theorem by grouping terms into only two sets of terms. However, we have to decide how to split it into the a and b pieces. There are two obvious ways: either cut the expression in half, so that a and b will be of equal size, or split off one component at a time. The latter method is more efficient in most cases. In other words, an expression $t_1 + t_2 + t_3 + \cdots + t_k$ will be treated as the sum $a + b$ where $a = t_1$ and $b = t_2 + t_3 + \cdots + t_k$, i.e.

$$(t_1 + t_2 + t_3 + \cdots + t_k)^n = [t_1 + (t_2 + t_3 + \cdots + t_k)]^n$$

and the binomial expansion is applied. This is implemented in Product:: expand().

- coeff() returns the symbolic or numeric coefficients of an expression. It is overloaded in three different forms:

  - Symbolic::coeff(const Symbolic &s)
    returns the coefficient of the term s in the expression.

  - Symbolic::coeff(int n)
    returns the constant term in the expression divided by n.

  - Symbolic::coeff(double n)
    returns the constant term in the expression divided by n.

  - Symbolic::coeff(const Symbolic &s,int n)
    returns the coefficient of the term sⁿ in the expression. Thus the constant term with respect to s is obtained with $n = 0$.

- Symbolic::subst(const Symbolic &s1,const Symbolic &s2,int &n) replaces all the terms s1 in an expression by another term s2. The reference parameter n is set to the number of substitutions that were made and is optional.

The function proceeds by checking through each term of the expression. If a match with s1 is found, it replaces s1 with s2. This is a useful function that aids the simplification of expressions. For example, an expression with a mixture of $\sin^2 x$ and $\cos^2 x$ terms such as

$$10 * \sin^2 x + 5 * \cos^2 x$$

may be simplified to
$$5 * \sin^2 x + 5$$
by replacing all the $\cos^2 x$ term by $1 - \sin^2 x$ and combining like terms. Each class implements the relevant `subst` method. Of special note are

`Sum::subst(const Symbolic&,constSymbolic&,int&)`

and

`Product::subst(const Symbolic&,constSymbolic&,int&)`.

```
Symbolic a("a"), b("b"), c("c"), d("d");
(a+b+c).subst(a+c,d); // b+d
(a*b*c).subst(a*c,d); // d*b
```

These two substitutions take into account the commutativity of the arithmetic operations. Non-commutative algebras make substitutions more complicated. Consider the following example. We use the commutation relation (Bose operators – non-commutative algebra)

$$bb^\dagger = I + b^\dagger b$$

and β is a scalar (commutative variable). It follows that

$$(b\beta)^2 b^\dagger \equiv bb^\dagger b\beta^2 + b\beta^2.$$

```
Symbolic b("b"), bd("bd"), beta("beta");
b = ~b; bd = ~bd; // non-commutative
(((b*beta)^2)*bd).subst(b*bd==1+bd*b); // b*bd*b*beta^(2)+b*beta^(2)
```

1. Expand positive integer powers.
   `((b*beta)^2) * bd => b * beta * b * beta * bd`

2. Initialize paths beginning with the location of the first factor of x. There could be none, one or many such paths.

   ```
     b * beta * b * beta * bd
     ↑               ↑
   ```

3. Build each path by considering all possible locations of subsequent factors of x and creating all possible paths.

   ```
     b * beta * b * beta * bd
     ①                      ②
              ①             ②
   ```

4. Discard paths that are too short.

5. Search for a path which may be substituted

   (a) When consecutive values are in the wrong order check that they commute. *Correct the ordering for each path.*

(b) Search for a position (from left to right) that all values can commute to, and set that as the place for substitution.

```
b * beta * b * beta * bd
↑
```

(c) Replace the factors of x with z.

Example 1. We consider the exponentiation x^y.

```
// expand.cpp

#include <iostream>
#include "symbolicc++.h"
using namespace std;

int main(void)
{
 Symbolic a("a"), b("b"), c("c"), y, z;

 y = (Symbolic(7)^3); cout << " y = " << y << endl;
 y = (a^0);            cout << " y = " << y << endl;
 y = (a^3);            cout << " y = " << y << endl;
 cout << endl;

 y = ((a+b-c)^3);      cout << " y = " << y << endl;
 cout << endl;

 y = (a+b)*(a-c);      cout << " y = " << y << endl;
 cout << endl;

 y = a+b;
 z = (y^4);            cout << " z = " << z << endl;
 return 0;
}
```

The output is

```
y = 343
y = 1
y = a^(3)

y = a^(3)+3*a^(2)*b-3*a^(2)*c+3*a*b^(2)-6*a*b*c+3*a*c^(2)+b^(3)-3*b^(2)*c
    +3*b*c^(2)-c^(3)

y = a^(2)-a*c+b*a-b*c

z = a^(4)+4*a^(3)*b+6*a^(2)*b^(2)+4*a*b^(3)+b^(4)
```

Example 2. We consider the derivatives of

$$y(x) = \frac{1}{1-x} + 2x^3 - z$$

with respect to x. Notice that $dz/dx = 0$ because the system assumes no dependency for any two variables unless specified otherwise. In the second part of this example, we consider the derivatives of the expression

$$u(v) = \frac{1}{3}v^{3/5} - \frac{2}{7}v^{1/5} + \frac{1}{6}$$

with respect to v. The coefficients and the degrees of the variable v are rational
numbers.

```
// derivatv.cpp

#include <iostream>
#include "symbolicc++.h"
using namespace std;

int main(void)
{
   int i;
   Symbolic x("x"), y, z("z");

   y = 1/(1-x) + 2*(x^3) - z;
   cout << "y = " << y << endl;

   for(i=0;i<8;i++)
   {
    y = df(y,x);
     cout << "y = " << y << endl;
   }
   cout << endl;

   Symbolic u, v("v");

   u = (v^(Symbolic(3)/5))/3 - 2*(v^(Symbolic(1)/5))/7 + Symbolic(1)/6;
   cout << "u = " << u << endl;

   for(i=0;i<8;i++)
   {
    u = df(u,v);
     cout << "u = " << u << endl;
   }
   return 0;
}
```

The output is

```
y = (-x+1)^(-1)+2*x^(3)-z
y = (-x+1)^(-2)+6*x^(2)
y = 2*(-x+1)^(-3)+12*x
y = 6*(-x+1)^(-4)+12
y = 24*(-x+1)^(-5)
y = 120*(-x+1)^(-6)
y = 720*(-x+1)^(-7)
y = 5040*(-x+1)^(-8)
y = 40320*(-x+1)^(-9)

u = 1/3*v^(3/5)-2/7*v^(1/5)+1/6
u = 1/5*v^(-2/5)-2/35*v^(-4/5)
u = -2/25*v^(-7/5)+8/175*v^(-9/5)
u = 14/125*v^(-12/5)-72/875*v^(-14/5)
u = -168/625*v^(-17/5)+144/625*v^(-19/5)
u = 2856/3125*v^(-22/5)-2736/3125*v^(-24/5)
u = -62832/15625*v^(-27/5)+65664/15625*v^(-29/5)
u = 1696464/78125*v^(-32/5)-1904256/78125*v^(-34/5)
u = -54286848/390625*v^(-37/5)+64744704/390625*v^(-39/5)
```

Example 3. We investigate the properties of the df() operator. By default, any two variables are assumed to be independent of each other. Dependency can be created by depend() and removed by nodepend().

```cpp
// depend.cpp

#include <iostream>
#include "symbolicc++.h"
using namespace std;

int main(void)
{
  Symbolic a("a"), b("b"), u("u"), v("v"), x("x"), z("z"), y("y");

  cout << "System assumes no dependency by default" << endl;
  cout << "df(y,x) => " << df(y,x) << endl;

  cout << "y is dependent on x" << endl;
  y = y[x];
  cout << "df(y,x) => " << df(y,x) << endl;
  y = sin(x*x+5) + x; cout << "y = " << y << endl;
  cout << "df(y,x) => " << df(y,x) << endl;
  cout << endl;

  cout << "u depends on x" << endl;
  u = u[x];
  cout << "df(cos(u),x) => " << df(cos(u),x) << endl;

  cout << "u depends on x, and x depends on v" << endl;
  x = x[v];
  u = Symbolic("u")[x];
  cout << "df(cos(u),v) => " << df(cos(u),v) << endl;

  // example
  y = cos(x*x+5)-2*sin(a*z+b*x); cout << "y = " << y << endl;
  cout << "df(y,v) => " << df(y,v) << endl;
  cout << endl;

  cout << "remove dependency" << endl;
  y = Symbolic("y");
  cout << "df(y,v) => " << df(y,v) << endl;
  cout << endl;

  cout << "derivative of constants gives zero" << endl;
  cout << "df(5,x) => " << df(5,x) << endl;
  cout << endl;

  // renew the variable y
  y = Symbolic("y");
  y = y[x, z];
  cout << "df(y,x)+df(y,x) => " << df(y,x)+df(y,x) << endl;
  cout << "df(y,x)*df(y,x) => " << df(y,x)*df(y,x) << endl;
  cout << endl;

  cout << "Another example," << endl;
  x = Symbolic("x");
  y = Symbolic("y");
  u = Symbolic("u");
```

```
 v = v[x];
 u = u[v];
 y = y[u];
 y = df(u,x)*sin(x);
 cout << "y = " << y << endl;
 cout << "let u = " << 2*v*x << " then," << endl;
 cout << "y => " << y[u == 2*v*x] << endl;
 cout << "df(y[u == 2*v*x],x) => " << df(y[u == 2*v*x],x) << endl;
 return 0;
}
```

The output is

```
System assumes no dependency by default
df(y,x) => 0
y is dependent on x
df(y,x) => df(y[x],x)
y = sin(x^(2)+5)+x
df(y,x) => 2*cos(x^(2)+5)*x+1

u depends on x
df(cos(u),x) => -sin(u[x])*df(u[x],x)
u depends on x, and x depends on v
df(cos(u),v) => -sin(u[x[v]])*df(u[x[v]],x[v])*df(x[v],v)
y = cos(x[v]^(2)+5)-2*sin(a*z+b*x[v])
df(y,v) => -2*sin(x[v]^(2)+5)*x[v]*df(x[v],v)-2*cos(a*z+b*x[v])*b*df(x[v],v)

remove dependency
df(y,v) => 0

derivative of constants gives zero
df(5,x) => 0

df(y,x)+df(y,x) => 2*df(y[x[v],z],x[v])
df(y,x)*df(y,x) => df(y[x[v],z],x[v])^(2)

Another example,
y = df(u[v[x]],v[x])*df(v[x],x)*sin(x)
let u = 2*v[x]*x then,
y => 2*x*df(v[x],x)*sin(x)
df(y[u == 2*v*x],x) => 2*df(v[x],x)*sin(x)+2*x*df(v[x],x,x)*sin(x)
                       +2*x*df(v[x],x)*cos(x)
```

Examples 4, 5 and 6 demonstrate how different forms of `coeff()` could be used to extract coefficients of expressions.

Example 4.

```
// coeff1.cpp

#include <iostream>
#include "symbolicc++.h"
using namespace std;

int main(void)
{
   Symbolic a("a"), b("b"), c("c");
```

```
   cout << (2*a-3*b*a-2+c).coeff(a) << endl;
   cout << (2*a-3*b*a-2+c).coeff(b*a) << endl;
   cout << (2*a-3*b*a-2+c).coeff(b) << endl;
   cout << endl;

   cout << (b*b+c-3).coeff(a,0) << endl;
   cout << (-b).coeff(a,0) << endl;
   return 0;
}
```

The output is

```
-3*b+2
-3
-3*a

b^(2)+c-3
-b
```

Example 5.

```
// coeff2.cpp

#include <iostream>
#include "symbolicc++.h"
using namespace std;

int main(void)
{
   Symbolic a("a"), b("b"), c("c"), d("d");
   Symbolic y, z;

   y = -a*5*a*b*b*c+4*a-2*a*b*c*c+6-2*a*b+3*a*b*c-8*c*c*b*a+4*c*c*c*a*b-3*b*c;
   cout << "y = " << y << endl;
   cout << endl;

   z = a;          cout << y.coeff(z) << endl;
   z = b;          cout << y.coeff(z) << endl;
   z = c;          cout << y.coeff(z) << endl;
   z = a*a;        cout << y.coeff(z) << endl;
   z = a*b;        cout << y.coeff(z) << endl;
   z = d;          cout << y.coeff(z) << endl;

   cout << endl;

   // find coefficients of the constant term
   z = a;
   cout << y.coeff(z,0) << endl;
   cout << y.coeff(1) << endl;

   return 0;
}
```

The output is

```
y = -5*a^(2)*b^(2)*c+4*a-10*a*b*c^(2)-2*a*b+3*a*b*c+4*c^(3)*a*b-3*b*c+6

-10*b*c^(2)-2*b+3*b*c+4*c^(3)*b+4
-10*a*c^(2)-2*a+3*a*c+4*c^(3)*a-3*c
```

```
-5*a^(2)*b^(2)+3*a*b-3*b
-5*b^(2)*c
-10*c^(2)+3*c+4*c^(3)-2
0

-3*b*c+6
6
```

Example 6.

```cpp
// coeff3.cpp

#include <iostream>
#include "symbolicc++.h"
using namespace std;

int main(void)
{
   int i;
   Symbolic x("x"), p("p");
   Symbolic w = 0;

   for(i=-5;i<=5;i++) w += (6+i)*(p^(5-i))*(x^i)/(6-i);

   cout << "w = " << w << endl; cout << endl;

   cout << "The coefficients are" << endl;
   for(i=-5;i<=5;i++) cout << w.coeff(x,i) << endl;
   return 0;
}
```

The output is

```
w = 1/11*p^(10)*x^(-5)+1/5*p^(9)*x^(-4)+1/3*p^(8)*x^(-3)+1/2*p^(7)*x^(-2)
    +5/7*p^(6)*x^(-1)+p^(5)+7/5*p^(4)*x+2*p^(3)*x^(2)+3*p^(2)*x^(3)+5*p*x^(4)
    +11*x^(5)

The coefficients are
1/11*p^(10)
1/5*p^(9)
1/3*p^(8)
1/2*p^(7)
5/7*p^(6)
p^(5)
7/5*p^(4)
2*p^(3)
3*p^(2)
5*p
11
```

Example 7. This example illustrates the use of subst() (using the indexing operator) and how identities can be used for simplification.

```cpp
// put.cpp

#include <iostream>
#include "symbolicc++.h"
using namespace std;
```

```
int main(void)
{
    Symbolic a("a"), b("b"), c("c"), w, y;

    // Test (1)
    y = (a+b)*(a+sin(cos(a)^2))*b;
    cout << "y = " << y << endl;

    y = y[b==c+c]; cout << "y = " << y << endl;

    y = y[cos(a)*cos(a)==1-sin(a)*sin(a)];
    cout << "y = " << y << endl; cout << endl;

    // Test (2)
    y = sin(cos(a)^2) + b; cout << "y = " << y << endl;

    y = y[cos(a)*cos(a)==1-sin(a)*sin(a)];
    cout << "y = " << y << endl; cout << endl;

    // Test (3)
    a = a[c]; b = b[c];
    y = 2*a*df(b*a*a,c)+a*b*c; cout << "y = " << y << endl;

    y = y[a==cos(c)]; cout << "y = " << y << endl;

    w = df(b,c); y = y[w==exp(a)]; cout << "y = " << y << endl;

    return 0;
}
```

The output is

```
y = a^(2)*b+a*sin(cos(a)^(2))*b+b^(2)*a+b^(2)*sin(cos(a)^(2))
y = 2*a^(2)*c+2*a*sin(cos(a)^(2))*c+4*c^(2)*a+4*c^(2)*sin(cos(a)^(2))
y = 2*a^(2)*c+2*a*sin(-sin(a)^(2)+1)*c+4*c^(2)*a+4*c^(2)*sin(-sin(a)^(2)+1)

y = sin(cos(a)^(2))+b
y = sin(-sin(a)^(2)+1)+b

y = 2*a[c]^(3)*df(b[c],c)+4*a[c]^(2)*b[c]*df(a[c],c)+a[c]*b[c]*c
y = 2*cos(c)^(3)*df(b[c],c)-4*cos(c)^(2)*b[c]*sin(c)+cos(c)*b[c]*c
y = 2*cos(c)^(3)*e^a[c]-4*cos(c)^(2)*b[c]*sin(c)+cos(c)*b[c]*c
```

Example 8. We integrate the expression

$$y(x) = (1 - x)^2 + \cos(x) + xe^x + c + z$$

with respect to x. We declare z to be dependent on x. Since the system assumes no dependency for any two variables we can use c as a symbolic constant. We also make use of integration by parts

$$\int xe^x dx = xe^x - \int e^x dx.$$

```
// integration.cpp

#include <iostream>
#include "symbolicc++.h"
using namespace std;

int main(void)
{
  Symbolic x("x"), c("c"), z("z"), y;

  z = z[x];
  y = (1-x)*(1-x)+cos(x)+x*exp(x)+c+z;

  cout << "y = " << y << endl;

  for(int i=0;i<3;i++)
  {
    y = integrate(y,x);
    y = y[integrate(x*exp(x),x) == x*exp(x) - exp(x)];
    cout << "y = " << y << endl;
  }

  return 0;
}
```

The output is

```
y = x^(2)-2*x+cos(x)+x*e^x+c+z[x]+1
y = 1/3*x^(3)-x^(2)+sin(x)+x*e^x-e^x+c*x+int(z[x],x)+x
y = 1/12*x^(4)-1/3*x^(3)-cos(x)+x*e^x-2*e^x+1/2*c*x^(2)+int(z[x],x,x)+1/2*x^(2)
y = 1/60*x^(5)-1/12*x^(4)-sin(x)+x*e^x-3*e^x+1/6*c*x^(3)+int(z[x],x,x,x)+1/6*x^(3)
```

8.6.1 Functions

We have described the availability of special functions such as sine and cosine in our symbolic system. All the operations available can be applied to these functions, e.g. the differentiation operator df().

A function is usually derived from Symbol. Since Symbol can already represent dependent variables with a parameter list, any derived class can also have a parameter list. For example Sin uses a single parameter while Power uses a list of two parameters.

There are 6 special functions classes derived from Symbol:

$$\text{Power, Sinh, Cosh, Sin, Cos, Log}$$

There are 14 special functions implemented in symbolicc++.h

$$\text{exp(), sinh(), cosh(), sin(), cos(), sqrt(), ln(),}$$
$$\text{tan(), cot(), sec(), csc(), log(), pow(), sqrt()}$$

Member functions that are associated with them include:

- The differentiation rules:

$$\frac{d}{dx}\exp(u) = \frac{du}{dx}\exp(u), \quad \frac{d}{dx}\sinh u = \frac{du}{dx}\cosh u, \quad \frac{d}{dx}\cosh u = \frac{du}{dx}\sinh u,$$

$$\frac{d}{dx}\sin u = \frac{du}{dx}\cos u, \frac{d}{dx}\cos u = -\frac{du}{dx}\sin u, \frac{d}{dx}\sqrt{u} = \frac{du}{dx}\frac{1}{2\sqrt{u}}, \frac{d}{dx}\ln u = \frac{du}{dx}\frac{1}{u}.$$

- The value for special arguments:

$$\exp(0) = 1, \qquad \sinh(0) = 0, \qquad \cosh(0) = 1, \qquad \sin(0) = 0,$$

$$\cos(0) = 1, \qquad \sqrt{1} = 1, \qquad \sqrt{0} = 0, \qquad \ln(1) = 0.$$

8.7 Simplification

Simplification plays a vital role in a symbolic system. It produces an equivalent but simpler form of a given expression. The following are some examples of simplification of algebraic expressions

```
2+5             ->   7
3*x-2*x         ->   x
u+0             ->   u
sin(2*x-x-x)    ->   0
```

where the symbol -> means "simplifies to" or "transforms to". Consider another expression,

```
8*a*a + b*c*(a+b-a-2*b+b) + 2 - power(a,2)*3 + cos(x-x).
```

There exists some obvious redundancy. By applying the algebraic simplification, a much more "clean" and "useful" result can be obtained, such as the simplified form of the expression above,

```
5*a^2 + 3.
```

Expression simplification is based on the existence of simplification rules for each type of mathematical expression. For example, simplification of the expression above involved the following algebraic manipulations:

```
    8*a*a + b*c*(a+b-a-2*b+b) + 2 - (a^2)*3 + cos(x-x)
→   8*a^2 + b*c*0 + 2 - 3*a^2 + cos(0)
→   5*a^2 + 0 + 2 + 1
→   5*a^2 + 3
```

In fact, there is no finite set of simplification rules that could simplify all kinds of algebraic expressions. This means that it is hard to find an algorithm that can completely simplify an arbitrary algebraic expression. However, with a suitable finite set of simplification rules, very good results can be achieved in practice.

8.7.1 Canonical Forms

Mathematical expressions may exist in several different but equivalent symbolic representations. A *canonical form* is a designated or "standard" representation for such expressions. For example, polynomials are usually represented by a series of terms with decreasing exponents, with each term preceded by a numerical coefficient. Following the "standard" convention, the conversion below would be appropriate:

$$-5 * x^3 + x^6 * 8 + 4 - 2 * x^2 \rightarrow 8 * x^6 - 5 * x^3 - 2 * x^2 + 4\,.$$

A conversion to canonical form may increase or decrease the complexity of an expression, or simply result in the rearrangement of terms. Quite often, such a conversion is to improve the readability. The most important application is the determination of two distinct symbolic representations for their "equivalence". Therefore, we need some rules that permit equivalent transformation from one representation to another. However, it is difficult to obtain the complete set of correct rules.

When a canonical form of an expression is carefully defined, transformation to the canonical form of two equivalent expressions will produce identical, or nearly identical representations. This will greatly reduce the complexity for comparison.

A mathematical expression is represented by a tree structure for our symbolic system. The simplification process consists of two major parts: One being the transformation to the canonical form, and the other being the reduction to simplified form based on some known rules. For example, `sin(a*(-1))` could be rewritten as `-1*sin(a)`. Its symbolic simplification is as follows:

```
                    sin(a*(-1))
    [Step 1]  →  sin(-1*a)
    [Step 2]  →  -1*sin(a)
```

The process involves one steps of canonical transformations `[Step 1]` (implemented in `Product::simplify()`) and one step of rule-based simplification `[Step 2]` (implemented in `Sin::simplify()`). The rule applied in `[Step 2]` is achieved simply due to the fact that the product `-1*a` is written in a canonical form, i.e. numerical factors appear first.

This example illustrates that a canonical form is a way of restricting the forms of expression which are allowed. Hence the simplification rules need only deal with a restricted class of expressions.

8.7.2 Simplification Rules and Member Functions

In this section, we describe in greater detail the simplification rules and the transformation to canonical forms. Suppose e, e_1, e_2, e_3, e_4 are arbitrary algebraic expressions and c, c_1, c_2 denote positive numerical constants. Some sample rules of simplification for each operator are listed in Figure 8.5.

Operations	Functions
$c_1 + c_2 \rightarrow c$	Sum::simplify()
$e + 0 \rightarrow e$	Sum::simplify()
$c_1 * c_2 \rightarrow c$	Product::simplify()
$0 * e \rightarrow 0$	Product::simplify()
$1 * e \rightarrow e$	Product::simplify()
$0^c \rightarrow 0$	Power::simplify()
$1^c \rightarrow 1$	Power::simplify()
$e^0 \rightarrow 1$	Power::simplify()

Figure 8.5: *Simplification rules.*

The subtraction operator (-) is implemented in terms of addition and multiplication, i.e. $a - b = a + (-1) * b$, while the division operator (/) is implemented in terms of powers and multiplication, i.e. $a/b = a * (b^{-1})$. Meanwhile, the transformation rules to canonical forms are given in Figure 8.6.

Operations	Functions
$(c_1 * e_1) * (c_2 * e_2) \rightarrow c * (e_1 * e_2)$ where $c - c_1 * c_2$	Product::simplify()
$((e_1 + e_2) + e_3) + e_4 \rightarrow e_1 + e_2 + e_3 + e_4$	Sum::simplify()
$((e_1 * e_2) * e_3) * e_4 \rightarrow e_1 * e_2 * e_3 * e_4$	Product::simplify()
$e_1 + (e_2) \rightarrow e_1 + e_2$	Sum::simplify()

Figure 8.6: *Canonical conversion rules.*

Finally, individual terms are grouped, summed or multiplied if they belong to the same class. For example,

```
2*a + 3*a        ->   5*a
exp(z) + exp(z)  ->   2*exp(z)
p^(-2) * p^(5)   ->   p^(3)
cos(x) * cos(x)  ->   (cos(x))^2
```

and the final simplified expression is obtained. Terms in a sum are considered the same if they are the same up to the numerical coefficient in the canonical form. With this limited set of simplification rules, could an expression be simplified to its simplest possible form? The answer to this question is not obvious. However, transformation to canonical form does play an important role in the process as it reduces the number of rule-based simplifications and also the complexity of the process.

In our canonical form for product expressions with numeric operands, only the first operand may be numeric. This implies that

$$(e * c) \rightarrow (c * e)$$

which is just the application of commutative law for multiplication. Hence the following rules are equivalent and they are not shown in Figure 8.5

$$e * 1 \longleftrightarrow 1 * e$$
$$e * 0 \longleftrightarrow 0 * e$$
$$e * c \longleftrightarrow c * e.$$

Similarly, the commutative law for addition may be employed

$$(e + c) \rightarrow (c + e).$$

This eliminates many unnecessary addition rules. Another important canonical conversion rule is the outward propagation multiplication rule. For example,

$$(c_1 * e_1) * (c_2 * e_2) \longleftrightarrow (c_1 * c_2) * (e_1 * e_2)$$
$$(c_1 * e_1)/(c_2 * e_2) \longleftrightarrow (c_1/c_2) * (e_1/e_2).$$

These rules do not result in simplifications, but convert expressions to canonical forms which may result in simplifications later.

8.8   Commutativity

So far we have assumed the commutative law holds for multiplication of symbols. However, in some applications this is not necessarily true. For example,

$$A * B \neq B * A$$

in general if A, B are $n \times n$ matrices. Thus we need a non-commutative operator. In fact, a large branch of mathematics called Lie Algebras involves operators which are non-commutative. Therefore, non-commutative operators must be built into a symbolic system.

For our computer algebra system, we have made the system commutative by default. To specify a non-commutative algebra, use the member function `commutative(int t)`. This will return an expression where a symbol, or all the entries of a vector or matrix of symbols, are marked as commutative or not commutative. The argument `t` specifies the choice of commutativity. If `t=0`, non-commutativity prevails. Otherwise, commutativity prevails. Another option is to use the ~ operator which returns a symbol (or vector, or matrix as the case may be) with the opposite commutativity to the given symbol.

Example. Consider the following program

```cpp
// commute.cpp

#include <iostream>
#include "symbolicc++.h"
using namespace std;

int main(void)
{
   Symbolic a("a"), b("b"), y;

   cout << "Commutative Algebra" << endl;
   cout << "===================" << endl;

   // The system is commutative by default
   y = a*b*a;        cout << " y = " << y << endl;
   y = a*b-b*a;      cout << " y = " << y << endl;

   y = ((a-b)^2);  cout << " y = " << y << endl;
   cout << endl;

   cout << "Non Commutative Algebra" << endl;
   cout << "=======================" << endl;
   a = ~a; b = ~b;

   y = a*b*a;        cout << " y = " << y << endl;
   y = a*b-b*a;      cout << " y = " << y << endl;

   y = ((a-b)^2);  cout << " y = " << y << endl;
   cout << endl;

   cout << "Commutative Algebra" << endl;
   cout << "===================" << endl;
   cout << " y = " << y[a == ~a, b == ~b] << endl;

   return 0;
}
```

The output is

```
Commutative Algebra
===================
 y = a^(2)*b
 y = 0
 y = a^(2)-2*a*b+b^(2)

Non Commutative Algebra
=======================
 y = a*b*a
 y = a*b-b*a
 y = a^(2)-a*b-b*a+b^(2)

Commutative Algebra
===================
 y = a^(2)-2*a*b+b^(2)
```

8.9   Symbolic and Numeric Interface

When a symbolic algebraic expression is evaluated, its numerical algebraic value is sometimes also needed. Determining the numerical value of an expression amounts to substituting numerical values for all the variables occurring in the expression. If values are not available for all variables in the result then a partial result will follow, i.e. the result will be a symbolic expression and not a simple number. Thus to calculate a numerical value we proceed as follows.

1. Substitute numerical values for variables using subst(), subst_all or the indexing operator [].

2. Cast the result to double or int to obtain a numerical result.

Example. The different properties are demonstrated in the following program.

```cpp
// setvalue.cpp

#include <iostream>
#include "symbolicc++.h"
using namespace std;

int main(void)
{
   Symbolic x("x"), y("y"), z("z");
   double   c1 = 0.5, c2 = 1.2;

   y = x*x+z/2.0;
   cout << "y = " << y << endl;

   cout << "Put x = " << c1 << ", z = " << c2 << endl;
   cout << "The value of y = " << y
        << " is " << double(y[x==c1,z==c2]) << endl;
   cout << endl;

   y = x*x*z + 0.7*z - x*z;
   cout << "y = " << y << endl;

   cout << "Put x = " << c1 << endl;
   cout << "The value of y is " << y[x==c1] << endl;
   cout << endl;

   cout << "Put z = " << c2 << endl;
   cout << "The value of y is "
        << double(y[x==c1, z==c2]) << endl;

   return 0;
}
```

The output is

```
y = x^(2)+0.5*z
Put x = 0.5, z = 1.2
The value of y = x^(2)+0.5*z is 0.85

y = x^(2)*z+0.7*z-x*z
Put x = 0.5
```

```
The value of y is 0.45*z

Put z = 1.2
The value of y is 0.54
```

8.10 Example Computation

Consider the simple program

```
#include <iostream>
#include <symbolicc++.h>
using namespace std;

int main(void)
{
 Symbolic a("a"), b("b");
 cout << ( (a+b)^2 ) << endl;
 return 0;
}
```

The line

```
Symbolic a("a");
```

1. calls the constructor `Symbolic("a")` which

2. calls the constructor `SymbolicProxy(Symbol("a").simplify())`,
   where `Symbol("a").simplify()`

   (a) calls the constructor `Symbol("a",1)` which creates a new instance of
       `Symbol`, and

   (b) the default constructor of the parent class `CloningSymbolicInterface`
       is called which involves

   (c) calling the default constructor of `SymbolicInterface` which sets its
       members `simplified` and `expanded` to zero, and

   (d) calling the default constructor of `Cloning` which sets its member
       `refcount` to zero, which returns to `Symbol("a",1)` and

   (e) sets the data members `Symbol::name` set to "a" and `Symbol::commutes`
       set to 1, while the default constructor is used for the `list` parameters

   (f) calls the member function `simplify()` which returns `*this` (i.e. the
       same `Symbol`) as the argument to the constructor
       `Simplified(const CloningSymbolicInterface &s)` which

   (g) calls the constructor `CastPtr<CloningSymbolicInterface>(s)` which

   (h) calls the constructor `CloningPtr(const Cloning &c,int=0)` which

   (i) calls the member function `Cloning::reference(0)` which

   (j) calls `Symbol::clone()` which

(k) calls `Cloning::clone<Symbol>(*this)` which

(l) calls the copy constructor for `Symbol` which

    i. which creates a new (dynamically allocated) instance of `Symbol`, and

    ii. the copy constructor of the parent class `CloningSymbolicInterface` is called which involves

    iii. calling the copy constructor of `SymbolicInterface` which copies its members `simplified` and `expanded`, and

    iv. calling the copy constructor of `Cloning` which sets its member `refcount` to zero, which returns the copy constructor for `Symbol` and

    v. copies the data members

        `Symbol::name`, `Symbol::commutes` and `parameters`

    (`parameters` is an empty list)

(m) returning to `Cloning::reference`, the member `refcount` is set to 1 in the dynamically allocated instance of `Symbol` and

(n) the data member `value` is set to resulting `Cloning*` pointing to the newly (dynamically) allocated instance if `Symbol`. Thus the derived class `SymbolicProxy` also points to this instance

(o) and sets `SymbolicInterface::simplified` to 1

and then calls the constructor

3. `SymbolicProxy(const SymbolicProxy &s)` where the first parameter `s` is set to the simplified symbol created above, which

4. calls the constructor `CastPtr<CloningSymbolicInterface>(s)` which

5. calls the constructor `CloningPtr(const CloningPtr&,int)` which

6. calls `Cloning::reference()` for the dynamically allocated `Symbol` which increases its `refcount` to 2 and

7. assigns the resulting pointer (a copy of the pointer to the dynamically allocated symbol) to the member `value`.

Notice how we have combined the convenience of creating new variable instances on the stack with dynamic allocation to create persistent data as required for mathematical objects.

As the various stack allocated objects go out of scope:

1. The destructor for `Symbol` is called.

2. The destructor for `SymbolicInterface` is called.

3. The destructor for `Cloning` is called.

4. The destructor for `Simplified` is called.

5. The destructor for `SymbolicProxy` is called.

6. The destructor for `CastPtr<CloningSymbolicInterface>` is called.

7. The destructor for `CloningPtr` is called for the stack allocated instance of `Symbolic`, which does nothing since `refcount==0`.

8. The destructor for `CloningPtr` is called, which decrements `refcount` in the dynamically allocated instance of `Symbolic`. Thus we end with `refcount==1`, which is referenced via the variable `a` of type `Symbolic`.

The declaration of `b`

```
Symbolic b("b");
```

proceeds similarly.

We consider the operations described in

```
cout << ( (a+b)^2 ) << endl;
```

according to the precedence order that C++ uses:

1. The overloaded operator `operator+(a,b)` is called which

   (a) calls the constructor `Sum(a,b)` which adds a and b to the list of summands which is returned as a

   (b) `Symbolic(Sum(a,b))`, i.e. the constructor

      $$\text{Symbolic(const CloningSymbolicInterface\&)}$$

      which calls `Sum::expand()` which returns the same sum to the constructor
      `Expanded(const CloningSymbolicInterface&)` which

   (c) calls the constructor
      `SymbolicProxy(const CloningSymbolicInterface&)` which

      i. calls the constructor `CastPtr<CloningSymbolicInterface>(s)` which
      ii. calls the constructor `CloningPtr(const Cloning &c,int=0)` which
      iii. calls the member function `Cloning::reference(0)` which
      iv. calls `Symbol::clone()` which
      v. calls `Cloning::clone<Symbol>(*this)` which
      vi. calls the copy constructor for `Sum` which

         A. which creates a new (dynamically allocated) instance of `Sum`, and
         B. the copy constructor of the parent class `CloningSymbolic-Interface` is called which involves
         C. calling the copy constructor of `SymbolicInterface` which copies its members `simplified` and `expanded`, and

      D. calling the copy constructor of `Cloning` which sets its member `refcount` to zero, which returns the copy constructor for `Symbol` and

      E. copies the data member `Sum::summands`, which requires that each summand (`a` and `b`) be copied as a new element in the new list: i.e. calling the assignment operators for `SymbolicProxy`, `CastPtr<CloningSymbolicInterface>(s)`, and `CloningPtr` which results in the `refcount` of `a` and `b` being set to 2

    vii. returning to `Cloning::reference()`, the member `refcount` is set to 1 in the dynamically allocated instance of `Sum` and

    viii. the data member `value` is set to the resulting `Cloning*` which points to the new dynamically allocated instance of `Sum`. Thus the derived class `SymbolicProxy` also points to this instance

    ix. and sets `SymbolicInterface::expanded` to 1

  (d) `simplify()` is called for the resulting `Expanded` sum which essentially leaves the sum unchanged. This result is passed to the constructor `Simplified(const CloningSymbolicInterface&)`, and in turn to `SymbolicProxy(const CloningSymbolicInterface&)`,

`CastPtr<CloningSymbolicInterface>(const CloningSymbolicInterface&)`,

    and `CloningPtr(const Cloning&)` which calls `reference()` to increment the `refcount` for the sum

  (e) the above result is passed to the copy constructors for

    `SymbolicProxy, CastPtr<CloningSymbolicInterface>, CloningPtr`

    which calls `reference()` to increment the `refcount` of the sum. This yields the result of type `Symbolic` for the next operation.

2. the overloaded operator `operator^(sum,2)` (where `sum` is the result of the previous step) calls the constructor `Power(sum,Symbolic(2))` to yield a value of type `Symbolic` in a similar way to that described for the `Sum`, except that the integer 2 is first cast to `Symbolic(2)` in a similar way to the description for `Symbolic("a")` except that the result points to a `Number<int>`.

3. The overloaded operator `operator<<(cout,result)` (where `result` is the result from the previous step, namely the `Sum` $a^2 + 2ab + b^2$, calls the `print()` member function for the sum which recursively calls `print()` for each of the subexpressions.

Chapter 9

Applications

The functionalities of a computer algebra system are best illustrated by applications. In this chapter, we present the applications of the classes introduced in Chapters 7 and 8. The problems presented here relate to number theory, nonlinear dynamics, special functions in mathematics and physics, etc. In general, applications are categorized under different classes. In each application, the mathematical background is first described, followed by the proposed solution to the problem and then the excerpt of a program that solves the problem symbolically. There are thirty-one applications presented in this chapter

9.1   Bitset Class

We apply the `bitset` class of the Standard Template Library to number theory
– the *prime numbers*. A prime number is a positive integer $p > 1$ such that no
other integer divides p except 1 and p. The sequence formed by the prime numbers
is perhaps the most famous sequence in number theory. We generate the prime
number sequence using the "sieve of Eratosthenes" described in Section 2.2.

To implement the *sieve algorithm*, one must decide how numbers are stored and
crossed out in the sequence. We could maintain an array of flags with each entry
corresponding to a number in the sequence. The array is initialized by setting all
the flags *true*. The crossing out of multiples is equivalent to setting the respective
flag to *false*. When we reach a prime p with $p^2 > N$, all the primes below N have
been found, then the array entries with flags equal to *true* will correspond to all the
prime numbers $\leq N$.

Usually, an array of flags is represented by a bit field. It is simply a vector of 0/1
values. The field is can be maintained by a vector of **unsigned char**. Since each
character is composed of 8 bits, it can store 8 flags. This data structure is very
effective in terms of storage space.

Actually, we only need to store flags for the odd numbers, because we know that all
even numbers besides 2 are not prime. In this way, we need $1000000/(8 \times 2) = 62500$
bytes to store the table of primes less than one million. The space complexity of
the algorithm is $O(N)$.

We list the class **Prime** which generates the prime number sequence. The header
file **bitset** of STL is included.

```
// prime.h

#ifndef PRIME_H
#define PRIME_H

#include <cassert>
#include <iostream>
#include <bitset>
using namespace std;

template <size_t n>
class Prime
{
private:
   // Data Fields
   unsigned int max_num, max_index, index, p, q, current;
   bitset<(n+1)/2 - 1> bvec;

public:
   // Constructors
   Prime();
   Prime(unsigned int num);
```

```cpp
    void reset();
    int  step();
    void run();
    int  is_prime(unsigned int) const;
    unsigned int current_prime() const;

    unsigned int operator () (unsigned int) const;
};

template <size_t n> Prime<n>::Prime()
    : max_num(n), max_index((n+1)/2 - 1), index(0),
      p(3), q(3), current(2) { bvec.set(); }

template <size_t n> void Prime<n>::reset()
{
    assert(n > 1);
    max_num = n;
    max_index = (n+1)/2 - 1;
    index = 0; p = 3; q = 3; current = 2;
    bvec.set();
}

template <size_t n> int Prime<n>::step()
{
    if(index < max_index)
    {
        while(!bvec.test(index))
        {
            ++index;
            p += 2;
            if(q < max_index) q += p+p-2;
            if(index > max_index) return 0;
        }
        current = p;

        if(q < max_index)
        {
            // cross out all odd multiples of p, starting with p^2
            // k = index of p^2
            unsigned int k = q;
            while(k < max_index) { bvec.reset(k); k += p; }
            ++index;
            p += 2;
            q += p+p - 2;
            return 1;
        }
        // p^2 > n, so bvec has all primes <= n recorded
        else // to next odd number
        {
            p += 2;
            ++index;
            return 2;
        }
    }
    else return 0;
}

template <size_t n> void Prime<n>::run() { while(step() == 1); }
```

```
template <size_t n>
int Prime<n>::is_prime(unsigned int num) const
{
    if(!(num%2)) return 0;
    if(bvec.test((num-3)/2)) return 1;
    return 0;
}

template <size_t n>
unsigned int Prime<n>::current_prime() const
{  return current; }

template <size_t n>
unsigned int Prime<n>::operator () (unsigned int idx) const
{
    unsigned int i;
    if(idx == 0) return 2;
    for(i=0;idx && i<max_index;i++)
        if(bvec.test(i)) --idx;
    return 2*i + 1;
}
#endif
```

The constructor `Prime(unsigned int N)` specifies the upper limit N of the prime number sequence $\{p_k\}$ where $1 < p_k \leq N$, sets up the bit vector and turns all the bits on.

Suppose `p` is an instance of the class `Prime`, the six member functions work as follows

- `step()` finds out the next prime number, step by step, starting from 3 and crosses out all the multiples of that prime number.

- `run()` repeatedly executes the function `step()` until all the prime numbers less than N are found.

- `reset()` resets the prime number sequence to the beginning 2, 3.

- `is_prime(unsigned int num)` checks if `num` is a prime number. It works provided the bit table has already been built.

- `current_prime()` returns the current prime number being iterated by the function `step()`.

- The subscript operator `p(unsigned int M)` returns the M-th prime number on the sequence.

Let us look at a simple program that calculates the total number of primes below 100, 1000, 10000, 100000, 1000000 and displays the first 20 prime numbers. It also demonstrates the ability to access the i-th prime number.

```
// sprime.cpp

#include <iostream>
#include "prime.h"
using namespace std;

int main(void)
{
    const size_t max0 = 1000000;
    unsigned int i, j, count = 0;

    Prime<max0> p;      // specifies the upper limit of the sequence
    p.run();            // generates the prime number sequence

    // for all odd numbers greater than 3, check if they are prime
    for(i=3,j=100;i<=max0;i+=2)
    {
        if(p.is_prime(i)) count++;

        // sum the number of primes below 100, 1000, ..., 1000000
        if(i==j-1)
        {
            j *= 10;
            cout << "There are " << count + 1
                 << " primes below " << i+1 << endl;
        }
    }
    cout << endl;

    // print the first 20 primes
    for(i=0;i<20;i++) cout << p(i) << " ";
    cout << endl << endl;

    // randomly pick some primes
    cout << "The   100th prime is " <<   p(99)   << endl;
    cout << "The   200th prime is " <<   p(199)  << endl;
    cout << "The  3000th prime is " <<   p(2999) << endl;
    cout << "The 10000th prime is " <<   p(9999) << endl;
    cout << "The 12500th prime is " << p(12499) << endl;

    return 0;
}
```

9.2   Verylong Class

9.2.1   Big Prime Numbers

In this section, we make use of the Verylong class to test whether a large positive integer number is a *prime number*. A prime number is a positive integer $p > 1$ such that no other integer divides p except 1 and p.

The function is_prime(unsigned int num) which we described in Section 9.1 can be used for checking whether num is a prime. It works provided the bit table

268   CHAPTER 9.   APPLICATIONS

has already been built. For a large number greater than 2^{32}, the built-in type
unsigned int is too small. The Verylong integer can be used to overcome this
problem. However, the main problem with this algorithm is the huge space requirement for the bit table. For large numbers such as a 20-digit number, it becomes
highly impractical.

We describe another algorithm that tests the primality for a positive integer with
little memory requirement. It is an obvious algorithm for a primality test. For
a given positive integer N, we divide N by successive primes $p = 2, 3, 5, \ldots$ until
a smallest p for which $N \bmod p = 0$, then N is not prime and has a factor p. If
at any stage we find that $N \bmod p \neq 0$ but $\lfloor N/p \rfloor \leq p$, we conclude that N is prime.

Let us restate the arguments above:

Given a positive integer N, we divide N by a sequence of "trial divisors"

$$2 = d_0 < d_1 < d_2 < d_3 < \cdots < d_n$$

which includes all prime numbers $\leq \sqrt{N}$ (the sequence may also include some non-
prime numbers if it is convenient). The last divisor in the sequence d_n is the smallest
number $\geq \sqrt{N}$. If, at any stage, $N \bmod d_i = 0$ for $0 \leq i \leq n$, then N is not prime.
Otherwise, it is prime.

The trial divisors sequence $d_0, d_1, d_2, \ldots, d_n$ works best when it contains only prime
numbers. Sometimes it is convenient to include some non-prime numbers as well
because such a sequence is usually easier to generate. One good sequence can be
generated using the following recurrence relation

$$d_0 = 2, \quad d_1 = 3, \quad d_2 = 5,$$
$$d_k = d_{k-1} + 2 \quad \text{for odd } k > 2$$
$$d_k = d_{k-1} + 4 \quad \text{for even } k > 2.$$

The first few trial divisors generated using the relation above are as follows:

$$2 \;\; 3 \;\; 5 \;\; 7 \;\; 11 \;\; 13 \;\; 17 \;\; 19 \;\; 23 \;\; 25 \;\; 29 \;\; 31 \;\; 35 \;\; 37 \;\; 41 \;\; 43 \;\; \ldots$$

This sequence contains no multiples of 2 or 3 and the first 9 numbers are all prime.
However, it also includes some non-prime numbers such as 25, 35, 49, etc.

```
// isprime.cpp

#include <iostream>
#include "verylong.h"
using namespace std;

template <class T> int is_prime(T p)
{
    T j(2), zero(0), one(1), two(2), four(4);
    T limit = T(sqrt(p)) + one;

    if(j < limit && p%j == zero) return 0;
```

```
    j++;

    if(j < limit && p%j == zero) return 0;
    j += two;
    while(j < limit)
    {
        if(p%j == zero) return 0;
        j += two;
        if(p%j == zero) return 0;
        j += four;
    }
    return 1;
}

int main(void)
{
    Verylong x;
    cout << "Please enter a positive integer number : ";
    cin >> x;
    if(is_prime(x)) cout << "The number " << x << " is prime " << endl;
    else            cout << "The number " << x << " is not prime " << endl;
    return 0;
}
```

Although this is not the best way to check for primality, it illustrates how the Verylong class could be incorporated into number theory. Sometimes, it is good to have the prime numbers table (see Section 9.1) as part of the program. For example, if the table contains all the 78498 prime numbers less than one million, we could test the primality of N less than 10^{12}. Such a table could be built by the class Prime which we have discussed in Section 9.1.

There are no known efficiently methods to test for the primality of large numbers. However, there are some algorithms [32] that will perform the testing in a reasonable amount of time.

An interesting extension of the program would be to find all the prime twins, for example $21, 23$. Why are there no prime triplets other than $2, 3, 5$?

The numbers $2^k - 1$ ($k \in \mathbf{N}$) are called *Mersenne numbers*. For $k = 2, 3, 5, 7, 13$, $19, 31, 61, 89, 107, 127, 521, 607, 1279, 2203, 2281, 3217, 4253, 4423, 9689, 9941$, $11213, \ldots$, we obtain prime numbers. Big prime numbers also play an important role in data encryption and network security.

The prime number $2^k - 1$ ($k = 11213$) can be found using the following small program. It is a large number consisting of 2816 digits.

```
// mersenne.cpp

#include <iostream>
#include "verylong.h"
using namespace std;

int main(void)
```

```
{
    Verylong p, one("1"), two("2"), mersenne("11213");
    p = pow(two,mersenne) - one;
    cout << p << endl;
    return 0;
}
```

9.2.2   Gödel Numbering

We can work with an alphabet which contains only a single letter, e.g. the letter
|.   The words constructed from this alphabet (apart from the empty word) are:
|, ||, |||, etc. These words can, in a trivial way, be identified with the natural numbers 0, 1, 2, ... Such an extreme standardization of the "material" is advisable
for some considerations. On the other hand, it is often convenient to disperse the
diversity of an alphabet consisting of several elements.

The use of an alphabet consisting of *one element* does not imply any essential
limitation. We can associate the words W over an alphabet **A** consisting of N
elements with natural numbers $G(W)$, in such a way that each natural number is
associated with at most one word. Similar arguments apply to words of an alphabet
consisting of *one element*. Such a representation of G is called a *Gödel numbering*
[17] (also called *arithmetization*) and $G(W)$ is the *Gödel number* of the the word W
with respect to G. The following are the requirements for an arithmetization of W:

1. If $W_1 \neq W_2$ then $G(W_1) \neq G(W_2)$.

2. There exists an algorithm such that for any given word W, the corresponding
   natural number $G(W)$ can be computed in a finite number of steps.

3. For any natural number n, it can be decided whether n is the Gödel number
   of a word W over **A** in a finite number of steps.

4. There exists an algorithm such that if n is the Gödel number of a word W over
   A, then this word W (which is unique by argument (1)) can be constructed
   in a finite number of steps.

Here is an example of a Gödel numbering. Consider the alphabet with the letters
a, b, c. A word is constructed by any finite concatenation of these – that is, a
placement of these letters side by side in a line. For example, *abcbba* is a word. We
can then number the words as follows:

Given a word $x_1 x_2 \ldots x_n$ where each x_i is a, b or c, we assign to it the number

$$2^{d_0} * 3^{d_1} * \cdots * p_n^{d_n}$$

where p_i is the i-th prime number (and 2 is the 0-th prime) and

$$d_i := \begin{cases} 1 \text{ if } x_i \text{ is } a \\ 2 \text{ if } x_i \text{ is } b \\ 3 \text{ if } x_i \text{ is } c \end{cases} .$$

The empty word is given the number 0.

For example, the word *acbc* has number $2^1 * 3^3 * 5^2 * 7^3 = 463050$, and *abc* has the number $2^1 * 3^2 * 5^3 = 2250$. The number 7350 represents *aabb* because $7350 = 2^1 * 3^1 * 5^2 * 7^2$.

To show that this numbering satisfies the criteria given above, we use the *fundamental theorem of arithmetic*:

> *Any natural number ≥ 2 can be represented as a product of primes, and that product is, except for the order of the primes, unique.*

We may number all kinds of objects, not just alphabets. In general, the criteria for a numbering to be useful are:

1. No two objects have the same number.

2. Given any object, we can "effectively" find the number that corresponds to it.

3. Given any number, we can "effectively" find if it is assigned to an object and, if so, to which object.

We list the `Goedel` class as described above. It uses `Prime.h` implemented in Section 9.1:

```
// goedel.h

#ifndef GOEDEL_H
#define GOEDEL_H

#include <cassert>
#include <iostream>
#include <string>
#include "prime.h"
#include "verylong.h"
using namespace std;

const size_t max_prime = 100000;

class Goedel
{
private:
   // Data Fields
   Verylong nvalue;
   string    wvalue;
   int       is_G;

public:
   // Constructors
   Goedel();
   Goedel(string);
   Goedel(Verylong);

   // Member Functions
   Verylong number() const;
   string   word() const;
```

```cpp
   int      is_goedel() const;

   void rename(string);
   void resize(Verylong);
};

Goedel::Goedel()
   : nvalue(Verylong("0")), wvalue(string("")), is_G(1) {}

Goedel::Goedel(string s)
   : nvalue(Verylong("1")), wvalue(s), is_G(1)
{
   static Prime<max_prime> p;
   Verylong temp, prim, zero("0"), one("1");
   p.run();

   for(int i=0;unsigned(i)<s.length();i++)
   {
      if(s[i]>='a' && s[i]<='c')
      {
         temp = one;
         prim = Verylong(p(i));
         for(char j='a';j<=s[i];j++) temp *= prim;
         nvalue *= temp;
      }
      else { nvalue = zero; break; }
   }
}

Goedel::Goedel(Verylong num) : nvalue(num), wvalue(string(""))
{
   static Prime<max_prime> p;
   static Verylong zero("0"), one("1");
   Verylong prim;
   int i=0, factor;

   p.run();

   while(num>one)
   {
      factor = 0; prim = Verylong(p(i));
      while(num % prim == zero) { num /= prim; ++factor; }
      switch(factor)
      {
         case 1: wvalue = wvalue + string("a"); break;
         case 2: wvalue = wvalue + string("b"); break;
         case 3: wvalue = wvalue + string("c"); break;
         default : is_G = 0; return;
      }
      ++i;
   }
   is_G = 1;
}

Verylong Goedel::number() const { return nvalue; }

string Goedel::word() const
{
```

```
      if(is_G) return wvalue;
      return string("");
}

int Goedel::is_goedel() const { return is_G; }

void Goedel::rename(string s)
{
   static Prime<max_prime> p;
   static Verylong zero("0"), one("1");
   Verylong temp, prim;
   p.run();
   nvalue = one;
   wvalue = s;
   is_G = 1;
   for(int i=0;unsigned(i)<s.length();i++)
   {
      if(s[i]>='a' && s[i]<='c')
      {
         temp = one;
         prim = Verylong(p(i));
         for(char j='a';j<=s[i];j++) temp *= prim;
         nvalue *= temp;
      }
      else { nvalue = zero; break; }
   }
}

void Goedel::resize(Verylong num)
{
   static Verylong zero("0"), one("1");
   assert(num >= zero);
   if(num == one) { is_G = 0; return; }
   static Prime<max_prime> p;
   int i=0, factor;
   Verylong prim;
   p.run();
   nvalue = num; wvalue = string("");
   while(num>one)
   {
      factor = 0; prim = Verylong(p(i));
      while(num % prim == zero) { num /= prim; ++factor; }
      switch(factor)
      {
         case 1: wvalue = wvalue + string("a"); break;
         case 2: wvalue = wvalue + string("b"); break;
         case 3: wvalue = wvalue + string("c"); break;
         default : is_G = 0; return;
      }
      ++i;
   }
   is_G = 1;
}
#endif
```

Next we provide a program which calculates the corresponding Gödel number of
acbc and aabb. It also lists all the Gödel numbers below 1500:

```cpp
// sgoedel1.cpp

#include <iostream>
#include "goedel.h"
using namespace std;

int main(void)
{
   Goedel g("acbc");

   cout << "The word is " << g.word();
   cout << " and its corresponding Goedel number is "
        << g.number() << endl;

   g.rename("aabb");

   cout << "The word is " << g.word();
   cout << " and its corresponding Goedel number is "
        << g.number() << endl;

   g.rename("abcbc");

   cout << "The word is " << g.word();
   cout << " and its corresponding Goedel number is "
        << g.number() << endl;
   cout << endl;

   Goedel h;

   // List all the Goedel Numbers below 1500
   for(int i=0;i<1500;i++)
   {
      h.resize(i);
      if(h.is_goedel())
          cout << i << " => " << h.word()
               << " is a Goedel number" << endl;
   }
   return 0;
}
```

Let us consider another example. It calculates the Gödel numbers for the strings:
a, aa, aaa, aaaa, ... Some of the Gödel numbers exceed the limit of the built-in
unsigned long type.

```cpp
// sgoedel2.cpp

#include <iostream>
#include <string>
#include "goedel.h"
using namespace std;

int main(void)
{
   string s;
   Goedel g;
```

```
for(int i=1;i<=20;i++)
{
    s = s + 'a';
    g.rename(s);
    cout << i << "   " << g.word() << " => " << g.number() << endl;
}
return 0;
}
```

9.2.3 Inverse Map and Denumerable Set

The set $\mathbf{N} \times \mathbf{N}$ is *denumerable* because it is equipotent to the natural numbers $\mathbf{N}$. In other words, there exists a 1-1 map between $\mathbf{N}$ and $\mathbf{N} \times \mathbf{N}$.

Let us write the elements of $\mathbf{N} \times \mathbf{N}$ in the form of an array as follows

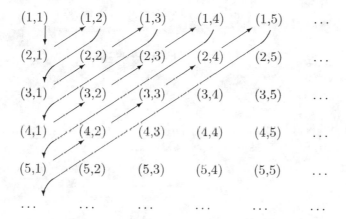

From the figure we see that we could arrange the elements of $\mathbf{N} \times \mathbf{N}$ into a linear sequence as indicated by the arrows, i.e.

$$(1,1),\ (2,1),\ (1,2),\ (3,1),\ (2,2),\ (1,3),\ (4,1),\ (3,2),\ \ldots$$

Therefore a 1-1 map between $\mathbf{N}$ and $\mathbf{N} \times \mathbf{N}$ exists. If $(m,n) \in \mathbf{N} \times \mathbf{N}$ we have

$$f(m,n) = \frac{1}{2}(m+n-1)(m+n-2) + n. \qquad (9.1)$$

The above relation can be obtained as follows. The pair (m,n) lies in the $(m+n-1)$-th diagonal stripe in the above figure and it is the n-th pair counting from the left of the stripe. The first stripe contains only one pair. In the second, there are two pairs and so on. Thus (m,n) is located at the position numbered by

$$n + \sum_{k=1}^{m+n-2} k$$

in the counting procedure. Thus (9.1) follows. How could we find the inverse of the problem? This means to find m and n if $f(m, n)$ is given.

To find the inverse map, we first have to find out in which diagonal stripe $f(m, n)$ lies. Note that $m = 1$ corresponds to the last element on the diagonal stripe. Let $k = m + n - 1$, thus for $m = 1$, $k = 1 + n - 1 = n$. By (9.1),

$$f(1, n) = \frac{1}{2}(n)(-1 + n) + n = \frac{n^2 + n}{2} = \frac{k^2 + k}{2}.$$

The last element in the k-th diagonal is given by

$$f(m, n) = \frac{k^2 + k}{2}$$

whereas the last element in the $(k - 1)$-th diagonal is given by

$$f(m, n) = \frac{(k - 1)^2 + (k - 1)}{2} = \frac{k^2 - k}{2}.$$

Thus, if $f(m, n)$ lies in the k-th diagonal stripe, then

$$\frac{k^2 - k}{2} < f(m, n) \le \frac{k^2 + k}{2}. \tag{9.2}$$

Let $x \in \mathbf{R}$ such that

$$\frac{x^2 + x}{2} = f(m, n) \tag{9.3}$$

then

$$x^2 + x - 2f(m, n) = 0 \quad \text{and} \quad x = \frac{-1 + \sqrt{1 + 8f(m, n)}}{2}$$

where we obviously choose the positive root. Using (9.2) and (9.3), we get $k - 1 < x \le k$. Thus, k is the smallest positive integer number greater than or equal to x. After we have obtained k, the values of m and n can readily be calculated by (9.1):

$$n = f(m, n) - \frac{k(k - 1)}{2}, \quad m = k - n + 1.$$

We implemented the inverse map using the algorithm described above. The program makes use of the **Verylong** class to handle very long integers. The **sqrt()** function used here is implemented in the **Verylong** class. It evaluates the integer square root of a **Verylong** number. An extension would be to find a 1-1 map between $\mathbf{N}$ and $\mathbf{N} \times \mathbf{N} \times \mathbf{N}$.

```cpp
// inverse.cpp

#include <iostream>
#include "verylong.h"
using namespace std;

void InverseMap(Verylong f,Verylong &m,Verylong &n)
```

```
{
    Verylong zero(0), one(1), two(2), eight(8), k;

    k = (sqrt(one+eight*f)-one)/two + one;
    n = f - k*(k-one)/two;
    if(n==zero) {--k; n += k;}
    m = k-n+one;
}

int main(void)
{
    Verylong one(1), two(2);
    Verylong f, m, n, result;

    cout << "Enter a number : ";
    cin  >> f;

    InverseMap(f,m,n);

    cout << "The corresponding m is " << m << " and n is " << n << endl;
    return 0;
}
```

9.3 Verylong and Rational Classes

9.3.1 Logistic Map

The *logistic map* $f : [0,1] \to [0,1]$ is given by

$$f(x) = 4x(1 - x).$$

It can also be written as the difference equation

$$x_{t+1} = 4x_t(1 - x_t), \qquad t = 0, 1, 2, \ldots, \quad x_0 \in [0,1]$$

where $x_t \in [0,1]$ for all $t \in \mathbf{N} \cup \{0\}$. Let $x_0 = 1/3$ be the initial value. Then

$$x_1 = \frac{8}{9}, \quad x_2 = \frac{32}{81}, \quad x_3 = \frac{6272}{6561}, \quad x_4 = \frac{7250432}{43046721}, \quad \ldots$$

This is a so-called *chaotic orbit* [53]. The exact solution of the logistic map is given by

$$x_t = \frac{1}{2} - \frac{1}{2}\cos(2^t \arccos(1 - 2x_0)).$$

Depending on the data type of x_0, we may get an approximate or exact orbit of the map. For an approximate orbit, we use the built-in data type double for x_0, whereas for the exact orbit of the logistic map, we use the Rational class that we have developed in Chapter 7.

We demonstrate the features of the Rational class as well as the coupling with the Verylong class. We compare Rational<long> and Rational<Verylong> in the iteration of the map.

We notice that the lengths of the numerator and denominator increase so fast that the data type `Rational<long>` could only hold the result up to the fourth iteration. Thus we have to use the data type `Rational<Verylong>`.

Quite often long fractions do not give a good idea of how large the value is; the floating point representation may be better. The conversion operator to `double` in the class provides such a function.

```cpp
// logistic.cpp

#include "rational.h"
#include "verylong.h"
using namespace std;

int main(void)
{
   Rational<int> a(4,1), b(1,1), x0(1,3); // initial value x0 = 1/3
   int i;

   cout << "x[0] = " << x0 << " or " << double(x0) << endl;

   for(i=1;i<=4;i++)      // cannot use higher values than 4
   {                      // out of range for data type int
      x0 = a*x0*(b - x0);
      cout << "x[" << i << "] = " << x0 << " or " << double(x0) << endl;
   }
   cout << endl;

   Rational<Verylong> c1("1"), c2("4/1"),
                 y0("1/3");        // initial value y0 = 1/3

   cout << "y[0] = " << y0 << " or " << double(y0) << endl;

   for(i=1;i<=6;i++)
   {
      y0 = c2*y0*(c1 - y0);
      cout << "y[" << i << "] = " << y0 << " or " << double(y0) << endl;
   }

   return 0;
}
```

9.3.2   Contracting Mapping Theorem

Let S be a closed set on a complete metric space X. A *contracting mapping* is a mapping $f : S \to S$ such that

$$d(f(x), f(y)) \le k d(x, y), \quad 0 \le k < 1, \quad d \text{ is the distance in } X.$$

One also says that "f is *lipschitzian* of order $k < 1$".

Contracting mapping theorem. A contracting mapping f has strictly one *fixed point*; i.e. there is one and only one point x^* such that

$$x^* = f(x^*).$$

The proof is by successive iteration. Let $x^* \in S$, then

$$f(x_0) \in S, \ldots, f^{(n)}(x_0) = f(f^{(n-1)}(x_0)) \in S$$

and

$$d(f^{(n)}(x_0), f^{(n-1)}(x_0)) \leq kd(f^{(n-1)}(x_0), f^{(n-2)}(x_0)) \leq \ldots \leq k^{n-1}d(f(x_0), x_0).$$

Since $k < 1$ the sequence $f^{(n)}(x_0)$ is a *Cauchy sequence* and it tends to a limit $x^* \in S$ when n tends to infinity

$$x^* = \lim_{n \to \infty} f^{(n)}(x_0) = \lim_{n \to \infty} f(f^{(n-1)}(x_0)) = f(x^*).$$

The uniqueness of x^* results from the defining property of contracting mappings. Assume that there is another point y^* such that $y^* = f(y^*)$, then

$$d(f(y^*), f(x^*)) = d(y^*, x^*).$$

On the other hand $d(f(y^*), f(x^*)) \leq kd(y^*, x^*)$, where $k < 1$. Hence, $d(y^*, x^*) = 0$ and $y^* = x^*$.

In the following program, we consider the map $f : [0, 1] \to [0, 1]$

$$f(x) = \frac{5}{2}x(1 - x).$$

Obviously the map f has a stable fixed point $x^* = 3/5$. We apply the contracting mapping theorem with the initial value $x_0 = 9/10$.

A brief description of the program is given below:

- The constructor of the program reads in two arguments. The first argument is the function name of the mapping while the second argument is the initial value.

- The increment operator (++) applies the mapping on the current value to obtain the next iteration value.

- The function Is_FP() checks if the values have converged to a fixed point by considering the relative difference between the previous and current values. If the relative difference is less than 10^{-5}, we claim that a fixed point has been obtained.

From the result obtained, we notice that the numerical value of the mapping converges to 0.6, which agrees with our discussion. However, the rational values (exact values) of the iteration grow very quickly.

```cpp
// contract.cpp

#include <iostream>
#include "rational.h"
#include "verylong.h"
using namespace std;

class Map
{
private:
   Rational<Verylong> (*function)(Rational<Verylong>);
   Rational<Verylong> value;          // current iterated value

public:
   Map(Rational<Verylong> (*f)(Rational<Verylong>), Rational<Verylong>);
   void operator ++ ();               // next iteration
   void Is_FP();                      // Is there a fixed point?
};

Map::Map(Rational<Verylong> (*f)(Rational<Verylong>),
         Rational<Verylong> x0) : function(f), value(x0) {}

void Map::operator ++ () {  value = (*function)(value); }

// Is there a fixed point?
void Map::Is_FP()
{
   Rational<Verylong> temp,  // the value in previous step
                      dist;  // the relative difference
                             // between previous and current step

   do
   {
     temp = value;
     ++(*this);
     dist = abs((value-temp)/temp); // the relative difference
     cout << "Value : " << value << " or " << double(value) << endl;
     cout << endl;
   } while(double(dist) > 1e-5);
}

// f(x) = 5/2 x(1-x)
Rational<Verylong> mapping(Rational<Verylong> x)
{
   return (Rational<Verylong>("5/2")*x*(Rational<Verylong>("1")-x));
}

int main(void)
{
   // initial value = 9/10
   Map M(mapping, Rational<Verylong>(Verylong("9"),Verylong("10")));
   M.Is_FP();
   return 0;
}
```

9.3.3   Ghost Solutions

Consider the nonlinear ordinary differential equation

$$\frac{du}{dt} = u(1 - u)$$

with initial condition $u(0) = u_0 > 0$. The fixed points are given by

$$u^* = 0, \quad u^* = 1 \,.$$

The fixed point $u^* = 1$ is asymptotically stable. The exact solution of the differential equation is given by

$$u(t) = \frac{u_0 e^t}{1 - u_0 + u_0 e^t} \quad \text{for} \ \ 0 < u_0 < 1 \,.$$

The exact solution starting from initial value $u_0 = 99/100 = 0.99$ is monotonously increasing and it converges to 1 as t tends to $+\infty$. For $t = \ln 99 \approx 4.5951$, we find that

$$u(t) = 0.9999 \,.$$

So, $u(t)$ is already quite near the asymptotically stable fixed point $u^* = 1$. In order to integrate this equation by a finite-difference scheme, we apply the *central difference scheme*

$$\frac{u_{n+1} - u_{n-1}}{2h} = u_n(1 - u_n) \qquad n = 1, 2, 3, \ldots \tag{9.4}$$

with initial conditions $u_0 = x_0$, $u_1 = x_0 + hx_0(1 - x_0)$. This second order difference equation can be rewritten as a system of two first order difference equations

$$v_{n+1} = u_n$$
$$u_{n+1} = v_n + 2hu_n(1 - u_n) \quad \text{for} \ \ n = 1, 2, 3, \ldots$$

with $v_1 = u_0 = x_0$ and $u_1 = x_0 + hx_0(1 - x_0)$. We iterate the system of nonlinear difference equations using the **Rational** and **Verylong** class with initial value $x_0 = 99/100$ and time-mesh length $h = 0.1$. This system does not converge to the fixed point $u^* = 1$ and we find oscillating behavior. Such a phenomenon is called the *ghost solution* (also called *spurious solution*).

```
// ghost.cpp

#include <fstream>
#include <math.h>
#include "rational.h"
#include "verylong.h"
using namespace std;

const Rational<Verylong> h("1/10");      // time-mesh length
const Rational<Verylong> x0("99/100"); // initial value
```

```
int main(void)
{
    Rational<Verylong> u, v, u1, twoh, t,
                    zero("0"), one("1"), two("2"), fifty("50");
    ofstream sout("ghosts.dat"); // contain the rational number values

    // initial values
    u = x0; v = u + h*u*(one-u);
    t = zero; twoh = two*h;

    while(t <= fifty)
    {
        u1 = u; u = v;
        v = u1 + twoh*u*(one-u);
        t += h;
        sout << t << " " << v << endl;
    }

    return 0;
}
```

For $0 < t < 1.4$, the numerical solution gives a good approximation to the true solution. The solution u_n increases monotonously approaching 1. After $t = 1.5$, the numerical solution is no longer monotonous. At $t = 3.0$, the value u_n becomes greater than 1 for the first time and the solution begins to oscillate thereafter. The amplitude of the oscillation grows larger and larger. The growth of this amplitude is geometric and the rate of growth is such that the amplitude is multiplied by about $e \approx 2.71828$ for each unit t increment. When $t = 10$, the oscillation loses its symmetry with respect to $u^* = 1$. The repetition of such cycles seems to be nearly periodic. The ghost solutions also appear even if h is quite small. One of the reasons for this phenomenon is that the central difference scheme is a second order difference scheme and that the instability enters at $u^* = 1$ and $u^* = 0$. The global behavior of the solution computed is very sensitive to the initial condition and the time-mesh length. To show that the behavior of (9.4) is not caused by the finite precision used in a digital computer we iterate the equations using exact arithmetic (`ghost.cpp`) to avoid rounding-off error. The `Rational` and `Verylong` class have been employed. We see that the oscillating behavior exists even for the exact arithmetics. The behavior is an inherited property of the difference equation.

9.3.4   Iterated Function Systems

A *hyperbolic iterated function* [7] system consists of a complete metric space $(\mathbf{X}, d)$ together with a finite set of contraction mappings $w_n : \mathbf{X} \to \mathbf{X}$, with *contractivity factors* s_n, for $n = 1, 2, \ldots, N$. The notation for the iterated function system is $\{\, \mathbf{X}\,;\, w_n,\ n = 1, 2, \ldots, N \,\}$ and its contractivity factor is

$$s = \max\{\, s_n : n = 1, 2, \ldots, N \,\}.$$

Let $(\mathbf{X}, d)$ be a complete metric space and $(\mathcal{H}(\mathbf{X}), h(d))$ denotes the corresponding space of non-empty compact subsets, with the Hausdorff metric $h(d)$. The following theorem summarizes the facts about a hyperbolic iterated function system:

Theorem. Let

$$\{\,\mathbf{X}\,;\,w_n,\ n=1,2,\ldots,N\,\}$$

be a hyperbolic iterated function system with contractivity factor s, then the transformation $W : \mathcal{H}(\mathbf{X}) \to \mathcal{H}(\mathbf{X})$ defined by

$$W(B) := \bigcup_{n=1}^{N} w_n(B)$$

for all $B \in \mathcal{H}(\mathbf{X})$, is a contraction mapping on the complete metric space $(\mathcal{H}(\mathbf{X}), h(d))$ with contractivity factor s, i.e.

$$h(W(B), W(C)) \le s \cdot h(B, C)$$

for all $B, C \in \mathcal{H}(\mathbf{X})$. Its unique fixed point, $A \in \mathcal{H}(\mathbf{X})$, obeys

$$A := W(A) = \bigcup_{n=1}^{N} w_n(A)$$

and is given by

$$A := \lim_{n \to \infty} W^{(n)}(B)$$

for any $B \in \mathcal{H}(\mathbf{X})$. The fixed point $A \in \mathcal{H}(\mathbf{X})$ is called the attractor of the iterated function system. For the proof we refer to the excellent book of Barnsley [7].

Next, we present an example in fractals – the standard *Cantor set*.

The standard Cantor set is described by the following iterated function system

$$\{\,\mathbf{R} : w_1, w_2\,\} \quad \text{where} \quad w_1(x) := \frac{1}{3}x \quad \text{and} \quad w_2(x) := \frac{1}{3}x + \frac{2}{3}\,.$$

This is an iterated function system with contractivity factor $s = 1/3$. Suppose

$$B_0 = [0, 1] \quad \text{and} \quad B_n = W^{(n)}(B_0)\,.$$

Then $B = \lim_{n \to \infty} B_n$ is the standard *Cantor set*. It is also called the Cantor middle third set or ternary Cantor set. The sets B_0, B_1, B_2, ... are given by

$$B_0 = [0, 1]$$
$$B_1 = \left[0, \frac{1}{3}\right] \cup \left[\frac{2}{3}, 1\right]$$
$$B_2 = \left[0, \frac{1}{9}\right] \cup \left[\frac{2}{9}, \frac{1}{3}\right] \cup \left[\frac{2}{3}, \frac{7}{9}\right] \cup \left[\frac{8}{9}, 1\right]$$
$$\vdots \qquad\qquad \vdots$$

This means we remove the open middle third, i.e. the interval $(\frac{1}{3}, \frac{2}{3})$ for the first step and remove the pair of intervals $(\frac{1}{9}, \frac{2}{9})$ and $(\frac{7}{9}, \frac{8}{9})$ in the second step. Continuing to remove the middle thirds in this fashion, we arrive at the Cantor set as

$n \to \infty$.

The standard Cantor set has cardinal c, is perfect, nowhere dense and has Lebesgue measure zero. Every $x \in B$ can be written as

$$x = a_1 3^{-1} + a_2 3^{-2} + a_3 3^{-3} + \cdots \qquad \text{where} \quad a_j \in \{0, 2\}.$$

The corresponding Cantor function is called the *Devil's staircase*.

Next, we iterate the Cantor set B_n for $n = 1, 2, 3, 4, 5, 6$ with the `Verylong` and `Rational` class:

```cpp
// cantor.cpp

#include <iostream>
#include "rational.h"
#include "verylong.h"
#include "vector.h"
using namespace std;

const Rational<Verylong> a = Rational<Verylong>("0"); // lower limit
const Rational<Verylong> b = Rational<Verylong>("1"); // upper limit

class Cantor
{
  private:
    Vector<Rational<Verylong> > CS;
    int currentSize;
  public:
    Cantor(int);
    Cantor(const Cantor&);  // copy constructor
    int step();
    void run();
    friend ostream& operator << (ostream&,const Cantor&);
};

Cantor::Cantor(int numStep) : CS((int)pow(2.0,numStep+1)), currentSize(2)
   { CS[0] = a; CS[1] = b; }

Cantor::Cantor(const Cantor& s)
   : CS(s.CS), currentSize(s.currentSize) { }

int Cantor::step()
{
  int i, newSize;
  static Rational<Verylong> three(3), tt(2,3);
  static int maxSize = CS.size();

  if(currentSize < maxSize)
  {
  for(i=0;i<currentSize;i++) CS[i] /= three;
  newSize = currentSize + currentSize;
  for(i=currentSize;i<newSize;i++) CS[i] = CS[i-currentSize] + tt;
  currentSize = newSize;
  return 1;
```

```
    }
    return 0;
}

void Cantor::run() {  while(step() != 0); }

ostream& operator << (ostream& s,const Cantor& c)
{
    for(int i=0;i<c.currentSize;i+=2)
    {
    s << "[" << c.CS[i] << " ";
    s << c.CS[i+1] << "] ";
    }
    return s;
}

int main(void)
{
    const int N = 6;
    Cantor C(N);
    cout << C << endl;
    for(int i=0;i<N;i++) { C.step(); cout << C << endl; }
    return 0;
}
```

9.4   Verylong, Rational and Derive Classes

9.4.1   Logistic Map and Ljapunov Exponent

We calculate the *Ljapunov exponent* of the *logistic map* [53] which is given by

$$x_{t+1} = 4x_t(1 - x_t)$$

where $t = 0, 1, 2, \ldots$ and $x_0 \in [0, 1]$. The *variational equation* of the logistic map (also called the *linearized equation*) is defined by

$$y_{t+1} = \frac{df(x_t)}{dx} y_t$$

where

$$f(x) = 4x(1 - x).$$

Since

$$\frac{df}{dx} = 4 - 8x$$

it follows that

$$y_{t+1} = (4 - 8x_t)y_t$$

with $y_0 \neq 0$. The *Ljapunov exponent* λ is defined as

$$\lambda(x_0, y_0) := \lim_{T \to \infty} \frac{1}{T} \ln \left| \frac{y_T}{y_0} \right|, \qquad y_0 \neq 0.$$

For almost all initial values we find that the Ljapunov exponent is given by

$$\lambda = \ln 2 \,.$$

We iterate the logistic map and the variational equation to find the Ljapunov exponent. The program makes use of the `Verylong`, `Rational` and `Derive` classes to approximate the Ljapunov exponent. The `Derive` class will do the differentiation. Thus the variational equation is obtained via exact differentiation.

```cpp
// ljapunov.cpp
// (1) Iteration of logistic equation and variational equation
// (2) Variational equation obtained via exact differentiation
// (3) Determination of Ljapunov exponent

#include <iostream>
#include <math.h>
#include "verylong.h"
#include "rational.h"
#include "derive.h"
using namespace std;

int main(void)
{
   int N = 100;
   double x1, x = 1.0/3.0;
   double y = 1.0;
   Derive<double> C1(1.0);   // constant 1.0
   Derive<double> C4(4.0);   // constant 4.0
   Derive<double> X;

   cout << "i = 0   x = " << x << "    " << "y = " << y << endl;
   for(int i=1;i<=N;i++)
   {
      x1 = x;
      x = 4.0*x1*(1.0 - x1);

      X.set(x1);
      Derive<double> Y = C4*X*(C1 - X);

      y = df(Y)*y;
      cout << "i = " << i << "    "
           << "x = " << x << "    " << "y = " << y << endl;
   }
   double lam = log(fabs(y))/N;
   cout << "Approximation value for lambda = " << lam << endl;
   cout << endl;

   int M = 9;
   Rational<Verylong> u1;
   Rational<Verylong> u("1/3"), v("1");
   Rational<Verylong> K1("1");
   Rational<Verylong> K2("4");
   Derive<Rational<Verylong> > D1(K1); // constant 1
   Derive<Rational<Verylong> > D4(K2); // constant 4
   Derive<Rational<Verylong> > U;

   cout << "j = 0   u = " << u << "    " << "v = " << v << endl;
```

```
for(int j=1;j<=M;j++)
{
   u1 = u;
   u = K2*u1*(K1 - u1);
   U.set(Rational<Verylong>(u1));
   Derive<Rational<Verylong> > V = D4*U*(D1 - U);

   v = df(V)*v;
   cout << "j = " << j << "   "
        << "u = " << u << "    " << "v = " << v << endl;
}
lam = log(fabs(double(v)))/M;
cout << "Approximation value for lambda = " << lam << endl;
return 0;
}
```

9.5 Verylong, Rational and Complex Classes

9.5.1 Mandelbrot Set

Suppose $\mathbf{C}$ is the complex plane, then the *Mandelbrot set* M is defined by

$$M := \left\{ c \in \mathbf{C} \ : \ c, \ c^2 + c, \ (c^2 + c)^2 + c, \ \ldots \not\to \infty \right\}.$$

To find the Mandelbrot set we study the recursion relation

$$z_{t+1} = z_t^2 + c, \qquad t = 0, 1, 2, \ldots$$

with initial value $z_0 = 0$. Since $z := x + iy$ and $c := c_1 + ic_2$ with x, y, c_1, $c_2 \subset \mathbf{R}$, we can write the recursion relation as

$$x_{t+1} = x_t^2 - y_t^2 + c_1$$
$$y_{t+1} = 2x_t y_t + c_2$$

with the initial value $(x_0, y_0) = (0, 0)$. Let us consider some points in the complex plane $\mathbf{C}$ and determine whether they belong to the Mandelbrot set M:

- $(c_1, c_2) = (0, 0)$ belongs to M

- $(c_1, c_2) = (\frac{1}{4}, \frac{1}{4})$ belongs to M

- $(c_1, c_2) = (\frac{1}{2}, 0)$ does not belong to M, because

$$z_0 = 0, \ \ z_1 = \frac{1}{2}, \ \ z_2 = \frac{3}{4}, \ \ z_3 = \frac{17}{16}, \ \ \ldots \to \infty.$$

We iterate the recursion relation with the starting point

$$z = \frac{1}{8} + i\frac{1}{3}.$$

Here, we have coupled three classes: the `Verylong`, `Rational` and `complex` classes. We wrap up this section by comparing the virtues of the two data types:

complex<Rational<long> > and complex<Rational<Verylong> >.

It is obvious that Rational<long> has limited application since it is restricted by the precision of the built-in data type long. However, Rational<Verylong> removes this restriction and it can be applied to problems that require multi-precision.

```cpp
// mandel.cpp

#include <iostream>
#include <complex>
#include "rational.h"
#include "verylong.h"
using namespace std;

int main(void)
{
   complex<Rational<long> > c(Rational<long>(1,8),Rational<long>(1,3)),
                     z0(Rational<long>(0,1),Rational<long>(0,1));

   cout << "Using data type Rational<long>" << endl;
   for(int i=1;i<=3;i++)
   {
      z0 = z0*z0 + c;
      cout << "z[" << i << "] = " << z0 << endl;
   }
   cout << endl;

   complex<Rational<Verylong> >
      d(Rational<Verylong>(1,8),Rational<Verylong>(1,3)),
      w0(Rational<Verylong>(0,1),Rational<Verylong>(0,1));

   cout << "Using data type Rational<Verylong>" << endl;
   for(int j=1;j<=5;j++)
   {
      w0 = w0*w0 + d;
      cout << "w[" << j << "] = " << w0 << endl;
   }
   return 0;
}
```

9.6   Symbolic Class

9.6.1   Polynomials

Legendre Polynomials

The *Legendre differential equation* is given by

$$\left((1-x^2)\frac{d^2}{dx^2} - 2x\frac{d}{dx} + n(n+1) \right) P_n(x) = 0 \qquad \text{where } n \text{ is a constant.}$$

This equation arises in the solution of problems in mechanics, quantum mechanics, etc. The solution to this differential equation is called the *Legendre polynomials* which can be represented by *Rodrigue's formula*

$$P_0(x) := 1, \quad P_n(x) := \frac{1}{2^n n!} \frac{d^n}{dx^n}(x^2 - 1)^n, \quad n = 1, 2, \ldots$$

The above equation can be rewritten as

$$P_n(x) = \frac{2^n}{n!} \frac{d^n}{dx^n}\left(\left(\frac{x-1}{2}\right)^n \left(\frac{x+1}{2}\right)^n\right).$$

Now, if we apply the well-known *Leibniz rule* for differentiating products, we get

$$P_n(x) = \sum_{k=0}^{n} \binom{n}{k}^2 \left(\frac{x-1}{2}\right)^{n-k} \left(\frac{x+1}{2}\right)^{k}.$$

The Legendre polynomials form a set of orthonormal functions on $(-1, 1)$, that is

$$\int_{-1}^{+1} P_m(x) P_n(x) dx = \frac{2}{2n+1} \delta_{mn}.$$

Furthermore, we have the recursion relation

$$(n + 1)P_{n+1}(x) = (2n + 1)xP_n(x) - nP_{n-1}(x), \quad n = 1, 2, \ldots$$

with $P_0(x) = 1$ and $P_1(x) = x$. The *generating function* for the Legendre polynomials is given by

$$\frac{1}{\sqrt{1 - 2tx + t^2}} = \sum_{n=0}^{\infty} P_n(x)t^n.$$

In the program we make use of the recursion relation to generate the first few Legendre polynomials

$$P_0(x) = 1, \quad P_1(x) = x, \quad P_2(x) = \frac{1}{2}(3x^2 - 1),$$

$$P_3(x) = \frac{1}{2}(5x^3 - 3x), \quad P_4(x) = \frac{1}{8}(35x^4 - 30x^2 + 3).$$

We also show that the Legendre differential equation holds for $n = 4$.

```
// legendre.h
// Recursion formula (n+1)P_{n+1}(x) = (2n+1)xP_n(x)-nP_{n-1}(x)

#include <iostream>
#include <assert.h>
#include "symbolicc++.h"
using namespace std;

class Legendre
{
 private:
   int maxTerm, currentStep;
   Symbolic P, Q;
   const Symbolic &x;
```

```
 public:
   Legendre(int,const Symbolic&);
   int step();
   void run();
   void reset();
   Symbolic current() const;
   Symbolic operator () (int);
   friend ostream & operator << (ostream &,const Legendre&);
};

Legendre::Legendre(int n, const Symbolic &kernal)
   : maxTerm(n), currentStep(0), P(1), x(kernal) {}

int Legendre::step()
{
   int prev = currentStep;
   Symbolic R, one(1);
   ++currentStep;
   if(currentStep==1) { Q = one; P = x; return 1; }
   if(currentStep <= maxTerm)
   {
      R = (prev+currentStep)*x*P/currentStep-prev*Q/currentStep;
      Q = P; P = R;
      return 1;
   }
   return 0;
}

void Legendre::run() { while(step()); }

void Legendre::reset() { currentStep = 0; P = 1; }

Symbolic Legendre::current() const { return P; }

Symbolic Legendre::operator () (int m)
{
   assert(m <= maxTerm);
   reset();
   for(int i=0;i<m;i++) step();
   return P;
}

ostream & operator << (ostream &s,const Legendre &L)
{ return s << L.P; }
```

```
// legendre.cpp
// Recursion formula (n+1)P_{n+1}(x) = (2n+1)xP_n(x)-nP_{n-1}(x)

#include <iostream>
#include "symbolicc++.h"
#include "legendre.h"
using namespace std;

int main(void)
{
   int n=4;
   Symbolic x("x");
```

```
Legendre P(n,x);

// Calculate the first few Legendre polynomials
cout << "P(0) = " << P << endl;
for(int i=1;i<=n;i++)
{
    P.step();
    cout << "P("<< i << ") = " << P << endl;
}
cout << endl;

// Another way to access the Legendre polynomial
cout << "P(1) = " << P(1) << endl;
cout << "P(2) = " << P(2) << endl;
cout << "P(3) = " << P(3) << endl;
cout << "P(4) = " << P(4) << endl; cout << endl;

// Show that the Legendre differential equation is satisfied for n = 4
Symbolic result;
result = df((1-x*x)*df(P.current(),x),x) + (n*(n+1))*P.current();
cout << result << endl;  // ==> 0
return 0;
}
```

Associated Legendre Functions

Having solved the Legendre differential equation, we now obtain the solution of

$$\left((1-x^2)\frac{d^2}{dx^2} - 2x\frac{d}{dx} + n(n+1) - \frac{m^2}{1-x^2} \right) P_n^m(x) = 0$$

where m, n are constants and m is not necessarily equal to zero. The solution to this equation $P_n^{|m|}(x)$ is called the *associated Legendre functions* of degree n ($n = 0, 1, 2, \ldots$) and order $|m| \leq n$. It is defined by the relation

$$P_n^{|m|}(x) := (1-x^2)^{|m|/2} \frac{d^{|m|}}{dx^{|m|}} P_n(x), \qquad |m| = 0, 1, 2, \ldots, \leq n. \qquad (9.5)$$

For $m = 0$, we have $P_n^0(x) = P_n(x)$ which are the Legendre polynomials. The functions $P_n^{|m|}$ satisfy the recurrence relation

$$(2n+1)x P_n^{|m|}(x) = (n - |m| + 1)P_{n+1}^{|m|}(x) + (n + |m|)P_{n-1}^{|m|}(x)$$

and the orthogonality relation

$$\int_{-1}^{+1} P_n^{|m|}(x) P_{n'}^{|m|}(x) dx = \frac{2}{2n+1} \frac{(n+|m|)!}{(n-|m|)!} \delta_{nn'}.$$

Finally, it can be shown that the $P_n^{|m|}$ form a complete set in the Hilbert space $L_2(-1, 1)$. The first few associated Legendre functions are given by

$$P_1^1(x) = (1-x^2)^{1/2}, \quad P_2^1(x) = 3x(1-x^2)^{1/2}, \quad P_2^2(x) = 3(1-x^2),$$

$$P_3^1(x) = \frac{3}{2}(1 - x^2)^{1/2}(5x^2 - 1), \quad P_3^2(x) = 15x(1 - x^2), \quad P_3^3(x) = 15(1 - x^2)^{3/2}.$$

In the following program, we make use of (9.5) and the Legendre polynomials developed in the previous section to construct the associated Legendre polynomials.

```cpp
// asslegendre.h
// Associate Legendre Polynomial
// P_l^|m|(w) = (1-w^2)^(|m|/2) (d/dw)^|m| P_l(w)
// Recursion formula (n+1)P_{n+1}(x) = (2n+1)xP_n(x)-nP_{n-1}(x)

#include <iostream>
#include "symbolicc++.h"
#include "legendre.h"
using namespace std;

class AssLegendre
{
private:
   Symbolic P;
   const Symbolic &x;

public:
   AssLegendre(const Symbolic &);
   AssLegendre(int,int,const Symbolic &);

   void redefine(int,int);
   Symbolic current() const;

   friend ostream &operator << (ostream &,const AssLegendre &);
};

AssLegendre::AssLegendre(const Symbolic &kernal) : x(kernal) {}

AssLegendre::AssLegendre(int l,int m,const Symbolic &kernal) : x(kernal)
{
   int i, absm = abs(m);
   Legendre L(l,x);
   P = L(l);
   for(i=0;i<absm;i++) P = df(P,x);
   P *= (1 - x*x)^(absm/Symbolic(2));
}

void AssLegendre::redefine(int l,int m)
{
   int absm = abs(m);
   Legendre L(l,x);
   P = L(l);
   for(int i=0;i<absm;i++) P = df(P,x);
   P *= (1 - x*x)^(absm/Symbolic(2));
}

Symbolic AssLegendre::current() const
{ return P; }

ostream & operator << (ostream & s,const AssLegendre &L)
{  return s << L.P; }
```

```
// asslegendre.cpp
// Associate Legendre Polynomial
// P_l^|m|(w) = (1-w^2)^(|m|/2) (d/dw)^|m| P_l(w)
// Recursion formula (n+1)P_{n+1}(x) = (2n+1)xP_n(x)-nP_{n-1}(x)

#include <iostream>
#include "symbolicc++.h"
#include "asslegendre.h"
using namespace std;

int main(void)
{
   int n=5;
   Symbolic x("x") ,Y;
   AssLegendre P(x);

   for(int i=0;i<n;i++)
   {
      for(int j=0;j<=i;j++)
      {
         P.redefine(i,j);
         Y = P.current();
         cout << "P(" << i << "," << j << ") = " << Y << endl;
      }
      cout << endl;
   }
   return 0;
}
```

Laguerre Polynomials

The *Laguerre polynomials* may be defined by *Rodrigue's formula*:

$$L_n(x) := e^x \frac{d^n}{dx^n}(x^n e^{-x}), \qquad n = 0, 1, \ldots$$

where $L_0(x) = 1$. Using the *Leibniz rule*, we obtain

$$L_n(x) = n! \sum_{m=0}^{n} (-1)^m \binom{n}{m} \frac{x^m}{m!}.$$

The Laguerre polynomials L_n are the solutions of the linear second order differential equation

$$x\frac{d^2 L_n}{dx^2} + (1 - x)\frac{dL_n}{dx} + nL_n = 0. \tag{9.6}$$

They obey the recursion relation

$$L_{n+1}(x) = (2n + 1 - x)L_n(x) - n^2 L_{n-1}(x)$$

where $L_0(x) = 1$ and $L_1(x) = 1 - x$. The Laguerre polynomials can also be defined by the generating function

$$\frac{1}{1 - t} \exp\left(\frac{-xt}{1 - t}\right) = \sum_{n=0}^{\infty} \frac{L_n(x)t^n}{n!}.$$

The orthogonality relation of the Laguerre polynomials is given by

$$\int_0^\infty e^{-x} L_m(x) L_n(x) dx = (n!)^2 \delta_{mn}.$$

Furthermore we have

$$\int_0^x L_n(t) dt = L_n(x) - \frac{L_{n+1}(x)}{n+1}.$$

In the program we make use of the recursion relation to generate the first few Laguerre polynomials. We also show that (9.6) holds for $n = 4$. The first few Laguerre polynomials are given by

$$L_0(x) = 1, \quad L_1(x) = 1 - x, \quad L_2(x) = 2 - 4x + x^2,$$

$$L_3(x) = 6 - 18x + 9x^2 - x^3, \quad L_4(x) = 24 - 96x + 72x^2 - 16x^3 + x^4.$$

```cpp
// laguerre.cpp
// Recursion formula L_{n+1}(x) = (2n+1-x)L_n(x)-n^2L_{n-1}(x)

#include <iostream>
#include "symbolicc++.h"
using namespace std;

class Laguerre
{
private:
   int maxTerm, currentStep;
   Symbolic P, Q;
   const Symbolic &x;

public:
   Laguerre(int,const Symbolic&);

   int step();
   void run();
   Symbolic current() const;
   friend ostream & operator << (ostream&,const Laguerre&);
};

Laguerre::Laguerre(int n,const Symbolic &kernal)
   : maxTerm(n), currentStep(0), P(1), x(kernal) {}

int Laguerre::step()
{
   int prev = currentStep;
   Symbolic R;
   ++currentStep;
   if(currentStep==1) { Q = 1; P = 1-x; return 1; }
   if(currentStep <= maxTerm)
   {
      R = (2*prev+1-x)*P-prev*prev*Q;
      Q = P; P = R;
      return 1;
   }
}
```

```
      return 0;
}

void Laguerre::run() {  while(step()) ; }

Symbolic Laguerre::current() const { return P; }

ostream & operator << (ostream & s, const Laguerre & L)
{  return s << L.P; }

int main(void)
{
   int n=4;
   Symbolic x("x");
   Laguerre L(n,x);

   // Calculate the first few Laguerre polynomials
   cout << "L(0) = " << L << endl;

   for(int i=1;i<=n;i++)
   {
      L.step();
      cout << "L("<< i << ") = " << L << endl;
   }
   cout << endl;

   // Show that the Laguerre differential equation is satisfied for n = 4.
   Symbolic result;
   result = x*df(L.current(),x,2) + (1-x)*df(L.current(),x)+n*L.current();
   cout << result << endl;
   return 0;
}
```

Hermite Polynomials

The *Hermite polynomials* are defined by

$$H_n(x) := (-1)^n e^{x^2} \frac{d^n}{dx^n} e^{-x^2}$$

where $n = 0, 1, 2, \ldots$ It can be proved that they satisfy the linear differential equation

$$\frac{d^2 H_n}{dx^2} - 2x \frac{dH_n}{dx} + 2n H_n = 0. \tag{9.7}$$

Using the differential equation, we can prove that the Hermite polynomials are orthogonal on $(-\infty, \infty)$ with respect to the weight function e^{-x^2}. We have

$$\int_{-\infty}^{\infty} e^{-x^2} H_n(x) H_m(x) dx = \sqrt{\pi} 2^n n! \delta_{nm}$$

where $n, m = 0, 1, 2, \ldots$ and δ_{nm} is the *Kronecker delta*. The recursion formula takes the form

$$H_{n+1}(x) = 2x H_n(x) - 2n H_{n-1}(x)$$

where $H_0(x) = 1$ and $H_1(x) = 2x$. The Hermite polynomials can also be defined by the generating function

$$e^{2tx-t^2} = \sum_{n=0}^{\infty} \frac{H_n(x)t^n}{n!}.$$

Furthermore we have

$$\int_0^x H_n(t)dt = \frac{H_{n+1}(x)}{2(n+1)} - \frac{H_{n+1}(0)}{2(n+1)}.$$

In the program, we make use of the recursion relation to generate the first few Hermite polynomials, and show that (9.7) holds for $n = 4$:

$$H_0(x) = 1, \quad H_1(x) = 2x, \quad H_2(x) = 4x^2 - 2, \quad H_3(x) = 8x^3 - 12x,$$

$$H_4(x) = 16x^4 - 48x^2 + 12, \quad H_5(x) = 32x^5 - 160x^3 + 120x.$$

```cpp
// hermite.cpp
// Recursion formula H_{n+1}(x) = 2xH_n(x)-2nH_{n-1}(x)

#include <iostream>
#include "symbolicc++.h"
using namespace std;

class Hermite
{
 private:
    int maxTerm, currentStep;
    Symbolic P, Q;
    const Symbolic &x;
 public:
    Hermite(int,const Symbolic&);
    int step();
    void run();
    Symbolic current() const;
    friend ostream & operator << (ostream&,const Hermite&);
};

Hermite::Hermite(int n,const Symbolic &kernal)
   : maxTerm(n), currentStep(0), P(1), x(kernal) {}

int Hermite::step()
{
    int prev = currentStep;
    Symbolic R;
    ++currentStep;

    if(currentStep==1)
    { Q = 1; P = 2*x; return 1; }

    if(currentStep <= maxTerm)
    {
        R = 2*x*P-2*prev*Q;
        Q = P;
```

```
      P = R;
      return 1;
   }
   return 0;
}

void Hermite::run() { while(step()); }

Symbolic Hermite::current() const { return P; }

ostream & operator << (ostream &s,const Hermite &H)
{ return s << H.P; }

int main(void)
{
   int n=4;
   Symbolic x("x");
   Hermite H(n,x);

   // Calculate the first few Hermite polynomials
   cout << "H(0) = " << H << endl;
   for(int i=1;i<=n;i++)
   {
      H.step();
      cout << "H("<< i << ") = " << H << endl;
   }
   cout << endl;

   // Show that the Hermite differential equation is satisfied for n=4.
   Symbolic result;

   result = df(H.current(),x,2)-2*x*df(H.current(),x)+2*n*H.current();

   cout << result << endl;
   return 0;
}
```

Chebyshev Polynomials

The *Chebyshev polynomials* are defined by the relation

$$T_n(x) := \cos(n \arccos x), \qquad n = 0, 1, 2, \ldots$$

Note that $T_{-n}(x) = T_n(x)$ and from the trigonometric formulae we get

$$T_{n+m}(x) + T_{n-m}(x) = 2T_n(x)T_m(x).$$

For $m = 1$, we obtain the recursion relation

$$T_{n+1}(x) = 2xT_n(x) - T_{n-1}(x)$$

where $T_0(x) = 1$ and $T_1(x) = x$. Thus we can successively compute all $T_n(x)$. Let $x := \cos\theta$, then

$$y = T_n(x) = \cos(n\theta)$$

298 CHAPTER 9. APPLICATIONS

$$\frac{dy}{dx} = \frac{n\sin(n\theta)}{\sin\theta}$$

$$\frac{d^2y}{dx^2} = \frac{-n^2\cos(n\theta) + n\sin(n\theta)\cot\theta}{\sin^2\theta}$$

$$= -\frac{n^2 y}{1-x^2} + \frac{x}{1-x^2}\frac{dy}{dx}.$$

Thus the polynomials $T_n(x)$ satisfy the second order linear ordinary differential equation

$$(1-x^2)\frac{d^2y}{dx^2} - x\frac{dy}{dx} + n^2 y = 0.$$

The first few polynomials are

$$T_0(x) = 1, \quad T_1(x) = x, \quad T_2(x) = 2x^2 - 1,$$

$$T_3(x) = 4x^3 - 3x, \quad T_4(x) = 8x^4 - 8x^2 + 1.$$

```cpp
// chebyshev.cpp
// Recursion formula T_{n+1}(x) = 2xT_n(x)-T_{n-1}(x)

#include <iostream>
#include "symbolicc++.h"
using namespace std;

class Chebyshev
{
private:
   int maxTerm, currentStep;
   Symbolic P, Q;
   const Symbolic &x;
public:
   Chebyshev(int,const Symbolic&);
   int step();
   void run();
   Symbolic current() const;
   friend ostream & operator << (ostream&,const Chebyshev&);
};

Chebyshev::Chebyshev(int n,const Symbolic &kernal)
   : maxTerm(n), currentStep(0), P(1), x(kernal) {}

int Chebyshev::step()
{
 Symbolic R;
 ++currentStep;

 if(currentStep==1) { Q = 1; P = x; return 1; }

 if(currentStep <= maxTerm)
 {
   R = 2*x*P-Q; Q = P; P = R; return 1;
 }
 return 0;
}
```

```
void Chebyshev::run() { while(step()); }

Symbolic Chebyshev::current() const { return P; }

ostream & operator << (ostream &s,const Chebyshev &T)
{ return s << T.P; }

int main(void)
{
  int n=4;
  Symbolic x("x");
  Chebyshev T(n,x);

  // Calculate the first few Chebyshev polynomials
  cout << "T(0) = " << T << endl;
  for(int i=1;i<=n;i++)
  {
   T.step();
   cout << "T("<< i << ") = " << T << endl;
  }
  cout << endl;

  // Show that the Chebyshev differential equation is satisfiedi for n=4.
  Symbolic result;

  result = (1-x*x)*df(T.current(),x,2)-x*df(T.current(),x)+n*n*T.current();
  cout << result << endl;
  return 0;
}
```

9.6.2 Cumulant Expansion

Suppose $x, a_n, b_n \in \mathbf{R}$ and

$$\exp\left[\sum_{n=1}^{\infty} \frac{b_n x^n}{n!}\right] = \sum_{n=0}^{\infty} \frac{a_n x^n}{n!}$$

with $a_0 = 1$. We determine the relation between the coefficients a_n and b_n. The k-th term of the exponential function on the left hand side is given by

$$\frac{1}{k!}\left(\sum_{n=1}^{\infty} \frac{b_n x^n}{n!}\right)^k = \frac{1}{k!}\left(\sum_{n_1=1}^{\infty} \frac{b_{n_1} x^{n_1}}{n_1!}\right) \cdots \left(\sum_{n_k=1}^{\infty} \frac{b_{n_k} x^{n_k}}{n_k!}\right)$$

$$= \frac{1}{k!}\sum_{n_1=1}^{\infty}\sum_{n_2=1}^{\infty} \cdots \sum_{n_k=1}^{\infty} \frac{b_{n_1} b_{n_2} \cdots b_{n_k} x^{n_1+n_2+\cdots+n_k}}{n_1!n_2!\cdots n_k!}.$$

Therefore,

$$\exp\left[\sum_{n=1}^{\infty} \frac{b_n x^n}{n!}\right] = 1 + \sum_{n=1}^{\infty} \frac{b_n x^n}{n!} + \frac{1}{2!}\sum_{n_1=1}^{\infty}\sum_{n_2=1}^{\infty} \frac{b_{n_1} b_{n_2} x^{n_1+n_2}}{n_1!n_2!} + \cdots$$

$$+ \frac{1}{k!}\sum_{n_1=1}^{\infty}\sum_{n_2=1}^{\infty} \cdots \sum_{n_k=1}^{\infty} \frac{b_{n_1} b_{n_2} \cdots b_{n_k} x^{n_1+n_2+\cdots+n_k}}{n_1!n_2!\cdots n_k!} + \cdots$$

$$= \sum_{n=0}^{\infty} \frac{a_n x^n}{n!}.$$

Equating terms of the same order in x, we can obtain the relation between a_n and b_n for all positive integers n. The first three terms are

$$x^1 : a_1 = b_1, \quad x^2 : a_2 = b_2 + b_1^2, \quad x^3 : a_3 = b_3 + 3b_2 b_1 + b_1^3.$$

It follows that

$$b_1 = a_1, \quad b_2 = a_2 - a_1^2, \quad b_3 = a_3 - 3a_2 a_1 + 2a_1^3.$$

In the program, we repeat the process of deriving the coefficients. The process shows that a computer algebra system can obtain the coefficients efficiently.

```cpp
// cumu.cpp

#include <iostream>
#include "symbolicc++.h"
using namespace std;

Symbolic Taylor(Symbolic u,const Symbolic &x,int n)
{
   Symbolic series = u[x==0];
   int fac = 1;

   for(int j=1;j<=n;j++)
   {
      u = df(u,x); fac = fac * j;
      series += (u[x==0]*(x^j)/fac);
   }
   return series;
}

int main(void)
{
   int fac, n=5;
   Symbolic x("x"), a("a",5), b("b",5);
   Symbolic y, P, Q;

   fac = 1; P = a(0); Q = 0;
   for(int i=1;i<n;i++)
   {
      fac *= i;
      P += a(i)*(x^i)/fac;
      Q += b(i)*(x^i)/fac;
   }

   cout << "P = " << P << endl;
   cout << "Q = " << Q << endl; cout << endl;
   y = Taylor(exp(Q),x,5);

   cout << "Taylor series expansion of exp(Q) = " << endl;
   cout << y << endl;
   cout << endl;
```

```
    cout << "Coefficient of x :"<< endl;
    cout << "exp(Q) => " << y.coeff(x,1) << endl;
    cout << "  P    => " << P.coeff(x,1) << endl;
    cout << endl;

    cout << "Coefficient of x^2 :"<< endl;
    cout << "exp(Q) => " << y.coeff(x,2) << endl;
    cout << "  P    => " << P.coeff(x,2) << endl;
    cout << endl;

    cout << "Coefficient of x^3 :"<< endl;
    cout << "exp(Q) => " << y.coeff(x,3) << endl;
    cout << "  P    => " << P.coeff(x,3) << endl;
    cout << endl;

    cout << "Coefficient of x^4 :"<< endl;
    cout << "exp(Q) => " << y.coeff(x,4) << endl;
    cout << "  P    => " << P.coeff(x,4) << endl;

    return 0;
}
```

9.6.3 Exterior Product

In this section, we give an implementation of the *exterior product* (also called wedge product or Grassmann product) described in Section 2.17. In the program the exterior product $\wedge$ is written as $*$. We evaluate the determinant of the 4×4 matrix

$$A := \begin{pmatrix} 1 & 2 & 5 & 2 \\ 0 & 1 & 2 & 3 \\ 1 & 0 & 1 & 0 \\ 0 & 3 & 0 & 7 \end{pmatrix}$$

by calculating the exterior product

$$\begin{pmatrix} 1 \\ 0 \\ 1 \\ 0 \end{pmatrix} \wedge \begin{pmatrix} 2 \\ 1 \\ 0 \\ 3 \end{pmatrix} \wedge \begin{pmatrix} 5 \\ 2 \\ 1 \\ 0 \end{pmatrix} \wedge \begin{pmatrix} 2 \\ 3 \\ 0 \\ 7 \end{pmatrix}.$$

The exterior product gives $24\mathbf{e}_1 \wedge \mathbf{e}_2 \wedge \mathbf{e}_3 \wedge \mathbf{e}_4$, where $\mathbf{e}_1, \mathbf{e}_2, \mathbf{e}_3, \mathbf{e}_4$ is the standard basis in $\mathbf{R}^4$. Thus, we find that $\det A = 24$.

```
// grass.cpp

#include <iostream>
#include "symbolicc++.h"
using namespace std;

int main(void)
{
    int i,j,n=4;
    Symbolic e("e",n);
    Symbolic y,result;
```

```
Symbolic A("A",n,n);

// non-commutative e0, e1, e2, ... , en
e = ~e;

cout << e << endl;

A(0,0) = 1; A(0,1) = 2; A(0,2) = 5; A(0,3) = 2;
A(1,0) = 0; A(1,1) = 1; A(1,2) = 2; A(1,3) = 3;
A(2,0) = 1; A(2,1) = 0; A(2,2) = 1; A(2,3) = 0;
A(3,0) = 0; A(3,1) = 3; A(3,2) = 0; A(3,3) = 7;

result = 1;
for(i=0;i<n;i++) result *= A.row(i)*e;

cout << result << endl; cout << endl;

Equations rules;

// set e(i)*e(i) to 0
for(i=0;i<n;i++) rules = (rules,e(i)*e(i) == 0);

// for all i>j, set e(i)*e(j) to -e(j)*e(i)
for(i=0;i<n;i++)
    for(j=0;j<i;j++)
        rules = (rules,e(i)*e(j) == -e(j)*e(i));

result = result.subst_all(rules);
cout << "result = " << result << endl;
return 0;
}
```

9.7   Symbolic Class and Symbolic Differentiation

9.7.1   First Integrals

Consider an autonomous system of first order ordinary differential equations

$$\frac{d\mathbf{u}}{dt} = V(\mathbf{u})$$

where $\mathbf{u} = (u_1, u_2, \ldots, u_n)$ and V_j are smooth functions of $u_1, u_2, \ldots, u_n$. A smooth function $I(\mathbf{u})$ is called a *first integral* of the differential equations if

$$\frac{d}{dt}I(\mathbf{u}(t)) \equiv \sum_{j=1}^{n} \frac{\partial I}{\partial u_j}\frac{du_j}{dt} \equiv \sum_{j=1}^{n} \frac{\partial I}{\partial u_j}V_j(\mathbf{u}) = 0.$$

A smooth function $I(\mathbf{u}(t), t)$ is called an explicitly *time-dependent first integral* of the differential equations if

$$\frac{d}{dt}I(\mathbf{u}(t), t) = \frac{\partial I}{\partial t} + \sum_{j=1}^{n} \frac{\partial I}{\partial u_j}V_j(\mathbf{u}(t), t) = 0.$$

Example 1. Consider the system of differential equations

$$\frac{du_1}{dt} = u_2 u_3, \qquad \frac{du_2}{dt} = u_1 u_3, \qquad \frac{du_3}{dt} = u_1 u_2.$$

We show that I_1, I_2 and I_3 where

$$I_1(\mathbf{u}) = u_1^2 - u_2^2, \qquad I_2(\mathbf{u}) = u_3^2 - u_1^2, \qquad I_3(\mathbf{u}) = u_2^2 - u_3^2$$

are first integrals of the system.

```cpp
// first1.cpp

#include <iostream>
#include "symbolicc++.h"
using namespace std;

int main(void)
{
   int N=3;
   Symbolic term, sum=0, u("u",N), v("",N), I("",N);

   v(0) = u(1)*u(2); v(1) = u(0)*u(2); v(2) = u(0)*u(1);
   I(0) = u(0)*u(0)-u(1)*u(1);
   I(1) = u(2)*u(2)-u(0)*u(0);
   I(2) = u(1)*u(1)-u(2)*u(2);

   for(int i=0;i<N;i++)
   {
      for(int j=0;j<N;j++)
      {
         term = df(I(i),u(j)) * v(j);
         sum += term;
         cout << "Partial Term " << j << " = " << term << endl;
      }
      cout << "Sum = " << sum << endl;
      cout << endl;
   }
   return 0;
}
```

Example 2. Consider the *Lorenz model*

$$\frac{du_1}{dt} = -\sigma u_1 + \sigma u_2, \qquad \frac{du_2}{dt} = -u_1 u_3 + r u_1 - u_2, \qquad \frac{du_3}{dt} = u_1 u_2 - b u_3.$$

For certain values of the bifurcation parameters σ, b and r the system admits explicitly the time-dependent first integrals. We insert the ansatz

$$I(\mathbf{u}(t), t) = (u_2^2 + u_3^2) \exp(2t)$$

into the system in order to find the conditions on the coefficients σ, r, b such that $I(\mathbf{u}(t), t)$ is a first integral.

```
// first2.cpp

#include <iostream>
#include "symbolicc++.h"
using namespace std;

int main(void)
{
  Symbolic u("u",3), v("v",4);
  Symbolic term, sum, I, R1, R2;
  Symbolic t("t"), s("s"), b("b"), r("r");

  // Lorenz Model
  v(0) = s*u(1)-s*u(0);
  v(1) = -u(1)-u(0)*u(2)+r*u(0);
  v(2) = u(0)*u(1)-b*u(2);
  v(3) = 1;

  // The ansatz
  I = (u(1)*u(1)+u(2)*u(2))*exp(2*t);
  sum = 0;

  for(int i=0;i<3;i++) sum += v(i)*df(I,u(i));

  sum += v(3)*df(I,t);
  cout << "sum = " << sum << endl; cout << endl;
  R1 = sum.coeff(u(2),2);
  R1 = R1/(exp(2*t));
  cout << "R1 = " << R1 << endl;

  R2 = sum.coeff(u(0),1); R2 = R2.coeff(u(1),1);
  R2 = R2/(exp(2*t));
  cout << "R2 = " << R2 << endl;
  return 0;
}
```

It is obvious that for R1, R2 to be equal to zero, we need $b = 1$ and $r = 0$.

9.7.2   Spherical Harmonics

Spherical harmonics are complex, single-valued functions on the surface of the unit sphere, i.e. functions of two real-valued arguments $0 \leq \theta \leq \pi$, $0 \leq \phi < 2\pi$. They are defined by

$$Y_{lm}(\theta,\phi) := (-1)^m \left[\frac{(2l+1)(l-m)!}{4\pi(l+m)!} \right]^{\frac{1}{2}} P_l^m(\cos\theta)e^{im\phi}, \quad m \geq 0$$

where $P_l^m(w)$ is the associated Legendre functions defined as

$$P_l^m(w) := (1-w^2)^{|m|/2} \frac{d^{|m|}}{dw^{|m|}} P_l(w)$$

with $l = 0, 1, 2, \ldots$ and $m = -l, -l+1, \ldots, l$. For negative values of m we have

$$Y_{lm}(\theta,\phi) = (-1)^m Y_{l,-m}^*(\theta,\phi)$$

where the asterisk (*) denotes the complex conjugate. The spherical harmonics satisfy the orthonormality relations

$$\int Y_{l'm'}^*(\theta,\phi)Y_{lm}(\theta,\phi)d\Omega = \int_0^{2\pi} d\phi \int_0^\pi d\theta \sin\theta\, Y_{l'm'}^*(\theta,\phi)Y_{lm}(\theta,\phi)$$
$$= \delta_{ll'}\delta_{mm'}$$

where we have written $d\Omega \equiv \sin\theta d\theta d\phi$ and $\int d\Omega$ means that we integrate over the full range of the angular variables (θ,ϕ), namely

$$\int d\Omega \equiv \int_0^{2\pi} d\phi \int_0^\pi d\theta \sin\theta.$$

The spherical harmonics form a basis in the Hilbert space $L_2(S^2)$ where

$$S^2 := \left\{ (x_1,x_2,x_3) : x_1^2 + x_2^2 + x_3^2 = 1 \right\}.$$

Thus any function in the Hilbert space $L_2(S^2)$ can be written as a linear combination of the spherical harmonics. The first few spherical harmonics are given by

$$Y_{00}(\theta,\phi) = \frac{1}{\sqrt{4\pi}}, \quad Y_{11}(\theta,\phi) = -\sqrt{\frac{3}{8\pi}}\sin\theta\, e^{i\phi},$$

$$Y_{10}(\theta,\phi) = \sqrt{\frac{3}{4\pi}}\cos\theta, \quad Y_{1-1}(\theta,\phi) = \sqrt{\frac{3}{8\pi}}\sin\theta\, e^{-i\phi}.$$

```cpp
// spheric.cpp

#include <iostream>
#include "symbolicc++.h"
#include "asslegendre.h"
using namespace std;

Symbolic PI("PI"), I("I");

int factorial(int n)
{
   int result=1;
   for(int i=2;i<=n;i++) result *= i;
   return result;
}

Symbolic Y(int l,int m,const Symbolic &phi,const Symbolic &w)
{
   Symbolic a, b, u("u"), result;
   int absm = abs(m);
   AssLegendre A(l,m,u);
   a = A.current(); a = a[u == cos(w)];
   if(m>0 && m%2) a = -a;
   b = sqrt((2*l+1)*factorial(l-absm)/(4*factorial(l+absm)*PI));
   result = a*b*exp(I*m*phi);
   return result;
}
```

```
int main(void)
{
    int n=3;
    Symbolic phi("phi"), w("w"), result;

    for(int i=0;i<=n;i++)
    {
        for(int j=-i;j<=i;j++)
        {
            result = Y(i,j,phi,w);
            result = result[cos(w)*cos(w) == 1-sin(w)*sin(w)];
            cout << "Y(" << i << "," << j << ") = " << result << endl;
        }
        cout << endl;
    }
    return 0;
}
```

9.7.3 Nambu Mechanics

In *Nambu mechanics* the phase space is spanned by an n-tuple of dynamical variables u_i, for $i = 1, \ldots, n$. The equations of motion of Nambu mechanics (i.e., the autonomous system of first order ordinary differential equations) can be constructed as follows. Let $I_k : \mathbf{R}^n \to \mathbf{R}$, for $k = 1, \ldots, n-1$ be smooth functions, then

$$\frac{du_i}{dt} = \frac{\partial(u_i, I_1, \ldots, I_{n-1})}{\partial(u_1, u_2, \ldots, u_n)}$$

where $\partial(u_i, I_1, \ldots, I_{n-1})/\partial(u_1, u_2, \ldots, u_n)$ denotes the *Jacobian*. Consequently, the equations of motion can also be written as (summation convention)

$$\frac{du_i}{dt} = \epsilon_{ijk\ldots l} \partial_j I_1 \ldots \partial_l I_{n-1}$$

where $\epsilon_{ijk\ldots l}$ is the generalized *Levi-Civita symbol* and $\partial_j \equiv \partial/\partial u_j$. The proof that $I_1, \ldots, I_{n-1}$ are first integrals of the system is as follows. Since (summation convention)

$$\frac{dI_i}{dt} = \frac{\partial I_i}{\partial u_j} \frac{du_j}{dt}$$

we have

$$\frac{dI_i}{dt} = (\partial_j I_i)\epsilon_{jl_1\ldots l_{n-1}}(\partial_{l_1} I_1) \ldots (\partial_{l_{n-1}} I_{n-1}) = \frac{\partial(I_i, I_1, \ldots, I_{n-1})}{\partial(u_1, \ldots, u_n)} = 0 \,.$$

Since the Jacobian matrix has two equal rows, it is singular. If the first integrals are polynomials, then the dynamical system is algebraically completely integrable. For the case $n = 3$, we obtain the equations of motion

$$\frac{du_1}{dt} = \frac{\partial I_1}{\partial u_2} \frac{\partial I_2}{\partial u_3} - \frac{\partial I_1}{\partial u_3} \frac{\partial I_2}{\partial u_2}, \quad \frac{du_2}{dt} = \frac{\partial I_1}{\partial u_3} \frac{\partial I_2}{\partial u_1} - \frac{\partial I_1}{\partial u_1} \frac{\partial I_2}{\partial u_3}, \quad \frac{du_3}{dt} = \frac{\partial I_1}{\partial u_1} \frac{\partial I_2}{\partial u_2} - \frac{\partial I_1}{\partial u_2} \frac{\partial I_2}{\partial u_1}.$$

Example. We consider the following first integrals

$$I_1 = u_1 + u_2 + u_3 \quad \text{and} \quad I_2 = u_1 u_2 u_3.$$

The program uses the first equation to construct the 3×3 matrices and then calculate their determinants

$$\frac{du_i}{dt} = \det \begin{pmatrix} \dfrac{\partial u_i}{\partial u_1} & \dfrac{\partial I_1}{\partial u_1} & \dfrac{\partial I_2}{\partial u_1} \\ \dfrac{\partial u_i}{\partial u_2} & \dfrac{\partial I_1}{\partial u_2} & \dfrac{\partial I_2}{\partial u_2} \\ \dfrac{\partial u_i}{\partial u_3} & \dfrac{\partial I_1}{\partial u_3} & \dfrac{\partial I_2}{\partial u_3} \end{pmatrix}$$

for $i = 1, 2, 3$.

```cpp
// nambu.cpp

#include <iostream>
#include "symbolicc++.h"
using namespace std;

void nambu(const Symbolic &I,const Symbolic &u,int n)
{
   int i, j;
   Symbolic J("J",n,n);

   for(i=0;i<n;i++)
      for(j=1;j<n;j++)
         J(i,j) = df(I(j-1),u(i));

   for(i=0;i<n;i++)
   {
      for(j=0;j<n;j++) J(j,0) = df(u(i),u(j));
      cout << "du(" << i << ")/dt = " << det(J) << endl;
   }
}

int main(void)
{
   Symbolic u("u",3), I("I",2);
   I(0) = u(0)+u(1)+u(2);
   I(1) = u(0)*u(1)*u(2);
   cout << "The equations are " << endl;
   nambu(I,u,3);

   return 0;
}
```

9.7.4   Taylor Expansion of Differential Equations

Consider the first order ordinary differential equation

$$\frac{du}{dx} = f(x, u), \quad u(x_0) = u_0$$

where f is an analytic function of x and u. The *Taylor series expansion* gives an approximate solution about $x = x_0$ and the solution of the differential equation is given by

$$u(x_0 + h) = u(x_0) + h \left(\frac{du}{dx} \right)_{x=x_0} + \frac{h^2}{2!} \left(\frac{d^2 u}{dx^2} \right)_{x=x_0} + \cdots$$

The derivatives can be obtained by the successive differentiation of the differential equation,

$$\frac{d^2 u}{dx^2} = \frac{\partial f}{\partial x} + \frac{\partial f}{\partial u} \frac{du}{dx} = \frac{\partial f}{\partial x} + \frac{\partial f}{\partial u} f$$

$$\frac{d^3 u}{dx^3} = \frac{\partial^2 f}{\partial x^2} + 2 \frac{\partial^2 f}{\partial x \partial u} f + \frac{\partial^2 f}{\partial u^2} f^2 + \frac{\partial f}{\partial u} \left(\frac{\partial f}{\partial x} + \frac{\partial f}{\partial u} f \right)$$

$$\vdots$$

The formulae for higher derivatives become more involved.

Example. Consider the *Riccati differential equation*

$$\frac{du}{dx} = u^2 + x, \qquad u(0) = 1.$$

This equation has no solution in terms of elementary functions. *Bessel functions* are needed to solve it. Suppose the solution may be represented by a Taylor series expansion,

$$u(x) = 1 + \sum_{j=1}^{\infty} \frac{x^j}{j!} \left(\frac{d^j u}{dx^j} \right)_{x=x_0}.$$

We compute all the successive derivatives

$$\frac{d^2 u}{dx^2} = 2u \frac{du}{dx} + 1, \quad \frac{d^3 u}{dx^3} = 2 \left(\frac{du}{dx} \right)^2 + 2u \frac{d^2 u}{dx^2}, \quad \cdots$$

These derivatives can be found using computer algebra. The initial value is then inserted and we obtain the Taylor series expansion up to order n. We solve the Riccati differential equation using the Taylor series method up to order $n = 4$, with initial value $u(0) = 1$. The solution is

$$u(x) = 1 + x + \frac{3}{2} x^2 + \frac{4}{3} x^3 + \frac{17}{12} x^4 + \cdots$$

```
// taylor1.cpp

#include <iostream>
#include "symbolicc++.h"
using namespace std;

int factorial(int N)
```

```
{
   int result=1;
   for(int i=2;i<=N;i++) result *= i;
   return result;
}

int main(void)
{
   int i, j, n=4;
   Symbolic u("u"), x("x"), result;
   Symbolic u0("",n), y("",n);
   u = u[x];
   u0(0) = u*u+x;
   for(j=1;j<n;j++) u0(j) = df(u0(j-1),x);

   cout << u0 << endl;

   // initial condition u(0)=1
   u0(0) = u0(0)[u==1,x==0];

   y(0) = u;
   for(i=1;i<n;i++) y(i) = df(y(i-1),x);

   // substitution of initial conditions
   for(i=1;i<n;i++)
      for(j=i;j>0;j--) u0(i) = u0(i)[y(j)==u0(j-1)];
   cout << u0 << endl;

   for(i=0;i<n;i++) u0(i) = u0(i)[u == 1];
   cout << u0 << endl;

   // Taylor series expansion
   result = 1;
   for(i=0;i<n;i++)
      result += (Symbolic(1)/factorial(i+1))*u0(i)*(x^(i+1));
   cout << "u(x) = " << result << endl;
   return 0;
}
```

Example. Consider the nonlinear initial value problem

$$\frac{du}{dx} = x^2 + xu - u^2, \quad \text{where} \quad u(0) = 1.$$

It is straightforward, though tedious, to compute all the derivatives

$$\frac{d^2u}{dx^2} = 2x + u + x\frac{du}{dx} - 2u\frac{du}{dx}$$

$$\frac{d^3u}{dx^3} = 2 + 2\frac{du}{dx} + x\frac{d^2u}{dx^2} - 2\left(\frac{du}{dx}\right)^2 - 2u\frac{d^2u}{dx^2}$$

$$\frac{d^4u}{dx^4} = 3\frac{d^2u}{dx^2} + x\frac{d^3u}{dx^3} - 6\frac{du}{dx}\frac{d^2u}{dx^2} - 2u\frac{d^3u}{dx^3}$$

$$\vdots$$

At the point $x = 0$, $u(x = 0) = 1$,

$$\left(\frac{du}{dx}\right)_{x=0} = -1, \quad \left(\frac{d^2u}{dx^2}\right)_{x=0} = 3, \quad \left(\frac{d^3u}{dx^3}\right)_{x=0} = -8, \quad \left(\frac{d^4u}{dx^4}\right)_{x=0} = 43, \quad \cdots$$

Finally, the solution $u(x)$ of the initial value problem near $x = 0$ is given by

$$u(x) = 1 - x + \frac{3}{2}x^2 - \frac{4}{3}x^3 + \frac{43}{24}x^4 + \cdots$$

We reproduce the procedure of the calculation using the algebra system.

```
// taylor2.cpp

#include <iostream>
#include "symbolicc++.h"
using namespace std;

int factorial(int N)
{
    int result=1;
    for(int i=2;i<=N;i++) result *= i;
    return result;
}

int main(void)
{
    int i, j, n=4;
    Symbolic x("x"), u("u"), result;
    Symbolic u0("",n), w("",n);
    u = u[x];
    u0(0) = x*x + x*u - u*u;
    for(i=1;i<n;i++) u0(i) = df(u0(i-1),x);
    cout << u0 << endl;

    // initial condition u(0)=1
    u0(0) = u0(0)[u==1,x==0];

    w(0) = u;
    for(i=1;i<n;i++) w(i) = df(w(i-1),x);

    // substitution of initial conditions
    for(i=1;i<n;i++)
        for(j=i;j>0;j--) u0(i) = u0(i)[w(j)==u0(j-1)];
    cout << u0 << endl;

    for(i=0;i<n;i++) u0(i) = u0(i)[u==1,x==0];
    cout << u0 << endl;

    // Taylor series expansion
    result = 1;
    for(i=0;i<n;i++) result += u0(i)/factorial(i+1)*(x^(i+1));
    cout << "u = " << result << endl;
    return 0;
}
```

9.7.5 Commutator of Two Vector Fields

Consider an autonomous system of first order ordinary differential equations

$$\frac{d\mathbf{x}}{dt} = V(\mathbf{x})$$

where $V_j : \mathbf{R}^n \to \mathbf{R}$ and $j = 1, 2, \ldots, n$. We assume that the functions V_j are analytic. We can associate the vector field

$$V = \sum_{j=1}^{n} V_j(\mathbf{x}) \frac{\partial}{\partial x_j}$$

with the differential equation. Let W be another vector field

$$W = \sum_{k=1}^{n} W_k(\mathbf{x}) \frac{\partial}{\partial x_k}.$$

We define the *commutator* of V and W as follows

$$[V, W] := \sum_{k=1}^{n} \sum_{j=1}^{n} \left(V_j \frac{\partial W_k}{\partial x_j} - W_j \frac{\partial V_k}{\partial x_j} \right) \frac{\partial}{\partial x_k}.$$

We find that the commutator satisfies the following properties

- $[V, W] = -[W, V]$

- $[V, U + W] = [V, U] + [V, W]$

- the *Jacobi identity* $[[V, W], U] + [[U, V], W] + [[W, U], V] = 0$

where U is an analytic vector field on $\mathbf{R}^n$. The analytic vector fields define a *Lie algebra*.

In the program we consider three vector fields V, W, U and show that they satisfy the three properties listed above.

```cpp
// comm.cpp

#include <iostream>
#include "symbolicc++.h"
using namespace std;

const int n = 3;

Symbolic x("x",n);

Symbolic commutator(const Symbolic &V,const Symbolic &W)
{
   Symbolic U(Symbolic(0),V.rows());

   for(int k=0;k<n;k++)
      for(int j=0;j<n;j++)
```

```
        U(k) += (V(j)*df(W(k),x(j))-W(j)*df(V(k),x(j)));

   return U;
}


int main(void)
{
   Symbolic V("",n), W("",n), U("",n), Y("",n);

   V(0) = x(0)*x(0); V(1) = x(1)*x(2); V(2) = x(2)*x(2);
   W(0) = x(2);      W(1) = x(0)-x(1); W(2) = x(1)*x(2);
   U(0) = x(1)*x(0); U(1) = x(1);      U(2) = x(0)-x(2);

   // [V,W] = -[W,V]
   Y = commutator(V,W) + commutator(W,V); cout << Y << endl;
   cout << endl;

   // [V,U+W] = [V,U] + [V,W]
   Y = commutator(V,U+W) - (commutator(V,U)+commutator(V,W));
   cout << Y << endl; cout << endl;

   // Jacobian Identity [[V,W],U] + [[U,V],W] + [[W,U],V] = 0
   Y =   commutator(commutator(V,W),U)
       + commutator(commutator(U,V),W)
       + commutator(commutator(W,U),V);
   cout << Y << endl;
   return 0;
}
```

9.7.6   Lie Derivative and Killing Vector Field

Let M be a Riemannian manifold with *metric tensor field*

$$g = \sum_{j=1}^{4} \sum_{k=1}^{4} g_{jk}(\mathbf{x}) dx_j \otimes dx_k$$

and V be a smooth vector field defined on M. The vector field V is called a *Killing vector field* if

$$L_V g = 0$$

where $L_V g$ denotes the *Lie derivative* of g with respect to V. The Lie derivative is linear, i.e.

$$L_V(T + S) = L_V T + L_V S$$

where T and S are (r,s) tensor fields. Furthermore the Lie derivative obeys the product rule

$$L_V(S \otimes T) = (L_V S) \otimes T + S \otimes (L_V T)$$

where S is an (r,s) tensor field and T is a (p,q) tensor field. Finally, we have in local coordinates

$$L_V dx_j = d(V \lrcorner dx_j) = dV_j = \sum_{k=1}^{4} \frac{\partial V_j}{\partial x_k} dx_k$$

where V is given by

$$V = \sum_{j=1}^{4} V_j(\mathbf{x}) \frac{\partial}{\partial x_j}.$$

Consequently, we obtain

$$L_V g = \sum_{j=1}^{4} \sum_{k=1}^{4} \sum_{l=1}^{4} \left(V_l \frac{\partial g_{jk}}{\partial x_l} + g_{lk} \frac{\partial V_l}{\partial x_j} + g_{jl} \frac{\partial V_l}{\partial x_k} \right) dx_j \otimes dx_k.$$

Example. We consider the *Gödel metric tensor field*

$$g = dx_1 \otimes dx_1 - \frac{1}{2} \exp(2x_1) \ dx_2 \otimes dx_2 + dx_3 \otimes dx_3 - dx_4 \otimes dx_4$$
$$- \exp(x_1) \ dx_2 \otimes dx_4 - \exp(x_1) \ dx_4 \otimes dx_2.$$

The metric tensor field can also be written in matrix form

$$(g_{jk}) = \begin{pmatrix} 1 & 0 & 0 & 0 \\ 0 & -\exp(2x_1)/2 & 0 & -\exp(x_1) \\ 0 & 0 & 1 & 0 \\ 0 & -\exp(x_1) & 0 & -1 \end{pmatrix}.$$

We show that the vector field

$$V = x_2 \frac{\partial}{\partial x_1} + \left(\exp(-2x_1) - \frac{x_2^2}{2} \right) \frac{\partial}{\partial x_2} - 2 \exp(-x_1) \frac{\partial}{\partial x_4}$$

is the Killing vector field of g.

```
// kill.cpp

#include <iostream>
#include "symbolicc++.h"
using namespace std;

int main(void)
{
   int N=4;
   Symbolic g("g",N,N), Lg("Lg",N,N), V("V",N), x("x",N);

   // The Goedel metric
   g(0,0) = 1; g(0,1) = 0;                g(0,2) = 0; g(0,3) = 0;
   g(1,0) = 0; g(1,1) = -exp(2*x(0))/2;   g(1,2) = 0; g(1,3) = -exp(x(0));
   g(2,0) = 0; g(2,1) = 0;                g(2,2) = 1; g(2,3) = 0;
   g(3,0) = 0; g(3,1) = -exp(x(0));       g(3,2) = 0; g(3,3) = -1;

   // The Killing vector field of the Goedel metric
   V(0) = x(1); V(1) = exp(-2*x(0)) - x(1)*x(1)/2;
   V(2) = 0;    V(3) = -2*exp(-x(0));

   // The Lie derivative
   for(int j=0;j<N;j++)
      for(int k=0;k<N;k++)
      {
```

```
        Lg(j,k) = 0;
        for(int l=0;l<N;l++)
            Lg(j,k) += V(l)*df(g(j,k),x(l)) + g(l,k)*df(V(l),x(j))
                       + g(j,l)*df(V(l),x(k));
    }

  cout << "The Goedel Metric, g\n" << g << endl;
  cout << "The Killing vector field of the Goedel metric, V\n"
       << V << endl; cout << endl;
  cout << "The Lie derivative of g with respect to V, Lg\n"
       << Lg << endl;
  return 0;
}
```

9.8   Matrix Class

9.8.1   Hilbert-Schmidt Norm

Let A and B be two arbitrary $n \times n$ matrices over $\mathbf{R}$. Define

$$(A, B) := \operatorname{tr}(AB^T)$$

where B^T denotes the transpose of B and tr() denotes the trace. We find that

$$(A, A) \geq 0$$
$$(A, B) = (B, A)$$
$$(cA, B) = c(A, B)$$
$$(A_1 + A_2, B) = (A_1, B) + (A_2, B)$$

where $c \in \mathbf{R}$. Thus (A, B) defines a scalar product for $n \times n$ matrices over $\mathbf{R}$. The scalar product induces a norm, which is given by

$$\|A\| := \sqrt{(A, A)} = \sqrt{\sum_{i=1}^{n} \sum_{j=1}^{n} |a_{ij}|^2}.$$

The norm is called the *Hilbert-Schmidt norm.* The results can be extended to infinite dimensions when we impose the condition

$$\sum_{i=1}^{\infty} \sum_{j=1}^{\infty} |a_{ij}|^2 < \infty.$$

If an infinite-dimensional matrix A satisfies this condition, we call A a *Hilbert-Schmidt operator.*

In the program we consider a 2×2 symbolic matrix

$$A := \begin{pmatrix} a & b \\ b & a \end{pmatrix}.$$

The norm is implemented in the `Matrix` class which has been described in Chapter 7.

```
// hilbert.cpp

#include <iostream>
#include "matrix.h"
#include "matnorm.h"
#include "symbolicc++.h"
using namespace std;

int main(void)
{
    int n = 2;
    Symbolic a("a"), b("b"), y1;
    Matrix<Symbolic> A(n,n);
    A[0][0] = a; A[0][1] = b;
    A[1][0] = b; A[1][1] = a;

    cout << "The " << n << "x" << n << " matrix A is \n" << A << endl;

    y1 = normH(A);
    cout << "The Hilbert-Schmidt norm of matrix A is " << y1 << endl;
    cout << endl;

    //a = 2.0; b = 3.0;
    cout << "Put a = " << 2.0 << " and b = " << 3.0 << endl;
    cout << "The Hilbert-Schmidt norm of matrix A is "
         << y1[a==2,b==3] << " or "
         << y1[a==2.0,b==3.0] << endl;
    return 0;
}
```

9.8.2 Lax Pair and Hamilton System

Consider the *Hamilton function* (*Toda lattice*)

$$H(\mathbf{p}, \mathbf{q}) = \frac{1}{2}(p_1^2 + p_2^2 + p_3^2) + \exp(q_1 - q_2) + \exp(q_2 - q_3) + \exp(q_3 - q_1)$$

where $\mathbf{p}$, $\mathbf{q}$ are the momentum and position vectors, respectively. The *Hamilton equations of motion* are given by

$$\frac{dp_j}{dt} = -\frac{\partial H}{\partial q_j}, \qquad \frac{dq_j}{dt} = \frac{\partial H}{\partial p_j}, \qquad \text{for} \quad j = 1, 2, 3.$$

Introducing the quantities

$$a_j := \frac{1}{2}\exp\left(\frac{1}{2}(q_j - q_{j+1})\right), \qquad b_j := \frac{1}{2}p_j$$

and cyclic boundary conditions (i.e. $q_4 \equiv q_1$), we find that the Hamilton equations of motion take the form (with $a_3 - 0$)

$$\frac{da_j}{dt} = a_j(b_j - b_{j+1}), \quad \frac{db_1}{dt} = -2a_1^2, \quad \frac{db_2}{dt} = 2(a_1^2 - a_2^2), \quad \frac{db_3}{dt} = 2a_2^2$$

where $j = 1, 2$. We introduce the matrices

$$L := \begin{pmatrix} b_1 & a_1 & 0 \\ a_1 & b_2 & a_2 \\ 0 & a_2 & b_3 \end{pmatrix}, \quad A := \begin{pmatrix} 0 & -a_1 & 0 \\ a_1 & 0 & -a_2 \\ 0 & a_2 & 0 \end{pmatrix}.$$

The eigenvalues of the matrix L are constants of motion and thus the coefficients of the characteristic polynomial are also constants of motion. The equations of motion can be rewritten as the *Lax representation*

$$\frac{dL}{dt} = [A, L](t).$$

It can be shown that the first integrals of the system take the forms

$$\mathrm{tr}(L^n) \quad \text{where } n = 1, 2, \ldots$$

In the program we demonstrate that the equations of motion and the Lax representation are equivalent. In the second part of the program, we show that

$$\mathrm{tr}L = b_1 + b_2 + b_3 \quad \text{and} \quad \mathrm{tr}(L^2) = b_1^2 + b_2^2 + b_3^2 + 2a_1^2 + 2a_2^2$$

are the first integrals. The determinant of L,

$$\det(L) = b_1 b_2 b_3 - b_1 a_2^2 - b_3 a_1^2$$

is also a first integral of the system. Can $\det(L)$ be expressed in the form $\mathrm{tr}(L^n)$ for some $n > 0$? Notice that $\det(L)$ is the product of the eigenvalues of L and $\mathrm{tr}(L)$ is the sum of the eigenvalues of L.

```cpp
// lax.cpp

#include <iostream>
#include "symbolicc++.h"
using namespace std;

int main(void)
{
   Symbolic L("L",3,3), A("A",3,3), Lt("Lt",3,3); // Lt=dL/dt
   Symbolic a1("a1"), a2("a2"), b1("b1"), b2("b2"), b3("b3"),
            a1t, a2t, b1t, b2t, b3t;

   L(0,0) = b1; L(0,1) = a1;  L(0,2) = 0;
   L(1,0) = a1; L(1,1) = b2;  L(1,2) = a2;
   L(2,0) = 0;  L(2,1) = a2;  L(2,2) = b3;
   A(0,0) = 0;  A(0,1) = -a1; A(0,2) = 0;
   A(1,0) = a1; A(1,1) = 0;   A(1,2) = -a2;
   A(2,0) = 0;  A(2,1) = a2;  A(2,2) = 0;

   Lt = A*L - L*A;
   cout << "Lt =\n" << Lt << endl;

   b1t = Lt(0,0); b2t = Lt(1,1); b3t = Lt(2,2);
   a1t = Lt(0,1); a2t = Lt(1,2);
   cout << "b1t = " << b1t << ", b2t = " << b2t
```

```
           << ", b3t = " << b3t << endl;
   cout << "a1t = " << a1t << ", a2t = " << a2t << endl;
   cout << endl;

   // Show that I(0),I(1),I(2) are first integrals
   int n = 3;
   Symbolic result;
   Symbolic I("I",n);

   I(0) = L.trace();        cout << "I(0) = " << I(0) << endl;
   I(1) = (L*L).trace();    cout << "I(1) = " << I(1) << endl;
   I(2) = L.determinant(); cout << "I(2) = " << I(2) << endl;
   cout << endl;

   for(int i=0;i<n;i++)
   {
     result = b1t*df(I(i),b1) + b2t*df(I(i),b2) + b3t*df(I(i),b3)
            + a1t*df(I(i),a1) + a2t*df(I(i),a2);

     cout << "result" << i+1 << " = " << result << endl;
   }
   return 0;
}
```

In the program, Lt corresponds to dL/dt and b1t, b2t, b3t, a1t, a2t correspond to db_1/dt, db_2/dt, db_3/dt, da_1/dt, da_2/dt, respectively.

The output shows that the equations of motion are equivalent to the Lax representation by comparing db_1/dt to the $(0,0)$-th, db_2/dt to the $(1,1)$-th, ...entry of the matrix L. Here result1 = 0, result2 = 0, result3 = 0 indicate that I[0], I[1] and I[2] are first integrals of the system.

9.8.3 Padé Approximant

When a power series of a function f diverges, the function has singularities in a certain region. A *Padé approximant* is a ratio of polynomials that contains the same information as the power series over an interval, often with information about whether singularities exist. In the $[N, M]$ Padé approximant the numerator has degree M and the denominator has degree N. The coefficients are determined by equating like powers of x in the equation

$$f(x)Q(x) - P(x) = Ax^{M+N+1} + Bx^{M+N+2} + \cdots \quad \text{with } Q(0) = 1$$

where $P(x)/Q(x)$ is the $[N, M]$ Padé approximant to f.

Suppose the solution to a differential equation can be expressed by a k-th order Taylor series

$$f(x) = a_0 + a_1x + a_2x^2 + \cdots + a_kx^k.$$

The $[N, M]$ Padé approximant $P_N^M(x)$ is given explicitly in terms of the coefficients a_j

$$[N,M](x) := \cfrac{\det \begin{vmatrix} a_{M-N+1} & a_{M-N+2} & \cdots & a_{M+1} \\ \vdots & \vdots & & \vdots \\ a_M & a_{M+1} & \cdots & a_{M+N} \\ \sum_{j=N}^{M} a_{j-N}x^j & \sum_{j=N-1}^{M} a_{j-N+1}x^j & \cdots & \sum_{j=0}^{M} a_j x^j \end{vmatrix}}{\det \begin{vmatrix} a_{M-N+1} & a_{M-N+2} & \cdots & a_{M+1} \\ \vdots & \vdots & & \vdots \\ a_M & a_{M+1} & \cdots & a_{M+N} \\ x^N & x^{N-1} & \cdots & 1 \end{vmatrix}}$$

with $N + M + 1 = k$. Note that $a_j \equiv 0$ if $j < 0$, and the sums for which the initial value is larger than the final value are taken to be zero. It often happens that $P_N^M(x)$ converges to the true solution of the differential equation as $N, M \to \infty$ even when the Taylor series solution diverges. Usually we only consider the convergence of the Padé sequence

$$\left\{ P_0^J(x), P_1^{J+1}(x), P_2^{J+2}(x), \dots \right\}$$

having $M = N + J$ and J is held constant while $N \to \infty$. The special sequence with $J = 0$ is called the *diagonal sequence*.

In the program we calculate the Padé approximant $[1,1]$, $[2,2]$ and $[3,3]$ for

$$f(x) = \sin(x) = \sum_{k=0}^{\infty} \frac{(-1)^k x^{2k+1}}{(2k+1)!} = x - \frac{1}{6}x^3 + \cdots .$$

Specifically, $[2,2]$ can be calculated as follows

$$[2,2] = \cfrac{\det \begin{vmatrix} a_1 & a_2 & a_3 \\ a_2 & a_3 & a_4 \\ a_0 x^2 & a_0 x + a_1 x^2 & a_0 + a_1 x + a_2 x^2 \end{vmatrix}}{\det \begin{vmatrix} a_1 & a_2 & a_3 \\ a_2 & a_3 & a_4 \\ x^2 & x & 1 \end{vmatrix}}$$

with $a_0 = 0, a_1 = 1, a_2 = 0, a_3 = -1/6$ and $a_4 = 0$. We find that

$$[2,2] = \frac{x}{1 + \dfrac{x^2}{6}} .$$

```
// pade.cpp

#include <iostream>
#include "symbolicc++.h"
using namespace std;

Symbolic Taylor(Symbolic u,Symbolic &x,int n)
```

```
{
   Symbolic series = u[x==0];
   int fac = 1;

   for(int j=1;j<=n;j++)
   {
      u = df(u,x); fac = fac*j;
      series += u[x==0]/fac*(x^j);
   }
   return series;
}

Symbolic Pade(const Symbolic &f,Symbolic &x,int N,int M)
{
   int i, j, k, N1 = N+1, M1 = M+1, n = M + N1;
   Symbolic y, z;
   Symbolic a("a",n);
   Symbolic P(Symbolic(0),N1,N1), Q(Symbolic(0),N1,N1);

   y = Taylor(f,x,n);
   for(i=0;i<n;i++) a(i) = y.coeff(x,i);

   for(i=0;i<N;i++)
      for(j=0;j<N1;j++)
      {
         k = M-N+i+j+1;
         if(k >= 0) P(i,j) = Q(i,j) = a(k);
         else       P(i,j) = Q(i,j) = 0;
      }

   for(i=0,i<N1;i++)
   {
      for(j=N-i;j<M1;j++)
      {
         k = j-N+i;
         if(k >= 0) P(N,i) += a(k)*(x^j);
      }
      Q(N,i) = x^(N-i);
   }
   y = det(P); z = det(Q);
   return y/z;
}

int main(void)
{
   Symbolic x("x"), f;
   f = sin(x);
   cout << Pade(f,x,1,1) << endl; // => x
   cout << Pade(f,x,2,2) << endl; // => -1/6*x*(-1/6-1/36*x^2)^(-1)
   cout << Pade(f,x,3,3) << endl;
       // => (-7/2160*x+1589885/3306816*x^3)*(-7/2160-7/43200*x^2)^(-1)
   return 0;
}
```

9.9   Array and Symbolic Classes

9.9.1   Pseudospherical Surfaces and Soliton Equations

In Section 3.2.2, we described how to find the *sine-Gordon equation* from a metric tensor field. An implementation using REDUCE was given. Here we give an implementation using SymbolicC++. In this case u is declared as a variable dependent on x1 and x2. The operator df() denotes differentiation. Since terms of the form $\cos^2(x)$ and $\sin^2(x)$ result from our calculation, we have included the identity

$$\sin^2(u) + \cos^2(u) \equiv 1$$

to simplify the expressions. This is done using the subst() function (via the C++ indexing operator []) in symbolicc++.h.

```cpp
// tensor.cpp

#include <iostream>
#include "array.h"
#include "symbolicc++.h"
using namespace std;

int main(void)
{
   int a,b,m,n,c,K=2;
   Symbolic g("g",K,K), g1("g1",K,K);    // inverse of g
   Array<Symbolic,2> Ricci(K,K), Ricci1(K,K);
   Array<Symbolic,3> gamma(K,K,K);
   Array<Symbolic,4> R(K,K,K,K);
   Symbolic u("u"), x("x",2);
   Symbolic sum, RR;

   // u depends on x1 and x2
   u = u[x(0),x(1)];

   g(0,0) = 1;          g(0,1) = cos(u);
   g(1,0) = cos(u);   g(1,1) = 1;

   g1 = g.inverse();

   for(a=0;a<K;a++)
      for(m=0;m<K;m++)
         for(n=0;n<K;n++)
         {
            sum = 0;
            for(b=0;b<K;b++)
               sum += g1(a,b)*(df(g(b,m),x(n))+df(g(b,n),x(m))
                       -df(g(m,n),x(b)));
            gamma[a][m][n] = sum / 2;

            cout << "gamma(" << a << "," << m << "," << n << ") = "
                  << gamma[a][m][n] << endl;
         }
   cout << endl;

   for(a=0;a<K;a++)
```

```
        for(m=0;m<K;m++)
          for(n=0;n<K;n++)
            for(b=0;b<K;b++)
              {
                R[a][m][n][b] = df(gamma[a][m][b],x(n))
                                -df(gamma[a][m][n],x(b));

                for(c=0;c<K;c++)
                  {
                    R[a][m][n][b] += gamma[a][c][n]*gamma[c][m][b]
                                   - gamma[a][c][b]*gamma[c][m][n];
                  }
                R[a][m][n][b] = R[a][m][n][b].subst(cos(u)*cos(u),
                                             1-sin(u)*sin(u));

              }

  for(m=0;m<K;m++)
    for(n=0;n<K;n++)
      {
        Ricci[m][n] = 0;
        for(b=0;b<K;b++) Ricci[m][n] += R[b][m][b][n];

        cout << "Ricci(" << m << "," << n << ") = "
             << Ricci[m][n] << endl;
      }
  cout << endl;

  for(m=0;m<K;m++)
    for(n=0;n<K;n++)
      {
        Ricci1[m][n] = 0;
        for(b=0;b<K;b++) Ricci1[m][n] += g1(m,b)*Ricci[n][b];
      }

  RR = 0;
  for(b=0;b<K;b++) RR += Ricci1[b][b];
  RR = RR[cos(u)*cos(u)==1-sin(u)*sin(u)];
  cout << "R = " << RR << endl;
  return 0;
}
```

9.10 Polynomial and Symbolic Classes

9.10.1 Picard's Method

We consider *Picard's method* to approximate a solution to the differential equation

$$\frac{dy}{dx} = f(x, y)$$

with initial condition $y(x_0) = y_0$, where f is an analytic function of x and y. Integrating both sides yields

$$y(x) = y_0 + \int_{x_0}^{x} f(s, y(s))ds.$$

Now starting with y_0 this formula can be used to approach the exact solution iteratively if the procedure converges. The next approximation is given by

$$y_{n+1}(x) = y_0 + \int_{x_0}^{x} f(s, y_n(s)) ds.$$

The example approximates the solution of the linear differential equation

$$\frac{dy}{dx} = x + y$$

and the nonlinear differential equation

$$\frac{dy}{dx} = x + y^2$$

using five and four steps of Picard's method. The initial conditions are $y(x = 0) = 1$. We also give the value $y(x = 2)$ for these approximations. In the first program picard.cpp we use the Polynomial class and integration from this class. In the second program spicard.cpp we use integration from the Symbolic class.

```cpp
// picard.cpp

#include <iostream>
#include <string>
#include "polynomial.h"
#include "rational.h"
using namespace std;

int main(void)
{
 Polynomial<Rational<Verylong> > x("x");
 Polynomial<Rational<Verylong> > pic(x);
 Rational<Verylong> zero(string("0")), one(string("1")), two(string("2"));
 int i;

 cout << endl << "x+y up to fifth approximation :" << endl;
 pic = one;
 cout << pic << endl;
 for(i=1;i<=5;i++)
 {
   //integrate and evaluate at the boundaries x and zero
   pic = one + Int(x+pic,"x") - (Int(x+pic,"x"))(zero);
   cout << pic << endl;
 }
 cout << "The approximation at x=2 gives " << pic(two) << endl;

 cout << endl << "x+y^2 up to fourth approximation :" << endl;
 pic = one;
 cout << pic << endl;
 for(i=1;i<=4;i++)
 {
   //integrate and evaluate at the boundaries x and zero
   pic = one + Int(x+(pic^2),"x") - (Int(x+(pic^2),"x"))(zero);
   cout << pic << endl;
 }
```

```
 cout << "The approximation at x=2 gives " << pic(two) << endl;

 return 0;
}
```

```
// spicard.cpp

#include <iostream>
#include "symbolicc++.h"
using namespace std;

int main(void)
{
 Symbolic x("x"), pic;
 int i;

 cout << endl << "x+y up to fifth approximation :" << endl;
 pic = 1;
 cout << pic << endl;

 for(i=1;i<=5;i++)
 {
  //integrate and evaluate at the boundaries x and zero
  pic = 1+integrate(x+pic,x)-integrate(x+pic,x)[x==0];
  cout << pic << endl;
 }

 cout << "The approximation at x=2 gives " << pic[x==2]
      << " (" << pic[x==2.0] << ")" << endl;

 cout << endl << "x+y^2 up to fourth approximation :" << endl;
 pic = 1;
 cout << pic << endl;

 for(i=1;i<=4;i++)
 {
  //integrate and evaluate at the boundaries x and zero
  pic = 1+integrate(x+pic*pic,x)-integrate(x+pic*pic,x)[x==0];
  cout << pic << endl;
 }
 cout << "The approximation at x=2 gives " << pic[x==2]
      << " (" << pic[x==2.0] << ")" << endl;

 return 0;
}
```

9.11   Lie Series Techniques

Let us consider an autonomous system of ordinary differential equations

$$\frac{d\mathbf{x}}{dt} = \mathbf{f}(\mathbf{x}), \quad \mathbf{x}(t=0) \equiv \mathbf{x}_0$$

where $\mathbf{x} = (x_1, \ldots, x_n)^T$ and $\mathbf{x}_0$ is the initial value at $t = 0$. Let f_j be analytic functions defined on $\mathbf{R}^n$. We consider the analytic vector field

$$V := \sum_{j=1}^{n} f_j(\mathbf{x}) \frac{\partial}{\partial x_j}.$$

Then the solution of the initial value problem, for sufficiently small t, can be written as

$$x_j(t) = \exp(tV) x_j|_{\mathbf{x}=\mathbf{x}(0)}$$

where $j = 1, 2, \ldots, n$. Expanding the exponential function yields

$$x_j(t) = x_j(0) + tV(x_j)|_{\mathbf{x}=\mathbf{x}(0)} + \frac{t^2}{2} V(V(x_j))|_{\mathbf{x}=\mathbf{x}(0)} + \cdots.$$

This method is called the *Lie series technique*. Let us consider the *Lorenz model* which is given by

$$\frac{dx_1}{dt} = \sigma(x_2 - x_1)$$

$$\frac{dx_2}{dt} = -x_1 x_3 + r x_1 - x_2$$

$$\frac{dx_3}{dt} = x_1 x_2 - b x_3$$

where σ, b and r are positive constants. The vector field is given by

$$V = \sigma(x_2 - x_1) \frac{\partial}{\partial x_1} + (-x_1 x_3 + r x_1 - x_2) \frac{\partial}{\partial x_2} + (x_1 x_2 - b x_3) \frac{\partial}{\partial x_3}.$$

Hence the solution for the system is

$$\mathbf{x}(t) = \begin{pmatrix} e^{tV} x_1 \\ e^{tV} x_2 \\ e^{tV} x_3 \end{pmatrix}_{\mathbf{x}=\mathbf{x}(0)} = \begin{pmatrix} x_1(0) + tV(x_1)|_{\mathbf{x}=\mathbf{x}(0)} + \dfrac{t^2}{2} V(V(x_1))|_{\mathbf{x}=\mathbf{x}(0)} + \cdots \\ x_2(0) + tV(x_2)|_{\mathbf{x}=\mathbf{x}(0)} + \dfrac{t^2}{2} V(V(x_2))|_{\mathbf{x}=\mathbf{x}(0)} + \cdots \\ x_3(0) + tV(x_3)|_{\mathbf{x}=\mathbf{x}(0)} + \dfrac{t^2}{2} V(V(x_3))|_{\mathbf{x}=\mathbf{x}(0)} + \cdots \end{pmatrix}$$

where

$$V(x_1) = \sigma(x_2 - x_1)$$
$$V(x_2) = -x_2 - x_1 x_3 + r x_1$$
$$V(x_3) = x_1 x_2 - b x_3$$

and

$$V(V(x_1)) = -\sigma^2(x_2 - x_1) + \sigma(-x_2 - x_1 x_3 + r x_1)$$
$$V(V(x_2)) = \sigma(x_2 - x_1)(-x_3 + r) + x_2 + x_1 x_3 - r x_1 - x_1(x_1 x_2 - b x_3)$$
$$V(V(x_3)) = \sigma x_2(x_2 - x_1) + x_1(-x_2 - x_1 x_3 + r x_1) - b(x_1 x_2 - b x_3).$$

The Lorenz model possesses a number of interesting properties. The divergence is $-(\sigma + b + 1)$. Hence each small volume element shrinks to zero as $t \to \infty$, at a rate independent of x_1, x_2, x_3. However, this does not imply that each volume element shrinks to a point. For certain parameter values (for example $\sigma = 10$, $b = 8/3$ and $r = 28$) we find that nearby trajectories separate exponentially. The system shows chaotic behavior.

In the following program, we apply the Lie series techniques to the Lorenz model. Here, we have expanded the exponential function up to second order. In the second part of the program, we iterate the Lie Series solution with $\sigma = 10$, $b = 8/3$ and $r = 28$.

```cpp
// lie.cpp

#include <iostream>
#include "symbolicc++.h"
using namespace std;

const int N = 3;

Symbolic u("u",N), ut("ut",N);

// The vector field V
template <class T> T V(const T& ss)
{
   T sum(0);
   for(int i=0;i<N;i++) sum += ut(i)*df(ss,u(i));
   return sum;
}

int main(void)
{
   int i,j;
   Symbolic t("t"), s("s"), b("b"), r("r");
   Symbolic us("",N);
   Equations values;

   cout << u << endl;
   // Lorenz model
   ut(0) = s*(u(1)-u(0));
   ut(1) = -u(1)-u(0)*u(2)+r*u(0);
   ut(2) = u(0)*u(1)-b*u(2);

   // Taylor series expansion up to order 2
   for(i=0;i<N;i++)
    us(i) = u(i)+t*V(u(i))+0.5*t*t*V(V(u(i)));
   cout << "us =\n" << us << endl;

   // Evolution of the approximate solution
   values = (t==0.01,r==40.0,s==16.0,b==4.0,u(0)==0.8,u(1)==0.8,u(2)==0.8);

   for(j=0;j<100;j++)
   {
    Equations newvalues = (t==0.01,r==40.0, s==16.0,b==4.0);
    for(i=0;i<N;i++)
    {
```

```
   newvalues = (newvalues,u(i)==us(i)[values]);
   cout << newvalues.back() << endl;
  }
  values = newvalues;
 }
 return 0;
}
```

9.12   Spectra of Small Spin Clusters

Consider the spin Hamilton operator

$$\hat{H} = a \sum_{j=1}^{3} \sigma_3(j)\sigma_3(j+1) + b \sum_{j=1}^{3} \sigma_1(j)$$

where a, b are real constants and

$$\sigma_i(1) = \sigma_i \otimes I_2 \otimes I_2$$

$$\sigma_i(2) = I_2 \otimes \sigma_i \otimes I_2$$

$$\sigma_i(3) = I_2 \otimes I_2 \otimes \sigma_i$$

where $\otimes$ denotes the Kronecker product, σ_i are the Pauli matrices

$$\sigma_1 := \begin{pmatrix} 0 & 1 \\ 1 & 0 \end{pmatrix}, \qquad \sigma_2 := \begin{pmatrix} 0 & -i \\ i & 0 \end{pmatrix}, \qquad \sigma_3 := \begin{pmatrix} 1 & 0 \\ 0 & -1 \end{pmatrix}$$

and I_2 is the 2×2 identity matrix. We adopt the cyclic boundary conditions, i.e., $\sigma_3(4) \equiv \sigma_3(1)$. The matrix representation of the first term on the right hand side is a diagonal matrix

$$\sum_{j=1}^{3} \sigma_3(j)\sigma_3(j+1) = \mathrm{diag}(3a, -a, -a, -a, -a, -a, -a, 3a).$$

The second term leads to non-diagonal terms. The 8×8 symmetric matrix for $\hat{H}$ becomes

$$\begin{pmatrix}
3a & b & b & 0 & b & 0 & 0 & 0 \\
b & -a & 0 & b & 0 & b & 0 & 0 \\
b & 0 & -a & b & 0 & 0 & b & 0 \\
0 & b & b & -a & 0 & 0 & 0 & b \\
b & 0 & 0 & 0 & -a & b & b & 0 \\
0 & b & 0 & 0 & b & -a & 0 & b \\
0 & 0 & b & 0 & b & 0 & -a & b \\
0 & 0 & 0 & b & 0 & b & b & 3a
\end{pmatrix}.$$

In the program we calculate the spin Hamilton operator symbolically. We determine the trace and determinant of the matrix symbolically. Then we substitute

$$a = \frac{1}{2}, \qquad b = \frac{1}{3}$$

into the matrix and the determinant to get a matrix with numerical values.

```cpp
// spins3.cpp

#include <iostream>
#include <fstream>
#include "symbolicc++.h"
using namespace std;

ofstream fout("spinS3.dat");

Symbolic sigma(char coord,int index,int N)
{
  int i;
  Symbolic I("I",2,2);
  I = I.identity();
  Symbolic result, s("s",2,2);

  if(coord=='x')
  { s(0,0) = 0; s(0,1) = 1; s(1,0) = 1; s(1,1) = 0; }
  else // 'z'
  { s(0,0) = 1; s(0,1) = 0; s(1,0) = 0; s(1,1) = -1; }

  if(index==0)
  {
   result = s;
   for(i=1;i<N;i++) result = kron(result,I);
  }
  else
  {
   result = I;
   for(i=1;i<index;i++)  result = kron(result,I);
   result = kron(result,s);
   for(i=index+1;i<N;i++) result = kron(result,I);
  }
  return result;
}

Symbolic H(int N,Symbolic a,Symbolic b)
{
  int i;
  Symbolic result, part, sigmaX("X",N), sigmaZ("Z",N);

  for(i=0;i<N;i++) sigmaX(i) = sigma('x',i,N);
  for(i=0;i<N;i++) sigmaZ(i) = sigma('z',i,N);

  part = 0;
  for(i=1;i<=N;i++) part += sigmaZ(i-1)*sigmaZ(i%N);
  result = a*part;

  part = 0;
  for(i=0;i<N;i++) part += sigmaX(i);
  result += b*part;

  return result;
}

int main(void)
{
```

```
int N=3;
Symbolic c = Symbolic(1)/2, d = Symbolic(1)/3;
Symbolic a("a"), b("b"), p("p"), det;
Symbolic result;

result = H(N,a,b);
fout << result << endl;
fout << "trace = " << result.trace() << endl;

det = result.determinant();
fout << "determinant = " << det << endl;
fout << endl;

// assigning numerical values
fout << "Put a = " << c << " and b = " << d << endl;
fout << endl;

fout << "The matrix becomes:" << endl;
fout << result[a == c, b == d] << endl;
fout << "determinant = " << det[a == c, b == d] << endl;
return 0;
}
```

9.13   Nonlinear Maps and Chaotic Behavior

We consider the map

$$x_{t+1} = r(3y_t + 1)x_t(1 - x_t)$$
$$y_{t+1} = r(3x_t + 1)y_t(1 - y_t)$$

which shows the Ruelle–Takens–Newhouse transition into chaos. The bifurcation parameter is r. We calculate the variational equation symbolically

$$u_{t+1} = r(3y_t + 1)(1 - 2x_t)u_t + 3rx_t(1 - x_t)v_t$$
$$v_{t+1} = 3ry_t(1 - y_t)u_t + r(3x_t + 1)(1 - 2y_t)v_t$$

and then iterate these four equations using the data type double. The largest Ljapunov exponent is calculated approximately for $r = 1.0834$

$$\lambda \approx \frac{1}{T} \ln \left(|u_T| + |v_T| \right)$$

where T is large. The fixed points of the map are given as the solutions of

$$r(3y^* + 1)x^*(1 - x^*) = x^*, \qquad r(3x^* + 1)y^*(1 - y^*) = y^*.$$

We find

$$x_1^* = \frac{1}{3r} \left(-\sqrt{4r^2 - 3r} + r \right), \; y_1^* = \frac{1}{3r} \left(-\sqrt{4r^2 - 3r} + r \right),$$
$$x_2^* = \frac{1}{3r} \left(\sqrt{4r^2 - 3r} + r \right), \; y_2^* = \frac{1}{3r} \left(\sqrt{4r^2 - 3r} + r \right),$$

$$x_3^* = \frac{r-1}{r}, \ y_3^* = 0,$$
$$x_4^* = 0, \ y_4^* = 0,$$
$$x_5^* = 0, \ y_5^* = \frac{r-1}{r}.$$

The fixed points (x_1^*, y_1^*) and (x_2^*, y_2^*) exist only for $r \geq 3/4$.

In the program we calculate symbolically the variational equations of the map described above. Then, we calculate the Ljapunov exponent with $r = 1.0834$ and the initial conditions

$$x_0 = 0.3, \qquad y_0 = 0.4, \qquad u_0 = 0.5, \qquad v_0 = 0.6.$$

We find that the Ljapunov exponent for the map is approximately 0.22.

```
// var.cpp

#include <iostream>
#include <cmath>
#include "symbolicc++.h"
using namespace std;

Symbolic f(Symbolic &x,Symbolic &y,Symbolic &r)
{ return r*(3.0*y+1.0)*x*(1.0-x); }

Symbolic g(Symbolic &x,Symbolic &y,Symbolic &r)
{ return r*(3.0*x+1.0)*y*(1.0-y); }

int main(void)
{
   int T, N = 1000;
   double x2, y2, u2, v2;
   Symbolic x("x"), x1("x1"), y("y"), y1("y1"), r("r"),
            u("u"), u1("u1"), v("v"), v1("v1");

   x1 = f(x,y,r);
   y1 = g(x,y,r);

   cout << "x1 = " << x1 << endl;
   cout << "y1 = " << y1 << endl;

   // Variational Equation
   u1 = df(x1,x)*u + df(x1,y)*v;
   v1 = df(y1,y)*v + df(y1,x)*u;

   cout << "u1 = " << u1 << endl;
   cout << "v1 = " << v1 << endl;
   cout << endl;

   // Calculation of the Ljapunov exponent by iterating the four equations.

   // Initial values
   Equations values = (x==0.3,y==0.4,r==1.0834,u==0.5,v==0.6);

   for(T=1;T<N;T++)
```

```
{
    x2 = x1[values]; y2 = y1[values]; u2 = u1[values]; v2 = v1[values];

    values = (r==1.0834,x==x2,y==y2,u==u2,v==v2);
    cout << "The Ljapunov exponent for T = " << T << " is "
         << log(fabs(double(rhs(values,u)))
                 +fabs(double(rhs(values,v))))/T << endl;
}
  return 0;
}
```

9.14   Numerical-Symbolic Application

Consider the equation

$$f(x) = 0$$

where it is assumed that f is at least twice differentiable. Let I be some interval containing a root of f. A method that approximates the root of f can be derived by taking the tangent line to the curve $y = f(x)$ at the point $(x_n, f(x_n))$ corresponding to the current estimate, x_n, of the root. The intersection of this line with the x-axis gives the next estimate to the root, x_{n+1}. The gradient of the curve $y = f(x)$ at the point $(x_n, f(x_n))$ is $f'(x_n)$. The tangent line at this point has the form $y = f'(x_n) * x + b$. Since this passes through $(x_n, f(x_n))$ we see that $b = f(x_n) - x_n * f'(x_n)$. Therefore the tangent line is

$$y = f'(x_n) * x + f(x_n) - x_n * f'(x_n).$$

To determine where this line cuts the x-axis we set $y = 0$. Taking the point of intersection as the next estimate, x_{n+1}, to the root, we have $0 = f'(x_n) * x_{n+1} + f(x_n) - x_n * f'(x_n)$. It follows that

$$x_{n+1} = x_n - \frac{f(x_n)}{f'(x_n)}, \quad n = 0, 1, 2, \ldots$$

This is the *Newton-Raphson scheme* [20], which has the following form

next-estimate = current estimate + correction term.

The correction term is $-f(x_n)/f'(x_n)$ and this must be small when x_n is close to the root if convergence is to be achieved. This depends on the behavior of $f'(x)$ near the root and, in particular, difficulty will be encountered when $f'(x)$ and $f(x)$ have roots close together. The Newton–Raphson method is of the form $x_{n+1} = g(x_n)$ with $g(x) := x - f(x)/f'(x)$. The order of the method can be examined. Differentiating this equation leads to $g'(x) = (f(x)f''(x))/((f'(x))^2)$. For convergence we require that

$$\left| \frac{f(x)f''(x)}{(f'(x))^2} \right| < 1$$

for all x in some interval I containing the root. Since $f(\alpha) = 0$, the above condition is satisfied at the root $x = \alpha$ provided that $f'(\alpha) \neq 0$. Then, provided that $g(x)$

is continuous, an interval I must exist in the neighborhood of the root over which the condition is satisfied. Difficulty is sometimes encountered when the interval I is small because the initial guess must be taken within this interval. This usually arises when $f(x)$ and $f'(x)$ have roots close together since the correction term is inversely proportional to $f'(x)$.

We use of the method described above to find the root of the function

$$f(x) = 4x - \cos(x).$$

In the program, we compare the built-in data type `double` and the symbolic iteration with numerical substitution at each step. The sequence converges to the value $x = 0.2426746806$.

```cpp
// newton.cpp

#include <iostream>
#include <iomanip>    // for setprecision()
#include "symbolicc++.h"
using namespace std;

template <class T> T f(T x)   // f(x)
{ return 4*x-cos(x); }

int main(void)
{
   int N=7;
   double y, u0 = -1.0;
   Symbolic x("x"), ff, ff1, ff2, C;
   Equation value = (x==u0);
   ff = f(x); ff1 = df(ff,x); ff2 = df(ff1,x);

   // Set numerical precision to 10 decimal places
   cout << setprecision(10);
   cout << "f(x) = " << ff << endl;
   cout << "f'(x) = " << ff1 << endl;
   cout << "f''(x) = " << ff2 << endl;
   cout << endl;

   // ======= Condition for convergence =======
   C = ff*ff2/ff1;
   cout << "C = " << C << endl;
   cout << "|C(x=u0)| = " << fabs(double(C[x==u0])) << endl;
   cout << endl;

   // ======= Symbolic computation =========
   for(int i=0;i<N;i++)
   {
    y = double((x-ff/ff1)[value]);
    value = (x==y);
    cout << "x = " << x[value] << endl;
   }
   return 0;
}
```

9.15   Bose Systems

The algebra for Bose particles is described by the Bose creation and annihilation operators, $b^\dagger$ and b respectively, which obey the commutation relation

$$[b, b^\dagger] = b\,b^\dagger - b^\dagger b = I$$

where I is the identity operator. Obviously b and $b^\dagger$ are noncommuting operators. It follows that if f is an analytic function then

$$[b, f(b^\dagger)] = \frac{\partial}{\partial b^\dagger} f(b^\dagger), \qquad [f(b), b^\dagger] = \frac{\partial}{\partial b} f(b).$$

An expression $E(b, b^\dagger)$ in b and $b^\dagger$ is considered to be in a normal form when it is written in the form

$$E(b, b^\dagger) = \sum_{j=1}^{\infty} \sum_{k=0}^{\infty} E_{j,k} \left(b^\dagger\right)^j b^k$$

for some $E_{j,k} \in \mathbf{C}$. We find that the normal form for $[f(b), g(b^\dagger)]$ where f and g are analytic is given by [31]

$$[f(b), g(b^\dagger)] = \sum_{j=1}^{\infty} \frac{1}{j!} \left(\frac{\partial^j}{\partial b^{\dagger j}} g(b^\dagger)\right) \left(\frac{\partial^j}{\partial b^j} f(b)\right).$$

In the following program we consider $f(b) = b^m$ and $g(b^\dagger) = (b^\dagger)^m$ where $m = 0, 1, 2, 3, 4$. Thus we have three methods to obtain the normal form. We can apply the commutation relation $[b, b^\dagger] = I$ until it is no longer possible. Or we can apply the formula above, calculating the partial derivatives. Lastly we can calculate the partial derivatives explicitly to find the formula

$$[b^m, (b^\dagger)^m] = \sum_{j=1}^{m} j! \binom{m}{j}^2 (b^\dagger)^{m-j} b^{m-j}.$$

The program calculates the normal form using all three methods.

```
// bose.cpp

#include <iostream>
#include "symbolicc++.h"
using namespace std;

int C(int n, int r)
{
 int num = 1, den = 1;
 for(int i=n;i>r;i--) num *= i;
 for(int j=2;j<=(n-r);j++) den *= j;
 return num/den;
}

int main(void)
{
```

```
Symbolic b("b"), bd("b'");
b = ~b; bd = ~bd;

for(int m=0; m<5; m++)
{
 int fac = 1, j;
 Symbolic comm = (b^m)*(bd^m)-(bd^m)*(b^m);
 Symbolic res, res2;
 cout << m << ": " << comm << " = "
      << comm.subst_all(b*bd==1+bd*b) << endl;

 for(j=1; j<=m; j++)
 {
  fac *= j;
  res += df(bd^m,bd,j)*df(b^m,b,j)/fac;
  res2 += fac*C(m,j)*C(m,j)*(bd^(m-j))*(b^(m-j));
 }
 cout << "    " << comm << " = " << res << endl;
 cout << "    " << comm << " = " << res2 << endl;
}
return 0;
}
```

9.16   Grassman Product and Lagrange Multipliers

Consider optimization of the function $f(x, y)$ subject to the constraint $g(x, y) = 0$.
We assume that $f(x, y)$ and $g(x, y)$ are smooth functions. The *Lagrange multiplier*
method provides critical points for this problem when the partial derivatives of f
and g with respect to x and y exist and

$$\frac{\partial g}{\partial x} \neq 0, \qquad \frac{\partial g}{\partial y} \neq 0.$$

The method consists of solving the equations

$$\frac{\partial f}{\partial x} = \lambda \frac{\partial g}{\partial x}, \qquad \frac{\partial f}{\partial y} = \lambda \frac{\partial g}{\partial y}, \qquad g(x, y) = 0$$

to obtain the critical points where λ is the Lagrange multiplier. Since

$$df = \frac{\partial f}{\partial x} dx + \frac{\partial f}{\partial y} dy, \qquad dg = \frac{\partial g}{\partial x} dx + \frac{\partial g}{\partial y} dy$$

we find

$$df \wedge dg = \left(\frac{\partial f}{\partial x} \frac{\partial g}{\partial y} - \frac{\partial f}{\partial y} \frac{\partial g}{\partial x} \right) dx \wedge dy$$

where we used $dx \wedge dx = dy \wedge dy = 0$ and $dy \wedge dx = -dx \wedge dy$. At the critical points
we find

$$df \wedge dg = 0.$$

Thus to find the critical points it is sufficient to solve

$$df \wedge dg = 0, \qquad g(x, y) = 0.$$

Since $dx \wedge dy = -dy \wedge dx$ we treat dx and dy as noncommuting variables in the Grassman product. In the following program we determine the equations to find the critical points of $f(x, y) = 2x^2 + y^2$ subject to the constraint $x + y = 1$.

```cpp
// lagrange.cpp

#include <iostream>
#include "symbolicc++.h"
using namespace std;

int main(void)
{
 Symbolic x("x"), y("y"), dx("dx"), dy("dy");
 Symbolic f = 2*x*x+y*y;
 Symbolic g = x+y-1;

 // noncommutative
 dx = ~dx; dy = ~dy;

 cout << "f = " << f << endl;
 cout << "g = " << g << endl;
 Symbolic d_f = df(f,x)*dx+df(f,y)*dy;
 Symbolic d_g = df(g,x)*dx+df(g,y)*dy;
 cout << "d_f = " << d_f << endl;
 cout << "d_g = " << d_g << endl;
 Symbolic wedge = (d_f*d_g).subst_all((dx*dx==0,dy*dy==0,dy*dx==-dx*dy));
 cout << (wedge.coeff(dx*dy)==0) << endl;
 return 0;
}
```

9.17   Interpreter for Symbolic Computation

Often it is inconvenient to compile and run a program to perform a simple symbolic computation. Most computer algebra systems provide a special purpose programming language and interpreter so that the user can perform symbolic computations interactively. In the following program we provide a simple interpreter that uses SymbolicC++ for the underlying symbolic computation. Since SymbolicC++ provides all the symbolic computation functionality, the interpreter concentrates mostly on providing a simple syntax. We interpret statements that end with a semicolon (;). Assignment is performed using the equals symbol (=). Two example sessions are given below.

Example 1. First integral.

We consider first integrals of systems of differential equations (see 9.7.1). Consider the system of differential equations

$$\frac{du_1}{dt} = u_1 - u_1 u_2, \qquad \frac{du_2}{dt} = -u_2 + u_1 u_2.$$

We find that

$$f(t) := \ln\left(u_1(t)\right) + \ln\left(u_2(t)\right) - u_1(t) - u_2(t)$$

is a first integral. To declare u as a function of t we use

```
u = function(u,t);
```

and to perform substitutions $f(x)\big|_{x \to y}$ we use

```
subst(f,x,y);
```

The following interactive session verifies that

$$f(t) := \ln(u_1(t)) + \ln(u_2(t)) - u_1(t) - u_2(t)$$

is a first integral.

```
u1 = function(u1,t);
 -> u1[t]
u2 = function(u2,t);
 -> u2[t]
f = ln(u1) + ln(u2) - u1 - u2;
 -> ln(u1[t])+ln(u2[t])-u1[t]-u2[t]
r = df(f, t);
 -> df(u1[t],t)*u1[t]^(-1)+df(u2[t],t)*u2[t]^(-1)
    -df(u1[t],t)-df(u2[t],t)
r1 = subst(r,  df(u1,t),  u1-u1*u2);
 -> -u2[t]+df(u2[t],t)*u2[t]^(-1)-u1[t]
    +u1[t]*u2[t]-df(u2[t],t)+1
r2 = subst(r1, df(u2,t), -u2+u1*u2);
 -> 0
```

Example 2. Invariants.

Consider the function $f : \mathbf{R} \to \mathbf{R}$

$$f(x) := 2x^2 - 1$$

and the function $f : \mathbf{R} \times \mathbf{R} \to \mathbf{R}$

$$g(x_1, x_2) := x_2 - 2x_1^2 + 2x_2^2 + d(1 + x_2 - 2x_1^2).$$

The function f is invariant under g if

$$f(f(x)) \equiv g(x, f(x)).$$

The following interactive session determines whether f is invariant under g.

```
f = 2*(x^2) - 1;
 -> 2*x^(2)-1
g = x2 - 2*(x1^2) + 2*(x2^2) + d*(1 + x2 - 2*(x1^2));
 -> x2-2*x1^(2)+2*x2^(2)+d*x2-2*d*x1^(2)+d
f1 = subst(f,x,f);
 -> 8*x^(4)-8*x^(2)+1
g1 = subst(g, x1, x);
 -> x2-2*x^(2)+2*x2^(2)+d*x2-2*d*x^(2)+d
g2 = subst(g1,x2,f);
 -> -8*x^(2)+8*x^(4)+1
f1;
 -> 8*x^(4)-8*x^(2)+1
```

The interpreter has to decompose a string into a sequence of tokens, where each token represents an operator, function or value. The sequence of tokens are evaluated using the standard precedence rules for arithmetic, i.e., first parenthesis, functions, exponentiation, multiplication, division, and finally addition and subtraction.

The function `get_tokens()` separates a string into tokens according to `separator` (an array of symbols used to separate tokens and are tokens themselves) and `ignore` (an array of symbols used to separate tokens but are discarded). The function `evaluate_tokens()` evaluates the list of tokens in the order assignment, parenthesis, functions and finally standard arithmetic operators according to precedence.

```cpp
// interpreter.cpp

#include <iostream>
#include <fstream>
#include <string>
#include <vector>
#include <map>
#include <cstdlib>
#include <cmath>
#include "symbolicc++.h"
using namespace std;

double error(string s)
{ cerr << "Error: " << s << ", using 0." << endl; return 0.0; }

class token
{
  private:
    int is_value;
    Symbolic v;
    string t;
    static map<string,Symbolic> values;
  public:
    token() : is_value(0), v(0), t("") {};
    token(const Symbolic &s) : is_value(1), v(s), t("") {};
    token(const string &s) : is_value(0), v(0), t(s)  {};
    Symbolic value();
    string name() { return t; }
    int isvalue() { return is_value; }
    Symbolic set(const Symbolic &d) { return values[t]=d; }
    int operator==(string s) { return (!is_value) && (t == s); }
    friend ostream& operator << (ostream&,token);
};

map<string,Symbolic> token::values;

Symbolic token::value()
{
  if(is_value) return v;
  char *end;
  int vali=(int)strtol(t.c_str(),&end,10);
  if(*end == '\0') return Symbolic(vali);
  double vald=strtod(t.c_str(),&end);
  if(*end == '\0') return Symbolic(vald);
  if(values.count(t)>0) return values[t];
```

```
  return Symbolic(t);
}

ostream& operator << (ostream& o,token t)
{ if(t.is_value) o << t.v; else o << t.t; return o;}

vector<token>
get_tokens(string s,string separator[],string ignore[])
{
  int i = 0, j, istoken = 0;
  vector<token> v;
  string value = "";
  while(i<(int)s.length())
  {
   istoken = 0;
   for(j=0;ignore[j] != "" && i<(int)s.length();j++)
    if(s.substr(i,ignore[j].length()) == ignore[j])
     i += ignore[j].length(), j = -1, istoken = 1;
   for(j=0;separator[j] != "" && !istoken;j++)
    if(s.substr(i,separator[j].length()) == separator[j])
    {
     if(value != "") { v.push_back(token(value)); value = ""; }
     v.push_back(token(separator[j]));
     i += separator[j].length();
     istoken = 1;
    }
   if(!istoken) value += s[i++];
   else if(value!="") { v.push_back(token(value)); value = ""; }
  }
  if(value != "") v.push_back(token(value));
  return v;
}

Symbolic spow(const Symbolic &x,const Symbolic &y) { return (x^y); }
Symbolic smul(const Symbolic &x,const Symbolic &y) { return x*y;  }
Symbolic sdiv(const Symbolic &x,const Symbolic &y) { return x/y;  }
Symbolic sadd(const Symbolic &x,const Symbolic &y) { return x+y;  }
Symbolic ssub(const Symbolic &x,const Symbolic &y) { return x-y;  }

Symbolic ssqrt(const vector<Symbolic> &x)  { return sqrt(x[0]);          }
Symbolic scos(const vector<Symbolic> &x)   { return cos(x[0]);           }
Symbolic ssin(const vector<Symbolic> &x)   { return sin(x[0]);           }
Symbolic stan(const vector<Symbolic> &x)   { return tan(x[0]);           }
Symbolic sexp(const vector<Symbolic> &x)   { return exp(x[0]);           }
Symbolic sln(const vector<Symbolic> &x)    { return ln(x[0]);            }
Symbolic slog(const vector<Symbolic> &x)   { return log(x[0],x[1]);      }
Symbolic sdf(const vector<Symbolic> &x)    { return df(x[0],x[1]);       }
Symbolic ssubst(const vector<Symbolic> &x) { return x[0].subst(x[1],x[2]);}
Symbolic sfunc(const vector<Symbolic> &x)  { return x[0][x[1]];          }

Symbolic evaluate(token t);

struct function {
 string name;
 int    args;
 Symbolic (*impl)(const vector<Symbolic>&);
};
```

```
Symbolic evaluate_tokens(vector<token> v)
{
  vector<token> v2, v3;
  int parenthesis, i, j, k;
  // function names, arity, and their implementation
  function functions[] = {
    { "sqrt",     1, ssqrt  },
    { "cos",      1, scos   },
    { "sin",      1, ssin   },
    { "tan",      1, stan   },
    { "exp",      1, sexp   },
    { "ln",       1, sln    },
    { "log",      2, slog   },
    { "df",       2, sdf    },
    { "subst",    3, ssubst },
    { "function", 2, sfunc  },
    { "" } };

  // default left operands for binary operators
  double initleft[] = { 1.0, 1.0, 0.0 };
  // binary operators and their implementation
  string opnames[][4] = { { "^", "" }, { "*", "/", "" }, { "+", "-", "" } };
  Symbolic (*opimpl[][3])(const Symbolic&,const Symbolic&) =
      { { spow }, { smul, sdiv }, { sadd, ssub } };

  // check for the assignment statement
  if(v.size()>2 && v[1] == "=") {
    for(j=2;j<(int)v.size();j++) v2.push_back(v[j]);
    return v[0].set(evaluate_tokens(v2));
  }

  // evaluate parenthesis first
  for(j=0;j<(int)v.size();j++)
  {
   if(v[j] == ")") return error("unbalanced parenthesis");
   else if(v[j] == "(")
   {
    for(parenthesis=1,j++;parenthesis && j<(int)v.size();j++)
    {
     if(v[j] == "(") parenthesis++;
     if(v[j] == ")") parenthesis--;
     // artificially end the parenthesized expression
     if(v[j] == "," && parenthesis == 1)
     {
      v2.push_back(token(evaluate_tokens(v3)));
      v3.clear();
     }
     else if(parenthesis) v3.push_back(v[j]);
    }
    if(parenthesis) return error("unbalanced parenthesis");
    v2.push_back(token(evaluate_tokens(v3)));
    v3.clear(); j--;
   }
   else v2.push_back(v[j]);
  }

  // evaluate functions
  for(j=0,v.clear();j<(int)v2.size();j++)
```

```
    {
    for(i=0;functions[i].name!="";i++)
    if(v2[j] == functions[i].name)
    {
      if(j+functions[i].args<(int)v2.size())
      {
        vector<Symbolic> args;
        for(k=1;k<=functions[i].args;k++)
         args.push_back(evaluate(v2[j+k]));
        v.push_back(token(functions[i].impl(args)));
        j+=functions[i].args;
      }
      else return error(functions[i].name         +
                        " without "                +
                        char('0'+functions[i].args) +
                        " arguments");
      break;
    }
    if(functions[i].name=="") v.push_back(v2[j]);
  }
  // evaluate operators in order of precedence
  for(k=0,v2.clear();k<3;k++,v = v2,v2.clear())
  {
    token left(initleft[k]);
    for(j=0;j<(int)v.size();j++)
    {
      for(i=0;opnames[k][i]!="";i++)
      if(v[j] == opnames[k][i])
      {
        if(v2.size()) v2.pop_back();
        if(j+1<(int)v.size())
          v2.push_back(token(opimpl[k][i](evaluate(left),
                                          evaluate(v[++j]))));
        else return error(opnames[k][i]+" without second argument");
        break;
      }
      if(opnames[k][i]=="") v2.push_back(v[j]);
      left = v2.back();
    }
  }
  // check that evaluation gave a single result
  if(v.size() != 1)
  {
    for(j=0;j<(int)v.size();j++)
     cerr << "token " << j+1 << " : " << v[j] << endl;
    return error("could not evaluate expression");
  }
  return v[0].value();
}

Symbolic evaluate(token t)
{ vector<token> v; v.push_back(t); return evaluate_tokens(v); }

Symbolic evaluateformula(istream &s)
{
  char c;
  string expression;
  static string ws[] = { " ", "\t", "\n", "\r", "" };
```

```
    static string separator[] = { "=", "+", "-", "*", "/",
                                  "^", "(", ")", ",", "" };
    do if((c = s.get()) != ';' && !s.eof()) expression += c;
    while(c != ';' && !s.eof());
    if(c != ';') return error("formula not terminated");
    vector<token> v = get_tokens(expression,separator,ws);
    return evaluate_tokens(v);
}

int main(void)
{
    while(!cin.eof())
    cout << " -> " << evaluateformula(cin) << endl;
    return 0;
}
```

Chapter 10

LISP and Computer Algebra

10.1   Introduction

LISP is short for *List Processing*. It is a computer language that is used in many applications of artificial intelligence. One of its major qualities is that it can manipulate lists easily. LISP was developed in the 1950s by John McCarthy [36].

It is one of the most commonly used languages for writing a computer algebra system. Reduce, Macsyma, Derive and Axiom are based on LISP. The basic data types in LISP are atoms and dotted pairs. Lists are built from dotted pairs. Most of the functions in LISP operate on lists. Atoms can be used as symbolic variables and lists can encode symbolic expressions using atoms for the variables. LISP was originally conceived as a system for symbolic computation (computation over symbols) which resulted in LISP being relatively well suited for computer algebra. Moreover arithmetic operations are also possible in LISP.

In this chapter we show how a computer algebra system can be built using LISP. This includes how simplification, differentiation, and polynomials are handled. We make use of functions and rely strongly on recursion. There are different dialects in LISP. In this chapter we use *Common LISP*. The programs listed here also run under *Portable Standard LISP* with some small modifications. In Common LISP, a function is indicated by (defun), whereas in Portable Standard LISP a function is indicated by (de). A comment in Common LISP is written as

```
; This is a single line comment in Common LISP
```

whereas a comment line in Portable Standard LISP is indicated by %

```
% This is a single line comment in Portable Standard LISP
```

The function (mapcan) is also implemented differently in Common LISP and Portable Standard LISP. In the following we use Common LISP.

LISP is described in a number of excellent textbooks [19], [2], [63] [57]. Some of them also discuss the implementations of symbolic manipulations [10], [26], [40], [61].

10.2   Basic Functions of LISP

Values in LISP are termed *S-expressions*, a contraction for symbolic expressions. An S-expression may be either an *atom*, which is written as a symbol, such as

```
a
part2
```

or a *dotted pair*, written in the form

```
(s_1 . s_2)
```

where s_1 and s_2 stand for arbitrary S-expressions. A dotted pair is the external representation of a data structure with two components, denoted car and cdr. The name car originates from "Contents of the Address part of Register" and cdr from "Contents of the Decrement part of Register" due to the 36 bit register size on the IBM 704 on which LISP was first implemented. It had 15 bits called the address part of the register and 15 bits called the decrement part. Thus 15 bits were available as pointers to data in each of the address and decrement parts. Each pointer could point to other LISP data such as atoms or dotted pairs. Some examples of dotted pairs are

```
(a . b)
(a . (b1 . b2))
((u . v) . (x . (y . z)))
```

An important subset of the S-expressions is the *list*, which satisfies the following constraints

- An atom is a list if it is the atom nil.

- A dotted pair (s_1 . s_2) is a list if s_2 is a list.

The atom nil is regarded as the null list. Therefore, a list in LISP is a sequence whose components are S-expressions. The following S-expressions

```
nil
(part2 . nil)
(a . (b . (c . nil)))
((part2 . a2) . nil)
```

are all lists. Lists are more conveniently expressed in list notation

```
(s_1  s_2 ... s_n)
```

for n ≥ 0. This is an abbreviation of

```
(s_1 . (s_2 . ( ... (s_n . nil) ... )))
```

For example,

```
(a b c)
```

abbreviates

```
(a . (b . (c . nil)))
```

The list (a) abbreviates

```
(a . nil)
```

and the list ( ) (the so-called empty list) abbreviates

```
nil
```

Note the difference between the dotted pair (a . b), and the two-component list (a b) which may also be written as (a . (b . nil)). Obviously a and (a) are not equivalent, because the latter abbreviates the dotted pair (a . nil).

S-expressions are the values manipulated by LISP programs. However, LISP programs are also S-expressions. For example, the notation for literals in LISP is

```
(quote S)
```

where S is an S-expression. Instead of (quote S) we can also write 'S. Syntactically, this is just a two-component list. The first component is the atom quote and the second is an S-expression S. Semantically, its value (relative to any state) is the S-expression S. For example, the value of

```
(quote part2)
```

is the atom part2, and the value of

```
(quote (a . b))
```

is the dotted pair (a . b). In LISP, an unquoted atom used as an expression is an *identifier*, except when it is nil or t, which always denote themselves. nil represents both logical falsehood and the empty list. t represents logical truth.

In LISP we use the notation (f x) for $f(x)$ and (f x y) for $f(x, y)$.

LISP provides five primitive operations for constructing, selecting and testing S-expression values

cons car cdr atom null

The function cons constructs a dotted pair from its two arguments. For example, the value of

```
(cons (quote a) (quote b))
```

is (a . b). Of course, the actual parameters of an invocation need not be literals. For example, the value of

```
(cons (quote a) (cons (quote b) nil))
```

is (a b). The functions `car` and `cdr` require a dotted pair (s_1 . s_2) as an argument and return the components s_1 and s_2, respectively. For example, the values of

```
(car (quote (a . b)))        (cdr (quote (a . b)))
```

are a and b, respectively. Next we consider the results of the functions `cons`, `car` and `cdr` on list arguments. The value of

```
(cons (quote a) (quote (b c d)))
```

is the list (a b c d). The result of applying `car` and `cdr` to a list argument is the first component of the list and the rest of the list, respectively. For example, the values of

```
(car (quote (a b c d)))       (cdr (quote (a b c d)))
```

are a and (b c d), respectively. Note also that the values of

```
(cons (quote a) nil)       (car (quote (a)))        (cdr (quote (a)))
```

are (a), a, and nil, respectively. The remaining two primitive functions in LISP are predicates, for which the result is one of the atoms — nil (representing false) or t (representing true). The value of

```
(atom E)
```

is t if the S-expression E is an atom, and nil if it is a dotted pair. The function `null` takes one argument. It returns t if its argument evaluates to nil, and returns nil otherwise. An example is

```
(null nil)
```

where the return value is t. The function > takes one or more arguments and returns t if they are ordered in decreasing order numerically. Otherwise, the function returns nil. For example,

```
(> 9 4)    ; return value : t
```

The function `zerop` takes a numeric argument and returns t if the argument evaluates to 0. Otherwise, the function returns nil. For example,

```
(zerop 1)   ; return value : nil
```

The arithmetic function + adds the arguments. For example,

```
(+ 2 4 -7)  ; return value : -1
```

The arithmetic function * multiplies the arguments. For example,

```
(* 4 5 6 7) ; return value : 840
```

The function `list` accepts one or more arguments. It places all of its arguments in a list. This function can also be invoked with no arguments, in which case it returns nil. For example (make a list out of the arguments)

```
(list 'c '(d f))  ; returns (c (d f))
```

To test whether two objects are identical (the same object) we use **eq**. We use **equal** to test whether two objects are identical in the sense that they would print the same. For example

```
(eq 'a 'a)             ; t
(eq (list 'a) (list 'a))     ; nil
(equal (list 'a) (list 'a)) ; t
```

LISP also has shortcut evaluation for truth tests, namely **and** and **or**. LISP interprets **nil** as false and non-**nil** as true. The operation **and** evaluates each argument and returns the first **nil** value and the last value otherwise. The **or** operation returns the first non-**nil** value and **nil** otherwise.

```
(and 1 2 3)      ; 3
(or  1 2 3)      ; 1
(and 1 nil 3)    ; nil
(or  nil 2 nil)  ; 2
(or  nil nil)    ; nil
```

We can define our own functions in LISP using **defun**, which is followed by the function name and a list of argument names for the function, followed by a sequence of expressions to be evaluated whenever the function is called. Thus we can define the function **square**

```
(defun square (x) (* x x))
(square 2)                    ; 4
```

which squares its argument. As a two argument example we consider the sum of squares $x, y \rightarrow x^2 + y^2$

```
(defun sum-square (a b) (+ (* a a) (* b b)))
(sum-square 2 3) ; 13
```

Since we are squaring both **a** and **b** it may be useful to define this in general, i.e. given (2 3) we would like to be able to compute (4 9). Perhaps we need the same operation for three numbers (2 3 4) $\rightarrow$ (4 9 16), then we would like to be able to generalize the computation on any list of numbers. This can be achieved with the function **mapcar**.

To perform the same operation on each element of a list we use **mapcar**

```
(mapcar #'car '( (a b c) (d e f) (h i j) )) ; (a d h)
```

mapcar takes as second argument a function with the same arity (number of arguments) as the number of lists in the remaining arguments. Thus to add corresponding elements in two lists we could use

```
(mapcar #'+ '(1 2 3) '(4 5 6)) ; (5 7 9)
```

Now if we wish to map square on the list (1 2 3 4), we would use

```
(mapcar square '(1 2 3 4)) ; Error
```

which gives an error due to the fact that `square` has no value. In LISP symbols can refer to both functions and values, depending on the context.

```
(setq square "square")
square                     ; "square" - variable
(square 2)                 ; 4          - function
```

To obtain the function interpretation we use `function` (or the shorter form `#'`) and `funcall` to call functions

```
(mapcar (function square) '(1 2 3 4)) ; (1 4 9 16)
(mapcar #'square '(1 2 3 4))          ; (1 4 9 16)
(funcall #'square 2)                  ; 4
```

If we only intend to use `square` once, it is inconvenient to use the symbol `square` when we can define the operation x → (* x x) independent of its name. Thus we can use the `lambda` form, which provides us with anonymous functions. For example

```
(mapcar #'(lambda (x) (* x x)) '(1 2 3 4)) ; (1 4 9 16)
```

Returning to `sum-square` we can now define a more flexible version

```
(defun sum_square (&rest l) (reduce #'+ (mapcar #'square l)))
(sum-square 2 3)     ; 13
(sum-square 3 5 9)   ; 115
```

Here we used `&rest l` to denote that all remaining arguments should be placed in the list `l` and `reduce` to combine all elements of a list

```
(defun show-rest (a &rest b) (princ b))
(show-rest 1)                      ; nil
(show-rest 'a '(1 3) 5 'rest)      ; ((1 3) 5 rest)
(reduce #'+ '(1 2 3))              ; 6
(reduce #'append '((1 2) (3 4) (5 6)))  ; (1 2 3 4 5 6)
```

The function `mapcan` takes two arguments: a function and a list. It maps the function over the elements in the list. The function used here must return a list, and the lists from the mapping are destructively spliced together. For example in Common LISP we have

```
(mapcan #'cdr '((a b c) (d e)))  ; return value : (b c e)
```

In Portable Standard LISP this is expressed as

```
(mapcan '((a b c) (d e)) 'cdr)  ; return value : (b c e)
```

Another useful function is `max` which determines the maximum of all of its arguments

```
(max 3 5 -1 7 9 2) ; 9
```

If we want to calculate the maximum of some values in a list we can `apply max` to the elements of the list

```
(apply #'max '(3 5 -1 7 9 2)) ; 9
```

Note that `reduce` is different to `apply`

```
; (sum-square 1 2 3) == 1^2 + 2^2 + 3^2
(apply #'sum-square '(1 2 3))  ; 14
; (sum-square (sum-square 1 2) 3) == (1^2 + 2^2)^2 + 3^2
(reduce #'sum-square '(1 2 3)) ; 34
```

If we wish to remove the definition of a variable or a function we use `unintern`. This is not usually necessary, but is useful in some situations (for example if a LISP system generates warnings about a function being redefined). For example

```
(defun cube (x) (* x x x))
(cube 2)           ; 8
(unintern 'cube)
(cube 2)                ; Error
```

Three more forms of quoting are available in the form of the backquote `, the comma , and splicing ,@ which are used for template style programming. The backquote allows quoting to be suppressed by using a comma

```
(eq 'a 'a)  ; t
(setq a 5)
(a = ,a)   ; (a = 5)
```

If we want to use the values in a list as part of another list we use splicing

```
(setq l '(1 2 3 4))
'(the squares of ,l are ,(mapcar #'square l))
; (the squares of (1 2 3 4) are (1 4 9 16))
'(the squares of ,@l are ,@(mapcar #'square l))
; (the squares of 1 2 3 4 are 1 4 9 16)
```

If we wish to display a LISP value on the screen we can use the `princ` and `print` functions, combined with `progn` which allows us to group expressions (evaluated in left to right order)

```
(princ '(+ (* 3 a) (/ a b)))      ; displays (+ (* 3 a) (/ a b))
(print "print")                   ; displays "print"
(princ "print")                   ; displays print
(progn (print 1) (print 2))       ; displays 1 and 2
(defun print12 () (print 1) (print 2))
(print12)                         ; displays 1 and 2
```

However, the function `format` provides powerful display and formatting capabilities similar to the `printf` function of C. The function `format` takes a stream as the first argument, or `t` for the standard output or `nil` to return the result as a string. The second argument is a format string that indicates how values should be displayed. The format string is displayed as provided, except that the character ~ in the format string denotes a special action. For example, `~A` displays the next argument in the same way as `princ` and `~%` begins display on the next line

```
(format nil "~A" '((* 3 5) => ,(* 3 5)))   ; "((* 3 5) => 15)"
(format t "~A ~% => ~A~%" '(* 3 5) (* 3 5)) ; displays (* 3 5)
;                                                       => 15
```

To introduce local variables we use `let` and `let*`. The `let` form consists of a list of variables and their values followed by statements that should be evaluated. The difference between `let` and `let*` is that `let*` allows subsequent local variables to refer to previously defined variables in the `let*` list. For example

```
(setq a 2)
(setq b 3)
(let ( (a     3)   ; a <- 3
       (b (* a a))) ; b <- a * a = 2 * 2, uses a above
    (princ b) )    ; displays 4 (local b)
(let* ( (a     3)   ; a <- 3
        (b (* a a))); b <- a * a = 3 * 3, uses local a
     (princ b) )    ; displays 9 (local b)
```

Similarly we use `labels` to locally define functions

```
(labels ( (square (x)   (* x x))
          (add    (x y) (+ x y)) )
        (format t "~A~%" (square 4)) ; displays 16
        (format t "~A~%" (add 2 3)) ); displays 5
```

We can reuse any functions defined in a file `file.lisp` using the `load` function

```
(load "file.lisp")
```

The function `cond` is used for *conditional processing*. It takes zero or more arguments, called *cases*. Every case is a list, whose first element is a test and the remaining elements are actions. The cases in a `cond` are evaluated one at a time, from first to last. When one of the tests returns a non-`nil` value, the remaining cases are skipped. The actions in the case are evaluated and the result is returned. If none of the tests returns a non-`nil` value, `cond` returns `nil`. For example, the function `length_list`, defined as

```
(defun length_list (L)
   (cond
      ((null L) 0)
      (t (+ 1 (length_list (cdr L)))))))
```

can be applied as follows

```
(setq L '(1 2 3 4 5 xx))  ; return value : (1 2 3 4 5 xx)
(length_list L)           ; return value : 6
```

Many functions in LISP can be defined in terms of other functions. For example, `caar` can be defined in terms of `car`. It is, therefore, natural to ask whether there is a smallest set of primitives necessary to implement the language. In fact, there is no single "best" minimal set of primitives; it all depends on the implementation. One possible set of primitives might include `car`, `cdr` and `cons` for manipulation of S-expressions, `quote`, `atom`, `read`, `print`, `eq` for equality, `cond` for conditionals, `setq` for assignment, and `defun` for definitions.

Recursion is at the heart of LISP programming. A recursive function is a function that calls itself again. Obviously we also need a stopping condition. It is important that any recursive function has a case (called the base case) for which it will not be called again so that the function does eventually return.

Example 1. The function `list_set` converts a list to a set using recursion.

```
; listset.lisp

(defun list_set (lis)
 (cond
  ( (not (consp lis)) nil )
  ( (member (car lis) (cdr lis)) (list_set (cdr lis)) )
  ( t (cons (car lis) (list_set (cdr lis))) ) ) )

(list_set '(a b x a c d 1 3 c 3)) ; (b x a d 1 c 3)
```

The function `consp(A : any) : boolean` returns `t` (true) if `A` is a dotted pair.

Example 2. The function `countsublists` counts the number of sublists in a list using recursion.

```
; countsub.lisp

(defun countsublists (lis)
 (cond
  ( (null lis) 0 )
  ( (atom lis) 0 )
  ( (atom (car lis)) (countsublists (cdr lis)) )
  ( t (+ 1
         (countsublists (car lis))
         (countsublists (cdr lis)) ) ) ) )

(countsublists '(a (1 2) (r s) (a b 23) z)) ; 3
(countsublists '())                         ; 0
```

Example 3. We show how `mapcar` could be implemented. The function `mymapcar` calls itself again except when given an empty list as argument.

```
(defun mymapcar (function list)
  (if (null list) nil
    (cons (funcall function (car list))
          (mymapcar function (cdr list)) ) ) )

(mymapcar #'car '( (1 2) (3 4) (5 6) )) ; (1 3 5)
```

10.3 LISP Macros and Infix Notation

The `format` example in the previous section shows that we can display a LISP expression together with its value which is a useful debugging tool. Unfortunately this requires us to list the identical expression twice which obscures the underlying intent of showing the same expression in two different ways. It would be clearer to express this functionality as

```
(show (* 3 5))
```

This is achieved using a LISP macro. Common LISP macros are defined using `defmacro`. Macros are LISP functions operating on a program expression interpreted as data and returns an expression which will be evaluated. Thus we can define the function `show`

```
(defmacro show (a)
  '(format t "~A~% => ~A~%" (quote ,a) ,a))
```

The statement (show (* 3 5)) is transformed as follows

```
(show (* 3 5))
'(format t "~A~% => ~A~%" (quote ,a) ,a))     ; a -> (* 3 5)
(format t "~A~% => ~A~%" (quote (* 3 5)) (* 3 5))
```

The last expression is returned to LISP as data, then LISP evaluates the expression as part of the program and consequently outputs

```
(* 3 5)
 => 15
```

We use a modified version of this macro extensively in the following programs to illustrate the results of applying different functions.

LISP prefix notation avoids the need for operator precedence, however mathematical formulae are usually expressed with the infix notation. To convert these formulae to prefix notation can be error prone. Instead we can use a LISP macro to automatically convert an infix expression to a prefix expression before a statement is evaluated by the LISP system.

```
; infix.lisp

; define the operators in decreasing precedence levels
(defvar precedence '( (* /) (+ -) ))

; convert any infix expressions with binary operations from ops
; to prefix form, first? specifies if this is the first element
; in a list
(defun convert (x ops first?)
  (cond
    ; pull out the first element of a single element list
    ( (and (listp x) (= (length x) 1) first?)
      (convert (car x) ops t) )
    ; a list of at least 3 elements: a op b ... -> (op a b) ...
    ;  op in ops has greater or equal precedence to other infix operators
    ;  in the expression to be converted
    ( (and (listp x) (> (length x) 2) (member (cadr x) ops))
      (convert (cons (list (cadr x) (car x) (caddr x))
                     (cdddr x) ) ops nil) )
    ; a pair of two elements - try to convert each element
    ( (consp x)  (cons (convert (car x) ops t)
      (convert (cdr x) ops nil)) )
    ; not a list to be converted
    ( t x ) ) )

; apply the conversion procedure for operators
; of decreasing precedence
(defun apply-precedence (pl x)
  (if (null pl) (convert x pl t)
      ; else apply the next precedence level
      ; after conversion for the current
      (apply-precedence (cdr pl) (convert x (car pl) t)) ) )
```

```
; convert an infix expression to prefix
; x - a list representing the expression to convert
(defun infix->prefix (&rest x) (apply-precedence precedence x) )

; macro for use within LISP
(defmacro infix (&rest x) (infix->prefix x))

; examples of infix within LISP
(princ (infix   2 + 3 * 5)) (newline)
(princ (infix (2 + 3) * 5)) (newline)

; example showing that infix is context sensitive
(labels ( (+ (x y) (append x y))
          (* (x y) (mapcar #'(lambda (x) (cons x y)) x)) )
    (princ (infix '(a b c) + '(d e f) * '(g h i))) (newline)
    (princ (infix ('(a b c) + '(d e f)) * '(g h i))) (newline) )
```

10.4 Examples from Symbolic Computation

10.4.1 Polynomials

In this section we show how to add and multiply polynomials [15] of the form

$$p(x) = \sum_{j=0}^{n} a_j x^j.$$

First we have to give a representation for the polynomial p, using a list of dotted pairs. The first element of the dotted pair is the exponent and the second element is the factor. Thus the `car` of the dotted pair is the exponent and the `cdr` of the dotted pair is the coefficient, e.g.

```
(car '(3 . 5))  ; return value : 3
(cdr '(3 . 5))  ; return value : 5
```

Consider, for example, the polynomial

$$p(x) = 3x^2 + 7x + 1.$$

The representation as a list of dotted pairs is given by

```
((2 . 3) (1 . 7) (0 . 1))
```

Consider another example: the polynomial $3x^2 + 2$ is represented as

```
((2 . 3) (0 . 2))
```

Using this representation, the polynomial 0 can be represented by `nil` or `((0 . 0))`, whereas the polynomial 1 is represented by `((0 . 1))`.

We give an implementation for the addition and multiplication of two polynomials [15]. *Recursion* is used here (i.e. the function calls itself). If the polynomial P has

m terms and the polynomial Q has n terms, the calculation time (that is the number of LISP operations) for add is bounded by $O(m+n)$, and for multiply is bounded by $O(m^2 n)$. Roughly speaking, we ought to sort the terms of the product so that they appear in decreasing order, and make use of the add function, corresponding to a sorting algorithm by insertion. Of course, the use of a better sorting method (such as quicksort) offers a more efficient multiplication algorithm, say $O(mn \log m)$. But most systems use an algorithm similar to the procedure described.

```
; poly.lisp

(defun add (P Q)
   (cond
      ((null P) Q)
      ((null Q) P)
      ((> (caar P) (caar Q)) (cons (car P) (add (cdr P) Q)))
      ((> (caar Q) (caar P)) (cons (car Q) (add P (cdr Q))))
      ((zerop (+ (cdar P) (cdar Q))) (add (cdr P) (cdr Q)))
      (t (cons (cons (caar P) (+ (cdar P) (cdar Q)))
         (add (cdr P) (cdr Q)))))))

(defun negate (P)
   (mapcar #'(lambda (x) (cons (car x) (- (cdr x)))) P))

(defun subtract (P Q)
   (add P (negate Q)))

(defun multiply (P Q)
   (cond
      ((or (null P) (null Q)) nil)
      (t (cons (cons (+ (caar P) (caar Q)) (* (cdar P) (cdar Q)))
               (add (multiply (list (car P)) (cdr Q))
                  (multiply (cdr P) Q))))))

(defun degree (P)
   (if (null P) 0 (caar P)))
```

Applying these functions in the program

```
; poly_eg.lisp

(load "poly.lisp")

(defun show1 (a)
   `(format t "~A~% => ~A~%" (quote ,a) ,a))

(defmacro show (&rest a)
   `(progn ,@(mapcar #'show1 a)))

(show
   (add '((2 . 3) (1 . 7) (0 . 1)) '((2 . 4) (0 . 2)))
   (add '((0 . 0)) '((3 . 4) (1 . 2) (0 . 7)))
   (negate '((2 . 4) (0 . 2)))
   (subtract '((2 . 3) (1 . 7) (0 . 1)) '((2 . 4) (0 . 2)))
   (multiply '((3 . 3) (0 . 2)) '((2 . 4) (1 . 1)))
   (multiply '((3 . 3) (0 . 2)) '((0 . 0)))
   (degree '((5 . 3) (2 . 2) (0 . -1)))
)
```

yields the results

```
(add '((2 . 3) (1 . 7) (0 . 1)) '((2 . 4) (0 . 2)))
  => ((2 . 7) (1 . 7) (0 . 3))
(add '((0 . 0)) '((3 . 4) (1 . 2) (0 . 7)))
  => ((3 . 4) (1 . 2) (0 . 7))
(negate '((2 . 4) (0 . 2)))
  => ((2 . -4) (0 . -2))
(subtract '((2 . 3) (1 . 7) (0 . 1)) '((2 . 4) (0 . 2)))
  => ((2 . -1) (1 . 7) (0 . -1))
(multiply '((3 . 3) (0 . 2)) '((2 . 4) (1 . 1)))
  => ((5 . 12) (4 . 3) (2 . 8) (1 . 2))
(multiply '((3 . 3) (0 . 2)) '((0 . 0)))
  => ((3 . 0) (0 . 0))
(degree '((5 . 3) (2 . 2) (0 . -1)))
  => 5
```

In the first example, we add the two polynomials

$$(3x^2 + 7x + 1) + (4x^2 + 2) = 7x^2 + 7x + 3.$$

In the second example, we add the two polynomials

$$0 + (4x^3 + 2x + 7) = 4x^3 + 2x + 7.$$

In the fifth example, we multiply two polynomials

$$(3x^3 + 2)(4x^2 + x) = 12x^5 + 3x^4 + 8x^2 + 2x.$$

In the sixth example, we multiply two polynomials

$$(3x^3 + 2) \cdot 0 = 0.$$

The output of the last example is ((3 . 0) (0 . 0)) which is $0 \cdot x^3 + 0 \cdot x^0$. This simplifies to 0.

10.4.2 Simplifications

Here we show how simplification can be implemented in Common LISP. We assume that the mathematical expression is given in *prefix notation*. This means an expression is arranged in the way that the operator appears before its operands.

In `simp1.lisp` we implement the rules

```
+ x          =>   x
- 0          =>   0
exp(0)       =>   1
log(1)       =>   0
sin(0)       =>   0
cos(0)       =>   1
arcsin(0)    =>   0
arctan(0)    =>   0
sinh(0)      =>   0
cosh(0)      =>   1
```

Basically, the simplification is done by comparing the expression case by case. If there is a match, replace the expression by the simplified form. Notice that

```
(simp-unary '(+ (+ x)))
(simp-unary '(- (+ x)))
(simp-unary '(+ (exp 0)))
(simp-unary '(exp (sin 0)))
(simp-unary '(log (exp (sin 0))))
```

are not simplified completely. This indicates that a recursion process is needed for the simplification to apply on different levels. Furthermore, we find that

```
(simp-unary '(+ x y))
```

gives a wrong answer, and

```
(simp-unary 'x)
(simp-unary '(x))
```

give error messages. Thus extra attention has to be given to the number and types of the arguments for the function simp-unary. The next attempt (simp2.lisp) tries to overcome these problems.

```
; simp1.lisp

(defun simp-unary (f)
   (let* ( (op  (car f))
           (opd (cadr f))
           (z?  (and (numberp opd) (zerop opd)))    ; zero
           (o?  (equal 1 opd))                      ; one
           (p?  (equal 'pi opd)) )                  ; pi
      (cond
         ((eq op '+)                   opd)  ; + x      => x
         ((and (eq op '-)     z?)   0)  ; - 0      => 0
         ((and (eq op 'exp)   z?)   1)  ; exp(0)   => 1
         ((and (eq op 'log)   o?)   0)  ; log(1)   => 0
         ((and (eq op 'sin)   z?)   0)  ; sin(0)   => 0
         ((and (eq op 'cos)   z?)   1)  ; cos(0)   => 1
         ((and (eq op 'sin)   p?)   0)  ; sin(pi)  => 0
         ((and (eq op 'cos)   p?)  -1)  ; cos(pi)  =>-1
         ((and (eq op 'arcsin) z?)  0)  ; arcsin(0) => 0
         ((and (eq op 'arctan) z?)  0)  ; arctan(0) => 0
         ((and (eq op 'sinh)  z?)   0)  ; sinh(0)  => 0
         ((and (eq op 'cosh)  z?)   1)  ; cosh(0)  => 1
         (t (list op opd)))))
```

We test the simplification function with the program

```
; simp1_eg.lisp

(load "simp1.lisp")

(defun show1 (a)
   '(format t "~A~%  => ~A~%" (quote ,a) ,a))
```

```
(defmacro show (&rest a)
  '(progn ,@(mapcar #'show1 a)))


; Applications of the simplification

(show
  (simp-unary '(+ x))
  (simp-unary '(exp 0))
  (simp-unary '(log 1))
  (simp-unary '(cosh 0))
  (simp-unary '(exp 1))
  (simp-unary '(cos pi))
  (simp-unary '(- x))
  (simp-unary '(- 0))

  ; the following expressions are not simplified
  (simp-unary '(+ (+ x)))
  (simp-unary '(- (+ x)))
  (simp-unary '(+ (exp 0)))
  (simp-unary '(exp (sin 0)))
  (simp-unary '(log (exp (sin 0))))

  ; the last expression simplifies incorrectly
  (simp-unary '(+ x y))   ;                    <--- WRONG !

  ; the following expressions generate errors
  (simp-unary 'x))    ; An attempt was made to do car on 'x',
                      ; which is not a pair <--- ERROR !

  (simp-unary '(x))   ; An attempt was made to do car on 'nil',
                      ; which is not a pair <--- ERROR !
)
```

The output is

```
(simp-unary '(+ x))
  => x
(simp-unary '(exp 0))
  => 1
(simp-unary '(log 1))
  => 0
(simp-unary '(cosh 0))
  => 1
(simp-unary '(exp 1))
  => (exp 1)
(simp-unary '(cos pi))
  => -1
(simp-unary '(- x))
  => (- x)
(simp-unary '(- 0))
  => 0
(simp-unary '(+ (+ x)))
  => (+ x)
(simp-unary '(- (+ x)))
  => (- (+ x))
(simp-unary '(+ (exp 0)))
  => (exp 0)
```

```
(simp-unary '(exp (sin 0)))
  => (exp (sin 0))
(simp-unary '(log (exp (sin 0))))
  => (log (exp (sin 0)))
(simp-unary '(+ x y))
  => x
```

In the program `simp2.lisp`, we have successfully overcome all the problems existing in the previous version. In this version, simplification may be applied on

- atoms, for which the original expression is returned;

- a list with only one element, for which we apply the function `simp` again on the argument;

- a list with two elements, for which we apply the function `simp-unary`, with the first argument being the operator and the other being the operand;

- a list with more than two elements, for which the message `cannot_simplify` is printed, because this expression can only be simplified by a binary operator.

We have built one more level on top of the function `simp-unary`. The new function `simp` plays an important role in handling the arguments provided by the users. It checks for the types as well as the number of arguments. A different action is taken for each case.

In the function `simp-unary`, most of the statements are the same as the previous program `simp1.lisp`. We have changed only one statement (line number 5 of `simp1.lisp`) in this function to make it recursive

```
(opd (simp (cadr f)))
```

This statement applies the simplification on the second argument of the function, enabling the simplification to apply on different levels of the operand. Note that the statement

```
(simp '(- (- x)))
```

is not simplified because the program does not cater for the interaction between different levels of expressions, i.e. the `minus` operators do not interact and cancel out each other. Furthermore, we see that the statement

```
(simp '(+ x y))
```

prompts the message `cannot_simplify`. This is because we have not implemented the binary operators yet. In the next program `simp3.lisp`, we consider the simplification of binary operators like +, -, *, / and `power`.

The refined simplification functions are

```
; simp2.lisp

(defun simp (f)
   (cond
      ((atom f) f)                      ; f is an atom
      ((null (cdr f)) (simp (car f)))   ; f has only one element
      ((null (cddr f)) (simp-unary f))  ; f has two elements
      (t (quote cannot_simplify)) ))    ; f has more than two elements

(defun simp-unary (f)
   (let* ( (op  (car f))
           (opd (simp (cadr f)))        ; local variables : op, opd, zp
           (z?  (and (numberp opd) (zerop opd)))   ; zero
           (o?  (equal 1 opd))                      ; one
           (p?  (equal 'pi opd)) )                  ; pi
      (cond
         ((eq op '+)                 opd)   ; + x        => x
         ((and (eq op '-)    z?)  0)  ; - 0        => 0
         ((and (eq op 'exp)  z?)  1)  ; exp(0)     => 1
         ((and (eq op 'log)  o?)  0)  ; log(1)     => 0
         ((and (eq op 'sin)  z?)  0)  ; sin(0)     => 0
         ((and (eq op 'cos)  z?)  1)  ; cos(0)     => 1
         ((and (eq op 'sin)  p?)  0)  ; sin(pi)    => 0
         ((and (eq op 'cos)  p?) -1)  ; cos(pi)    =>-1
         ((and (eq op 'arcsin) z?) 0)  ; arcsin(0) => 0
         ((and (eq op 'arctan) z?) 0)  ; arctan(0) => 0
         ((and (eq op 'sinh) z?)  0)  ; sinh(0)    => 0
         ((and (eq op 'cosh) z?)  1)  ; cosh(0)    => 1
         (t (list op opd)))))
```

Testing some examples again

```
; simp2_eg.lisp

(load "simp2.lisp")

(defun show1 (a)
   '(format t "~A~%  => ~A~%" (quote ,a) ,a))

(defmacro show (&rest a)
   '(progn ,@(mapcar #'show1 a)))


; Applications of the simplification
(show
  (simp 'x)
  (simp '(x))

  (simp '(+ x))
  (simp '(+ (+ x)))
  (simp '(+ (- x)))
  (simp '(+ (- 0)))
  (simp '(- (+ x)))

  (simp '(- (sin 0)))
  (simp '(+ (exp 0)))
  (simp '(- (exp 0)))
  (simp '(+ (log 1)))
```

```
(simp '(+ (cosh 0)))
(simp '(exp (sin 0)))
(simp '(log (exp (sin 0))))

(simp '(- (- x)))
(simp '(+ x y))
)
```

yields the output

```
(simp 'x)
 => x
(simp '(x))
 => x
(simp '(+ x))
 => x
(simp '(+ (+ x)))
 => x
(simp '(+ (- x)))
 => (- x)
(simp '(+ (- 0)))
 => 0
(simp '(- (+ x)))
 => (- x)
(simp '(- (sin 0)))
 => 0
(simp '(+ (exp 0)))
 => 1
(simp '(- (exp 0)))
 => (- 1)
(simp '(+ (log 1)))
 => 0
(simp '(+ (cosh 0)))
 => 1
(simp '(exp (sin 0)))
 => 1
(simp '(log (exp (sin 0))))
 => 0
(simp '(- (- x)))
 => (- (- x))
(simp '(+ x y))
 => cannot_simplify
```

In the program `simp3.lisp`, on top of the unary operators simplification, we implement the following binary operators

$$+ \qquad - \qquad * \qquad / \qquad \text{power}$$

In the function `simp`, we change the statement in line number 8 of `simp2.lisp` from

```
(t (quote cannot_simplify))
```

to

```
(t (simp-binary f))
```

This statement simply means if the argument contains more than two elements, apply the binary operator simplification. In the function `simp-binary`, we consider the following simplification

- operation x + y

    if x = 0 return y
    if y = 0 return x

- operation x − y

    if x = 0,
        if y = 0 return 0
        else return −y
    if y = 0 return x
    if x = y return 0

- operation x ∗ y

    if x = 0 or y = 0 return 0
    if x = 1 return y
    if y = 1 return x

- operation x/y

    if x = 0 return 0
    if y − 0 return **infinity**
    if y = 1 return x

- operation x^y

    if x = 0 return 0
    if y = 0 or x = 1 return 1
    if y = 1 return x

The final revision is as follows

```
; simp3.lisp

(load "simp2.lisp")

; remove definition of simp -- avoid warnings about redefinition
(unintern 'simp)

; redefine simp
(defun simp (f)
  (cond
     ((atom f) f)                      ; f is an atom
     ((null (cdr f)) (simp (car f)))   ; f has only one element
     ((null (cddr f)) (simp-unary f))  ; f has two elements
     (t (simp-binary f)) ))            ; f has more than two elements

(defun simp-binary (f)
  (let* ( (op   (car f))
          (opd1 (simp (cadr f)))                ; simplify first operand
```

```
         (opd2 (simp (caddr f)))           ; simplify second operand
         (zp1  (and (numberp opd1) (zerop opd1)))
         (zp2  (and (numberp opd2) (zerop opd2))) )
      (cond
        ((and (eq op '+)                    ; operation: x + y
          (cond (zp1 opd2)                  ; if x=0 return y
                (zp2 opd1))))               ; if y=0 return x

        ((and (eq op '-)                    ; operation: x - y
          (cond (zp1                        ; if x=0,
                  (cond (zp2 0)             ;    if y=0 return 0
                        (t (list '- opd2))))) ;   else return -y
                (zp2 opd1)                  ; if y=0 return x
                ((equal opd1 opd2) 0))))    ; if x=y return 0

        ((and (eq op '*)                    ; operation: x * y
          (cond ((or zp1 zp2) 0)            ; if x=0 or y=0 return 0
                ((equal 1 opd1) opd2)       ; if x=1 return y
                ((equal 1 opd2) opd1))))    ; if y=1 return x

        ((and (eq op '/)                    ; operation: x / y
          (cond (zp1 0)                     ; if x=0 return 0
                (zp2 'infinity)             ; if y=0 return infinity
                ((equal 1 opd2) opd1))))    ; if y=1 return x

        ((and (eq op 'power)                ; operation: x^y
          (cond (zp1 0)                     ; if x=0 return 0
                ((or zp2 (onep opd1)) 1)    ; if y=0 or x=1 return 1
                ((equal 1 opd2) opd1))))    ; if y=1 return x

        (t (list op opd1 opd2)))))
```

The test program is given below.

```
; simp3_eg.lisp

(load "simp3.lisp")

(defun show1 (a)
   '(format t "~A~%  => ~A~%" (quote ,a) ,a))

(defmacro show (&rest a)
   '(progn ,@(mapcar #'show1 a)))


; Applications of the simplification

(show
  (simp '(+ (+ x)))
  (simp '(+ (- x)))
  (simp '(- (- x)))               ; not simplified !

  (simp '(+ (sin 0)))
  (simp '(+ (exp 0)))
  (simp '(+ (log 1)))
  (simp '(- (cosh 0)))
  (simp '(- (sin 0)))
  (simp '(- (arctan 0)))
```

```
  (simp '(+ x 0))
  (simp '(+ 0 x))
  (simp '(- x 0))
  (simp '(* x 0))
  (simp '(/ x 0))
  (simp '(/ 0 x))
  (simp '(- (/ x 0)))
  (simp '(power x 0))

  (simp '(+ x y))
  (simp '(+ (* x 0) (* x 1)))
  (simp '(- (+ 0 x) (* 1 x)))

  (simp '(- (+ x y) (- y x)))   ; not simplified !
)
```

Now we find

```
(simp '(+ (+ x)))
  => x
(simp '(+ (- x)))
  => (- x)
(simp '(- (- x)))
  => (- (- x))
(simp '(+ (sin 0)))
  => 0
(simp '(+ (exp 0)))
  => 1
(simp '(+ (log 1)))
  => 0
(simp '(- (cosh 0)))
  => (- 1)
(simp '(- (sin 0)))
  => 0
(simp '(- (arctan 0)))
  => 0
(simp '(+ x 0))
  => x
(simp '(+ 0 x))
  => x
(simp '(- x 0))
  => x
(simp '(* x 0))
  => 0
(simp '(/ x 0))
  => infinity
(simp '(/ 0 x))
  => 0
(simp '(- (/ x 0)))
  => (- cannot_simplify)
(simp '(power x 0))
  => 1
(simp '(+ x y))
  => (+ x y)
(simp '(+ (* x 0) (* x 1)))
  => x
(simp '(- (+ 0 x) (* 1 x)))
```

```
  => 0
(simp '(- (+ x y) (- y x)))
  => (- (+ x y) (- y x))
```

10.4.3   Differentiation

In this section, we consider the differentiation of algebraic (polynomial) expressions. The following rules are implemented

$$\frac{dc}{dx} = 0, \qquad c \text{ is a constant}$$

$$\frac{dx}{dx} = 1$$

$$\frac{d}{dx}(f(x) + g(x)) = \frac{df}{dx} + \frac{dg}{dx}$$

$$\frac{d}{dx}(f(x) - g(x)) = \frac{df}{dx} - \frac{dg}{dx}$$

$$\frac{d}{dx}(f(x) * g(x)) = f(x) * \frac{dg}{dx} + \frac{df}{dx} * g(x).$$

In the program `differ.lisp`, ex is the expression to be differentiated and v stands for the variable. The expression is given in *prefix notation*. A prefix notation is one in which the operator appears before its operands.

For example, the mathematical expression

$$(3 + x)(x - a),$$

when it is expressed in the prefix notation using LISP, becomes

$$(* (+ 3 x) (- x a))$$

Basically, the program proceeds by checking and matching the operators. When a match is found, it applies the corresponding differentiation rules on the expression. Sometimes, the expression becomes more complicated after the differentiation. Simplification of the expression becomes necessary. Therefore, we apply the simplification program `simp3.lisp`, which we have just developed, when such a case arises.

The implementation of the function `diff` together with some applications is as follows. We used `cond` extensively in the simplification algorithms to conditionally evaluate and transform expressions, however sometimes only a simple transformation is necessary. Consider the derivatives of the functions $\sin x$, $\cos x$ and $\exp x$. This can be simply represented as a nested list

```
( (exp exp) (sin cos) (cos -sin) )
```

This is an associative array, thus the derivative of $\sin x$ is found using `assoc`

```
(assoc 'sin '( (exp exp) (sin cos) (cos -sin) )) ;  (sin cos)
```

and we can use `cadr` to find `cos`. A better representation is

```
(assoc '(sin x) '( ((exp x)    (exp x))
                   ((sin x)    (cos x))
                   ((cos x) (- (sin x))) )) ;  nil
```

but `nil` indicates that no match was found. It is necessary to use `equal` instead of `eq`, specified by the `:test` key, for a structural comparison

```
(assoc '(sin x) '( ((exp x)    (exp x))
                   ((sin x)    (cos x))
                   ((cos x) (- (sin x))) ) :test #'equal)
; ((sin x) (cos x))
```

If we apply the chain rule to

$$\frac{d}{dx}\cos e^x = -e^x \sin e^x$$

we need the following entry in our associative list

```
( (cos (exp x)) (- (* (exp x) (sin (exp x)))) )
```

This is a more complicated version of the entry which we had for `cos` in order to take into account the chain rule for `exp`. It is not feasible to have an entry in the associative list for every possible function composition. Thus we search for $\cos x$ in the table to find `(- (sin x))` and then substitute e^x for x before multiplying with the derivative of e^x (apply the chain rule).

```
(subst '(exp x) 'x '(- (sin x)))     ; (- (sin (exp x)))
```

The function for differentiation is

```
; differ.lisp

(defun diff (ex v)                     ; d(ex)/dv
   (cond
      ((atom ex)
         (cond ((eq ex v) 1) (t 0)))   ; d(v)/dv = 1, d(constant)/dv = 0

      ((eq (car ex) '+)                ; d(a+b)/dv = da/dv + db/dv
         (list '+ (diff (cadr ex) v) (diff (caddr ex) v)))

      ((eq (car ex) '*)                ; d(a*b)/dv = da/dv * b + a * db/dv
         (list '+
            (list '* (diff (cadr ex) v) (caddr ex))
            (list '* (cadr ex) (diff (caddr ex) v))))

      ((eq (car ex) '/)                ; d(a/b)/dv = (da/dv)/b - (a/b*b)*db/dv
         (list '-
            (list '/ (diff (cadr ex) v) (caddr ex))
            (list '* '(/ ,(cadr ex) (* ,(caddr ex) ,(caddr ex)))
                  (diff (caddr ex) v))))

      ((eq (car ex) '-)                ; d(a-b)/dv = da/dv - db/dv
         (list '- (diff (cadr ex) v) (diff (caddr ex) v)))
```

```
        ((atom (car ex))                        ; d/dv f(a) = f'(a) da/dv
          (let ( (form (assoc (car ex) '( (exp     (exp x))
                                          (sin     (cos x))
                                          (cos (- (sin x)))
                                          (ln      (/ 1 x)) ))) )
              (list '* (if form (subst (cadr ex) 'x (cadr form))
                            '(diff ,(car ex)  ,v))
                    (diff (cadr ex) v)))))))
```

Below we consider some test cases.

```
; differ_eg.lisp

(load "differ.lisp")
(load "simp3.lisp")

(defun show1 (a)
  '(format t "~A~%  => ~A~%" (quote ,a) ,a))

(defmacro show (&rest a)
  '(progn ,@(mapcar #'show1 a)))

; Applications
(show
 ;simple-examples
 (diff 'x 'x)
 (diff '2 'x)
 (diff 2 'x)
 (diff 'x 'u)

 ;sum-rule
 (diff '(+ x x) 'x)

 ;product-rule
 (diff '(* x x) 'x)
 (simp (diff '(* x x) 'x))

 ;more-examples
 (diff '(* (+ 3 x) (- a x)) 'x)
 (simp (diff '(* (+ 3 x) (- a x)) 'x))
 (diff '(* (* x x) x) 'x)
 (simp (diff '(* (* x x) x) 'x))

 ;chain-rule
 (simp (diff '(cos (exp x)) 'x))
 (simp (diff '(/ (sin x) (cos x)) 'x))
 (simp (diff '(f (* (cos x) (exp x))) 'x))
)
```

The output is

```
(diff 'x 'x)
  => 1
(diff '2 'x)
  => 0
(diff 2 'x)
  => 0
(diff 'x 'u)
  => 0
```

```
(diff '(+ x x) 'x)
  => (+ 1 1)
(diff '(* x x) 'x)
  => (+ (* 1 x) (* x 1))
(simp (diff '(* x x) 'x))
  => (+ x x)
(diff '(* (+ 3 x) (- a x)) 'x)
  => (+ (* (+ 0 1) (- a x)) (* (+ 3 x) (- 0 1)))
(simp (diff '(* (+ 3 x) (- a x)) 'x))
  => (+ (- a x) (* (+ 3 x) (- 1)))
(diff '(* (* x x) x) 'x)
  => (+ (* (+ (* 1 x) (* x 1)) x) (* (* x x) 1))
(simp (diff '(* (* x x) x) 'x))
  => (+ (* (+ x x) x) (* x x))
(simp (diff '(cos (exp x)) 'x))
  => (* (- (sin (exp x))) (exp x))
(simp (diff '(/ (sin x) (cos x)) 'x))
  => (- (/ (cos x) (cos x))
        (* (/ (sin x) (* (cos x) (cos x))) (- (sin x))))
(simp (diff '(f (* (cos x) (exp x))) 'x))
  => (* (diff f x)
        (+ (* (- (sin x)) (exp x)) (* (cos x) (exp x))))
```

10.5 LISP, Haskell and Computer Algebra

10.5.1 A simple computer algebra system in LISP

We describe a very simple computer algebra system in LISP below. The system consists of symbols (LISP symbols), numbers, sums (represented by a list `(+ ...)` where the remaining list elements are to be summed) and products (represented by a list `(* ...)` where the remaining list elements are to be multiplied). Thus we define two basic operations add and mul to perform the addition and multiplication. The function expand shrinks sums of sums to a single sum, products of products to a single product and applied the distributive law. The function tostring converts the symbolic expression to a string for display.

```
; symbol.lisp

(load "infix.lisp")

(defun sum     (l) (cons '+ l))
(defun product (l) (cons '* l))

(defun symbol?  (s) (symbolp s))
(defun number?  (n) (numberp n))
(defun sum?     (s) (and (consp s) (equal (car s) '+)))
(defun product? (s) (and (consp s) (equal (car s) '*)))

(defun add (a b) '(+ ,a ,b))
(defun mul (a b) '(* ,a ,b))

(defun ifrec (again f)
  (if again #'(lambda (x) (funcall f nil x)) #'(lambda (x) x)) )
```

```lisp
(defun split (p l)
 (if (null l) (list nil nil nil)
     (if (funcall p (car l)) (list nil (car l) (cdr l))
         (let ( (s (split p (cdr l))) )
              (cons (cons (car l) (car s)) (cdr s)) ) ) ) )

(defun expand1 (again s)
 (cond ( (symbol? s) s )
       ( (number? s) s )
       ; sum or product with one element in the list
       ( (null (cddr s)) (expand (cadr s)) )
       ; expand each element in the sum, and absorb sums
       ( (sum? s)
         (funcall (ifrec again #'expand1) (expand2 s #'sum? #'sum)) )
       ; apply distributivity for product, when it contains a sum
       ; otherwise expand and absorb
       ( t (let* ( (s1 (split #'sum? (cdr s))          )
                   (a   (car s1)                       )
                   ; cadr -> sum, cdadr -> list from sum
                   (s2 (and (sum? (cadr s1)) (cdadr s1)) )
                   (b   (caddr s1)                     ) )
             (if (null s2)
                 ; no sum found
                 (funcall (ifrec again #'expand1)
                          (expand2 s #'product? #'product) )
                 ; sum found
                 (expand1 again
                    (sum (mapcar (if (and (null a) (null b))
                                     #'(lambda (x) x)
                                     #'(lambda (x)
                                         (product (append a (list x) b))))
                                 s2 ) ) ) ) ) ) ) )

; implement expandabsorb for both sum and product
; using p? <- sum?, c <- sum or p? <- product?, c <- product
(defun expand2 (s p? c)
 (labels ( (absorb (x) (if (funcall p? x) (cdr x) (list x)))
           (expandabsorb (x) (absorb (expand x))) )
         (let ( (newlist (reduce #'append
                                 (mapcar #'expandabsorb
                                         (cdr s) ) ) ) )
              (funcall c newlist) ) ) )

(defun expand (s) (expand1 t s))

(defun tostring (s)
 (cond ( (symbol? s) (format nil "~A" s) )
       ( (number? s) (format nil "~A" s) )
       ( (null (cdr s)) "")
       ; sum or product of one symbol
       ( (and (symbol? (cadr s)) (null (cddr s)))
         (format nil "~A" (cadr s)) )
       ; sum or product of one number
       ( (and (number? (cadr s)) (null (cddr s)))
         (format nil "~A" (cadr s)) )
       ; sum or product of one element
       ( (and (null (cddr s)))
         (format nil "(~A)" (tostring (cadr s))) )
```

```
      ( (sum? s)
        (let ( (summand1 (tostring (sum (cons (cadr s) nil))))
               (summand2 (tostring (sum (cddr s)))) )
            (format nil "~A + ~A" summand1 summand2) ) )
      ( (product? s)
        (let ( (factor1 (tostring (cadr s)))
               (factor2 (tostring (product (cddr s)))) )
            (format nil "~A~A" factor1 factor2) ) ) ) )

; examples
(labels ( (+ (a b) (add a b))
          (* (a b) (mul a b))
          (print (a) (format t "~A~%" a)) )
        (defvar a 'a)
        (defvar b 'b)
        (defvar c (infix 3 * (a + b) * a))
        (defvar d 'd)
        (defvar e (infix 3.0 * d * (d + d)))
        (print (tostring c))
        (print (tostring (expand c)))
        (print (tostring (expand (infix a + b + a))))
        (print (tostring e)) )
```

10.5.2   A simple computer algebra system in Haskell

Haskell [30][23] is a general purpose, statically typed, purely functional programming language. Haskell typing is similar to C++ in that every variable and expression has a type at compile time (compared to LISP variables where the type
is determined at run time), but different in that the type can usually be inferred.
Haskell is also strongly typed whereas C++ is weakly typed. Thus the type of a
variable or function need not be explicitly specified. A statically typed language,
such as C++ or Haskell, specifies the type of a variable at compile time (i.e. in
the program source). A dynamically typed language, such as LISP and Smalltalk,
determines the type of a variable at run time, thus taking a car of a symbol causes
an error when the program is run. In C++ or Haskell the compiler would generate
an error before the program is run. C++ is weakly typed since, for example, we
can treat an integer directly as type double from

$$*(double*)\&x$$

where x is of type int. In Haskell this is not possible, a variable has a single type
and there is no means to interpret the variable's type in any other way (although
a function may convert a value from one type to another). Functions are usually
described using pattern matching. Comments begin with -- and continue until the
end of the line. Types can be specified with a type annotation, or can be inferred
by the compiler. A type annotation is a statement of the form

```
variable :: Type
```

for example

```
x :: Integer
```

LISP and Haskell share some concepts although with different syntax. Variables
and functions in Haskell begin with lower case letters while type constructors begin
with capital letters.

	LISP	Haskell
Comment	`; comment`	`-- comment`
Pair	`(cons a b)`	`a : b`
Empty list	`nil`	`[]`
List	`(cons a (cons b nil))`	`a : b : []`
List	`(list a b)`	`[ a, b ]`
List append	`(append a b)`	`a ++ b`
Variable	`(defvar a "a")`	`a = "a"`
Function	`(defun f (a b) ...)`	`f a b = ...`
Lambda	`(lambda (a) ...)`	`\a -> ...`
Lambda	`(lambda (a b) ...)`	`\a -> \b -> ...`
car	`(defun f (l) (car l))`	`f (a : l) = a`
cdr	`(defun f (l) (cdr l))`	`f (a : l) = l`
mapcar	`mapcar`	`map`
Output	`(format nil "~A" a)`	`show a`
Output	`(format t "~A" a)`	`putStr (show a)`
Let	`(let ((a 1) (b 2)) ...)`	`let {a=1 ; b=2} in ...`

In addition Haskell is indentation sensitive, has lazy evaluation, pattern matching
and

sections	`a + b` is equivalent to `(+) a b`
data types	`data Bool = True	False`
	`data Rational t = Ratio t t`
type annotations	`x :: Integer`
	`x = 3`
	`r :: Rational Integer`
	`r = Ratio x 5`

where `t` (lower case) is a type variable and `Ratio` (first letter is upper case) is a
type constructor which takes two parameters of type `t`, for example `Integer` in the
last example.

Below we reimplement the LISP computer algebra system from the previous section
in Haskell. The Haskell version is slightly different since sums and products must
always be treated separately, whereas in LISP we know that when the expression
is neither a number nor a symbol (i.e. sum or product) we can simplify the sum or
product because they share the same list structure.

```
-- symbol.hs

-- Every program has module Main
module Main where

-- A symbolic expression over a type a (eg. Double) is a
-- symbol, a number with value, a sum of symbolic expressions
```

```
-- or a product of symbolic expressions
data Symbolic a = Symbol String
                | Number a
                | Sum [ Symbolic a ]
                | Product [ Symbolic a ]

-- tests for the different cases

isSymbol  (Symbol x)  = True
isSymbol  x           = False
isNumber  (Number x)  = True
isNumber  x           = False
isSum     (Sum x)     = True
isSum     x           = False
isProduct (Product x) = True
isProduct x           = False


-- addition a + b
add a b = Sum [ a, b ]
-- multiplication a * b
mul a b = Product [ a, b ]


-- control for recursion
ifrec again f = if again then f False else (\x -> x)


-- expand flattens the tree for a symbolic expression
--   and applies the distributive property for multiplication
-- the arguments are a boolean value which specifies
--   whether the expansion should continue recursively
--   and an expression to expand

split p [] = ([], [], [])
split p (a : as) = if p a then ([], [a], as)
                   else (a : x, y, z) where (x, y, z) = split p as

expand1 again (Sum [s])     = expand s
expand1 again (Product [p]) = expand p
expand1 again (Sum summands)
 = (ifrec again expand1) (Sum newlist)
   where expandabsorb x = absorb (expand x)
         absorb (Sum s) = s
         absorb x       = [x]
         newlist        = foldl1 (++) (map expandabsorb summands)
expand1 again (Product factors)
 = if or (map isSum factors) then expand1 again (Sum (map mul s))
   else (ifrec again expand1) (Product newlist)
   where (a, [Sum s], b)  = split isSum factors
         mul              = if (null a) && (null b) then \x -> x
                            else \x -> Product (a++ [x] ++ b)
         expandabsorb x       = absorb (expand x)
         absorb (Product p) = p
         absorb x           = [x]
         newlist            = foldl1 (++) (map expandabsorb factors)
expand1 again x = x


expand x = expand1 True x

-- tostring converts a symbolic expression
```

```
-- to a string representation
tostring (Symbol s)              = s
tostring (Number n)              = show n
tostring (Sum      [])           = ""
tostring (Product [])            = ""
tostring (Sum      ((Symbol s) : [])) = s
tostring (Product ((Symbol s) : [])) = s
tostring (Sum      ((Number n) : [])) = (show n)
tostring (Product ((Number n) : [])) = (show n)
tostring (Sum      (a : []))     = "(" ++ (tostring a) ++ ")"
tostring (Product (a : []))      = "(" ++ (tostring a) ++ ")"
tostring (Sum      (first : rest))
 = summand1 ++ "+" ++ summand2
   where summand1 = tostring (Sum (first : []))
         summand2 = tostring (Sum rest)
tostring (Product (first : rest))
 = factor1 ++ factor2
   where factor1 = tostring (Product (first : []))
         factor2 = tostring (Product rest)


-- Examples

-- explicit type
a :: Symbolic Integer
a = Symbol "a"
-- inferred type b :: Symbolic a
b = Symbol "b"
-- inferred type c :: Symbolic Integer
c = let { (+) = add; (*) = mul } in (Number 3) * (a + b) * a
-- inferred type d :: Symbolic a
d = Symbol "d"
-- inferred type e :: Symbolic Double
e = let { (+) = add; (*) = mul } in (Number 3.0) * d * (d + d)

-- every program must have a main expression
-- that evaluates to some IO
main = let (+) = add
           (*) = mul
         in do putStrLn (tostring c)
               putStrLn (tostring (expand c))
               putStrLn (tostring (expand (a+b+a)))
               putStrLn (tostring e)
```

The output is

```
(3(a+b))a
(3aa)+(3ba)
a+b+a
(3.0d)(d+d)
```

Chapter 11

Lisp using C++

11.1 Lisp Operations in C++

In this section, we show how a Lisp system can be implemented using object-oriented programming with C++. In the program, we implement the following functions (described in Chapter 10):

car   cdr   atom   cons   cond   append

The basic data structures of a Lisp system are the atom and dotted pair. Lists are built from dotted pairs. In this case an *atom* is a C++ basic data type or abstract data type and a *dotted pair* is an abstract data type with two data members (a pointer to the `car` and a pointer to the `cdr` part of the dotted pair). A list may consist of basic data types (e.g. `int`, `double`) or abstract data types (e.g. `string`, `Verylong`, `Matrix`).

Basically, the program consists of three classes related to each other as shown in Figure 11.1.

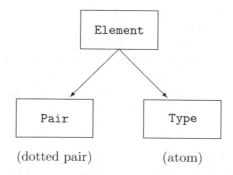

Figure 11.1: *Schematic diagram of the inheritance relationship.*

The inheritance hierarchy consists of an abstract base class Element and two derived classes: Pair and Type. The Pair class corresponds to the dotted pairs in Lisp, whereas the Type class corresponds to the atoms in Lisp.

The Element class, which is an abstract base class, specifies only the interface but not the implementation of the member functions available in the derived classes. All the functions in the class are declared as virtual. This allows the derived classes to override the definition of the functions. There are four member functions declared in the class:

- virtual Element *clone() const = 0;
  it declares a function that duplicates an Element.

- virtual void print(ostream&) const = 0;
  it declares a function that displays the content of an Element.

- virtual int atom() const = 0;
  it declares a function that checks if an Element is an atom.

- friend ostream& operator << (ostream&,const Element*);
  it declares an output stream function.

The structure and functionalities of the Pair class may be summarized as follows:

- Data field: The class contains only two data fields _car and _cdr. They represent the first and second elements of the dotted pair, respectively.

- Constructors and destructor:

  - Pair(): The default constructor assigns NULL to both _car and _cdr.

  - Pair(const Element *ee): it assigns ee to _car and NULL to _cdr.

  - Pair(const Element *e1,const Element *e2): it assigns e1 and e2 to the first and second elements of the dotted pair, respectively.

  - The copy constructor and destructor are overloaded as well.

- Member functions:

  - Element* car() const;
    it returns the first element of the dotted pair.

  - Element* cdr() const;
    it returns the second element of the dotted pair.

  - void car(const Element* ee);
    it sets the first element of the dotted pair to ee.

  - void cdr(const Element* ee);
    it sets the second element of the dotted pair to ee.

  - The virtual functions

```
Element* clone() const;
void print(ostream&) const;
int atom() const;
```

and the output stream operator

```
ostream& operator << (ostream&,const Pair*);
```

are overridden with new definitions for the class.

The **Type** class is implemented as a **template** class. Its structure and functionalities may be summarized as follows:

- Data field: The only data member in the class is **thing**. It represents the numerical value of an instance of the **Type** class.

- Constructors:

    - **Type()**: The default constructor sets the value of **thing** to zero.
    - **Type(T vv)**: It sets the value of **thing** to vv.
    - The copy constructor is overloaded.

- Member functions:

    - **const T& value() const;**
      it returns the value of **thing**.
    - The virtual functions

    ```
    Element* clone() const;
    void print(ostream&) const;
    int atom() const;
    ```

    and the output stream operator

    ```
    ostream& operator << (ostream&,const Type<T>*);
    ```

    are overridden with new definitions for the class.

A class has been derived from **Element** to provide the atoms **nil** and **t**. The atoms are passed as function parameters using the address of operator (**&**) for example **cons(&t,&nil);** Notice that only the function **atom()** is declared as a member function for the class. The rest of the functions which we intend to implement are declared as global functions:

- **Element* car(const Element*);**
  It applies on a dotted pair and returns the first element of the pair. An error message is reported if the argument is an atom.

- **Element* cdr(const Element*);**
  It applies on a dotted pair and returns the second element of the pair. If the argument is an atom, an error message is reported.

- `Pair* cons(const Element *p1,const Element *p2);`
  It forms a dotted pair with p1 and p2 as the first and second element of the pair, respectively.

- `Pair* append(const Element *p1,const Element *p2);`
  It replaces the last element of the dotted pair p1 with p2. Note that p2 may be an atom or dotted pair(s). An error message is reported if p1 is an atom.

- `Element* null(Element *e);`
  Returns &t if e points to the nil atom and &nil otherwise.

- `int is_lisp_list(Element *l);`
  Determines if l points to a valid Lisp list.

- `int is_nonempty_lisp_list(Element *l);`
  Determines if l points to a valid Lisp list which is not empty.

- `Element *cond(Element *c);`
  Implements the Lisp cond conditional statement. c must point to a non-empty list of lists where each list represents a condition to evaluate. cond returns &nil if no condition is satisfied.

In the following, we list the program of the implementation. The listing is followed by an application program lisprun.cpp, which makes use of the functions. The dotted pairs may contain elements of different types, such as Verylong and double.

```
// lisp.h

#ifndef __LISP_H
#define __LISP_H

#include <iostream>
#include "identity.h"
using namespace std;

// Abstract base class
class Element
{
public:
   virtual Element *clone() const = 0;
   virtual void print(ostream&) const = 0;
   virtual int atom() const = 0;
   virtual ~Element() {}
};

//unique class for atoms nil and t
class __nil_and_t_lisp_class : public Element
{
 private:
        int is_t;
 public:
        __nil_and_t_lisp_class(int i=0) : is_t(i) {}
        void print(ostream &s) const {s<<((is_t)?"t":"nil");}
        Element *clone(void) const {return (Element*)this;}
        int atom(void) const {return 1;}
```

```
}nil(0),t(1);

class Pair : public Element   // Dotted pair
{
private:
   Element *_car;              // First element of the dotted pair
   Element *_cdr;              // Second element of the dotted pair

public:
   // Constructors
   Pair();
   Pair(const Element*);
   Pair(const Element*,const Element*);
   Pair(const Pair&);
   virtual ~Pair();

   Pair & operator = (const Pair&);

   // Member functions
   Element *car() const;
   Element *cdr() const;

   void car(const Element*);
   void cdr(const Element*);

   Element *clone() const;
   void print(ostream&) const;
   int atom() const;
};

template <class T>
class Type : public Element   // Atom
{
private:
   T thing;

public:
   // Constructors
   Type();
   Type(T);
   Type(const Type<T>&);

   // Member functions
   const T &value() const;

   Type<T> &operator = (const Type<T>&);
   Element *clone() const;
   void print(ostream&) const;
   int atom() const;
};

// Global functions
Pair *cons(const Element*,const Element*);
Element *append(const Element*,const Element*);
Element *car(const Element*);
Element *cdr(const Element*);

// Implementation
```

```cpp
// class Pair
Pair::Pair() : _car(&nil), _cdr(&nil) {}

Pair::Pair(const Element *e) : _car(e->clone()), _cdr(&nil) {}

Pair::Pair(const Element *e1,const Element *e2) :
  _car(e1->clone()), _cdr(e2->clone()) {}

Pair::Pair(const Pair &p)
{
   _car = &nil; _cdr = &nil;
   *this = p;
}

Pair::~Pair()
{
   //don't delete constants of the system
   if((_car != &nil)&&(_car != &t)) delete _car; _car = &nil;
   if((_cdr != &nil)&&(_cdr != &t)) delete _cdr; _cdr = &nil;
}

Pair &Pair::operator = (const Pair &p)
{
   if(this != &p)
   {
      if((_car != &nil)&&(_car != &t)) delete _car;
      if((_car != &nil)&&(_car != &t)) delete _cdr;

      _car = (p.car())->clone();
      _cdr = (p.cdr())->clone();
   }
   return *this;
}

Element *Pair::car() const
{ return _car; }

Element *Pair::cdr() const
{ return _cdr; }

void Pair::car(const Element *e)
{ if(e != &nil) _car = e->clone(); }

void Pair::cdr(const Element *e)
{ if(e != &nil) _cdr = e->clone(); }

Element *Pair::clone() const
{
   Pair *p = new Pair(*this);
   return p;
}

void Pair::print(ostream &os) const
{ os << "(" << _car << " . " << _cdr << ")"; }

int Pair::atom() const { return 0; }
```

```
ostream &operator << (ostream &os,const Pair *p)
{
    if(p != NULL)
    { p->print(os); return os; }
    os << "nil";
    return os;
}

ostream &operator << (ostream &os,const Element *e)
{
    if(e != NULL)
    {
        e->print(os);
        return os;
    }
    os << "nil";
    return os;
}

template <class T> Type<T>::Type() : thing(zero(T())) {}

template <class T> Type<T>::Type(T t) : thing(t) {}

template <class T> Type<T>::Type(const Type<T> &t)
{ thing = t.value(); }

template <class T>
Type<T> &Type<T>::operator = (const Type<T> &t)
{
    if(this != &t) thing = t.value();
    return *this;
}

template <class T> const T &Type<T>::value() const { return thing; }

template <class T> int Type<T>::atom() const { return 1; }

template <class T>
Element *Type<T>::clone() const
{
    Type<T> *t = new Type<T>(*this);
    return t;
}

template <class T> void Type<T>::print(ostream &os) const { os << thing; }

template <class T>
ostream &operator << (ostream &os,const Type<T> *t)
{
    t->print(os);
    return os;
}

Pair *cons(const Element *e1,const Element *e2)
{
    Pair *p = new Pair;
    p->car(e1);
    p->cdr(e2);
```

```
      return p;
}

Element *append(const Element *e1,const Element *e2)
{
   if(! e1->atom())
   {
      Pair *p = new Pair(*(Pair *)e1);
      Pair *c = p;
      while(e->cdr() != &nil) e = (Pair *)e->cdr();
      e->cdr(e2);
      return p;
   }
   cerr << "\nFirst argument of append must be a list" << endl;
   return &nil;
}

Element *car(const Element *e)
{
   if(! e->atom()) return ((Pair*)e)->car();
   cerr << "\ncar: cannot take car of an atom!" << endl;
   return &nil;
}

Element *cdr(const Element *e)
{
   if(! e->atom()) return ((Pair*)e)->cdr();
   cerr << "\ncdr: cannot take cdr of an atom!" << endl;
   return &nil;
}

Element *null(Element *e)
{
 if(e==&nil) return &t;
 return &nil;
}

int is_lisp_list(Element *l)
{
 if(l==&nil) return 1;
 if(l->atom()) return 0;
 return is_lisp_list(((Pair*)l)->cdr());
}

int is_nonempty_lisp_list(Element *l)
{
 if(l->atom()) return 0;
 return is_lisp_list(((Pair*)l)->cdr());
}

Element *cond(Element *e)
{
 if(is_nonempty_lisp_list(e))
 {
  Element *firstcase=((Pair*)e)->car();
  if(is_nonempty_lisp_list(firstcase))
  {
   Element *condition=((Pair*)firstcase)->car();
```

```
    if(condition->atom())
    {
     if(condition==&nil)
      return cond(((Pair*)e)->cdr());
     else if(condition==&t)
      return ((Pair*)firstcase)->cdr();
     else cerr<<"cond expects a case first element to be nil or t."<<endl;
    }
    else cerr<<"cond expects a case first element to be nil or t."<<endl;
   }
   else cerr<<"cond expects a list for each case."<<endl;
  }
  else cerr<<"cond expects a list for evaluation."<<endl;
  return &nil;
}
#endif
```

The following program applies the classes in `lisp.h` which has just been developed.

```cpp
// lisprun.cpp

#include <iostream>
#include "lisp.h"
#include "verylong.h"
using namespace std;

typedef Type<int> l_int;
typedef Type<char> l_char;
typedef Type<double> l_double;
typedef Type<Verylong> l_verylong;

Element *l_plus(Element *e)
{
 if(is_nonempty_lisp_list(e))
 {
  if(is_nonempty_lisp_list(cdr(e)))
  {
   //assume int for calculations
   Element *param1,*param2;
   param1=car(e); param2=car(cdr(e));
   if(param2!=&nil)
    return l_int(((l_int*)param1)->value()
             +((l_int*)param2)->value()).clone();
  }
 }
 cerr<<"Plus takes a list of two integers as arguments"<<endl;
 return &nil;
}

Element* length_list(Element *e)
{
 l_int zero(0),one(1);

 //stop C++ recursion
 if(e==&nil) return cons(&zero,&nil);
                                 // Equivalent LISP code
 return cond(cons (                // (cond
  cons(null(e),cons(&zero,&nil)), // ((null e) 0)
```

```
   cons(                                      // (t (l_plus 1 (length_list (cdr e)))))
   cons(&t,cons(l_plus(cons(&one,length_list(cdr(e)))),&nil)),&nil)));
}

int main(void)
{
   // Define two "atoms"
   l_int a(1);      // 1
   l_char b('a'); // a

   cout << "Examples on the cons function:" << endl;

   Pair *A = cons(&a,&nil);  // in Lisp :  ( 1 )
   Pair *B = cons(&b,&nil);  // in Lisp :  ( a )

   Pair *C = cons(&a,&b);       // in Lisp : (cons '1 'a)
   cout << "(cons '1 'a)     => " << C << endl;

   C = cons(&a,B);              // in Lisp (cons '1 '(a))
   cout << "(cons '1 '(a))     => " << C << endl;

   C = cons(A,&b);              // in Lisp (cons '(1) 'a)
   cout << "(cons '(1) 'a     => " << C << endl;

   C = cons(A,B);               // in Lisp (cons '(1) '(a))
   cout << "(cons '(1) '(a)) => " << C << endl;
   cout << endl;

   cout << "Examples on the append function:" << endl;

   // (setq D1 '(1 2 3))
   Pair *D1 = cons(new l_int(1),cons(new l_int(2),cons(new l_int(3),&nil)));

   // (setq D2 '(a b c))
   Pair *D2 = cons(new l_char('a'),cons(new l_char('b'),
                   cons(new l_char('c'),&nil)));

   Pair *D = (Pair*)append(D1,D2);
   cout << "(append '(1 2 3) '(a b c)) => " << D << endl;
   cout << endl;

   cout << "Examples on the car and cdr functions:" << endl;

   // setq E '((1 2 3) (a b c))
   Pair *E = cons(D1,cons(D2,&nil));

   cout << "(car '((1 2 3) (a b c)))          => " << car(E)        << endl;
   cout << "(car (car '((1 2 3) (a b c))))  => " << car(car(E)) << endl;
   cout << "(cdr '((1 2 3) (a b c)))          => " << cdr(E)        << endl;
   cout << "(car (cdr '((1 2 3) (a b c))))  => " << car(cdr(E)) << endl;
   cout << "(cdr (cdr '((1 2 3) (a b c))))  => " << cdr(cdr(E)) << endl;
   cout << "(car 'a) => " << car(&a) << endl;
   cout << endl;

   // Abstract data types
   cout << "Applications with the abstract data type: Verylong" << endl;

   l_verylong v(Verylong("123456789012"));
```

```
    l_double r(3.14159);
    Pair *Very = cons(&v,cons(&r,&nil));
    cout << "(cons 'v '(r)) => " << Very << endl;

    cout << endl << "Applications on the cond function:" << endl;

    //((nil r) (t v))
    cout << "(cond (nil 3.14159) (t 123456789012)) => "
        << cond(cons(cons(&nil,cons(&r,&nil)),
                    cons(cons(&t,cons(&v,&nil)),&nil)))
        <<endl;
    //((t r) (nil v))
    cout << "(cond (t 3.14159) (nil 123456789012)) => "
        << cond(cons(cons(&t,cons(&r,&nil)),
                    cons(cons(&nil,cons(&v,&nil)),&nil)))
        <<endl;

    Pair *test=cons(&a,cons(&b,cons(&v,cons(&r,&nil))));
    cout << endl << "(a b v r) => " << test << endl;
    cout << "(length_list (a b v r)) => " << length_list(test) << endl;

    return 0;
}
```

The output is

```
Examples on the cons function:
(cons '1 'a)     => (0xbb806070 . 0xbb806078)
(cons '1 '(a))   => (0xbb806080 . 0xbb801160)
(cons '(1) 'a)   => (0xbb801180 . 0xbb806098)
(cons '(1) '(a)) => (0xbb8011a0 . 0xbb8011b0)

Examples on the append function:
(append '(1 2 3) '(a b c)) => (0xbb806140 . 0xbb801290)

Examples on the car and cdr functions:
(car '((1 2 3) (a b c)))       => (0xbb806188 . 0xbb801340)
(car (car '((1 2 3) (a b c)))) => 1
(cdr '((1 2 3) (a b c)))       => (0xbb801370 . 0x8054ca8)
(car (cdr '((1 2 3) (a b c)))) => (0xbb8061a0 . 0xbb801380)
(cdr (cdr '((1 2 3) (a b c)))) => nil

car: cannot take car of an atom!
(car 'a) => nil

Applications with the abstract data type: Verylong
(cons 'v '(r)) => (0xbb8013d0 . 0xbb8013e0)

Applications on the cond function:
(cond (nil 3.14159) (t 123456789012)) => (0xbb801550 . 0x8054ca8)
(cond (t 3.14159) (nil 123456789012)) => (0xbb801670 . 0x8054ca8)

(a b v r) => (0xbb8061c0 . 0xbb801780)
(length_list (a b v r)) => (0xbb806310 . 0x8054ca8)
```

11.2   λ-Calculus and C++

In the previous section we showed how LISP operations and data could be embedded in C++. However, some difficulty arises when expressions involving `if` or `cond` have to be evaluated since these control which subexpressions are to be evaluated. This is because we rely on the C++ compiler to determine which statements to evaluate, and not the LISP representation in C++. To closer implement LISP in C++ we need to write a LISP interpreter, i.e. a C++ program which given LISP statements as strings can evaluate those statements. LISP has some similarity to λ-calculus [5][6]. Although LISP was not originally a computer implementation of λ-calculus [37], it did use some concepts from λ-calculus as described below. The (`defun` ...) form is close to that of a λ abstraction. LISP even has a (`lambda` ...) form which is almost identical to that used in the λ-calculus. In this section we implement an interpreter for λ-calculus which could be used as the basis for a LISP interpreter.

11.2.1   λ-Calculus

The λ-calculus is essentially a substitution calculus. First we describe the terms in λ-calculus.

A term is

1. A variable such as x or y.

2. An abstraction $(\lambda x.f)$ where f is a term in λ-calculus.

3. An application $(f\ y)$ where f is an abstraction and y is a term in λ-calculus.

A variable is said to be *bound* if it appears after λ in an abstraction, for example x is bound in $(\lambda x.f)$. A variable which is not bound is said to be *free*. Thus y is free in $(\lambda x.(y\ x))$ whereas x is bound. It may be assumed that every free variable represents an unknown abstraction so that application is not restricted.

The semantics of λ-calculus are given by the following rules

1. α conversion: $(\lambda x.f) \rightarrow_\alpha (\lambda y.f_{x \rightarrow y})$ where y is not free in f and $f_{x \rightarrow y}$ denotes the replacement of every free occurrence of x in f by y.

2. β reduction: $((\lambda x.f)\ y) \rightarrow_\beta f_{x \rightarrow y}$.

3. η reduction: $((\lambda x.f)\ y) \rightarrow_\eta f$ for x not free in f.

η reduction can be considered a special case of β reduction. Thus substitution implements application.

We see that we can translate almost directly from λ-calculus to LISP:
$(\lambda x.(y\ x)) \rightarrow$ (`lambda` (x) (y x))

Now we can describe some of the concepts which are important in LISP programming.

$$first \equiv (\lambda x.(\lambda y.x))$$
$$second \equiv (\lambda x.(\lambda y.y))$$
$$cons \equiv (\lambda x.(\lambda y.(\lambda w.((w\ x)\ y)))).$$

The notation is a little clumsy, so we use left associativity for application $(w\ x\ y) \equiv ((w\ x)\ y)$ and right associativity for abstraction $(\lambda x \lambda y.x) \equiv (\lambda x.(\lambda y.x))$

$$first \equiv (\lambda x \lambda y.x)$$
$$second \equiv (\lambda x \lambda y.y)$$
$$cons \equiv (\lambda x \lambda y \lambda w.(w\ x\ y)).$$

Now we have

$$car \equiv (\lambda x.(x\ first))$$
$$cdr \equiv (\lambda x.(x\ second))$$

and consequently

$$
\begin{aligned}
(cons\ a\ b) \ &\equiv\ ((\lambda x \lambda y \lambda w.(w\ x\ y))\ a\ b) \\
&\rightarrow_\beta ((\lambda y \lambda w.(w\ a\ y))\ b) \\
&\rightarrow_\beta (\lambda w.(w\ a\ b)) \\
(car\ (cons\ a\ b)) \ &\equiv\ ((\lambda x.(x\ first))\ (cons\ a\ b)) \\
&\rightarrow_\beta ((cons\ a\ b)\ first) \\
&\rightarrow_\beta ((\lambda w.(w\ a\ b))\ first) \\
&\rightarrow_\beta (first\ a\ b)) \\
&\equiv\ ((\lambda x \lambda y.x)\ a\ b)) \\
&\rightarrow_\beta ((\lambda y.a)\ b) \\
&\rightarrow_\eta a.
\end{aligned}
$$

Similarly $(cdr\ (cons\ a\ b)) \rightarrow b$. In the above example we used reduction steps to simplify the expression. For the first reduction we applied β reduction for the application of *car*, however we could also have chosen to first apply β reduction for *cons*. Different reductions strategies have different properties. LISP would reduce b in $(a\ b)$ first (strict reduction) whereas Haskell would reduce a first (non-strict reduction) which avoids reducing terms that are never used. Both strategies have advantages and disadvantages. Strict evaluation requires if to be a special form, as is the case in LISP, but easily supports sequencing of input and output operations. Non-strict reduction allows if to be an abstraction term while sequencing for input and output must be handled specially.

To construct lists we need the empty list **nil**, and the corresponding test for an empty list, **null**.

$$nil \equiv (\lambda x.(x\ first\ first\ first\ first))$$
$$null \equiv (\lambda x.(x\ (\lambda a\lambda b.second)))$$

with

$$
\begin{aligned}
(null\ nil) \ &\equiv\ ((\lambda x.(x\ (\lambda a\lambda b.second)))\ nil) \\
&\to_\beta (nil\ (\lambda a\lambda b.second)) \\
&\equiv\ ((\lambda x.(x\ first\ first\ first\ first))\ (\lambda a\lambda b.second)) \\
&\to_\beta ((\lambda a\lambda b.second)\ first\ first\ first\ first) \\
&\to_\eta (second\ first\ first) \\
&\to_\beta first \\
(null\ (cons\ a\ b)) \ &\equiv\ ((\lambda x.(x\ (\lambda a\lambda b.second)))\ (cons\ a\ b)) \\
&\to_\beta ((cons\ a\ b)\ (\lambda a\lambda b.second)) \\
&\to_\beta ((\lambda w.(w\ a\ b))\ (\lambda a\lambda b.second)) \\
&\to_\beta ((\lambda a\lambda b.second)\ a\ b) \\
&\to_\eta second.
\end{aligned}
$$

Below we see that *first* and *second* can be considered as *true* and *false* values respectively. Thus we can implement LISP style lists.

In the following we consider the non-strict reduction strategy, the implementation that follows can easily be adapted to the strict reduction strategy. Thus we can implement boolean values and conditionals using

$$true \equiv first \equiv (\lambda x\lambda y.x)$$
$$false \equiv second \equiv (\lambda x\lambda y.y)$$
$$if \equiv (\lambda w\lambda x\lambda y.(w\ x\ y)).$$

Thus we have

$$
\begin{aligned}
(if\ true\ a\ b) \ &\equiv\ ((\lambda w\lambda x\lambda y.(w\ x\ y))\ true\ a\ b) \\
&\to_\beta ((\lambda x\lambda y.(true\ x\ y))\ a\ b) \\
&\to_\beta ((\lambda y.(true\ a\ y))\ b) \\
&\to_\beta (true\ a\ b) \\
&\to_\beta a
\end{aligned}
$$

and

$$(if\ false\ a\ b) \to b.$$

The λ-calculus has many more interesting properties, such as anonymous recursive functions and a representation of the natural numbers (Church numerals). See [5][6] for details.

11.2.2 C++ Implementation

The core of the C++ implementation resides in `expression.h`. The enumeration type `expression_type` describes λ-calculus terms with the extensions of basic list and numeric types.

We define the type `environment` as a mapping `map<string,expression>` from free variables to their definitions. This allows us to name functions using an entry in the environment. Similarly built in functions supported by the interpreter are stored in a `map<string,builtin>` where

```
typedef expression (*builtin)(expression&,environment&);
```

describes a C++ function which manipulates a given expression in a given environment.

The implementation is simple, thus strings are not supported, however we follow the convention that a free variable evaluates to itself and can be displayed. Thus free variables can, to some degree, fill the role of strings. To further enhance this role, symbols can be written between single quotes, for example `'free variable'`, so as to include white space. To obtain a single quote we use `''` (i.e. quote the open quotation) and `'` (outside of a quote).

Comments begin with `;` and expressions follow the parenthesis notation of LISP.

Interpreter

The main file `lambda.cpp` initializes basic commands and starts the interpreter. It is very easy, in principle, to add new commands. For example we could add new arithmetic operators `++` and `>=`. Built in commands access the internal structure allowing us to test the type of an expression `list?`, `lambda?` etc. In the interest of having a relatively concise implementation we have omitted these definitions.

```cpp
// lambda.cpp

#include <iostream>
#include <fstream>
#include <sstream>
#include "expression.h"
#include "builtin_arith.h"
#include "builtin_core.h"
#include "builtin_io.h"
#include "builtin_list.h"
#include "builtin_math.h"
using namespace std;

environment global;

int main(int argc,char *argv[])
{
  stringstream s;
```

```
// core
builtin_commands["apply"]      = builtin_apply;
builtin_commands["define"]      = builtin_define;
builtin_commands["defined?"]    = builtin_definedp;
builtin_commands["sequence"]    = builtin_sequence;
builtin_commands["sequence*"]   = builtin_sequence_star;
// io
builtin_commands["display"]     = builtin_display;
builtin_commands["import"]      = builtin_import;
builtin_commands["newline"]     = builtin_newline;
// lists
builtin_commands["car"]         = builtin_car;
builtin_commands["cdr"]         = builtin_cdr;
builtin_commands["cons"]        = builtin_cons;
builtin_commands["empty?"]      = builtin_emptyp;
builtin_commands["list"]        = builtin_list;
// arithmetic
builtin_commands["+"]           = builtin_add;
builtin_commands["-"]           = builtin_sub;
builtin_commands["*"]           = builtin_mul;
builtin_commands["/"]           = builtin_div;
builtin_commands["%"]           = builtin_mod;
builtin_commands["="]           = builtin_equality;
builtin_commands[">"]           = builtin_greater;
// math
builtin_commands["sin"]         = builtin_sin;
builtin_commands["cos"]         = builtin_cos;
builtin_commands["exp"]         = builtin_exp;
builtin_commands["log"]         = builtin_log;
builtin_commands["sqrt"]        = builtin_sqrt;

if(argc==1) interpreter(cin,global);
else
{
 s << "(import " << argv[1] << ")" << endl;
 interpreter(s,global);
}
return 0;
}
```

Expressions

The header file `expression.h` implements the internal structure for expressions and their environment and for built in expressions.

The header file provides a parser and evaluator for expressions such as

```
(lambda (x) x)      ; single argument
(lambda (x y) x)    ; multiple arguments
((lambda (x) x) y) ; application
```

Each expression has a type, namely an application, bound variable, free variable, λ abstraction, list or number. Bound variables are created using the method **bind** which searches through an expression for all variables with a given name and changes them to bound variables which point to an expression from which the value will be obtained. Numeric data is converted from a symbol on demand, i.e. `1` is a symbol until an expression requires a numeric value (for example `+`). In this case the method `numeric` attempts to extract the numeric value from the symbol. The symbol `1` does not have a numeric type (it is considered a free variable), but the results of numeric computations always have numeric type.

A method `copy` is provided to copy expressions, α-conversion is applied whenever an abstraction is copied.

The operator `>>` parses the given stream taking into account the rules for building symbols and ignoring comments. The operator `<<` is used to output expressions.

For convenience, a function `interpreter` is provided which reads from a stream and interprets all the expressions it finds. This function can be used to interpret the contents of a file.

The function `evaluate` evaluates an expression, which consists almost entirely of applying outermost β reduction.

The environment is manipulated with the function `define` and values are determined with `lookup`.

The STL `vector` class is used to store all complex expressions. In the case of an abstraction, the `vector tokens` has an abstraction at index 0 and the arguments in the remaining positions. In the case of a list, the `vector tokens` represents the contents of the list. In the case of an abstraction the `vector tokens` has a list (`vector`) of bound variables at index 0 and the expression for which substitution must be performed at index 1.

```
// expression.h

#ifndef EXPRESSION_H
#define EXPRESSION_H

#include <iostream>
#include <fstream>
#include <map>
#include <sstream>
#include <string>
#include <vector>
using namespace std;

// lambda terms, with extra support for lists and numbers
typedef enum {
  APPLICATION, BOUND_VARIABLE, FREE_VARIABLE, LAMBDA, LIST, NUMBER
} expression_type;

class expression;

typedef map<string,expression> environment;
```

```
void define(string&,expression&,environment&);
expression lookup(string&,environment&);
expression evaluate(expression,environment&);
void interpreter(istream&,environment&);

typedef expression (*builtin)(expression&,environment&);
map<string,builtin> builtin_commands;

class expression
{
 public:
  expression_type type;
  string identifier;
  double number;
  vector<expression> tokens;
  expression *reference;

  void bind(expression &variable,expression &body);
  expression &copy(const expression &src,expression &dest);
  double numeric(environment &);

  expression() : type(FREE_VARIABLE), identifier("?") {}
  expression(string s) : type(FREE_VARIABLE), identifier(s) {}
  expression(double d) : type(NUMBER), number(d) {}
  expression(const expression &e) { copy(e, *this); }
  expression(vector<expression> &v);

  expression &operator=(const expression &e)
  { return copy(e,*this); }

  friend ostream &operator<<(ostream &,const expression&);
  friend istream &operator>>(istream &,expression&);
};

expression::expression(vector<expression> &v) : tokens(v)
{
 vector<expression> sublambda(3), lambda_variables;

 if(v.size()==3 && v[0].type==FREE_VARIABLE && v[0].identifier=="lambda")
 {
  type = LAMBDA;
  switch(v[1].type)
  {
   case FREE_VARIABLE:
    cerr << "Error: free variable as lambda variable\n"; break;
   case BOUND_VARIABLE:
    cerr << "Error: bound variable as lambda variable\n"; break;
   case LAMBDA:
    cerr << "Error: lambda expression as lambda variable\n"; break;
   case APPLICATION:
     tokens[1].type = LIST;
     switch(tokens[1].tokens.size())
     {
      case 0:  break;
      case 1:  bind(tokens[1].tokens[0],tokens[2]);
               break;
      default: lambda_variables
               = vector<expression>(tokens[1].tokens.begin()+1,
```

```
                                    tokens[1].tokens.end());
                sublambda[0] = expression("lambda");
                sublambda[1] = expression(lambda_variables);
                sublambda[2] = expression(tokens[2]);
                tokens[1].tokens.erase(tokens[1].tokens.begin()+1,
                                       tokens[1].tokens.end());
                tokens[2] = expression(sublambda);
                bind(tokens[1].tokens[0],tokens[2]);
        }
    default:            break;
    }
 }
 else type = APPLICATION;
}

void expression::bind(expression &variable,expression &body)
{
 vector<expression>::iterator i;
 switch(body.type)
 {
  case FREE_VARIABLE: if(body.identifier==variable.identifier)
                      {
                       body.type = BOUND_VARIABLE;
                       body.reference = &variable;
                      }
                      break;
  case LAMBDA:        bind(variable,body.tokens[2]);
                      break;
  case APPLICATION:   for(i=body.tokens.begin();
                          i!=body.tokens.end();i++)
                        bind(variable,*i);
  default:            break;
 }
}

expression &expression::copy(const expression &src,expression &dest)
{
 if(&dest==&src) return dest;
 dest.type = src.type;
 expression *oldreference;
 static map<expression*, expression*> rebind;
 map<expression*, expression*>::iterator newbinding;

 switch(dest.type)
 {
  case FREE_VARIABLE:  dest.identifier = src.identifier; break;
  case NUMBER:         dest.number = src.number; break;
  // the next case is essentially alpha conversion
  // so that each abstraction has its own variable
  case BOUND_VARIABLE: dest.identifier = src.identifier;
                       newbinding = rebind.find(src.reference);
                       if(newbinding != rebind.end())
                        dest.reference = newbinding->second;
                       else copy(*(src.reference),dest);
                       break;
  // creates a new abstraction with its own variable
  // and applies alpha conversion
  case LAMBDA:         dest.tokens.resize(3);
```

```
                        copy(src.tokens[0],dest.tokens[0]);
                        copy(src.tokens[1],dest.tokens[1]);
                        oldreference
                          = (expression*)&(src.tokens[1].tokens[0]);
                        rebind[oldreference]
                          = &(dest.tokens[1].tokens[0]);
                        copy(src.tokens[2],dest.tokens[2]);
                        rebind.erase(oldreference);
                        break;
  case LIST:
  case APPLICATION:     dest.tokens = src.tokens; break;
  default:              break;
 }

 return dest;
}

double expression::numeric(environment &env)
{
 expression e = evaluate(*this,env);
 if(e.type==NUMBER) return e.number;
 if(e.type==FREE_VARIABLE)
 {
  double d;
  istringstream is(e.identifier);
  is >> d;
  return d;
 }
 cerr << "Error: Tried to use " << e << " as a number, using 0.0.\n";
 return 0.0;
}

void define(string &name,expression &definition,environment &env)
{env[name] = definition;}

expression lookup(string &name,environment &env)
{
 environment::iterator i = env.find(name);
 if(i != env.end()) return i->second;
 throw(string("not found"));
}

ostream &operator<<(ostream &o,const expression &e)
{
 vector<expression>::const_iterator i;
 switch(e.type)
 {
  case FREE_VARIABLE:  o << e.identifier; break;
  case BOUND_VARIABLE: o << *(e.reference); break;
  case NUMBER:         o << e.number; break;
  case LAMBDA:
  case LIST:
  case APPLICATION:    o << "(";
                       for(i=e.tokens.begin();
                             i!=e.tokens.end();i++)
                        o << " " << (*i);
                       o << " )";
  default:             break;
```

```
  }
 return o;
}

int isparen(char c) {return c == '(' || c== ')';}

void skipwhitespace(istream &in)
{
 char c = ' ';
 while(isspace(c) && !in.fail() && !in.eof()) c = in.get();
 if(!in.fail() && !in.eof()) in.unget();
}

istream &operator>>(istream &in,expression &expr)
{
 char c;
 string token;
 vector<expression> tokens;
 skipwhitespace(in); c = in.get();
 if(c==')')
 {
  cerr << "Error: in expression: mismatched parenthesis.\n";
  return in;
 }
 if(c=='(' && !in.fail() && !in.eof())
 {
  skipwhitespace(in); c = in.get();
  while(c!=')' && !in.fail() && !in.eof())
  {
   in.unget();
   tokens.push_back(expression());
   in >> tokens.back();
   skipwhitespace(in); c = in.get();
  }
  if(c!=')')
  {
   cerr << "Error: in expression: mismatched parenthesis.\n";
   return in;
  }
  expr = expression(tokens);
 }
 else
 {
  int quote = 0;
  int linecomment = 0;
  while(!in.fail() && !in.eof()
        && (quote || linecomment || (!isspace(c) && !isparen(c))) )
  {
   if(c=='\'' && !quote)           quote = 1;
   else if(c=='\'' && quote)       quote = 0;
   else if(c==';')                 linecomment = 1;
   else if(c=='\n' && linecomment) {linecomment = 0; in.unget();}
   else if(!linecomment) token = token + c;
   c = in.get();
  }
  if(isparen(c)) in.unget();
  expr = expression(token);
 }
```

```
 return in;
}

expression evaluate(expression e,environment &env)
{
 vector<string>::iterator i;
 vector<expression>::iterator j;
 map<string, builtin>::iterator b;
 switch(e.type)
 {
  case FREE_VARIABLE:   try {e = evaluate(lookup(e.identifier,env),env);}
                        catch(string error) { }
                        return e;
  case BOUND_VARIABLE: return evaluate(*(e.reference),env);
  case LIST:
  case NUMBER:
  case LAMBDA:          return e;
  case APPLICATION:
      e.tokens[0] = evaluate(e.tokens[0],env);
      // beta reduction
      if(e.tokens[0].type==LAMBDA && e.tokens.size()==2)
      {
       // for strict evaluation we would evaluate e.tokens[1] first
       e.tokens[0].tokens[1].tokens[0] = e.tokens[1];
       return evaluate(e.tokens[0].tokens[2],env);
      }
      // multiple lambda applications using left associativity
      else if(e.tokens[0].type==LAMBDA && e.tokens.size()>2)
      {
       vector<expression> v(e.tokens.begin(),e.tokens.begin()+2);
       e.tokens[1] = expression(v);
       e.tokens.erase(e.tokens.begin());
       return evaluate(e,env);
      }
      // builtin expression
      else if(e.tokens[0].type==FREE_VARIABLE
              && (b=builtin_commands.find(e.tokens[0].identifier))
                 !=builtin_commands.end())
       return evaluate((b->second)(e,env),env);
      else cerr << "Error: Invalid application:\n"
               << e.tokens[0] << endl;
      return expression("#error");
  default: break;
 }
 return expression("#ERROR");
}

void interpreter(istream &in,environment &env)
{
 expression expr;
 while(!in.fail() && !in.eof())
 { in >> expr; evaluate(expr,env); }
}
#endif
```

Built in Core

The commands implemented are illustrated in the following example.

```
(apply + (list 1 2 3))    ; => 6
(defined? id)             ; => false
(define id (lambda (x) x))
(defined? id)             ; => true
id                        ; => ( lambda ( x ) x )
(sequence (define x 1) x) ; => 1 : global define
x                         ; => 1
(sequence* (define x 2) x) ; => 2 : local define
x                         ; => 1
```

The implementation is as follows.

```cpp
// builtin_core.h

#ifndef BUILTIN_CORE_H
#define BUILTIN_CORE_H

#include "expression.h"

expression builtin_apply(expression &e,environment &env)
{
 expression e1 = e;
 if(e1.tokens.size()==3)
 {
  e1.tokens[1] = evaluate(e1.tokens[1], env);
  e1.tokens[2] = evaluate(e1.tokens[2], env);
 }
 if(e1.tokens.size()!=3 || e1.tokens[2].type!=LIST)
 {
  cerr << "Error: apply takes two arguments,"
       << " a function and a list\n";
  return expression("#error");
 }
 e1.tokens[2].tokens.insert(e1.tokens[2].tokens.begin(),
                            1, e1.tokens[1]);
 e1.tokens[2].type = APPLICATION;
 return e1.tokens[2];
}

expression builtin_define(expression &e,environment &env)
{
 if(e.tokens[1].type!=FREE_VARIABLE)
 {
  cerr << "Error: define must bind a free variable.\n";
  return expression("#error");
 }
 define(e.tokens[1].identifier,e.tokens[2],env);
 return expression("#definition:" + e.tokens[1].identifier);
}

expression builtin_definedp(expression &e,environment &env)
{
 if(e.tokens[1].type!=FREE_VARIABLE)
 {
  cerr << "Error: defined? takes a free variable as argument\n";
```

```
  return expression("#error");
}
try
{
 lookup(e.tokens[1].identifier,env);
 return expression("true");
}
catch(string error) {return expression("false");}
return expression("#error");
}

expression builtin_sequence(expression &e,environment &env)
{
 expression e1;
 vector<expression>::iterator j = e.tokens.begin() + 1;
 while(j!=e.tokens.end()) e1 = evaluate(*(j++),env);
 return e1;
}

expression builtin_sequence_star(expression &e,environment &env)
{
 expression e1;
 environment local = env;
 vector<expression>::iterator j = e.tokens.begin() + 1;
 while(j!=e.tokens.end()) e1 = evaluate(*(j++),local);
 return e1;
}
#endif
```

Built in Input and Output

The output functions are `display` and `newline`. The application (`newline`) prints
a new line character '\n'. `display` evaluates its argument and displays it.

```
(display id)               ; displays id
(define id (lambda (x) x))
(display id)               ; displays ( lambda ( x ) x )
```

The function (`import filename`) loads and interprets the file named `filename` in
the current environment.

```
// builtin_io.h

#ifndef BUILTIN_IO_H
#define BUILTIN_IO_H

expression builtin_display(expression &e,environment &env)
{
 int space = 0;
 vector<expression>::iterator i;
 for(i=e.tokens.begin()+1;i!=e.tokens.end();i++)
 {
  if(space) cout << " ";
  cout << evaluate(*i, env);
  space = 1;
 }
```

```
   return expression("#display");
}

expression builtin_import(expression &e,environment &env)
{
 vector<expression>::iterator i;
 for(i=e.tokens.begin()+1;i!=e.tokens.end();i++)
 {
  expression e1 = evaluate(*i, env);
  if(e1.type==FREE_VARIABLE)
  {
   ifstream f(e1.identifier.c_str());
   if(f.fail())
    cerr << "Error: (import " << e1 << ") failed to open file\n";
   else interpreter(f,env);
   f.close();
  }
  else
   cerr << "Error: (import " << e1 << ") is not a filename\n";
 }
 return expression("#import");
}

expression builtin_newline(expression &e,environment &env)
{cout << endl; return expression("#newline");}
#endif
```

Built in Lists

The basic operations for lists are supported, i.e. cons, car and cdr. One difference, for simplicity, is that (cons a b) requires that b is a list.

```
(list a b c)            ; ( a b c )
(cons a (list a b c))   ; ( a a b c )
(empty? (list a b c))   ; false
(empty? (list))         ; true
(car (list a b c))      ; a
(cdr (list a b c))      ; (b c)
```

```
// builtin_list.h

#ifndef BUILTIN_LIST_H
#define BUILTIN_LIST_H

#include "expression.h"

expression builtin_car(expression &e,environment &env)
{
 expression e1 = e;
 if(e1.tokens.size()==2) e1.tokens[1] = evaluate(e1.tokens[1],env);
 if(e1.tokens.size()!=2 || e1.tokens[1].type!=LIST
                        || e1.tokens[1].tokens.size()<1)
 {
  cerr << "Error: car only operates on lists"
       << " with at least one element.\n";
  return expression("#error");
```

```
}
 return e1.tokens[1].tokens[0];
}

expression builtin_cdr(expression &e,environment &env)
{
 expression e1 = e;
 if(e1.tokens.size()==2) e1.tokens[1] = evaluate(e1.tokens[1],env);
 if(e1.tokens.size()!=2 || e1.tokens[1].type!=LIST
                        || e1.tokens[1].tokens.size()<1)
 {
  cerr << "Error: cdr only operates on lists"
       << " with at least one element.\n";
  return expression("#error");
 }
 expression e2 = e1.tokens[1];
 e2.tokens.erase(e2.tokens.begin());
 return e2;
}

expression builtin_cons(expression &e,environment &env)
{
 expression e1 = e;
 if(e1.tokens.size()==3)
 {
  e1.tokens[1] = evaluate(e1.tokens[1],env);
  e1.tokens[2] = evaluate(e1.tokens[2],env);
 }
 if(e1.tokens.size()!=3 || e1.tokens[2].type!=LIST)
 {
  cerr << "Error: cons takes two arguments, the second a list.\n";
  return expression("#error");
 }
 e1.tokens[2].tokens.insert(e1.tokens[2].tokens.begin(),1,e1.tokens[1]);
 return e1.tokens[2];
}

expression builtin_emptyp(expression &e,environment &env)
{
 expression e1 = e;
 if(e1.tokens.size()==2) e1.tokens[1] = evaluate(e1.tokens[1],env);
 if(e1.tokens.size()!=2 || e1.tokens[1].type!=LIST)
 {
  cerr << "Error: empty? takes one argument of type list.\n";
  return expression("#error");
 }
 return expression((e1.tokens[1].tokens.size()==0)?"true":"false");
}

expression builtin_list(expression &e,environment &env)
{
 expression e1 = e;
 e1.tokens.erase(e1.tokens.begin());
 e1.type = LIST;
 return e1;
}
#endif
```

Built in Arithmetic

The basic C++ arithmetic operations are supported, with similar conventions to LISP:

```
(+ 1 2 3 4) ; 10 = 1 + 2 + 3 + 4
(- 1 2 3 4) ; -8 = 1 - 2 - 3 - 4
(- 2)       ; -2
(* 1 2 3 4) ; 24 = 1 * 2 * 3 * 4
(/ 1 2 3 4) ; 0.0416667 = 1 / 2 / 3 / 4 = 1 / 24
(/ 24)      ; 0.0416667 = 1 / 24
(% 53 21)   ; 11
(% 53 21 7) ; 4 = ((53 % 21) % 7)
(> 3 1)     ; true
(> 7 5 3)   ; true : (7 > 5) && (5 > 3)
(= 3 1)     ; false
(= 3 3 1)   ; false : (3 == 3) && (3 == 1)
```

```cpp
// builtin_arith.h

#ifndef BUILTIN_ARITH_H
#define BUILTIN_ARITH_H

#include "expression.h"

expression builtin_add(expression &e,environment &env)
{
 double acc = 0.0;
 vector<expression>::iterator j = e.tokens.begin() + 1;
 while(j != e.tokens.end()) acc += (j++)->numeric(env);
 return expression(acc);
}

expression builtin_sub(expression &e,environment &env)
{
 int first = 1;
 double acc = 0.0;
 vector<expression>::iterator j = e.tokens.begin() + 1;
 if(e.tokens.size()==2) first = 0;
 while(j!=e.tokens.end())
  {
   if(first) {acc = (j++)->numeric(env); first = 0;}
   else acc -= (j++)->numeric(env);
  }
 return expression(acc);
}

expression builtin_mul(expression &e,environment &env)
{
 double acc = 1.0;
 vector<expression>::iterator j = e.tokens.begin() + 1;
 while(j!=e.tokens.end()) acc *= (j++)->numeric(env);
 return expression(acc);
}

expression builtin_div(expression &e,environment &env)
{
 int first = 1;
```

```
 double acc = 1.0;
 vector<expression>::iterator j = e.tokens.begin() + 1;
 if(e.tokens.size()==2) first = 0;
 while(j!=e.tokens.end())
 {
  if(first) {acc = (j++)->numeric(env); first = 0;}
  else acc /= (j++)->numeric(env);
 }
 return expression(acc);
}

expression builtin_mod(expression &e,environment &env)
{
 int first = 1;
 double acc = 1.0;
 vector<expression>::iterator j = e.tokens.begin() + 1;
 while(j!=e.tokens.end())
 {
  if(first) {acc = (j++)->numeric(env); first = 0;}
  else acc = fmod(acc, (j++)->numeric(env));
 }
 return expression(acc);
}

expression builtin_equality(expression &e,environment &env)
{
 int holds = 1, first = 1;
 double i, i0;
 vector<expression>::iterator j = e.tokens.begin() + 1;
 while(j!=e.tokens.end())
 {
  i = (j++)->numeric(env);
  if(first) {i0 = i; first = 0;}
  else if(i!=i0) {holds = 0; break;};
 }
 return expression((holds)?"true":"false");
}

expression builtin_greater(expression &e,environment &env)
{
 int holds = 1, first = 1;
 double i, i0;
 vector<expression>::iterator j = e.tokens.begin() + 1;
 while(j!=e.tokens.end())
 {
  i = (j++)->numeric(env);
  if(first) {i0 = i; first = 0;}
  else if(i>=i0) {holds = 0; break;};
  i0 = i;
 }
 return expression((holds)?"true":"false");
}
#endif
```

Built in Transcendental Functions

There is support for a few transcendental functions, sin, cos, exp, log and sqrt.

```
(log (exp 1)) ; 1
```

Below is the listing for `sin`, the remaining functions are implemented almost identically.

```
// builtin_math.h

#ifndef BUILTIN_MATH_H
#define BUILTIN_MATH_H

expression builtin_sin(expression &e,environment &env)
{
 if(e.tokens.size()!=2)
 {
  cerr << "Error: incorrect number of arguments to sin\n";
  return expression("#error");
 }
 return expression(sin(e.tokens[1].numeric(env)));
}

...
```

Examples

Most programs use conditional statements. The implementation from λ-calculus is given below.

```
; logic.lambda :
;   Basic logic operations implemented in the lambda calculus

(define true  (lambda (x y) x))
(define false (lambda (x y) y))
(define if    (lambda (test consequent1 consequent2)
                      (test consequent1 consequent2) ) )

(define not (lambda (x) (if x false true)))

(define or  (lambda x
                    (if (empty? x) false
                        (if (car x) true (apply or (cdr x))) ) ) )
(define and (lambda x
                    (if (empty? x) true
                        (if (car x) (apply and (cdr x)) false)) ) )

(define cond (lambda x
                     (if (empty? x) false
                         (if (caar x) (cadar x) (apply cond (cdr x))) ) ) )
```

Note that the interpreter is not lexically scoped, thus the return value **true** depends on the current environment:

```
(> 5 3)                 ; true
(import logic.lambda)
(> 5 3)                 ; ( lambda ( x ) ( lambda ( y ) x ) )
```

As another example we give the implementation of `mapcar`

```
; mapcar.lambda

(import logic.lambda)

(define mapcar
 (lambda (function l)
  (if (empty? l) (list)
      (cons (function (car l)) (mapcar function (cdr l))) ) ) )

(mapcar (lambda (x) (+ x 1)) (list 1 2 3)) ; ( 2 3 4 )
```

The next example implements cons, car and cdr using the definitions we considered
in λ-calculus.

```
; mapcar2.lambda

(import logic.lambda)

(define first  (lambda (x y) x))
(define second (lambda (x y) y))
(define cons   (lambda (x y w) (w x y)))
(define car    (lambda (x) (x first)))
(define cdr    (lambda (x) (x second)))
(define cadr   (lambda (x) (car (cdr x))))
(define caddr  (lambda (x) (car (cdr (cdr x)))))
(define nil    (lambda (x) (x true true true true)))
(define null   (lambda (x) (x (lambda (a b) false))))

(define p (cons a b))
(display         p) (newline)        ; ( lambda ( w ) ( w a b ) )
(display (car p)) (newline)          ; a
(display (cdr p)) (newline)          ; b
(display (if (null nil) yes no)) ; yes
(newline)
(display (if (null   p) yes no)) ; no
(newline)

(define mapcar
 (lambda (function l)
  (if (null l) nil
      (cons (function (car l)) (mapcar function (cdr l))) ) ) )

(define ml (mapcar (lambda (x) (+ x 1))
                   (cons 1 (cons 2 (cons 3 nil))) ) )

(display (car   ml)) (newline) ; 2
(display (cadr  ml)) (newline) ; 3
(display (caddr ml)) (newline) ; 4
```

Chapter 12

Gene Expression Programming

12.1 Introduction

Gene expression programming (Ferreira [18], Hardy and Steeb [22]) is a genome/phe-nome genetic algorithm which combines the simplicity of genetic algorithms and the abilities of genetic programming. In a sense gene expression programming is a generalization of genetic algorithms and genetic programming. Gene expression programming is different from genetic programming because the expense of manag-ing a tree structure and ensuring correctness of programs is eliminated. We provide an introduction in the following.

A *gene* is a symbolic string with a head and a tail. Each symbol represents an operation. The operation + takes two arguments and adds them. For example, +x2 is the operation + with arguments x and 2 giving x+2. The operation * also takes two arguments and multiplies them. The operation x would evaluate to the value of the variable x. The tail consists only of operations which take no arguments. The string represents expressions in prefix notation, i.e. 5-3 would be stored as - 5 3. The reason for the tail is to ensure that the expression is always complete. Suppose the string has h symbols in the head which is specified as an input to the algorithm, and t symbols in the tail which is determined from h. Thus if n is the maximum number of arguments for an operation we must have

$$h + t - 1 = hn.$$

The left hand side is the total number of symbols except for the very first symbol. The right hand side is the total number of arguments required for all operations. We assume, of course, that each operation requires the maximum number of arguments so that any string of this length is a valid string for the expression. Thus the equation states that there must be enough symbols to serve as arguments for all operations. Now we can determine the required length for the tail $t = h(n-1) + 1$.

Example 1. Consider the symbols x, y and the operations $*$, $+$ which take two arguments. Let $h = 5$. Then $t = 6$ since $n = 2$. Thus a gene would be

+*xy2|xyy32y

The vertical line indicates the beginning of the tail. The expression is (x*y)+2. The tree structure would be

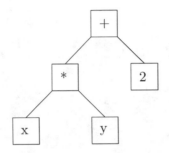

Example 2. Suppose we use $h = 8$, and $n = 2$ for arithmetic operations. Thus the tail length must be $t = 9$. So the total gene length is 17. We could then represent the expression

$$\cos(x^2 + 2) - \sin(x)$$

with the string

-c+*xx2s|x1x226x31

The vertical | is used to indicate the beginning of the tail. Here c represents cos and s represents sin. We can represent the expressions with trees. For the example above, the root of the tree would be '-' with branches for the parameters. Thus we could represent the expression as follows

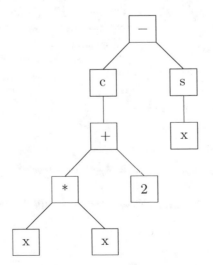

Note that not all of the symbols from the tail are used. Only one symbol from the tail is used in the present example.

A *chromosome* is a collection of genes. The genes combine to form an expression using some operation with the same number of arguments as genes in the chromosome. For example the expressions of genes of a chromosome may be added together. For operations applied to chromosomes we often concatenate the genes to obtain a single string of symbols. For example, suppose we have the following genes forming a chromosome

```
-c+*xx2s|x1x226x31
-c+*xx2s|x1x226x31
-c++x2*x|x1x226x31
```

we form the chromosome by concatenation

```
-c+*xx2s|x1x226x31|-c+*xx2s|x1x226x31|-c++x2*x|x1x226x31
```

where | indicates the beginning of the head and tail portions of the genes. We apply operations to chromosomes in a collection which is called the *population* of chromosomes.

A number of operations are applied to chromosomes.

- **Replication**. The chromosome is unchanged. The roulette wheel selection technique can be used to select chromosomes for replication.

- **Mutation**. Randomly change symbols in a chromosome. Symbols in the tail of a gene may not operate on any arguments. Typically 2 point mutations per chromosome is used. For example, let `+*xy2|xyy32y`. Then a mutation could yield `+*yy2|xyy32y` or `++xy2|xyy32y` or `+*xy-|xyy32y`.

- **Insertion**. A portion of a chromosome is chosen to be inserted in the head of a gene. The tail of the gene is unaffected. Thus symbols are removed from the end of the head to make room for the inserted string. Typically a probability of 0.1 of insertion is used. As an example suppose `+x2` is to be inserted into

  ```
  -c+*xx2s|x1x226x31
  ```

  at the fourth position in the head. We obtain

  ```
  -c++x2*x|x1x226x31
  ```

  which represents the expression

  $$\cos((x+2)+x^2) - 1.$$

- **Gene transposition**. One gene in a chromosome is randomly chosen to be the first gene. All other genes in the chromosome are shifted downwards in the chromosome to make place for the first gene.

- **Recombination**. The crossover operation. This can be one point (the chromosomes are split in two and corresponding sections are swapped), two point (chromosomes are split in three and the middle portion is swapped) or gene (one entire gene is swapped between chromosomes) recombination. Typically the sum of the probabilities of recombination is 0.7.

Another operation could be swapping. For example, consider again +*xy2|xyy32y. Then swapping the second and seventh position (counting from left and starting from zero) we obtain +*yy2|xyx32y.

In the following example we implement these operations. For simplicity we use only one gene in a chromosome and only one point recombination.

12.1.1   Example

In our example we consider smooth one-dimensional maps $f : \mathbf{R} \to \mathbf{R}$ having the properties

$$f(0) = 0, \quad f(1) = 0, \quad f\left(\frac{1}{2}\right) = 1$$

$$\frac{1}{2} < f\left(\frac{1}{4}\right) < 1, \qquad \frac{1}{2} < f\left(\frac{3}{4}\right) < 1.$$

For example the logistic map $g(x) := 4x(1-x)$ satisfies this set of properties. We use Gene Expression Programming to perform symbolic regression in order to obtain maps which satisfy the above properties. We expect that the logistic map should be found since we limit the functions in the implementation to polynomials.

We generalize the set of properties as follows. The points of evaluation are specified by a subset of $X \times Y$ where X and Y are given by $f : X \to Y$. Denote by $F_= \subset X \times Y$ the subset of all $(x, y) \in X \times Y$ such that we require of f that $f(x) = y$, by $F_> \subset X \times Y$ the subset of all $(x, y) \in X \times Y$ such that we require of f that $f(x) > y$, by $F_< \subset X \times Y$ the subset of all $(x, y) \in X \times Y$ such that we require of f that $f(x) < y$. Thus we can define the fitness function of a function g (where a smaller value indicates a higher fitness)

$$\text{fitness}(g) := \sum_{(x,y)\in F_=} |g(x) - y| + \sum_{(x,y)\in F_>} |g(x) - y| H(y - g(x))$$

$$+ \sum_{(x,y)\in F_<} |g(x) - y| H(g(x) - y)$$

where

$$H(x) := \begin{cases} 1 \, x > 0 \\ 0 \, \text{otherwise} \end{cases}$$

is the step function.

We apply gene expression programming until we find a function g such that

$$\text{fitness}(g) < \epsilon$$

for given $\epsilon > 0$. A typical value of ϵ is 0.001.

In the C++ program the function `evalr()` takes a character string and evaluates (i.e., returns type `double`) the corresponding function at the given point (the parameter `x` of type `double`) using recursion. The function `eval()` uses `evalr()` for evaluation without modifying the pointer argument. Similarly `printr()` and `print()` are responsible for output of the symbolic expressions in readable form. A character array and an array of type `double` are passed to the `fitness()` function. The array of `double` consists of triplets: the point of evaluation, the expected value and the comparison type. If the comparison type is zero then the goal is equality (i.e. we can use the absolute value of the difference between the expected value and the actual value), less than zero then the actual value should be less than the expected value, greater than zero then the actual value should be greater than the expected value. Finally, the function `gep()` implements the gene expression programming algorithm. As arguments `gep()` takes the `data` points (point of evaluation, and comparison with a given value), the number of data points `N`, the size of the population `P` and the desired fitness `eps`. The function `strncpy` of the string header file `cstring` is used to copy specific regions of the character strings representing the chromosomes. We need to specify the length since the chromosomes are not null terminated.

We use 10 symbols ($h = 10$) for the head portion of the representations, thus the total gene length is 21 (we use only addition, subtraction and multiplication). Since we only use x as a terminal symbol we obtain polynomials of order up to 11, i.e. the highest order polynomial supported by the representation is x^{11}. We randomly choose terminals (symbols which do not take arguments, such as x) and non-terminals (symbols which do take arguments, for example +) for the head and then terminals for the tail. We use 0.1 for the probability of mutation, 0.4 for the probability of insertion and 0.7 for the probability of recombination. At each iteration of the algorithm we eliminate the worst half of the population.

```
// gepchaos.cpp

#include <cstdlib>
#include <ctime>
#include <cmath>
#include <iostream>
#include <string>
using namespace std;

const double pi = 3.1415927;
const int nsymbols = 5;
// 2 terminal symbols (no arguments) x and 1
const int terminals = 2;
// terminal symbols first
const char symbols[nsymbols] = {'1','x','+','-','*'};
const int n = 2;    // for +,- and * which take 2 arguments
```

```
int h = 10;

double evalr(char *&e,double x)
{
 switch(*(e++))
 {
  case '1': return 1.0;
  case 'x': return x;
  case 'y': return pi*x;
  case 'c': return cos(evalr(e,x));
  case 's': return sin(evalr(e,x));
  case '+': return evalr(e,x)+evalr(e,x);
  case '-': return evalr(e,x)-evalr(e,x);
  case '*': return evalr(e,x)*evalr(e,x);
  default : return 0.0;
 }
}

double eval(char *e,double x)
{
 char *c = e;
 return evalr(c,x);
}

void printr(char *&e)
{
 switch(*(e++))
 {
  case '1': cout << '1'; break;
  case 'x': cout << 'x'; break;
  case 'y': cout << "pi*x"; break;
  case 'c': cout << "cos(";
            printr(e);
            cout << ")";
            break;
  case 's': cout << "sin(";
            printr(e);
            cout << ")";
            break;
  case '+': cout << '(';
            printr(e);
            cout << '+';
            printr(e);
            cout << ')';
            break;
  case '-': cout << '(';
            printr(e);
            cout << '-';
            printr(e);
            cout << ')';
            break;
  case '*': cout << '(';
            printr(e);
            cout << '*';
            printr(e);
            cout<<')';
            break;
 }
```

```
}

void print(char *e)
{
 char *c = e;
 printr(c);
}

double fitness(char *c,double *data,int N)
{
 double sum = 0.0;
 double d;

 for(int j=0;j<N;j++)
 {
  d=eval(c,data[3*j])-data[3*j+1];
  if(data[3*j+2] == 0) sum += fabs(d);
  else if(data[3*j+2] > 0) sum -= (d > 0.0)?0.0:d;
  else if(data[3*j+2] < 0) sum += (d < 0.0)?0.0:d;
 }
 return sum;
}

// N number of data points
// population of size P
// eps = accuracy required
void gep(double *data,int N,int P,double eps)
{
 int i,j,k,replace,replace2,rlen,rp;
 int t = h*(n-1)+1;
 int gene_len = h+t;
 int pop_len = P*gene_len;
 int iterations = 0;
 char *population = new char[pop_len];
 char *elim = new char[P];
 int toelim = P/2;
 double bestf, f;       // best fitness, fitness value
 double sumf = 0.0;     // sum of fitness values
 double pmutate = 0.1;  // probability of mutation
 double pinsert = 0.4;  // probability of insertion
 double precomb = 0.7;  // probability of recombination
 double r,lastf;        // random numbers and roulette wheel selection
 char *best = (char*)NULL; //best gene
 char *iter;            // iteration variable

 // initialize the population
 for(i=0;i < pop_len;i++)
  if(i%gene_len < h)
   population[i] = symbols[rand()%nsymbols];
  else
   population[i] = symbols[rand()%terminals];

 // initial calculations
 bestf = fitness(population,data,N);
 best = population;
 for(i=0,sumf=0.0,iter=population;i<P;i++,iter+=gene_len)
 {
  f = fitness(iter,data,N);
```

```
 sumf += f;
 if(f<bestf)
 {
  bestf = f;
  best = population+i*gene_len;
 }
}

while(bestf >= eps)
{
 // reproduction
 // roulette wheel selection
 for(i=0;i<P;i++) elim[i] = 0;
 for(i=0;i<toelim;i++)
 {
  r = sumf*(double(rand())/RAND_MAX);
  lastf = 0.0;
  for(j=0;j<P;j++)
  {
   f = fitness(population+j*gene_len,data,N);
   if((lastf<=r) && (r<f+lastf))
   {
    elim[j] = 1;
    j = P;
   }
   lastf += f;
  }
 }

 for(i=0;i < pop_len;)
 {
  if(population+i == best) i += gene_len; //never modify/replace best gene
  else for(j=0;j<gene_len;j++,i++)
  {
   // mutation or elimination due to failure in selection
   // for reproduction
   if((double(rand())/RAND_MAX < pmutate) || elim[i/gene_len])
   if(i%gene_len < h) population[i] = symbols[rand()%nsymbols];
   else population[i] = symbols[rand()%terminals];
  }

  // insertion
  if(double(rand())/RAND_MAX < pinsert)
  {
  // find a position in the head of this gene for insertion
  // -gene_len for the gene since we have already moved
  //  to the next gene
  replace = i-gene_len;
  rp = rand()%h;
  // a random position for insertion source
  replace2 = rand()%pop_len;
  // a random length for insertion from the gene
  rlen = rand()%(h-rp);
  // create the new gene
  char *c = new char[gene_len];
  // copy the shifted portion of the head
  strncpy(c+rp+rlen,population+replace+rp,h-rp-rlen);
  // copy the tail
```

```
    strncpy(c+h,population+replace+h,t);
    // copy the segment to be inserted
    strncpy(c+rp,population+replace2,rlen);
    // if the gene is fitter use it
    if(fitness(c,data,N) < fitness(population+replace,data,N))
     strncpy(population+replace,c,h);
    delete[] c;
   }

   // recombination
   if(double(rand())/RAND_MAX < precomb)
   {
    // find a random position in the gene for one point recombination
    replace = i-gene_len;
    rlen = rand()%gene_len;
    // a random gene for recombination
    replace2 = (rand()%P)*gene_len;
    // create the new genes
    char *c[5];
    c[0] = population+replace;
    c[1] = population+replace2;
    c[2] = new char[gene_len];
    c[3] = new char[gene_len];
    c[4] = new char[gene_len];
    strncpy(c[2],c[0],rlen);
    strncpy(c[2]+rlen,c[1]+rlen,gene_len-rlen);
    strncpy(c[3],c[1],rlen);
    strncpy(c[3]+rlen,c[0]+rlen,gene_len-rlen);
    // take the fittest genes
    for(j=0;j<4;j++)
    for(k=j+1;j<4;j++)
     if(fitness(c[k],data,N) < fitness(c[j],data,N))
     {
      strncpy(c[4],c[j],gene_len);
      strncpy(c[j],c[k],gene_len);
      strncpy(c[k],c[4],gene_len);
     }
    delete[] c[2];
    delete[] c[3];
    delete[] c[4];
   }
  }

 // fitness
 for(i=0,sumf=0.0,iter=population;i<P;i++,iter+=gene_len)
 {
  f = fitness(iter,data,N);
  sumf += f;
  if(f < bestf)
  {
   bestf = f;
   best = population+i*gene_len;
  }
 }
 iterations++;
}

print(best);
```

```
cout << endl;
cout << "Fitness of " << bestf << " after "
     << iterations << " iterations." << endl;

delete[] population;
delete[] elim;
}

int main(void)
{
 srand(time(NULL)); //set the seed for the random number generator

 double data[] = {0,0,0,1,0,0,0.5,1,0,0.25,0.5,1,0.75,0.5,1,
                  0.25,1,-1,0.75,1,-1};
 gep(data,7,30,0.001);
 cout << endl;

 return 0;
}
```

The output is

```
(((((1-x)*((((x+x)*1)+x)+x))-x)+x)
Fitness of 0 after 400 iterations.
```

We list some typical results below, all with fitness 0.

$$((1 - x) * ((x + (((x - x) + x) + x)) + x)) = 4x(1 - x)$$
$$((x + (((((((x - x) - x) - x) * x) - x) * x) + x)) + x) = x(1 - x)(2x + 3)$$
$$(x - (((x + x) * ((x - 1) + x)) - x)) = 4x(1 - x)$$
$$((((1 - x) * 1) * (((1 + 1) + 1) + 1)) * x) = 4x(1 - x)$$
$$((((1 + 1) - (((x + x) * x) + x)) * x) + x) = x(1 - x)(2x + 3)$$
$$((1 * ((x - (((((x + x) * x) + x) * x)) + x)) + x) = x(1 - x)(2x + 3)$$
$$((1 - x) * ((((((x + x) * x) + x) * 1) + x) + x)) = x(1 - x)(2x + 3)$$
$$(((((1 - (1 * (((x + x) * x) - 1)))) - x) + 1) * x) = x(1 - x)(2x + 3)$$

We find that most of the time either $4x(1 - x)$ or $x(1 - x)(2x + 3)$ is the fittest map. Of course we find the desired functions with less iterations if the population size is increased. The map

$$g(x) = x(1 - x)(2x + 3)$$

satisfies the conditions given above, but g has values greater than 1 on the interval

$$\left(\frac{1}{2}, \frac{\sqrt{5}}{2} - \frac{1}{2} \right).$$

Thus we find that almost all initial values $x_0 \in [0, 1]$ escape the interval $[0, 1]$ under iterations of the map. The set of all points whose iterate stay in $[0, 1]$ are of measure zero and form a Cantor set. We can also change the **symbols** array to include the symbols **c** for cosine, **s** for sine and **y** for πx. In this case we find the function $\sin(\pi x)$ nearly every time as the fittest map.

Another application of gene expression programming is to find boolean expressions. In boolean expressions we have the AND, OR, XOR and NOT operation. The NAND gate is an AND gate followed by a NOT gate. The NAND gate is universal gate, where all other gates can be built from.

As an example consider the truth table

a	b	O
0	0	1
0	1	1
1	0	0
1	1	1

where a and b are the inputs and O is the output.

Assume that $\cdot$ (in C++ the bitwise "&" operator) is the AND-operation, $+$ (in C++ the bitwise "|" operator) is the OR operation and $\bar{a}$ (in C++ the bitwise "~") is the NOT operation (of a). Using gene expression programming we can form the expressions

```
&a|1~b|aab10bb
|b~a|ababa
```

The first boolean expression would be $a \cdot (1 + \bar{b})$. The second boolean expression would be $b + \bar{a}$. In the first case the tree structure would be given by

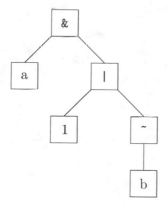

For the fitness function we compare the output for each entry from the given truth table with the output from the gene. For example, consider the expression written in gene expression programming $E(a,b) = $ ~&a|b0. The expression could also be written in standard notation

$$E(a, b) = \overline{a \cdot (b + 0)}.$$

Thus we find $E(0,0) = 1$, $E(0,1) = 1$, $E(1,0) = 1$, $E(1,1) = 0$. Therefore the fitness is 2, since at two entries the values differ. Of course the fitness we have to achieve is 0.

```cpp
// gepbool.cpp

#include <cstdlib>
#include <ctime>
#include <cmath>
#include <iostream>
#include <string>
using namespace std;

const int nsymbols = 7;
// 4 terminal symbols (no arguments) a, b, 0 and 1
const int terminals = 4;
// terminal symbols first
const char symbols[nsymbols] = {'0','1','a','b','|','~','&'};
const int n = 2;    // for +,- and * which take 2 arguments
int h = 10;

int evalr(char *&e,int a,int b)
{
 switch(*(e++))
  {
   case '0': return 0;
   case '1': return 1;
   case 'a': return a;
   case 'b': return b;
   case '|': return evalr(e,a,b) | evalr(e,a,b);
   case '~': return (~evalr(e,a,b)) & 1;
   case '&': return evalr(e,a,b) & evalr(e,a,b);
   default : return 0;
  }
}

int eval(char *e,int a,int b)
{
 char *c = e;
 return evalr(c,a,b);
}

void printr(char *&e)
{
 switch(*(e++))
  {
   case '0': cout << '0'; break;
   case '1': cout << '1'; break;
   case 'a': cout << 'a'; break;
   case 'b': cout << 'b'; break;
   case '|': cout << '(';
             printr(e);
             cout << '|';
             printr(e);
             cout << ')';
             break;
   case '~': cout << "~(";
             printr(e);
             cout << ')';
             break;
   case '&': cout << '(';
```

```
            printr(e);
            cout << '&';
            printr(e);
            cout << ')';
            break;
  }
}

void print(char *e)
{
 char *c = e;
 printr(c);
}

double fitness(char *c,int *data,int N)
{
 int d, sum = 0;

 for(int j=0;j<N;j++)
 {
  d=eval(c,data[3*j],data[3*j+1])^data[3*j+2];
  sum += d;
 }
 return sum;
}

// N number of data points
// population of size P
// eps = accuracy required
void gop(int *data,int N,int P,double eps)
{
 int i,j,k,replace,replace2,rlen,rp;
 int t = h*(n-1)+1;
 int gene_len = h+t;
 int pop_len = P*gene_len;
 int iterations = 0;
 char *population = new char[pop_len];
 char *elim = new char[P];
 int toelim = P/2;
 double bestf,f;        // best fitness, fitness value
 double sumf = 0.0;     // sum of fitness values
 double pmutate = 0.1;  // probability of mutation
 double pinsert = 0.4;  // probability of insertion
 double precomb = 0.7;  // probability of recombination
 double r,lastf;        // random numbers and roulette wheel selection
 char *best = (char*)NULL; //best gene
 char *iter;            // iteration variable

 // initialize the population
 for(i=0;i<pop_len;i++)
  if(i%gene_len < h)
   population[i] = symbols[rand()%nsymbols];
  else
   population[i] = symbols[rand()%terminals];

 // initial calculations
 bestf = fitness(population,data,N);
 best = population;
```

```
for(i=0,sumf=0.0,iter=population;i<P;i++,iter+=gene_len)
{
 f = fitness(iter,data,N);
 sumf += f;
 if(f<bestf)
 {
  bestf = f;
  best = population+i*gene_len;
 }
}

while(bestf >= eps)
{
 // reproduction
 // roulette wheel selection
 for(i=0;i<P;i++) elim[i] = 0;
 for(i=0;i<toelim;i++)
 {
  r = sumf*(double(rand())/RAND_MAX);
  lastf = 0.0;
  for(j=0;j<P;j++)
  {
   f = fitness(population+j*gene_len,N);
   if((lastf<=r) && (r<f+lastf))
   {
    elim[j] = 1;
    j = P;
   }
   lastf += f;
  }
 }

 for(i=0;i<pop_len;)
 {
  if(population+i == best) i += gene_len; //never modify/replace best gene
  else for(j=0;j<gene_len;j++,i++)
  {
   // mutation or elimination due to failure in selection
   // for reproduction
   if((double(rand())/RAND_MAX < pmutate) || elim[i/gene_len])
   if(i%gene_len < h) population[i] = symbols[rand()%nsymbols];
   else population[i] = symbols[rand()%terminals];
  }

  // insertion
  if(double(rand())/RAND_MAX < pinsert)
  {
   // find a position in the head of this gene for insertion
   //  -gene_len for the gene since we have already moved
   //   to the next gene
   replace = i-gene_len;
   rp = rand()%h;
   // a random position for insertion source
   replace2 = rand()%pop_len;
   // a random length for insertion from the gene
   rlen = rand()%(h-rp);
   // create the new gene
   char *c = new char[gene_len];
```

```
   // copy the shifted portion of the head
   strncpy(c+rp+rlen,population+replace+rp,h-rp-rlen);
   // copy the tail
   strncpy(c+h,population+replace+h,t);
   // copy the segment to be inserted
   strncpy(c+rp,population+replace2,rlen);
   // if the gene is fitter use it
   if(fitness(c,data,N) < fitness(population+replace,data,N))
    strncpy(population+replace,c,h);
   delete[] c;
  }


  // recombination
  if(double(rand())/RAND_MAX < precomb)
  {
   // find a random position in the gene for one point recombination
   replace = i-gene_len;
   rlen = rand()%gene_len;
   // a random gene for recombination
   replace2 = (rand()%P)*gene_len;
   // create the new genes
   char *c[5];
   c[0] = population+replace;
   c[1] = population+replace2;
   c[2] = new char[gene_len];
   c[3] = new char[gene_len];
   c[4] = new char[gene_len];
   strncpy(c[2],c[0],rlen);
   strncpy(c[2]+rlen,c[1]+rlen,gene_len-rlen);
   strncpy(c[3],c[1],rlen);
   strncpy(c[3]+rlen,c[0]+rlen,gene_len-rlen);
   // take the fittest genes
   for(j=0;j<4;j++)
   for(k=j+1;j<4;j++)
    if(fitness(c[k],data,N) < fitness(c[j],data,N))
    {
     strncpy(c[4],c[j],gene_len);
     strncpy(c[j],c[k],gene_len);
     strncpy(c[k],c[4],gene_len);
    }
   delete[] c[2];
   delete[] c[3];
   delete[] c[4];
  }
 }


// fitness
for(i=0,sumf=0.0,iter=population;i<P;i++,iter+=gene_len)
{
 f = fitness(iter,data,N);
 sumf += f;
 if(f < bestf)
 {
  bestf = f;
  best = population+i*gene_len;
 }
}
iterations++;
```

```
}

print(best);
cout << endl;
cout << "Fitness of " << bestf << " after "
     << iterations << " iterations." << endl;

delete[] population;
delete[] elim;
}

int main(void)
{
 srand(time(NULL)); //set the seed for the random number generator
 int data[] = {0,0,1, 0,1,1, 1,0,0, 1,1,1};
 gep(data,4,30,0.5);
 cout << endl;
 return 0;
}
```

An example output is

```
((b|0)|(~(a)&~(b)))
Fitness of 0 after 17 iterations.
```

Another example output is

```
(b|~(((a|a)&((a|b)|1))))
Fitness of 0 after 12 iterations.
```

The above expression simplifies to $\overline{a} + b$.

12.2   Multi Expression Programming

Multi Expression Programming [42] is a genetic programming variant which encodes multiple expressions in a single chromosome. The expressions are represented in a similar way to gene expression programming, except that expressions may refer to (contain) expressions that appear in earlier in the chromosome. The fitness of a chromosome is the best fitness over each of the expressions encoded in the chromosome. Crossing generally occurs at the level of expressions.

Example. The expression

$$\sin b + a * b + c * d + a - b$$

must be represented as a chromosome in multi expression programming which can be constructed in the following way

0:	a	The variable a
1:	b	The variable b
2:	c	The variable c
3:	d	The variable d
4:	$*, \mathbf{0}, \mathbf{1}$	The expression $a * b$
5:	$*, \mathbf{2}, \mathbf{3}$	The expression $c * d$
6:	$+, \mathbf{4}, \mathbf{5}$	The expression $(a * b) + (c * d)$
7:	$+, \mathbf{6}, \mathbf{0}$	The expression $(a * b) + (c * d) + a$
8:	$-, \mathbf{7}, \mathbf{1}$	The expression $((a * b) + (c * d) + a) - b$
9:	$\sin, \mathbf{1}$	The expression $\sin b$
10:	$+, \mathbf{9}, \mathbf{8}$	The expression $(\sin b) + (((a * b) + (c * d) + a) - b)$

The operation $-, x, y$ means $x - y$. The expression must be evaluated for given values of a, b, c and d. The numbers in bold are the expressions which can be referenced (backwards) from within other expressions. A number in bold appearing in an expression must be less than the number of the current expression. The following program constructs this chromosome and evaluates the expression for $a = 3$, $b = 4$, $c = 5$ and $d = 6$.

```cpp
// mep.cpp

#include <cctype>
#include <cstdlib>
#include <cmath>
#include <iostream>
#include <map>
#include <sstream>
#include <string>
#include <vector>
using namespace std;

typedef double (*op)(vector<double>);

// assume appropriate number of arguments
double add(vector<double> v)    { return v[0]+v[1]; }
double sub(vector<double> v)    { return v[0]-v[1]; }
double mul(vector<double> v)    { return v[0]*v[1]; }
double sine(vector<double> v)   { return sin(v[0]); }
double cosine(vector<double> v) { return cos(v[0]); }

string show(vector<string> e,string s = "")
{
  size_t i = 0, j = 0;
  string r;
  vector<string> operands;

  if(s == "") return show(e,e.back());

  while(i < s.size() && j != string::npos)
  {
    j = s.find(',',1);
    operands.push_back(s.substr(i,j-i));
    if(j != string::npos) i = j + 1;
  }
}
```

```
  if(operands.size() == 1)
  {
   if(isdigit(operands[0][0]))
   {
    int in;
    istringstream is(operands[0]);
    is >> in;
    return show(e,e[in]);
   }
   else return operands[0];
  }

  r = show(e,operands[0]);
  r += "(";

  for(i=1;i<operands.size()-1;i++)
  {
   r += show(e,operands[i]);
   r += ",";
  }

  r += show(e,operands[i]);
  r += ")";
  return r;
}

double evaluate(vector<string> e, map<string,double> values,
                map<string,op> ops, string s="")
{
  size_t i = 0, j = 0;
  string r;
  vector<string> operands;
  vector<double> operand_values;

  if(s == "") return evaluate(e,values,ops,e.back());

  while(i < s.size() && j != string::npos)
  {
   j = s.find(',',i);
   operands.push_back(s.substr(i,j-i));
   if(j != string::npos) i = j + 1;
  }

  if(operands.size() == 1)
  {
   if(isdigit(operands[0][0]))
   {
    int i;
    istringstream is(operands[0]);
    is >> i;
    return evaluate(e,values,ops,e[i]);
   }
   else return values[operands[0]];
  }

  r = show(e,operands[0]);
```

```
    for(i=1;i<operands.size();i++)
      operand_values.push_back(evaluate(e,values,ops,operands[i]));
    return ops[operands[0]](operand_values);
}

int main(void)
{
  map<string,op>      ops;
  map<string,double>  values;
  vector<string>      expression(11);

  // operators used and corresponding arity
  ops["+"] = add; ops["-"] = sub; ops["*"] = mul; ops["s"] = sine;

  // values for evaluation
  values["a"] = 3.0; values["b"] = 4.0; values["c"] = 5.0; values["d"] = 6.0;

  // (a*b) + (c*d) + a - b + sin(b)
  expression[0]  = "a";
  expression[1]  = "b";
  expression[2]  = "c";
  expression[3]  = "d";
  expression[4]  = "*,0,1";
  expression[5]  = "*,2,3";
  expression[6]  = "+,4,5";
  expression[7]  = "+,6,0";
  expression[8]  = "-,7,1";
  expression[9]  = "s,1";
  expression[10] = "+,9,8";

  cout << show(expression,"") << endl;
  cout << evaluate(expression, values, ops) << endl;
  return 0;
}
```

The output is

```
+(s(b),-(+(+(*(a,b),*(c,d)),a),b))
40.2432
```

Consider the chromosome from above

0:	a
1:	b
2:	c
3:	d
4:	$*, 0, 1$
5:	$*, 2, 3$
6:	$+, 4, 5$
7:	$+, 6, 0$
8:	$-, 7, 1$
9:	$\sin, 1$
10:	$+, 9, 8$

Mutation randomly changes a subexpression in the chromosome, for example mutating the first chromosome at position 6 might yield

$$
\begin{array}{ll}
\textbf{0:} & a \\
\textbf{1:} & b \\
\textbf{2:} & c \\
\textbf{3:} & d \\
\textbf{4:} & *, 0, 1 \\
\textbf{5:} & *, 2, 3 \\
\textbf{6:} & \sin, 4 \\
\textbf{7:} & +, 6, 0 \\
\textbf{8:} & -, 7, 1 \\
\textbf{9:} & \sin, 1 \\
\textbf{10:} & +, 9, 8
\end{array}
$$

In other words the entire subexpression $+, 4, 5$ is replaced by the new subexpression $\sin 4$.

Crossing is implemented at the subexpression level in the same way as for genetic algorithms and gene expression programming. Consider for example the one point crossing, at position 4, between the two chromosomes

$$
\begin{array}{llll}
\textbf{0:} & a & \textbf{0:} & a \\
\textbf{1:} & b & \textbf{1:} & b \\
\textbf{2:} & c & \textbf{2:} & *, 1, 0 \\
\textbf{3:} & d & \textbf{3:} & \cos, 0 \\
\textbf{4:} & *, 0, 1 & \textbf{4:} & +, 0, 3 \\
\textbf{5:} & *, 2, 3 & \textbf{5:} & *, 3, 3 \\
\textbf{6:} & +, 4, 5 & \textbf{6:} & +, 5, 2 \\
\textbf{7:} & +, 6, 0 & \textbf{7:} & \sin, 2 \\
\textbf{8:} & -, 7, 1 & \textbf{8:} & -, 7, 1 \\
\textbf{9:} & \sin, 1 & \textbf{9:} & *, 1, 1 \\
\textbf{10:} & +, 9, 8 & \textbf{10:} & +, 9, 8
\end{array}
$$

which yields the two children

$$
\begin{array}{llll}
\textbf{0:} & a & \textbf{0:} & a \\
\textbf{1:} & b & \textbf{1:} & b \\
\textbf{2:} & c & \textbf{2:} & *, 1, 0 \\
\textbf{3:} & d & \textbf{3:} & \cos, 0 \\
\textbf{4:} & +, 0, 3 & \textbf{4:} & *, 0, 1 \\
\textbf{5:} & *, 3, 3 & \textbf{5:} & *, 2, 3 \\
\textbf{6:} & +, 5, 2 & \textbf{6:} & +, 4, 5 \\
\textbf{7:} & \sin, 2 & \textbf{7:} & +, 6, 0 \\
\textbf{8:} & -, 7, 1 & \textbf{8:} & -, 7, 1 \\
\textbf{9:} & *, 1, 1 & \textbf{9:} & \sin, 1 \\
\textbf{10:} & +, 9, 8 & \textbf{10:} & +, 9, 8
\end{array}
$$

and the two point crossing between position 3 and 6 which yields the children

0:	a	**0:**	a
1:	b	**1:**	b
2:	c	**2:**	$*, 1, 0$
3:	$\cos, 0$	**3:**	d
4:	$+, 0, 3$	**4:**	$*, 0, 1$
5:	$*, 3, 3$	**5:**	$*, 2, 3$
6:	$+, 5, 2$	**6:**	$+, 4, 5$
7:	$+, 6, 0$	**7:**	$\sin, 2$
8:	$-, 7, 1$	**8:**	$-, 7, 1$
9:	$\sin, 1$	**9:**	$*, 1, 1$
10:	$+, 9, 8$	**10:**	$+, 9, 8$

Mutation and crossing could be refined to change the subexpressions, although this would be a little more involved.

Multi expression programming can also be applied to boolean expressions, and is particularly useful for circuit design. Consider for example the *full adder* which is given by the following truth table

a	b	c_{in}	s	c_{out}
0	0	0	0	0
0	0	1	1	0
0	1	0	1	0
0	1	1	0	1
1	0	0	1	0
1	0	1	0	1
1	1	0	0	1
1	1	1	1	1

where a and b are two bits that should be added, c_{in} is the carry bit from another adder (previous stage) which should be added to the sum of a and b, s is the resulting sum bit and c_{out} is the resulting carry bit (for the next stage).

One problem in circuit design is to find an efficient logic circuit that implements a boolean function using only NAND gates. The design consisting of the least number of NAND gates is considered optimum. Multi expression programming allows us to accomodate the fanout operation (reuse of a subexpression in more than one other subexpression) which can result in designs with less NAND gates. The fitness function should count the number of NAND gates, where less NAND gates used should be interpreted as greater fitness. The fitness function should only count those NAND gates that are actually used in the computation since, as with gene expression programming, part of the chromosome may not actually be used in the resulting expression.

The chromosome that can be used to implement the full adder is given by

0:	a
1:	b
2:	c
3:	NAND,0,1
4:	NAND,0,3
5:	NAND,1,3
6:	NAND,4,5
7:	NAND,2,6
8:	NAND,6,7
9:	NAND,2,7
10:	NAND,8,9
11:	NAND,3,7

Here subexpression **10** computes s and subexpression **11** computes c_{out}. Combining both output functions in a single chromosome allows the chromosome to share subexpressions for the implementation of each of the functions, thus using less NAND gates in total than if s and c_{out} each had their own circuit without sharing common NAND gates.

Chapter 13

Program Listing

This chapter contains the listing of the header files of the classes presented in Chapters 7 and 8. For each class, a brief description of the public member functions is given prior to the complete program listing.

13.1 Identities

The identity.h header file provides template definitions for the Identity elements 0 (zero(T)) and 1 (one(T)), as well as specializations for char, short, int, long, float, double and complex<T>.

For a detailed description of the functions please refer to Section 7.1.

```
#ifndef IDENTITY_H
#define IDENTITY_H

#include <iostream>
#include <cstdlib>
#include <complex>
using namespace std;

template <class T> T zero(T) { return T() - T(); }

template <class T> T one(T)
{
 cerr << "one() not implemented" << endl;
 abort();
}

template <> char zero(char) { return (char) 0; }
template <> char one(char) { return (char) 1; }

template <> short zero(short) { return (short) 0; }
template <> short one(short) { return (short) 1; }
```

```
template <> int zero(int) { return (int) 0; }
template <> int one(int) { return (int) 1; }

template <> long zero(long) { return (long) 0; }
template <> long one(long) { return (long) 1; }

template <> float zero(float) { return (float) 0.0; }
template <> float one(float) { return (float) 1.0; }

template <> double zero(double) { return (double) 0.0; }
template <> double one(double) { return (double) 1.0; }

template <class T>
complex<T> zero(complex<T>) { return complex<T>(zero(T())); }

template <class T>
complex<T> one(complex<T>) { return complex<T>(one(T())); }

#endif
```

13.2 Verylong Class

The public interface of the `Verylong` class:

- `Verylong(const string & = "0")` : Constructor.

- `Verylong(int)` : Constructor.

- `abs(const Verylong&)` : Absolute value function.

- `sqrt(const Verylong&)` : Integer square root function.

- `pow(const Verylong&,const Verylong&)` : Integer power function.

- `div(const Verylong&,const Verylong&)` : Double division function.

- Arithmetic operators : `++`, `--`, `-`(unary), `+`, `-`, `*`, `/`, `%`, `=`, `+=`, `-=`, `*=`, `/=`, `%=`

- Relational operators : `==`, `!=`, `<`, `<=`, `>`, `>=`

- Type conversion operators : `operator int()`, `operator double()`

- Stream operators : `>>`, `<<`

For a detailed description of the class structure and each member function, please refer to Section 7.2.

```
// verylong.h
// Very Long Integer Class

#ifndef VERYLONG_H
#define VERYLONG_H
```

```
#include <cassert>
#include <cctype>
#include <cmath>
#include <cstdlib>
#include <iomanip>
#include <iostream>
#include <limits>
#include <string>
#include "identity.h"
using namespace std;

class Verylong
{
   private:
      // Data Fields
      string vlstr;       // The string is stored in reverse order.
      int    vlsign;      // Sign of Verylong: +=>0; -=>1

      // Private member functions
      Verylong multdigit(int) const;
      Verylong mult10(int) const;

   public:
      // Constructors and destructor
      Verylong(const string&);
      Verylong(int);
      Verylong(const Verylong&);
      ~Verylong();

      // Conversion operators
      operator int () const;
      operator double () const;
      operator string () const;

      // Arithmetic operators and Relational operators
      const Verylong & operator = (const Verylong&);  // assignment operator
      Verylong operator - () const;       // negate  operator
      Verylong operator ++ ();            // prefix  increment operator
      Verylong operator ++ (int);         // postfix increment operator
      Verylong operator -- ();            // prefix  decrement operator
      Verylong operator -- (int);         // postfix decrement operator

      Verylong operator += (const Verylong&);
      Verylong operator -= (const Verylong&);
      Verylong operator *= (const Verylong&);
      Verylong operator /= (const Verylong&);
      Verylong operator %= (const Verylong&);

      friend Verylong operator + (const Verylong&,const Verylong&);
      friend Verylong operator - (const Verylong&,const Verylong&);
      friend Verylong operator * (const Verylong&,const Verylong&);
      friend Verylong operator / (const Verylong&,const Verylong&);
      friend Verylong operator % (const Verylong&,const Verylong&);

      friend int operator == (const Verylong&,const Verylong&);
      friend int operator != (const Verylong&,const Verylong&);
      friend int operator <  (const Verylong&,const Verylong&);
```

```
        friend int operator <= (const Verylong&,const Verylong&);
        friend int operator >  (const Verylong&,const Verylong&);
        friend int operator >= (const Verylong&,const Verylong&);

        // Other functions
        friend Verylong abs(const Verylong&);
        friend Verylong sqrt(const Verylong&);
        friend Verylong pow(const Verylong&,const Verylong&);
        friend double div(const Verylong&,const Verylong&);

        // Class Data
        static const Verylong zero;
        static const Verylong one;
        static const Verylong two;

        // I/O stream functions
        friend ostream & operator << (ostream&,const Verylong&);
        friend istream & operator >> (istream&,Verylong&);
};

// Class Data

const Verylong Verylong::zero = Verylong("0");

const Verylong Verylong::one = Verylong("1");

const Verylong Verylong::two = Verylong("2");

// Constructors, Destructors and Conversion operators.

Verylong::Verylong(const string &value = "0")
{
    string s = (value == "") ? "0" : value;

    vlsign = (s[0] == '-') ? 1:0;        // check for negative sign
    if(ispunct(s[0]))                    // if the first character
        vlstr = s.substr(1,s.length()-1); // is a punctuation mark.
    else vlstr = s;
}

Verylong::Verylong(int n)
{
    if(n < 0) { vlsign = 1; n = (-n); } // check for sign and convert the
    else vlsign = 0;                     // number to positive when negative

    if(n > 0)
        while(n >= 1)              // extract the number digit by digit and store
        {                          // internally
            vlstr = char(n%10 + '0') + vlstr;
            n /= 10;
        }
    else vlstr = string("0"); // else number is zero
}

Verylong::Verylong(const Verylong &x) : vlstr(x.vlstr),vlsign(x.vlsign) { }

Verylong::~Verylong() { }
```

```
Verylong::operator int() const
{
   int number, factor = 1;
   static Verylong max0(INT_MAX);
   static Verylong min0(INT_MIN+1);
   string::const_reverse_iterator j=vlstr.rbegin();

   if(*this > max0)
   {
     cerr << "Error: Conversion Verylong->integer is not possible" << endl;
     return INT_MAX;
   }
   else if(*this < min0)
   {
     cerr << "Error: Conversion Verylong->integer is not possible" << endl;
     return INT_MIN;
   }

   number = *j - '0';

   for(j++;j!=vlstr.rend();j++)
   {
     factor *= 10;
     number += (*j-'0') * factor;
   }

   if(vlsign) return -number;
   return number;
}

Verylong::operator double() const
{
   double sum, factor = 1.0;
   string::const_reverse_iterator i = vlstr.rbegin();

   sum = double(*i) - '0';
   for(i++;i!=vlstr.rend();i++)
   {
     factor *= 10.0;
     sum += double(*i-'0') * factor;
   }
   if(vlsign) return -sum;
   return sum;
}

Verylong::operator string () const
{
   if(vlstr.length() == 0) return string("0");
   return vlstr;
}

// Various member operators

const Verylong & Verylong::operator = (const Verylong &rhs)
{
   if(this == &rhs) return *this;

   vlstr = rhs.vlstr;
```

```cpp
    vlsign = rhs.vlsign;

    return *this;
}

// Unary - operator
Verylong Verylong::operator -() const
{
    Verylong temp(*this);
    if(temp != zero)  temp.vlsign = !vlsign;
    return temp;
}

// Prefix increment operator
Verylong Verylong::operator ++ ()
{ return *this = *this + one; }

// Postfix increment operator
Verylong Verylong::operator ++ (int)
{
    Verylong result(*this);
    *this = *this + one;
    return result;
}

// Prefix decrement operator
Verylong Verylong::operator -- ()
{ return *this = *this - one; }

// Postfix decrement operator
Verylong Verylong::operator -- (int)
{
    Verylong result(*this);
    *this = *this - one;
    return result;
}

Verylong Verylong::operator += (const Verylong &v)
{  return *this = *this + v; }

Verylong Verylong::operator -= (const Verylong &v)
{  return *this = *this - v; }

Verylong Verylong::operator *= (const Verylong &v)
{  return *this = *this * v; }

Verylong Verylong::operator /= (const Verylong &v)
{  return *this = *this / v; }

Verylong Verylong::operator %= (const Verylong &v)
{  return *this = *this % v; }

// Various friendship operators and functions.

Verylong operator + (const Verylong &u,const Verylong &v)
{
    char digitsum, d1, d2, carry = 0;
    string temp;
```

```
      string::const_reverse_iterator j, k;

   if(u.vlsign ^ v.vlsign)
   {
     if(u.vlsign == 0) return u-abs(v);
     else                    return v-abs(u);
   }

   for(j=u.vlstr.rbegin(),  k=v.vlstr.rbegin();
       j!=u.vlstr.rend() || k!=v.vlstr.rend();)
   {
     d1 = (j == u.vlstr.rend()) ? 0 : *(j++)-'0'; // get digit
     d2 = (k == v.vlstr.rend()) ? 0 : *(k++)-'0'; // get digit
     digitsum = d1 + d2 + carry;                  // add digits
     carry = (digitsum >= 10) ? 1 : 0;
     digitsum -= 10*carry;
     temp = char(digitsum+'0') + temp;
   }
   if(carry) temp = '1' + temp;  // if carry at end, last digit is 1
   if(u.vlsign) temp = '-' + temp;
   return Verylong(temp);
}


Verylong operator - (const Verylong &u,const Verylong &v)
{
   char d, d1, d2, borrow = 0;
   int negative;
   string temp, temp2;
   string::reverse_iterator i, j;
   if(u.vlsign ^ v.vlsign)
   {
     if(u.vlsign == 0) return u+abs(v);
     else                    return -(v+abs(u));
   }
   Verylong w, y;
   if(u.vlsign == 0)  // both u,v are positive
    if(u<v) { w=v; y=u; negative=1;}
    else    { w=u; y=v; negative=0;}
   else                // both u,v are negative
    if(u<v) { w=u; y=v; negative=1;}
    else    { w=v; y=u; negative=0;}

   for(i=w.vlstr.rbegin(),  j=y.vlstr.rbegin();
       i!=w.vlstr.rend() || j!=y.vlstr.rend();)
   {
     d1 = (i == w.vlstr.rend()) ? 0:*(i++)-'0';
     d2 = (j == y.vlstr.rend()) ? 0:*(j++)-'0';
     d = d1 - d2 - borrow;
     borrow = (d < 0) ? 1 : 0;
     d += 10 * borrow;
     temp = char(d+'0') + temp;
   }

   while(temp[0] == '0') temp = temp.substr(1);
   if(negative) temp = '-' + temp;
   return Verylong(temp);
}
```

```cpp
Verylong operator * (const Verylong &u,const Verylong &v)
{
    Verylong pprod("1"), tempsum("0");
    string::const_reverse_iterator r = v.vlstr.rbegin();

    for(int j=0;r!=v.vlstr.rend();j++,r++)
    {
      int digit = *r - '0';              // extract a digit
      pprod = u.multdigit(digit);        // multiplied by the digit
      pprod = pprod.mult10(j);           // "adds" suitable zeros behind
      tempsum = tempsum + pprod;         // result added to tempsum
    }
    tempsum.vlsign = u.vlsign^v.vlsign; // to determine sign
    return tempsum;
}

//  This algorithm is the long division algorithm.
Verylong operator / (const Verylong &u,const Verylong &v)
{
    int len = u.vlstr.length() - v.vlstr.length();
    string temp;
    Verylong w,y,b,c,d,quotient=Verylong::zero;

    if(v == Verylong::zero)
    {
      cerr << "Error : division by zero" << endl;
      return Verylong::zero;
    }

    w = abs(u); y = abs(v);
    if(w < y) return Verylong::zero;
    c = Verylong(w.vlstr.substr(0,w.vlstr.length()-len));

    for(int i=0;i<=len;i++)
    {
      quotient = quotient.mult10(1);
      b = d = Verylong::zero;            // initialize b and d to 0
      while(b < c) { b = b + y; d = d + Verylong::one; }
      if(c < b)                          // if b>c, then
      {                                  // we have added one count too many
        b = b - y;
        d = d - Verylong::one;
      }
      quotient = quotient + d;           // add to the quotient
      if(i < len)
      {
        // partial remainder * 10 and add to next digit
        c = (c-b).mult10(1);
        c += Verylong(w.vlstr[w.vlstr.length()-len+i]-'0');
      }
    }
    quotient.vlsign = u.vlsign^v.vlsign;  // to determine sign
    return quotient;
}

Verylong operator % (const Verylong &u,const Verylong &v)
{ return (u - v*(u/v)); }
```

```cpp
int operator == (const Verylong &u,const Verylong &v)
{ return (u.vlsign==v.vlsign && u.vlstr==v.vlstr); }

int operator != (const Verylong &u,const Verylong &v)
{ return !(u==v); }

int operator < (const Verylong &u,const Verylong &v)
{
    if      (u.vlsign < v.vlsign) return 0;
    else if(u.vlsign > v.vlsign) return 1;
    // exclusive or (^) to determine sign
    if      (u.vlstr.length() < v.vlstr.length()) return (1^u.vlsign);
    else if(u.vlstr.length() > v.vlstr.length()) return (0^u.vlsign);
    return (u.vlstr < v.vlstr && !u.vlsign) || (u.vlstr > v.vlstr && u.vlsign);
}

int operator <= (const Verylong &u,const Verylong &v)
{ return (u<v || u==v); }

int operator > (const Verylong &u,const Verylong &v)
{ return (!(u<v) && u!=v); }

int operator >= (const Verylong &u,const Verylong &v)
{ return (u>v || u==v); }

// Calculate the absolute value of a number
Verylong abs(const Verylong &v)
{
    Verylong u(v);
    if(u.vlsign) u.vlsign = 0;
    return u;
}

// Calculate the integer square root of a number
// based on the formula (a+b)^2 = a^2 + 2ab + b^2
Verylong sqrt(const Verylong &v)
{
    // if v is negative, error is reported
    if(v.vlsign) {cerr << "NaN" << endl; return Verylong::zero; }

    int j, k = v.vlstr.length()+1, num = k >> 1;
    Verylong y, z, sum, tempsum, digitsum;
    string temp, w(v.vlstr);

    k = 0;
    j = 1;

    // segment the number 2 digits by 2 digits
    if(v.vlstr.length() % 2) digitsum = Verylong(w[k++] - '0');
    else
    {
        digitsum = Verylong((w[k] - '0')*10 + w[k+1] - '0');
        k += 2;
    }

    // find the first digit of the integer square root
    sum = z = Verylong(int(sqrt(double(digitsum))));
    // store partial result
```

```
   temp = char(int(z) + '0');
   digitsum = digitsum - z*z;

   for(;j<num;j++)
   {
     // get next digit from the number
     digitsum = digitsum.mult10(1) + Verylong(w[k++] - '0');
     y = z + z;          // 2*a
     z = digitsum/y;
     tempsum = digitsum.mult10(1) + Verylong(w[k++] - '0');
     digitsum = -y*z.mult10(1) + tempsum - z*z;
     // decrease z by 1 and re-calculate when it is over-estimated.
     while(digitsum < Verylong::zero)
     {
       --z;
       digitsum = -y*z.mult10(1) + tempsum - z*z;
     }
     temp = temp + char(int(z) + '0');// store partial result
     z = sum = sum.mult10(1) + z;       // update value of the partial result
   }
   Verylong result(temp);
   return result;
}

// Raise a number X to a power of degree
Verylong pow(const Verylong &X,const Verylong &degree)
{
   Verylong N(degree), Y("1"), x(X);

   if(N == Verylong::zero) return Verylong::one;
   if(N < Verylong::zero) return Verylong::zero;

   while(1)
   {
     if(N%Verylong::two != Verylong::zero)
     {
       Y = Y * x;
       N = N / Verylong::two;
       if(N == Verylong::zero) return Y;
     }
     else  N = N / Verylong::two;
     x = x * x;
   }
}

// Double division function
double div(const Verylong &u,const Verylong &v)
{
   double qq = 0.0, qqscale = 1.0;
   Verylong w,y,b,c;
   int d, count;
   // number of significant digits
   int decno = numeric_limits<double>::digits;

   if(v == Verylong::zero)
   {
     cerr << "ERROR : Division by zero" << endl;
     return 0.0;
```

```
    }
    if(u == Verylong::zero) return 0.0;

    w=abs(u); y=abs(v);
    while(w<y) { w = w.mult10(1); qqscale *= 0.1; }

    int len = w.vlstr.length() - y.vlstr.length();
    string temp = w.vlstr.substr(0,w.vlstr.length()-len);
    c = Verylong(temp);

    for(int i=0;i<=len;i++)
    {
      qq *= 10.0;
      b = Verylong::zero; d = 0;   // initialize b and d to 0
      while(b < c) { b += y; d += 1;}

      if(c < b) { b -= y; d -= 1;} // if b>c, then we have added one too many
      qq += double(d);             // add to the quotient
      c = (c-b).mult10(1);         // the partial remainder * 10
      if(i < len)                  // and add to next digit
          c += Verylong(w.vlstr[w.vlstr.length()-len+i]-'0');
    }
    qq *= qqscale; count = 0;

    while(c != Verylong::zero && count < decno)
    {
      qqscale *= 0.1;
      b = Verylong::zero; d = 0;   // initialize b and d to 0
      while(b < c) { b += y; d += 1;}
      if(c < b) { b -= y; d -= 1;} // if b>c, then we have added one too many
      qq += double(d)*qqscale;
      c = (c-b).mult10(1);
      count++;
    }
    if(u.vlsign^v.vlsign) qq *= (-1.0); // check for the sign
    return qq;
}

ostream & operator << (ostream &s,const Verylong &v)
{
    if(v.vlstr.length() > 0) { if(v.vlsign) s << "-"; s << v.vlstr; }
    else s << "0";
    return s;
}

istream & operator >> (istream &s,Verylong &v)
{
    string temp(10000, ' ');
    s >> temp;
    v = Verylong(temp);
    return s;
}


// Private member functions: multdigit(), mult10().

// Multiply this Verylong number by num
Verylong Verylong::multdigit(int num) const
```

```
{
    int carry = 0;
    string::const_reverse_iterator r;

    if(num)
    {
      string temp;
      for(r=vlstr.rbegin();r!=vlstr.rend();r++)
      {
        int d1 = *r - '0',                   // get digit and multiplied by
              digitprod = d1*num + carry;    // that digit plus carry
        if(digitprod >= 10)                  // if there's a new carry,
        {
          carry = digitprod/10;              // carry is high digit
          digitprod -= carry*10;             // result is low digit
        }
        else carry = 0;                      // otherwise carry is 0
        temp = char(digitprod + '0') + temp; // insert char in string
      }
      if(carry) temp = char(carry + '0') + temp; //if carry at end,
      Verylong result(temp);
      return result;
    }
    else return zero;
}

// Multiply this Verylong number by 10*num
Verylong Verylong::mult10(int num) const
{
    if(*this != zero)
    {
      string temp;
      for(int j=0;j<num;j++) temp = temp + '0';
      Verylong result(vlstr+temp);
      if(vlsign) result = -result;
      return result;
    }
    else return zero;
}

template <> Verylong zero(Verylong) { return Verylong::zero; }
template <> Verylong one(Verylong) { return Verylong::one; }

#endif
```

13.3   Rational Class

The public interface of the Rational class:

- Rational() : Default constructor.

- Rational(T) : Integer constructor.

- Rational(T,T) : Rational number constructor.

- num() : Numerator of the rational number.

- den() : Denominator of the rational number.

- frac() : Fractional part of the rational number.

- normalize() : Normalization of the rational number.

- output_style(output_type) : The parameter is either Rational::decimal or Rational::fraction which determines whether to display rationals in decimal or as ratios.

- Arithmetic operators : -(unary), +, -, *, /, =, +=, -=, *=, /=

- Relational operators : ==, !=, <, <=, >, >=

- Type conversion operator : operator double()

- Stream operators : >>, <<

For a detailed description of the class structure and each member function, refer to Section 7.3.

```cpp
// rational.h
// Rational Numbers Class
#ifndef RATIONAL_H
#define RATIONAL_H

#include <iostream>
#include <sstream>
#include <string>
#include <cstdlib>
#include <vector>
#include <ctype.h>
#include "identity.h"
using namespace std;

template <class T>
class Rational
{
   public:
      enum output_type { decimal, fraction };

   private:
      // Data Fields : Numerator and Denominator
      T p,q;

      // Private member function
      T gcd(T,T);

      // output type : decimal or fraction
      static output_type output_preference;

   public:
      // Constructors and Destructor
      Rational();
```

```
        Rational(T);
        Rational(T,T);
        Rational(const Rational<T>&);
        Rational(const string&);
        Rational(const double&);
        ~Rational();

        // Conversion operator
        operator double () const;

        // Member functions
        T num() const;               // numerator of r
        T den() const;               // denominator of r
        Rational<T> frac() const; // fractional part of r
        void normalize();            // normalize the rational number

        // Arithmetic operators and Relational operators
        const Rational<T> &operator = (const Rational<T>&);
        Rational<T> operator - () const;
        const Rational<T> &operator += (const Rational<T>&);
        const Rational<T> &operator -= (const Rational<T>&);
        const Rational<T> &operator *= (const Rational<T>&);
        const Rational<T> &operator /= (const Rational<T>&);

        Rational<T> operator + (const Rational<T>&) const;
        Rational<T> operator - (const Rational<T>&) const;
        Rational<T> operator * (const Rational<T>&) const;
        Rational<T> operator / (const Rational<T>&) const;

        Rational<T> operator ++ ();
        Rational<T> operator ++ (int);
        Rational<T> operator -- ();
        Rational<T> operator -- (int);

        int operator == (const Rational<T>&) const;
        int operator != (const Rational<T>&) const;
        int operator >  (const Rational<T>&) const;
        int operator <  (const Rational<T>&) const;
        int operator >= (const Rational<T>&) const;
        int operator <= (const Rational<T>&) const;

        // I/O stream functions
        ostream &output(ostream&) const;
        istream &input(istream&);

        static void output_style(output_type);
};

template <class T>
typename Rational<T>::output_type
Rational<T>::output_preference = Rational<T>::fraction;

// Constructors, destructor and conversion operator.

template <class T> Rational<T>::Rational () : p(zero(T())),q(one(T())) {}

template <class T> Rational<T>::Rational (T N) : p(N),q(one(T())) {}
```

```
template <class T> Rational<T>::Rational (T N, T D) : p(N),q(D)
{
    if(D==zero(T()))
    {
        cerr << "Zero denominator in Rational Number " << endl;
        return;
    }
    if(q < zero(T())) {p = -p; q = -q;}
    normalize();
}

template <class T> Rational<T>::Rational(const Rational<T> &r)
    : p(r.p), q(r.q) {}

// for a string "a" or "a/b" or "a.b"
// spaces are allowed for "a / b"
template <class T> Rational<T>::Rational(const string &s)
{
  string s1 = s;
  string::size_type fraction, decimal, space;
  p = zero(T());
  q = one(T());
  // erase spaces
  space = s1.find(" ");
  while(space != string::npos)
  {
   s1.erase(space, 1);
   space = s1.find(" ");
  }
  fraction = s1.find("/");
  decimal = s1.find(".");

  if(fraction != string::npos)
  {
   istringstream istr(s1.substr(0,fraction));
   istr >> p;
   // the following line allows 1/2/3 == 1/6
   *this /= Rational<T>(s1.substr(fraction+1));
  }
  else if(decimal != string::npos)
  {
   string s2 = s1.substr(decimal+1);
   int power = s2.length(); // 10^power in denomenator
   // change aaa.bbbb -> aaa + bbbb/10000
   while(s2[0] == '0') s2.erase(0,1);
   *this += Rational<T>(s2 + "/" + "1" + string(power, '0'));
  }
  else
  {
   istringstream istr(s1);
   istr >> p;
  }
}

template <class T> Rational<T>::Rational(const double &d)
{
 static int base = numeric_limits<double>::radix;
 static vector<Rational<T> > digits(0);
```

```
static Rational<T> rbase("0");
Rational<T> rint("0"), rfrac("0"), rfact;
double integer, fraction;
int sign = (d<0);

// since we don't know the type T and base is usually small
// we can increment to find rbase
if(digits.size() == 0)
{
 digits.resize(base);
 for(int i=0;i< base;i++) digits[i] = rbase++;
}

if(sign) fraction = modf(-d,&integer);
else fraction = modf(d,&integer);

rfact = Rational<T>("1");
while(floor(integer) != 0.0)
{
 rint += digits[int(fmod(integer, base))];
 integer /= base;
 rint /= rbase;
 rfact *= rbase;
}

 rint *= rfact;
 rfact = Rational<T>("1");
 while(fraction != 0.0)
 {
  rfrac *= rbase;
  rfact /= rbase;
  fraction *= base;
  fraction = modf(fraction, &integer);
  rfrac += digits[int(integer)];
 }
 rfrac *= rfact;
 if(sign) *this = -rint - rfrac;
 else     *this =  rint + rfrac;
}

template <class T> Rational<T>::~Rational() {}

template <class T> Rational<T>::operator double() const
{ return double(p)/double(q); }

// Member functions
template <class T> T Rational<T>::num() const { return p; }

template <class T> T Rational<T>::den() const { return q; }

template <class T> Rational<T> Rational<T>::frac() const
{
    static Rational<T> zero(zero(T())), one(one(T()));
    Rational<T> temp(*this);

    if(temp < zero)
    {
        while(temp < zero) temp = temp + one;
```

```
         return temp - one;
      }
      else
      {
         while(zero < temp) temp = temp - one;
         return temp + one;
      }
}

template <class T> void Rational<T>::normalize()
{
   T t;
   if(p < zero(T())) t = -p;
   else              t = p;
   t = gcd(t, q);
   if(t > one(T()))   { p /= t; q /= t;}
}


// Various operators

template <class T>
const Rational<T> & Rational<T>::operator = (const Rational<T> &r)
{ p = r.p; q = r.q; return *this; }

template <class T> Rational<T> Rational<T>::operator - () const
{ return Rational<T>(-p,q); }

template <class T>
const Rational<T> & Rational<T>::operator += (const Rational<T> &r)
{ return *this = *this + r; }

template <class T>
const Rational<T> & Rational<T>::operator -= (const Rational<T> &r)
{ return *this = *this - r; }

template <class T>
const Rational<T> & Rational<T>::operator *= (const Rational<T> &r)
{ return *this = *this * r; }

template <class T>
const Rational<T> & Rational<T>::operator /= (const Rational<T> &r)
{ return *this = *this / r; }

template <class T>
Rational<T> Rational<T>::operator + (const Rational<T> &r2) const
{ return Rational<T> (p * r2.q + r2.p * q, q * r2.q); }

template <class T>
Rational<T> Rational<T>::operator - (const Rational<T> &r2) const
{ return Rational<T> (p * r2.q - r2.p * q, q * r2.q); }

template <class T>
Rational<T> Rational<T>::operator * (const Rational<T> &r2) const
{ return Rational<T> (p * r2.p, q * r2.q); }

template <class T>
Rational<T> Rational<T>::operator / (const Rational<T> &r2) const
{ return Rational<T> (p * r2.q, q * r2.p); }
```

```
template <class T>
Rational<T> Rational<T>::operator ++ () { p += q; return *this; }

template <class T>
Rational<T> Rational<T>::operator ++ (int)
{ Rational<T> r = *this; ++(*this); return r; }

template <class T>
Rational<T> Rational<T>::operator -- () { p -= q; return *this; }

template <class T>
Rational<T> Rational<T>::operator -- (int)
{ Rational<T> r = *this; --(*this); return r; }

template <class T>
int Rational<T>::operator == (const Rational<T> &r2) const
{ return (p * r2.q) == (r2.p * q); }

template <class T>
int Rational<T>::operator != (const Rational<T> &r2) const
{ return !(*this == r2); }

template <class T>
int Rational<T>::operator >  (const Rational<T> &r2) const
{ return (p*r2.q - r2.p*q > zero(T())); }

template <class T>
int Rational<T>::operator <  (const Rational<T> &r2) const
{ return (p*r2.q - r2.p*q < zero(T())); }

template <class T>
int Rational<T>::operator >= (const Rational<T> &r2) const
{ return (*this>r2) || (*this==r2); }

template <class T>
int Rational<T>::operator <= (const Rational<T> &r2) const
{ return (*this<r2) || (*this==r2); }

template <class T>
ostream & Rational<T>::output(ostream &s) const
{
   switch(output_preference)
   {
     case fraction:
       if(q == one(T())) return s << p;
       return s << p << "/" << q;
       break;
     case decimal:
       return s << double(*this);
       break;
     default:
       cerr << "Unknown output stylei in Rational::output." << endl;
       return s;
       break;
   }
}
```

```
template <class T>
ostream & operator << (ostream &s,const Rational<T> &r)
{ return r.output(s); }

template <class T>
istream & Rational<T>::input(istream &s)
{
   char c;
   T     n, d(one(T()));
   s.clear();                  // set stream state to good
   s >> n;                     // read numerator
   if(! s.good()) return s;    // can't get an integer, just return
   c = s.peek();               // peek next character
   if(c == '/')
   {
      c = s.get();             // clear '/'
      s >> d;                  // read denominator
      if(! s.good())
      {
         s.clear(s.rdstate() | ios::badbit);
         return s;
      }
   }
   *this = Rational<T>(n,d);
   return s;
}

template <class T>
istream & operator >> (istream &s,Rational<T> &r)
{ return r.input(s); }

template <class T>
void Rational<T>::output_style(output_type t)
{ output_preference = t; }

// Private member function: gcd()
template <class T> T Rational<T>::gcd (T a, T b)
{
   while(b > zero(T())) { T m = a % b; a = b; b = m; }
   return a;
}

// This function is a global function.
// It should be in a global header file.
template <class T> T abs(const T &x)
{
   if(x > zero(T())) return x;
   return -x;
}

#include "verylong.h"
template <>
Rational<Verylong>::operator double() const
{ return div(p,q); }

template <class T>
Rational<T> zero(Rational<T>) { return Rational<T>(zero(T())); }
```

```
template <class T>
Rational<T> one(Rational<T>) { return Rational<T>(one(T())); }
#endif
```

13.4   Quaternion Class

The public interface of the `Quaternion` class:

- `Quaternion()` : Default constructor.

- `Quaternion(T,T,T,T)` : Constructor.

- `sqr()` : Square of the quaternion.

- `conjugate()` : Conjugate of the quaternion.

- `inverse()` : Inverse of the quaternion.

- `magnitude()` : Magnitude of the quaternion.

- Arithmetic operators : -(unary), +, -, *, /, ~

- Stream operators : >>, <<

For a detailed description of the class structure and each member function, please refer to Section 7.4.

```
// quatern.h
// Template Class for Quaternions


#ifndef QUATERNION_H
#define QUATERNION_H

#include <iostream>
#include <cmath>          // for sqrt()
#include "identity.h"
using namespace std;

template <class T> class Quaternion
{
    private:
        // Data Fields
        T r, i, j, k;

    public:
        // Constructors
        Quaternion();
        Quaternion(T,T,T,T);
        Quaternion(const Quaternion<T>&);
        ~Quaternion();
```

```
      // Operators
      const Quaternion<T> &operator = (const Quaternion<T>&);
      Quaternion<T> operator + (const Quaternion<T>&);
      Quaternion<T> operator - (const Quaternion<T>&);
      Quaternion<T> operator - () const;
      Quaternion<T> operator * (const Quaternion<T>&);
      Quaternion<T> operator * (T);
      Quaternion<T> operator / (const Quaternion<T>&);
      Quaternion<T> operator ~ () const;

      // Member Functions
      Quaternion<T> sqr();
      Quaternion<T> conjugate() const;
      Quaternion<T> inverse() const;
      double magnitude() const;

      // Streams
      ostream &print(ostream &) const;
      istream &input(istream &s);
};

template <class T> Quaternion<T>::Quaternion()
  : r(zero(T())),i(zero(T())),j(zero(T())),k(zero(T())) {}

template <class T> Quaternion<T>::Quaternion(T r1,T i1,T j1,T k1)
  : r(r1),i(i1),j(j1),k(k1) {}

template <class T> Quaternion<T>::Quaternion(const Quaternion<T> &arg)
  : r(arg.r),i(arg.i),j(arg.j),k(arg.k) {}

template <class T> Quaternion<T>::~Quaternion() {}

template <class T>
const Quaternion<T> &Quaternion<T>::operator = (const Quaternion<T> &rvalue)
{
   r = rvalue.r; i = rvalue.i; j = rvalue.j; k = rvalue.k;
   return *this;
}

template <class T>
Quaternion<T> Quaternion<T>::operator + (const Quaternion<T> &arg)
{ return Quaternion<T>(r+arg.r,i+arg.i,j+arg.j,k+arg.k); }

template <class T>
Quaternion<T> Quaternion<T>::operator - (const Quaternion<T> &arg)
{ return Quaternion<T>(r-arg.r,i-arg.i,j-arg.j,k-arg.k); }

template <class T>
Quaternion<T> Quaternion<T>::operator - () const
{ return Quaternion<T>(-r,-i,-j,-k); }

template <class T>
Quaternion<T> Quaternion<T>::operator * (const Quaternion<T> &arg)
{
   return Quaternion<T>(r*arg.r - i*arg.i - j*arg.j - k*arg.k,
                        r*arg.i + i*arg.r + j*arg.k - k*arg.j,
                        r*arg.j + j*arg.r + k*arg.i - i*arg.k,
                        r*arg.k + k*arg.r + i*arg.j - j*arg.i);
```

```
}

template <class T>
Quaternion<T> Quaternion<T>::operator * (T arg)
{ return Quaternion<T>(r*arg,i*arg,j*arg,k*arg); }

template <class T>
Quaternion<T> Quaternion<T>::operator / (const Quaternion<T> &arg)
{ return *this * arg.inverse(); }

// Normalize Quaternion
template <class T>
Quaternion<T> Quaternion<T>::operator ~ () const
{
   Quaternion<T> result;
   double length = magnitude();
   result.r = r/length; result.i = i/length;
   result.j = j/length; result.k = k/length;
   return result;
}

template <class T> Quaternion<T> Quaternion<T>::sqr()
{
   Quaternion<T> result;
   T temp;
   T two = one(T()) + one(T());
   temp = two*r;
   result.r = r*r - i*i - j*j - k*k;
   result.i = temp*i; result.j = temp*j; result.k = temp*k;
   return result;
}

template <class T> Quaternion<T> Quaternion<T>::conjugate() const
{ return Quaternion<T>(r,-i,-j,-k); }

template <class T> Quaternion<T> Quaternion<T>::inverse() const
{
   Quaternion<T> temp1(conjugate());
   T            temp2 = r*r + i*i + j*j + k*k;
   return Quaternion<T>(temp1.r/temp2,temp1.i/temp2,
                        temp1.j/temp2,temp1.k/temp2);
}

template <class T> double Quaternion<T>::magnitude() const
{ return sqrt(r*r + i*i + j*j + k*k); }

template <class T> ostream &Quaternion<T>::print(ostream &s) const
{
   s << "(" << r << "," << i << ","
            << j << "," << k << ")";
   return s;
}

template <class T> istream &Quaternion<T>::input(istream &s)
{
   s >> r >> i >> j >> k;
   return s;
}
```

```
template <class T> Quaternion<T> operator * (T factor,Quaternion<T> &arg)
{ return arg * factor; }

template <class T>
ostream & operator << (ostream &s,const Quaternion<T> &arg)
{ return arg.print(s); }

template <class T>
istream & operator >> (istream &s,Quaternion<T> &arg)
{ return arg.input(s); }

#endif
```

13.5 Derive Class

The public interface of the **Derive** class:

- **Derive()** : Default constructor.

- **Derive(const T)** : Constructor.

- **set(const T)** : Specifies the point where the derivative takes place.

- **exp(const Derive<T>&)** : Exponential function.

- **cos(const Derive<T>&)** : Cosine function.

- **sin(const Derive<T>&)** : Sine function.

- **df(const Derive<T>&)** : Exact derivative of an expression.

- Arithmetic operators : -(unary), +, -, *, /, +=, -=, *=, /=

- Stream operator : <<

For a detailed description of the class structure and each member function, please refer to Section 7.5.

```
// derive.h
// The Derivation Class

#ifndef DERIVE_H
#define DERIVE_H

#include <iostream>
#include <cmath>
#include "identity.h"
using namespace std;

template <class T> class Derive
{
```

```
   private:
       // Data Field
       T u, du;

       // Private Constructor
       Derive(const T&,const T&);

   public:
       // Constructors
       Derive();
       Derive(const T&);
       Derive(const Derive<T>&);

       // Member Function
       void set(const T);

       // Arithmetic Operators
       Derive<T> operator - () const;
       Derive<T> operator += (const Derive<T>&);
       Derive<T> operator -= (const Derive<T>&);
       Derive<T> operator *= (const Derive<T>&);
       Derive<T> operator /= (const Derive<T>&);
       Derive<T> operator += (const T&);
       Derive<T> operator -= (const T&);
       Derive<T> operator *= (const T&);
       Derive<T> operator /= (const T&);

       Derive<T> operator + (const Derive<T>&) const;
       Derive<T> operator - (const Derive<T>&) const;
       Derive<T> operator * (const Derive<T>&) const;
       Derive<T> operator / (const Derive<T>&) const;
       Derive<T> operator + (const T&) const;
       Derive<T> operator - (const T&) const;
       Derive<T> operator * (const T&) const;
       Derive<T> operator / (const T&) const;
       Derive<T> exp() const;
       Derive<T> sin() const;
       Derive<T> cos() const;
       T df() const;
       ostream &output(ostream&) const;
};

template <class T> Derive<T>::Derive() : u(zero(T())),du(one(T())) {}

template <class T> Derive<T>::Derive(const T &v) : u(v),du(zero(T())) {}

template <class T> Derive<T>::Derive(const T &v,const T &dv) : u(v),du(dv) {}

template <class T> Derive<T>::Derive(const Derive<T> &r) : u(r.u),du(r.du) {}

template <class T> void Derive<T>::set(const T v) { u = v; }

template <class T> Derive<T> Derive<T>::operator - () const
{ return Derive<T>(-u,-du); }

template <class T> Derive<T> Derive<T>::operator += (const Derive<T> &r)
{ return *this = *this + r; }
```

```
template <class T> Derive<T> Derive<T>::operator -= (const Derive<T> &r)
{ return *this = *this - r; }

template <class T> Derive<T> Derive<T>::operator *= (const Derive<T> &r)
{ return *this = *this * r; }

template <class T> Derive<T> Derive<T>::operator /= (const Derive<T> &r)
{ return *this = *this / r; }

template <class T> Derive<T> Derive<T>::operator += (const T &r)
{ return *this = *this + r; }

template <class T> Derive<T> Derive<T>::operator -= (const T &r)
{ return *this = *this - r; }

template <class T> Derive<T> Derive<T>::operator *= (const T &r)
{ return *this = *this * r; }

template <class T> Derive<T> Derive<T>::operator /= (const T &r)
{ return *this = *this / r; }

template <class T>
Derive<T> Derive<T>::operator + (const Derive<T> &y) const
{ return Derive<T>(u+y.u,du+y.du); }

template <class T>
Derive<T> Derive<T>::operator - (const Derive<T> &y) const
{ return Derive<T>(u-y.u,du-y.du); }

template <class T>
Derive<T> Derive<T>::operator * (const Derive<T> &y) const
{ return Derive<T>(u*y.u,y.u*du+u*y.du); }

template <class T>
Derive<T> Derive<T>::operator / (const Derive<T> &y) const
{ return Derive<T>(u/y.u,(y.u*du-u*y.du)/(y.u*y.u)); }

template <class T>
Derive<T> Derive<T>::operator + (const T &y) const
{ return *this + Derive<T>(y); }

template <class T>
Derive<T> Derive<T>::operator - (const T &y) const
{ return *this - Derive<T>(y); }

template <class T>
Derive<T> Derive<T>::operator * (const T &y) const
{ return *this * Derive<T>(y); }

template <class T>
Derive<T> Derive<T>::operator / (const T &y) const
{ return *this / Derive<T>(y); }

template <class T>
Derive<T> operator + (const T &y,const Derive<T> &d)
{ return Derive<T>(y)+d; }

template <class T>
```

```
Derive<T> operator - (const T &y,const Derive<T> &d)
{ return Derive<T>(y)-d; }

template <class T>
Derive<T> operator * (const T &y,const Derive<T> &d)
{ return Derive<T>(y)*d; }

template <class T>
Derive<T> operator / (const T &y,const Derive<T> &d)
{ return Derive<T>(y)/d; }

template <class T>
Derive<T> Derive<T>::exp() const { return Derive<T>(exp(u),du*exp(u)); }

template <class T>
Derive<T> exp(const Derive<T> &x) { return x.exp(); }

template <class T>
Derive<T> Derive<T>::sin() const { return Derive<T>(sin(u),du*cos(u)); }

template <class T>
Derive<T> sin(const Derive<T> &x) { return x.sin(); }

template <class T>
Derive<T> Derive<T>::cos() const { return Derive<T>(cos(u),-du*sin(u)); }

template <class T>
Derive<T> cos(const Derive<T> &x) { return x.cos(); }

template <class T> T Derive<T>::df() const { return du; }

template <class T> T df(const Derive<T> &x) { return x.df(); }

template <class T>
ostream &Derive<T>::output(ostream &s) const { return s << u; }

template <class T>
ostream &operator << (ostream &s,const Derive<T> &r)
{ return r.output(s); }

#endif
```

13.6   Vector Class

The public interface of the Vector class:

- Vector() : Default constructor.

- Vector(int) : Constructor.

- Vector(int,const T&) : Constructor.

- reset(int n) : Resets the vector to size n.

- reset(int n,T v) : Resets the vector to size n and initializes entries to v.
- Arithmetic operators : +(unary), -(unary), +, -, *, /, =, +=, -=, *=, /=
- Relational operators : ==, !=
- Subscript operator : []
- Dot product : |
- Cross product : %
- Stream operators : >>, <<

The class Vector<T> is derived from the STL class vector<T> and consequently inherits vector<T> members such as size and resize.

Auxiliary functions in VecNorm.h:

- norm1(const Vector&) : One-norm of the vector.
- norm2(const Vector&) : Two-norm of the vector.
- normI(const Vector&) : Infinite-norm of the vector.
- normalize(const Vector&) : Normalization of the vector.

For a detailed description of the class structure and each member function, please refer to Section 7.6.

13.6.1 Vector Class

```
// vector.h
// Vector class

#ifndef MVECTOR_H
#define MVECTOR_H

#include <cassert>
#include <cmath>
#include <iostream>
#include <vector>
#include "identity.h"
using namespace std;

// definition of class Vector
template <class T> class Vector: public vector<T>
{
  public:
    // Constructors
    Vector();
```

```
        Vector(int);
        Vector(int,const T&);
        Vector(const Vector<T>&);
        ~Vector();

        // Member Functions
        T& operator [] (int);
        const T& operator [] (int) const;
        void reset(int);
        void reset(int,const T&);

        // Arithmetic Operators
        const Vector<T>& operator = (const T&);
        Vector<T> operator + () const;
        Vector<T> operator - () const;

        Vector<T> operator += (const Vector<T>&);
        Vector<T> operator -= (const Vector<T>&);
        Vector<T> operator *= (const Vector<T>&);
        Vector<T> operator /= (const Vector<T>&);
        Vector<T> operator +  (const Vector<T>&) const;
        Vector<T> operator -  (const Vector<T>&) const;
        Vector<T> operator *  (const Vector<T>&) const;
        Vector<T> operator /  (const Vector<T>&) const;

        Vector<T> operator += (const T&);
        Vector<T> operator -= (const T&);
        Vector<T> operator *= (const T&);
        Vector<T> operator /= (const T&);
        Vector<T> operator +  (const T&) const;
        Vector<T> operator -  (const T&) const;
        Vector<T> operator *  (const T&) const;
        Vector<T> operator /  (const T&) const;

        T operator | (const Vector<T>&) const; // Dot product / Inner product
        Vector<T> operator % (const Vector<T>&) const; // Cross product

        ostream &output(ostream&) const;
        istream &input(istream&);
};

// implementation of class Vector
template <class T> Vector<T>::Vector() : vector<T>() { }

template <class T> Vector<T>::Vector(int n) : vector<T>(n) { }

template <class T> Vector<T>::Vector(int n,const T &value)
 : vector<T>(n,value) { }

template <class T> Vector<T>::Vector(const Vector<T> &v) : vector<T>(v) { }

template <class T> Vector<T>::~Vector() { }

template <class T> void Vector<T>::reset(int length)
{ reset(length, zero(T())); }

template <class T> T& Vector<T>::operator [] (int i)
{ return vector<T>::at(i); }
```

```cpp
template <class T> const T& Vector<T>::operator [] (int i) const
{ return vector<T>::at(i); }

template <class T> void Vector<T>::reset(int length, const T &value)
{
    vector<T>::resize(length);
    for(int i=0;i<length;i++) vector<T>::at(i) = value;
}

template <class T> const Vector<T> & Vector<T>::operator = (const T &value)
{
    int length = vector<T>::size();
    for(int i=0;i<length;i++) vector<T>::at(i) = value;
    return *this;
}

template <class T> Vector<T> Vector<T>::operator + () const
{ return *this; }

template <class T> Vector<T> Vector<T>::operator - () const
{ return *this * T(-1); }

template <class T> Vector<T> Vector<T>::operator += (const Vector<T> &v)
{
    int length = vector<T>::size();
    assert(vector<T>::size()==v.size());
    for(int i=0;i<length;i++) vector<T>::at(i) += v[i];
    return *this;
}

template <class T> Vector<T> Vector<T>::operator -= (const Vector<T> &v)
{
    int length = vector<T>::size();
    assert(vector<T>::size()==v.size());
    for(int i=0;i<length;i++) vector<T>::at(i) -= v[i];
    return *this;
}

template <class T> Vector<T> Vector<T>::operator *= (const Vector<T> &v)
{
    int length = vector<T>::size();
    assert(vector<T>::size()==v.size());
    for(int i=0;i<length;i++) vector<T>::at(i) *= v[i];
    return *this;
}

template <class T> Vector<T> Vector<T>::operator /= (const Vector<T> &v)
{
    int length = vector<T>::size();
    assert(vector<T>::size()==v.size());
    for(int i=0;i<length;i++) vector<T>::at(i) /= v[i];
    return *this;
}

template <class T>
Vector<T> Vector<T>::operator + (const Vector<T> &v) const
{
```

```cpp
      Vector<T> result(*this);
      return result += v;
}

template <class T>
Vector<T> Vector<T>::operator - (const Vector<T> &v) const
{
   Vector<T> result(*this);
   return result -= v;
}

template <class T>
Vector<T> Vector<T>::operator * (const Vector<T> &v) const
{
   Vector<T> result(*this);
   return result *= v;
}

template <class T>
Vector<T> Vector<T>::operator / (const Vector<T> &v) const
{
   Vector<T> result(*this);
   return result /= v;
}

template <class T> Vector<T> Vector<T>::operator += (const T &c)
{
   int length = vector<T>::size();
   for(int i=0;i<length;i++) vector<T>::at(i) += c;
   return *this;
}

template <class T> Vector<T> Vector<T>::operator -= (const T &c)
{
   int length = vector<T>::size();
   for(int i=0;i<length;i++) vector<T>::at(i) -= c;
   return *this;
}

template <class T> Vector<T> Vector<T>::operator *= (const T &c)
{
   int length = vector<T>::size();
   for(int i=0;i<length;i++) vector<T>::at(i) *= c;
   return *this;
}

template <class T> Vector<T> Vector<T>::operator /= (const T &c)
{
   int length = vector<T>::size();
   for(int i=0;i<length;i++) vector<T>::at(i) /= c;
   return *this;
}

template <class T> Vector<T> Vector<T>::operator + (const T &c) const
{
   Vector<T> result(*this);
   return result += c;
}
```

```
template <class T> Vector<T> Vector<T>::operator - (const T &c) const
{
    Vector<T> result(*this);
    return result -= c;
}

template <class T> Vector<T> Vector<T>::operator * (const T &c) const
{
    Vector<T> result(*this);
    return result *= c;
}

template <class T> Vector<T> Vector<T>::operator / (const T &c) const
{
    Vector<T> result(*this);
    return result /= c;
}

template <class T> Vector<T> operator + (const T &c, const Vector<T> &v)
{ return v+c; }

template <class T> Vector<T> operator - (const T &c, const Vector<T> &v)
{ return -v+c; }

template <class T> Vector<T> operator * (const T &c, const Vector<T> &v)
{ return v*c; }

template <class T> Vector<T> operator / (const T &c, const Vector<T> &v)
{
    int length = v.size();
    Vector<T> result(v.size());
    for(int i=0;i<length;i++) result[i] = c/v[i];
    return result;
}

// Dot Product / Inner Product
template <class T> T Vector<T>::operator | (const Vector<T> &v) const
{
    int length = vector<T>::size();
    assert(vector<T>::size() == v.size());
    T result(zero(T()));
    for(int i=0;i<length;i++) result = result + vector<T>::at(i)*v[i];
    return result;
}

// Cross Product
template <class T>
Vector<T> Vector<T>::operator % (const Vector<T> &v) const
{
    assert(vector<T>::size() == 3 && v.size() == 3);
    Vector<T> result(3);
    result[0] = vector<T>::at(1)*v[2]-v[1]*vector<T>::at(2);
    result[1] = v[0]*vector<T>::at(2)-vector<T>::at(0)*v[2];
    result[2] = vector<T>::at(0)*v[1]-v[0]*vector<T>::at(1);
    return result;
}
```

```
template <class T> ostream& Vector<T>::output(ostream &s) const
{
   int lastnum = vector<T>::size();
   for(int i=0;i<lastnum;i++) s << "[" << vector<T>::at(i) << "]" << endl;
   return s;
}

template <class T> ostream& operator << (ostream &s,const Vector<T> &v)
{ return v.output(s); }

template <class T> istream& Vector<T>::input(istream &s)
{
   int i, num;
   s.clear();                  // set stream state to good
   s >> num;                   // read size of Vector
   if(! s.good()) return s;  // can't get an integer, just return
   vector<T>::resize(num);
   for(i=0;i<num;i++)
   {
      s >> vector<T>::at(i); // read in entries
      if(! s.good())
      {
         s.clear(s.rdstate() | ios::badbit);
         return s;
      }
   }
   return s;
}

template <class T> istream & operator >> (istream &s,Vector<T> &v)
{ return v.input(s); }
#endif
```

13.6.2   Vector Norms

```
// vecnorm.h
// Norms of Vectors

#ifndef MVECNORM_H
#define MVECNORM_H

#include <iostream>
#include <math.h>
using namespace std;

template <class T> T norm1(const Vector<T> &v)
{
   T result(0);
   for(int i=0;i<int(v.size());i++) result = result + abs(v[i]);
   return result;
}

double norm1(const Vector<double> &v)
{
   double result(0);
   for(int i=0;i<int(v.size());i++) result = result + fabs(v[i]);
```

```
      return result;
}

template <class T> double norm2(const Vector<T> &v)
{
   T result(0);
   for(int i=0;i<int(v.size()));i++) result = result + v[i]*v[i];
   return sqrt(double(result));
}

template <class T> T normI(const Vector<T> &v)
{
   T maxItem(abs(v[0])), temp;
   for(int i=1;i<int(v.size()));i++)
   {
      temp = abs(v[i]);
      if(temp > maxItem) maxItem = temp;
   }
   return maxItem;
}

double normI(const Vector<double> &v)
{
   double maxItem(fabs(v[0])), temp;
   for(int i=1;i<int(v.size()));i++)
   {
      temp = fabs(v[i]);
      if(temp > maxItem) maxItem = temp;
   }
   return maxItem;
}

template <class T> Vector<T> normalize(const Vector<T> &v)
{
   Vector<T> result(v.size());
   double length = norm2(v);
   for(int i=0;i<int(v.size()));i++) result[i] = v[i]/length;
   return result;
}
#endif
```

13.7   Matrix Class

The public interface of the Matrix class:

- Matrix() : Default constructor.

- Matrix(int,int) : Constructor.

- Matrix(int,int,const T&) : Constructor.

- Matrix(const Vector<T>&) : Constructor.

- identity() : Creates an identity matrix.

- `transpose()` : Transpose of the matrix.

- `inverse()` : Inverse of the matrix.

- `trace()` : Trace of the matrix.

- `determinant()` : Determinant of the matrix.

- `rows()` : Number of rows of the matrix.

- `cols()` : Number of columns of the matrix.

- `resize(int,int)` : Resizes the matrix.

- `resize(int,int,const T&)` : Resizes the matrix and initializes the rest of the entries.

- `vec(const Matrix<T>&)` : Vectorize operator.

- `kron(const Matrix<T>&,const Matrix<T>&)` : Kronecker product.

- `hadamard(const Matrix<T>&,const Matrix<T>&)` : Hadamard product.

- `dsum(const Matrix<T>&,const Matrix<T>&)` : Direct sum.

- `fill(const T &v)` : Fills the matrix with the value v.

- Arithmetic operators : `+`(unary), `-`(unary), `+`, `-`, `*`, `/`, `=`, `+=`, `-=`, `*=`, `/=`

- Row vector operator : `[]`

- Column vector operator : `()`

- Stream operators : `>>`, `<<`

Auxiliary functions in `MatNorm.h`:

- `norm1(const Matrix&)` : One-norm of the matrix.

- `normI(const Matrix&)` : Infinite-norm of the matrix.

- `normH(const Matrix&)` : Hilbert-Schmidt norm of the matrix.

For a detailed description of the class structure and each member function, please refer to Section 7.7.

13.7.1   Matrix Class

```cpp
// matrix.h
// Matrix class

#ifndef MATRIX_H
#define MATRIX_H

#include <iostream>
#include <cmath>
#include <cassert>
#include <string>
#include <utility>
#include "identity.h"
#include "vector.h"
using namespace std;

// definition of class Matrix
template <class T> class Matrix
{
   protected:
      // Data Fields
      int rowNum, colNum;
      Vector<Vector<T> > mat;

   public:
      // Constructors
      Matrix();
      Matrix(int,int);
      Matrix(int,int,const T&);
      Matrix(const Vector<T>&);
      Matrix(const Matrix<T>&);
      ~Matrix();

      // Member Functions
      Vector<T>& operator [] (int);
      const Vector<T>& operator [] (int) const;
      Vector<T>  operator () (int) const;

      Matrix<T> identity();
      Matrix<T> transpose() const;
      Matrix<T> inverse() const;
      T trace() const;
      T determinant() const;

      int rows() const;
      int cols() const;
      void resize(int,int);
      void resize(int,int,const T&);
      void fill(const T&);

      // Arithmetic Operators
      const Matrix<T>& operator = (const Matrix<T>&);
      const Matrix<T>& operator = (const T&);

      Matrix<T> operator + () const;
      Matrix<T> operator - () const;
      Matrix<T> operator += (const Matrix<T>&);
```

```cpp
     Matrix<T> operator -= (const Matrix<T>&);
     Matrix<T> operator *= (const Matrix<T>&);
     Matrix<T> operator +  (const Matrix<T>&) const;
     Matrix<T> operator -  (const Matrix<T>&) const;
     Matrix<T> operator *  (const Matrix<T>&) const;
     Vector<T> operator *  (const Vector<T>&) const;

     Matrix<T> operator += (const T&);
     Matrix<T> operator -= (const T&);
     Matrix<T> operator *= (const T&);
     Matrix<T> operator /= (const T&);
     Matrix<T> operator +  (const T&) const;
     Matrix<T> operator -  (const T&) const;
     Matrix<T> operator *  (const T&) const;
     Matrix<T> operator /  (const T&) const;

     Vector<T> vec() const;
     Matrix<T> kron(const Matrix<T>&) const;
     Matrix<T> dsum(const Matrix<T>&) const;
     Matrix<T> hadamard(const Matrix<T>&) const;
     pair<Matrix<T>, Matrix<T> > LU() const;

     ostream &output(ostream&) const;
     istream &input(istream&);
};

template <class T> T tr(const Matrix<T> &m) { return m.trace(); }
template <class T> T det(const Matrix<T> &m) { return m.determinant(); }

// implementation of class Matrix
template <class T> Matrix<T>::Matrix()
   : rowNum(0), colNum(0), mat() {}

template <class T> Matrix<T>::Matrix(int r,int c)
   : rowNum(r), colNum(c), mat(r)
{ for(int i=0;i<r;i++) mat[i].resize(c); }

template <class T> Matrix<T>::Matrix(int r,int c,const T &value)
   : rowNum(r), colNum(c), mat(r)
{ for(int i=0;i<r;i++) mat[i].resize(c,value); }

template <class T> Matrix<T>::Matrix(const Vector<T> &v)
   : rowNum(v.size()), colNum(1), mat(rowNum)
{ for(int i=0;i<rowNum;i++) mat[i].resize(1,v[i]); }

template <class T> Matrix<T>::Matrix(const Matrix<T> &m)
   : rowNum(m.rowNum), colNum(m.colNum), mat(m.mat)
{ }

template <class T> Matrix<T>::~Matrix() { }

template <class T> Vector<T> & Matrix<T>::operator [] (int index)
{
   assert(index>=0 && index<rowNum);
   return mat[index];
}


template <class T>
```

```
const Vector<T> & Matrix<T>::operator [] (int index) const
{
   assert(index>=0 && index<rowNum);
   return mat[index];
}


template <class T> Vector<T> Matrix<T>::operator () (int index) const
{
   assert(index>=0 && index<colNum);
   Vector<T> result(rowNum);
   for(int i=0;i<rowNum;i++) result[i] = mat[i][index];
   return result;
}


template <class T> Matrix<T> Matrix<T>::identity()
{
   for(int i=0;i<rowNum;i++)
      for(int j=0;j<colNum;j++)
         if(i==j) mat[i][j] = one(T());
         else     mat[i][j] = zero(T());
   return *this;
}


template <class T> Matrix<T> Matrix<T>::transpose() const
{
   Matrix<T> result(colNum,rowNum);
   for(int i=0;i<rowNum;i++)
      for(int j=0;j<colNum;j++) result[j][i] = mat[i][j];
   return result;
}

// Symbolical Inverse using Leverrier's Method
template <class T> Matrix<T> Matrix<T>::inverse() const
{
   assert(rowNum == colNum);
   Matrix<T> B(*this), D, I(rowNum, colNum);
   T c0(B.trace()), c1, j(one(T()));
   int i;
   I.identity();
   for(j++,i=2;i<rowNum;i++,j++)
   {
      B = *this*(B-c0*I);
      c0 = B.trace()/j;
   }
   D = *this*(B-c0*I);
   c1 = D.trace()/j;
   return (B-c0*I)/c1;
}


template <class T> T Matrix<T>::trace() const
{
   assert(rowNum == colNum);
   T result(zero(T()));
   for(int i=0;i<rowNum;i++) result += mat[i][i];
   return result;
}
```

```cpp
// Symbolical determinant
template <class T> T Matrix<T>::determinant() const
{
   assert(rowNum==colNum);
   Matrix<T> B(*this), I(rowNum, colNum, zero(T()));
   T c(B.trace());
   int i;
   for(i=0;i<rowNum;i++) I[i][i] = one(T());
   // Note that determinant of int-type gives zero
   // because of division by T(i)
   for(i=2;i<=rowNum;i++)
   {
      B = *this * (B-c*I);
      c = B.trace()/T(i);
   }
   if(rowNum%2) return c;
   return -c;
}

template <class T> int Matrix<T>::rows() const
{ return rowNum; }

template <class T> int Matrix<T>::cols() const
{ return colNum; }

template <class T> void Matrix<T>::resize(int r,int c)
{ resize(r, c, zero(T())); }

template <class T> void Matrix<T>::resize(int r,int c,const T &value)
{
   mat.resize(r);
   for(int i=0;i<r;i++) mat[i].resize(c,value);
   rowNum = r; colNum = c;
}

template <class T> void Matrix<T>::fill(const T &value)
{
   for(int i=0;i<rowNum;i++)
      for(int j=0;j<colNum;j++) mat[i][j] = value;
}

template <class T>
const Matrix<T> & Matrix<T>::operator = (const Matrix<T> &m)
{
   if(this == &m) return *this;
   rowNum = m.rowNum; colNum = m.colNum;
   mat = m.mat;
   return *this;
}

template <class T>
const Matrix<T> & Matrix<T>::operator = (const T &value)
{
   for(int i=0;i<rowNum;i++) mat[i] = value;
   return *this;
}

template <class T> Matrix<T> Matrix<T>::operator + () const
```

```
{ return *this; }

template <class T> Matrix<T> Matrix<T>::operator - () const
{ return *this * T(-1); }

template <class T> Matrix<T> Matrix<T>::operator += (const Matrix<T> &m)
{ return *this = *this + m; }

template <class T> Matrix<T> Matrix<T>::operator -= (const Matrix<T> &m)
{ return *this = *this - m; }

template <class T> Matrix<T> Matrix<T>::operator *= (const Matrix<T> &m)
{ return *this = *this * m; }

template <class T>
Matrix<T> Matrix<T>::operator + (const Matrix<T> &m) const
{
    assert(rowNum == m.rowNum && colNum == m.colNum);
    Matrix<T> result(*this);
    for(int i=0;i<rowNum;i++) result[i] += m[i];
    return result;
}

template <class T>
Matrix<T> Matrix<T>::operator - (const Matrix<T> &m) const
{
    assert(rowNum == m.rowNum && colNum == m.colNum);
    Matrix<T> result(*this);
    for(int i=0;i<rowNum;i++) result[i] -= m[i];
    return result;
}

template <class T>
Matrix<T> Matrix<T>::operator * (const Matrix<T> &m) const
{
    assert(colNum == m.rowNum);
    Matrix<T> result(rowNum, m.colNum, zero(T()));
    for(int i=0;i<rowNum;i++)
      for(int j=0;j<m.colNum;j++)
        for(int k=0;k<colNum;k++)
            result[i][j] += mat[i][k]*m[k][j];
    return result;
}

template <class T>
Vector<T> Matrix<T>::operator * (const Vector<T> &v) const
{
    assert(colNum == v.size());
    Vector<T> result(rowNum);
    // dot product | is used
    for(int i=0;i<rowNum;i++) result[i] = (mat[i] | v);
    return result;
}

template <class T> Matrix<T> Matrix<T>::operator += (const T &c)
{
    assert(rowNum == colNum);
    for(int i=0;i<rowNum;i++) mat[i][i] += c;
```

```
    return *this;
}

template <class T> Matrix<T> Matrix<T>::operator -= (const T &c)
{
    assert(rowNum == colNum);
    for(int i=0;i<rowNum;i++) mat[i][i] -= c;
    return *this;
}

template <class T> Matrix<T> Matrix<T>::operator *= (const T &c)
{
    for(int i=0;i<rowNum;i++) mat[i] *= c;
    return *this;
}

template <class T> Matrix<T> Matrix<T>::operator /= (const T &c)
{
    for(int i=0;i<rowNum;i++) mat[i] /= c;
    return *this;
}

template <class T>
Matrix<T> Matrix<T>::operator + (const T &value) const
{
    assert(rowNum == colNum);
    Matrix<T> result(*this);
    return result += value;
}

template <class T>
Matrix<T> Matrix<T>::operator - (const T &value) const
{
    assert(rowNum == colNum);
    Matrix<T> result(*this);
    return result -= value;
}

template <class T>
Matrix<T> Matrix<T>::operator * (const T &value) const
{
    Matrix<T> result(*this);
    return result *= value;
}

template <class T>
Matrix<T> Matrix<T>::operator / (const T &value) const
{
    Matrix<T> result(*this);
    return result /= value;
}

template <class T>
Matrix<T> operator + (const T &value,const Matrix<T> &m)
{ return m + value; }

template <class T>
Matrix<T> operator - (const T &value,const Matrix<T> &m)
```

```
{ return -m + value; }

template <class T>
Matrix<T> operator * (const T &value,const Matrix<T> &m)
{
 int i, j;
 Matrix<T> m1(m);
 for(i=0;i<m1.rows();i++)
  for(j=0;j<m1.cols();j++)
   m1[i][j] = value*m1[i][j];
 return m1;
}

template <class T>
Matrix<T> operator / (const T &value,const Matrix<T> &m)
{
   Matrix<T> result(m.rows(),m.cols());
   for(int i=0;i<result.rows();i++) result[i] = value/m[i];
   return result;
}

// Vectorize operator
template <class T> Vector<T> Matrix<T>::vec() const
{
   int i=0, j, k, size = rowNum*colNum;
   Vector<T> result(size);
   for(j=0;j<colNum;j++)
      for(k=0;k<rowNum;k++) result[i++] = mat[k][j];
   return result;
}

template <class T> Vector<T> vec(const Matrix<T> &m)
{ return m.vec(); }

// Kronecker Product
template <class T> Matrix<T> Matrix<T>::kron(const Matrix<T> &m) const
{
   int size1 = rowNum*m.rowNum,
       size2 = colNum*m.colNum,
       i, j, k, p;
   Matrix<T> result(size1, size2);
   for(i=0;i<rowNum;i++)
      for(j=0;j<colNum;j++)
         for(k=0;k<m.rowNum;k++)
            for(p=0;p<m.colNum;p++)
               result[k+i*m.rowNum][p+j*m.colNum]
                  = mat[i][j]*m.mat[k][p];
   return result;
}

template <class T>
Matrix<T> kron(const Matrix<T> &s,const Matrix<T> &m)
{ return s.kron(m); }

// Direct Sum
template <class T> Matrix<T> Matrix<T>::dsum(const Matrix<T> &m) const
{
   int size1 = rowNum+m.rowNum,
```

```cpp
      size2 = colNum+m.colNum;
   Matrix<T> result(size1, size2);
   for(int i=0;i<size1;i++)
      for(int j=0;j<size2;j++)
      {
         if(i < rowNum && j < colNum)
           result[i][j] = mat[i][j];
         else if(i >= rowNum && j >= colNum)
           result[i][j] = m.mat[i-rowNum][j-colNum];
         else
           result[i][j] = zero(T());
      }
   return result;
}


template <class T>
Matrix<T> dsum(const Matrix<T> &s,const Matrix<T> &m)
{ return s.dsum(m); }


// Hadamard product
template <class T> Matrix<T> Matrix<T>::hadamard(const Matrix<T> &m) const
{
   assert(rowNum == m.rowNum && colNum == m.colNum);
   Matrix<T> result(rowNum, colNum, zero(T()));
   for(int i=0;i<rowNum;i++)
      for(int j=0;j<m.colNum;j++)
         result[i][j] = mat[i][j]*m[i][j];
   return result;
}


template <class T>
Matrix<T> hadamard(const Matrix<T> &s,const Matrix<T> &m)
{ return s.hadamard(m); }


template <class T>
pair<Matrix<T>, Matrix<T> > Matrix<T>::LU() const
{
 assert(rowNum == colNum);
 Matrix<T> L(rowNum,colNum,zero(T()));
 Matrix<T> U(*this);
 for(int i=0;i<rowNum;i++)
 {
  assert(U[i][i] != zero(T()));
  L[i][i] = U[i][i];
  U[i] /= L[i][i];
  U[i][i] = one(T());
  for(int j=i+1;j<colNum;j++)
  {
   L[j][i] = U[j][i];
   U[j] -= L[j][i]*U[i];
   U[j][i] = zero(T());
  }
 }
 return make_pair(L, U);
}


template <class T>
pair<Matrix<T>, Matrix<T> > LU(const Matrix<T> &m)
```

```
{ return m.LU(); }

template <class T>
int operator == (const Matrix<T> &m1,const Matrix<T> &m2)
{
   if(m1.rows() != m2.rows()) return 0;
   for(int i=0;i<m1.rows();i++)
      if(m1[i] != m2[i]) return 0;
   return 1;
}


template <class T>
int operator != (const Matrix<T> &m1,const Matrix<T> &m2)
{ return !(m1==m2); }

template <class T> ostream & Matrix<T>::output(ostream &s) const
{
   int t = colNum-1, maxwidth=0, i, j, k, l;
   vector<string> m(rowNum*colNum);
   for(i=0,k=0;i<rowNum;i++)
   {
      for(j=0;j<colNum;j++,k++)
      {
      // strore the string representation for each
      // element so that we can compute the maximum
      // string length and then center each element
      // in its column
      ostringstream os;
      os << mat[i][j];
      m[k] = os.str();
      if(maxwidth < (int)m[k].length()) maxwidth = m[k].length();
      }
   }
   for(i=0,k=0;i<rowNum;i++)
   {
      s << "[";
      for(j=0;j<t;j++,k++)
      {
      // add spaces around the string to center it
      l = maxwidth-m[k].length();
      if(l%2) m[k] = " " + m[k];
      for(l=l/2;l>0;l--) m[k] = " " + m[k] + " ";
      // output the centered string
      s << m[k] << " ";
      }
      // add spaces around the string to center it
      l = maxwidth-m[k].length();
      if(l%2) m[k] = " " + m[k];
      for(l=l/2;l>0;l--) m[k] = " " + m[k] + " ";
      // output the centered string
      s << m[k++] << "]" << endl;
   }
   return s;
}

template <class T> ostream & operator << (ostream &s,const Matrix<T> &m)
{ return m.output(s); }
```

```
template <class T> istream & Matrix<T>::input(istream &s)
{
   int i, j, num1, num2;
   s.clear();                    // set stream state to good
   s >> num1;                    // read in row number
   if(! s.good()) return s;      // can't get an integer, just return
   s >> num2;                    // read in column number
   if(! s.good()) return s;      // can't get an integer, just return
   resize(num1,num2);            // resize to Matrix into right order
   for(i=0;i<num1;i++)
      for(j=0;j<num2;j++)
      {
         s >> mat[i][j];
         if(! s.good())
         {
            s.clear(s.rdstate() | ios::badbit);
            return s;
         }
      }
   return s;
}

template <class T> istream & operator >> (istream &s,Matrix<T> &m)
{ return m.input(s); }
#endif
```

13.7.2   Matrix Norms

```
// matnorm.h
// Norms of Matrices

#ifndef MATNORM_H
#define MATNORM_H

#include <iostream>
#include <math.h>
#include "vector.h"
#include "vecnorm.h"
#include "identity.h"
using namespace std;

template <class T> T norm1(const Matrix<T> &m)
{
   T maxItem(0), temp;
   int i,j;
   for(i=0;i<m.rows();i++) maxItem += m[i][0];
   for(i=1;i<m.cols();i++)
   {
      temp = zero(T());
      for(j=0;j<m.rows();j++) temp += abs(m[j][i]);
      if(temp > maxItem) maxItem = temp;
   }
   return maxItem;
}

template <class T> T normI(const Matrix<T> &m)
```

```
{
    T maxItem(norm1(m[0]));
    for(int i=1;i<m.rows();i++)
        if(norm1(m[i]) > maxItem) maxItem = norm1(m[i]);
    return maxItem;
}

template <class T> T normH(const Matrix<T> &m)
{ return sqrt((m*(m.transpose())).trace()); }
#endif
```

13.8 Array Class

The public interface of the `Array<T,n>` class:

- `Array<T,n>(int, ...)` : Constructor for an n-dimensional array with the sizes of each dimension as arguments.

- `Array<T,n>(const vector<int>&)` : Constructor for an n-dimensional array with the sizes of each dimension stored in a vector.

- `Array<T,n>(const vector<int>&,const T&t)` : Constructor for an n-dimensional array with the sizes of each dimension stored in a vector and all elements initialized to the value `t`.

- `resize(int, ...)` : Resizes the array.

- `resize(const vector<int>&)` : Resizes the array.

- `resize(const vector<int>&,const T&t)` : Resizes the array and sets all elements to the value `t`.

- Arithmetic operators : `+, -, =, +=, -=, *=`

- Stream operator : `<<`

The class `Array<T,n>` is derived from the STL class `vector<T>` and consequently inherits `vector<T>` members such as `size` and `resize`.

For a detailed description of the classes and each member function, please refer to Section 7.8.

```
// array.h
// The Array Class

#ifndef ARRAY_H
#define ARRAY_H

#include <iostream>
#include <cassert>
```

```cpp
#include <cstdarg>
#include <vector>
using namespace std;

template <int d> vector<int> dimensions(int d1, ...)
{
   vector<int> dim;
   va_list ap;

   dim.push_back(d1);
   // collect the remaining sizes in dim
   va_start(ap,d1);
    for(int i=1;i<d;++i) dim.push_back(va_arg(ap,int));
   va_end(ap);
   return dim;
}

template <class T,int d>
class Array : public vector<Array<T,d-1> >
{
   public:
      // Constructors
      Array();
      Array(int,...);
      Array(vector<int>);
      Array(vector<int>,const T&);
      Array(const Array<T,d>&);
      ~Array();

      // Member Functions
      void resize(int,...);
      void resize(vector<int>);
      void resize(vector<int>,const T&);

      // Arithmetic Operators
      const Array<T,d> & operator = (const T&);
      Array<T,d> operator *= (const T&);
      Array<T,d> operator += (const Array<T,d>&);
      Array<T,d> operator -= (const Array<T,d>&);
      Array<T,d> operator +  (const Array<T,d>&);
      Array<T,d> operator -  (const Array<T,d>&);
};    // end declaration class Array<T,d>

// Constructor, destructor and copy constructor.

template <class T,int d>
Array<T,d>::Array() : vector<Array<T,d-1> >() {}

template <class T,int d>
Array<T,d>::Array(int r, ...) : vector<Array<T,d-1> >(r)
{
   vector<int> dim;
   va_list ap;

   dim.push_back(r);
   // collect the remaining sizes in dim
   va_start(ap,r);
    for(int i=0;i<d-1;++i) dim.push_back(va_arg(ap,int));
```

```
   va_end(ap);
   // resize the arrays with sizes in dim
   resize(dim);
}

template <class T,int d>
Array<T,d>::Array(vector<int> r)
{
   if(!r.empty())
   {
    vector<Array<T,d-1> >::resize(r.front());
    vector<int> n(r.begin()+1, r.end());
    // resize the arrays with sizes in r
    for(int i=0;i<r.front();++i)
     vector<Array<T,d-1> >::at(i).resize(n);
   }
}

template <class T,int d>
Array<T,d>::Array(vector<int> r,const T &num)
{
   if(!r.empty())
   {
    vector<Array<T,d-1> >::resize(r.front());
    vector<int> n(r.begin()+1,r.end());
    // resize the arrays with sizes in r
    for(int i=0;i<r.front();++i)
     vector<Array<T,d-1> >::at(i).resize(n,num);
   }
}

template <class T,int d>
Array<T,d>::Array(const Array<T,d> &m) : vector<Array<T,d-1> >(m) { }

template <class T,int d> Array<T,d>::~Array() { }

// Member functions

template <class T,int d>
void Array<T,d>::resize(int r, ...)
{
   vector<int> dim;
   va_list ap;
   dim.push_back(r);
   // collect the remaining sizes in dim
   va_start(ap,r);
    for(int i=0;i<d-1;++i) dim.push_back(va_arg(ap,int));
   va_end(ap);
   resize(dim);
}

template <class T,int d>
void Array<T,d>::resize(vector<int> r)
{
   if(!r.empty())
   {
    vector<Array<T,d-1> >::resize(r.front());
    vector<int> n(r.begin()+1, r.end());
```

```cpp
    // resize the arrays with sizes in r
    for(int i=0;i<r.front();++i)
     vector<Array<T,d-1> >::at(i).resize(n);
  }
}

template <class T,int d>
void Array<T,d>::resize(vector<int> r,const T &value)
{
  if(!r.empty())
  {
   vector<Array<T,d-1> >::resize(r.front());
   vector<int> n(r.begin()+1, r.end());
   // resize the arrays with sizes in r
   for(int i=0;i<r.front();++i)
    vector<Array<T,d-1> >::at(i).resize(n,value);
  }
}

// Various member operators

template <class T,int d>
const Array<T,d> & Array<T,d>::operator = (const T &num)
{
   int length = vector<Array<T,d-1> >::size();
   for(int i=0;i<length;i++) vector<Array<T,d-1> >::at(i) = num;
   return *this;
}

template <class T,int d>
Array<T,d> Array<T,d>::operator *= (const T &num)
{
   int length = vector<Array<T,d-1> >::size();
   for(int i=0;i<length;i++) vector<Array<T,d-1> >::at(i) *= num;
   return *this;
}

template <class T,int d>
Array<T,d> Array<T,d>::operator += (const Array<T,d> &m)
{ return *this = *this + m; }

template <class T,int d>
Array<T,d> Array<T,d>::operator -= (const Array<T,d> &m)
{ return *this = *this - m; }

template <class T,int d>
Array<T,d> Array<T,d>::operator + (const Array<T,d> &m)
{
   int length = vector<Array<T,d-1> >::size();
   assert(m.size() == (vector<Array<T,d-1> >::size()));
   Array<T,d> temp(*this);
   for(int i=0;i<length;i++)
    temp[i] = vector<Array<T,d-1> >::at(i)+m[i];
   return temp;
}

template <class T,int d>
Array<T,d> Array<T,d>::operator - (const Array<T,d> &m)
```

```
{
    int length = vector<Array<T,d-1> >::size();
    assert(m.size() == (vector<Array<T,d-1> >::size()));
    Array<T,d> temp(*this);
    for(int i=0;i<length;i++)
      temp[i] = vector<Array<T,d-1> >::at(i)-m[i];
    return temp;
}

template <class T,int d>
ostream & operator << (ostream &s,const Array<T,d> &m)
{
    for(int i=0;i<int(m.size());i++) s << m[i] << endl;
    return s;
}


//Override the template definition for the 1-dimensional case

template <class T>
class Array<T,1> : public vector<T>
{
    public:
        // Constructors
        Array(int = 0);
        Array(vector<int>);
        Array(int,const T&);
        Array(vector<int>,const T&);
        Array(const Array<T,1>&);
        ~Array();

        // Member Functions
        void resize(int);
        void resize(vector<int>);
        void resize(int,const T&);
        void resize(vector<int>,const T&);

        // Arithmetic Operators
        const Array<T,1> & operator = (const T&);
        Array<T,1> operator *= (const T&);
        Array<T,1> operator += (const Array<T,1>&);
        Array<T,1> operator -= (const Array<T,1>&);
        Array<T,1> operator +  (const Array<T,1>&);
        Array<T,1> operator -  (const Array<T,1>&);
};      // end declaration class Array<T,1>

// Constructors, destructors and copy constructor.

template <class T>
Array<T,1>::Array(int n) : vector<T>(n) { }

template <class T>
Array<T,1>::Array(vector<int> n)
{ if(!n.empty()) vector<T>::resize(n.front()); }

template <class T>
Array<T,1>::Array(int n,const T &num) : vector<T>(n,num) { }

template <class T>
```

```
Array<T,1>::Array(vector<int> n,const T &num)
{ if(!n.empty()) vector<T>::resize(n.front(),num); }

template <class T>
Array<T,1>::Array(const Array<T,1> &v) : vector<T>(v) { }

template <class T> Array<T,1>::~Array() { }

// Member functions

template <class T> void Array<T,1>::resize(int n)
{ vector<T>::resize(n); }

template <class T> void Array<T,1>::resize(vector<int> n)
{ if(!n.empty()) vector<T>::resize(n.front()); }

template <class T>
void Array<T,1>::resize(int n,const T &value)
{ vector<T>::resize(n,value); }

template <class T>
void Array<T,1>::resize(vector<int> n,const T &value)
{ if(!n.empty()) vector<T>::resize(n.front(),value); }

// Various member operators

template <class T>
const Array<T,1> & Array<T,1>::operator = (const T &num)
{
    int length = vector<T>::size();
    for(int i=0;i<length;i++) vector<T>::at(i) = num;
    return *this;
}

template <class T>
Array<T,1> Array<T,1>::operator *= (const T &num)
{
    int length = vector<T>::size();
    for(int i=0;i<length;i++) vector<T>::at(i) *= num;
    return *this;
}

template <class T>
Array<T,1> Array<T,1>::operator += (const Array<T,1> &v)
{ return *this = *this + v; }

template <class T>
Array<T,1> Array<T,1>::operator -= (const Array<T,1> &v)
{ return *this = *this - v; }

template <class T>
Array<T,1> Array<T,1>::operator + (const Array<T,1> &v)
{
    int length = vector<T>::size();
    assert(vector<T>::size() == v.size());
    Array<T,1> temp(length);
    for(int i=0;i<length;i++) temp[i] = vector<T>::at(i)+v[i];
    return temp;
```

```
}

template <class T>
Array<T,1> Array<T,1>::operator - (const Array<T,1> &v)
{
    int length = vector<T>::size();
    assert(vector<T>::size() == v.size());
    Array<T,1> temp(length);
    for(int i=0;i<length;i++) temp[i] = vector<T>::at(i)-v[i];
    return temp;
}

template <class T>
ostream & operator << (ostream &s,const Array<T,1> &v)
{
    s << "[";
    for(int i=0;i<int(v.size())-1;i++) s << v[i] << " ";
    s << v.back() << "]";
    return s;
}

#endif
```

13.9 Polynomial Class

The public interface of the Polynomial class:

- Polynomial() : Default constructor.

- Polynomial(string) : Constructor giving the variable name.

- Polynomial(const T&) : Constant polynomial constructor.

- Polynomial<T> karatsuba(const Polynomial<T>&,const Polynomial<T>&) : Karatsuba multiplication.

- Polynomial<T> newton(const Polynomial<T>&,const Polynomial<T>&) : Newton iteration for division.

- Polynomial<T> gcd(const Polynomial<T>&,const Polynomial<T>&) : Greatest common divisor.

- int constant() : returns 1 if the polynomial is constant and 0 otherwise.

- Polynomial<T> Diff(const string &x) : Differentiation with respect to the variable named x.

- Polynomial<T> Int(const string &x) : Integration with respect to the variable named x.

- Polynomial<T> reverse() : Reverse the polynomial.

- list<Polynomial<T> > squarefree() : Calculate the square free decomposition.

- Arithmetic operators : +(unary), –(unary), +, -, *, /, %, ^, =, +=, -=, *=, /=

- Relational operators : ==, !=

- Function operator : ()

- Stream operator : <<

For a detailed description of the class structure and each member function, please refer to Section 7.9.

```cpp
// polynomial.h

#ifndef _POLYNOMIAL
#define _POLYNOMIAL

#include <cassert>
#include <iostream>
#include <list>
#include <string>
#include <utility>  // for pair
#include "identity.h"
using namespace std;

//Polynomial class

template <class T>
class Polynomial
{
 public:
   static int Karatsuba, Newton;

   Polynomial() {}
   Polynomial(const T&);
   Polynomial(string x);
   Polynomial(const Polynomial<T> &p)
     : variable(p.variable), terms(p.terms) {}

   Polynomial<T>& operator=(const T&);
   Polynomial<T>& operator=(const Polynomial<T>&);

   Polynomial<T> operator+() const;
   Polynomial<T> operator-() const;

   Polynomial<T> operator+(const Polynomial<T>&) const;
   Polynomial<T> operator-(const Polynomial<T>&) const;
   Polynomial<T> operator*(const Polynomial<T>&) const;
   Polynomial<T> operator/(const Polynomial<T>&) const;
   Polynomial<T> operator%(const Polynomial<T>&) const;

   Polynomial<T> operator^(unsigned int) const;

   Polynomial<T> operator+(const T&) const;
   Polynomial<T> operator-(const T&) const;
   Polynomial<T> operator*(const T&) const;
   Polynomial<T> operator/(const T&) const;
   Polynomial<T> operator%(const T&) const;
```

```
    Polynomial<T>& operator+=(const Polynomial<T>&);
    Polynomial<T>& operator-=(const Polynomial<T>&);
    Polynomial<T>& operator*=(const Polynomial<T>&);
    Polynomial<T>& operator/=(const Polynomial<T>&);
    Polynomial<T>& operator%=(const Polynomial<T>&);

    Polynomial<T>& operator+=(const T&);
    Polynomial<T>& operator-=(const T&);
    Polynomial<T>& operator*=(const T&);
    Polynomial<T>& operator/=(const T&);
    Polynomial<T>& operator%=(const T&);

    Polynomial<T> karatsuba(const Polynomial<T>&,
                            const Polynomial<T>&,int) const;
    Polynomial<T> newton(const Polynomial<T>&,const Polynomial<T>&) const;
    Polynomial<T> gcd(const Polynomial<T>&,const Polynomial<T>&) const;
    Polynomial<T> Diff(const string &) const;
    Polynomial<T> Int(const string &) const;
    Polynomial<T> reverse() const;
    list<Polynomial<T> > squarefree() const;

    int operator==(const Polynomial<T>&) const;
    int operator!=(const Polynomial<T>&) const;

    int operator==(const T&) const;
    int operator!=(const T&) const;

    T operator()(const T&) const;
    ostream &output1(ostream &) const;
    ostream &output2(ostream &) const;
    int constant() const;
  protected:
    void remove_zeros(void);
    string variable;
    list<pair<T,int> > terms;
};

// multiply using Karatsuba algorithm
template <class T> int Polynomial<T>::Karatsuba = 1;
// inversion by Newton iteration for division
template <class T> int Polynomial<T>::Newton = 1;

// additional functions that are not members of the class
template <class T>
Polynomial<T> operator+(const T&,const Polynomial<T>&);
template <class T>
Polynomial<T> operator-(const T&,const Polynomial<T>&);
template <class T>
Polynomial<T> operator*(const T&,const Polynomial<T>&);
template <class T>
Polynomial<T> operator/(const T&,const Polynomial<T>&);
template <class T>
Polynomial<T> operator%(const T&,const Polynomial<T>&);
template <class T>
int operator==(T,const Polynomial<T>&);
template <class T>
int operator!=(T,const Polynomial<T>&);
```

```cpp
// implementation

template <class T>
Polynomial<T>::Polynomial(const T &c) : variable("")
{ if(c!=zero(T())) terms.push_back(pair<T,int>(c,0)); }

template <class T>
Polynomial<T>::Polynomial(string x): variable(x)
{ terms.push_back(pair<T,int>(one(T()),1)); }

template <class T>
Polynomial<T>& Polynomial<T>::operator=(const T &c)
{ return *this = Polynomial<T>(c); }

template <class T>
Polynomial<T>& Polynomial<T>::operator=(const Polynomial<T> &p)
{
 if(this == &p) return *this;
 variable = p.variable;
 terms = p.terms;
 return *this;
}

template <class T>
Polynomial<T> Polynomial<T>::operator+() const
{ return *this; }

template <class T>
Polynomial<T> Polynomial<T>::operator-() const
{
 Polynomial<T> p2(*this);
 typename list<pair<T,int> >::iterator i = p2.terms.begin();
 for(;i!=p2.terms.end();i++) i->first = -(i->first);
 return p2;
}

template <class T>
Polynomial<T>
Polynomial<T>::operator+(const Polynomial<T> &p) const
{
 Polynomial<T> p2;
 typename list<pair<T,int> >::const_iterator i = terms.begin();
 typename list<pair<T,int> >::const_iterator j = p.terms.begin();

 if(constant()) p2.variable = p.variable;
 else if(p.constant()) p2.variable = variable;
 else assert((p2.variable = variable) == p.variable);

 while(i!=terms.end() && j!=p.terms.end())
 {
  if(i->second < j->second)      p2.terms.push_back(*(j++));
  else if(i->second > j->second) p2.terms.push_back(*(i++));
  else  // i->second == j->second
  {
   p2.terms.push_back(pair<T,int>(i->first + j->first,i->second));
   i++; j++;
```

```
  }
 }

 while(i != terms.end())   p2.terms.push_back(*(i++));
 while(j != p.terms.end()) p2.terms.push_back(*(j++));

 p2.remove_zeros();
 return p2;
}


template <class T>
Polynomial<T>
Polynomial<T>::operator-(const Polynomial<T> &p) const
{ return (*this) + (-p); }


template <class T>
Polynomial<T>
Polynomial<T>::operator*(const Polynomial<T> &p) const
{
 Polynomial<T> p2;
 typename list<pair<T,int> >::const_iterator i, j;

 if(constant()) p2.variable = p.variable;
 else if(p.constant()) p2.variable = variable;
 else assert((p2.variable=variable) == p.variable);

 if(!Karatsuba) // conventional multiplication
 for(i=terms.begin();i!=terms.end();i++)
 {
  Polynomial<T> t;
  t.variable = p2.variable;
  for(j=p.terms.begin();j!=p.terms.end();j++)
   t.terms.push_back(pair<T,int>(i->first*j->first,i->second+j->second));

  p2 += t;
 }
 else // Karatsuba multiplication
 {
  int degree = 1;
  int degree1 = (terms.empty()) ? 0 : terms.front().second;
  int degree2 = (p.terms.empty()) ? 0 : p.terms.front().second;
  while(degree <= degree1 || degree <= degree2) degree *= 2;
  Polynomial<T> k = karatsuba(*this,p,degree);
  k.variable = p2.variable;
  return k;
 }
 return p2;
}


template <class T>
Polynomial<T>
Polynomial<T>::operator/(const Polynomial<T> &p) const
{
 Polynomial<T> p2, p3(*this);
 typename list<pair<T,int> >::const_iterator i;
 typename list<pair<T,int> >::iterator j;

 if(constant()) p2.variable = p.variable;
```

```
  else if(p.constant()) p2.variable = variable;
  else assert((p2.variable=variable) == p.variable);

  assert(p != zero(T()));

  if(!Newton)  // long division
  while(p3.terms.size() > 0 &&
        p3.terms.front().second >= p.terms.front().second)
  {
   Polynomial<T> t;
   t.variable = p2.variable;
   t.terms.push_back(
     pair<T,int>(p3.terms.front().first/p.terms.front().first,
                 p3.terms.front().second-p.terms.front().second));
   p2 += t;
   p3 -= t * p;
  }
  else        // Newton iteration
   return newton(*this,p);

  return p2;
}

template <class T>
Polynomial<T>
Polynomial<T>::operator%(const Polynomial<T> &p) const
{ return *this - p * (*this/p); }

template <class T>
Polynomial<T> Polynomial<T>::operator^(unsigned int n) const
{
 Polynomial<T> result(one(T())), factor(*this);
 while(n>0)
 {
  if(n%2 == 1) result *= factor;
  factor *= factor;
  n /= 2;
 }
 return result;
}

template <class T>
Polynomial<T> Polynomial<T>::operator+(const T &c) const
{ return (*this) + Polynomial<T>(c); }

template <class T>
Polynomial<T> Polynomial<T>::operator-(const T &c) const
{ return (*this) - Polynomial<T>(c); }

template <class T>
Polynomial<T> Polynomial<T>::operator*(const T &c) const
{ return (*this) * Polynomial<T>(c); }

template <class T>
Polynomial<T> Polynomial<T>::operator/(const T &c) const
{ return (*this) / Polynomial<T>(c); }

template <class T>
```

```
Polynomial<T> Polynomial<T>::operator%(const T &c) const
{ return (*this) / Polynomial<T>(c); }

template <class T>
Polynomial<T>& Polynomial<T>::operator+=(const Polynomial<T> &p)
{ return *this = *this + p; }

template <class T>
Polynomial<T>& Polynomial<T>::operator-=(const Polynomial<T> &p)
{ return *this = *this - p; }

template <class T>
Polynomial<T>& Polynomial<T>::operator*=(const Polynomial<T> &p)
{ return *this = *this * p; }

template <class T>
Polynomial<T>& Polynomial<T>::operator/=(const Polynomial<T> &p)
{ return *this = *this / p; }

template <class T>
Polynomial<T>& Polynomial<T>::operator%=(const Polynomial<T> &p)
{ return *this = *this % p; }

template <class T>
Polynomial<T> &Polynomial<T>::operator+=(const T &c)
{ return *this += Polynomial<T>(c); }

template <class T>
Polynomial<T> &Polynomial<T>::operator-=(const T &c)
{ return *this -= Polynomial<T>(c); }

template <class T>
Polynomial<T> &Polynomial<T>::operator*=(const T &c)
{ return *this *= Polynomial<T>(c); }

template <class T>
Polynomial<T> &Polynomial<T>::operator/=(const T &c)
{ return *this /= Polynomial<T>(c); }

template <class T>
Polynomial<T> &Polynomial<T>::operator%=(const T &c)
{ return *this %= Polynomial<T>(c); }

template <class T>
T Diff(const T &t,const string &x)
{ return zero(T()); }

// partial template specialization for polynomials
template <class T>
Polynomial<T> Diff(const Polynomial<T>&p,const string &x)
{ return p.Diff(x); }

template <class T>
Polynomial<T> Polynomial<T>::karatsuba(const Polynomial<T> &p1,
                                       const Polynomial<T> &p2,int n) const
{
 typename list<pair<T,int> >::const_iterator i;
 typename list<pair<T,int> >::iterator j;
```

```
  int n2 = n/2;
  Polynomial<T> f0;
  Polynomial<T> f1;
  Polynomial<T> g0;
  Polynomial<T> g1;

  if(n == 1)
  {
   Polynomial<T> p;
   if(p1.terms.empty() || p2.terms.empty()) return p;
   T t1 = p1.terms.front().first;
   T t2 = p2.terms.front().first;
   if(t1*t2 == zero(T())) return p;
   p.terms.push_back(make_pair(t1*t2,0));
   return p;
  }

  for(i=p1.terms.begin();i!=p1.terms.end() && i->second>=n2;i++)
   f1.terms.push_back(make_pair(i->first, i->second - n2));
  f0.terms.insert(f0.terms.end(),i,p1.terms.end());

  for(i=p2.terms.begin();i!=p2.terms.end() && i->second>=n2;i++)
   g1.terms.push_back(make_pair(i->first, i->second - n2));
  g0.terms.insert(g0.terms.end(),i,p2.terms.end());

  Polynomial<T> r1 = karatsuba(f1,g1,n2);
  Polynomial<T> r2 = karatsuba(f0,g0,n2);
  Polynomial<T> r3 = karatsuba(f0+f1,g0+g1,n2);
  Polynomial<T> r4 = r3 - r1 - r2;

  // multiply by x^(n/2)
  for(j=r4.terms.begin();j!=r4.terms.end();j++) j->second += n2;
  // multiply by x^n
  for(j=r1.terms.begin();j!=r1.terms.end();j++) j->second += n;

  return r1 + r4 + r2;
}

template <class T>
Polynomial<T> Polynomial<T>::newton(const Polynomial<T> &p1,
                                    const Polynomial<T> &p2) const
{
 T two = one(T()) + one(T());
 Polynomial<T> g = one(T());
 Polynomial<T> f = p2.reverse();

 g.variable = f.variable;

 int n = (p1.terms.empty()) ? 0 : p1.terms.front().second;
 int m = (p2.terms.empty()) ? 0 : p2.terms.front().second;
 int i, j, r;

 if(n<m) return Polynomial<T>(zero(T()));

 T lc = (p2.terms.empty()) ? one(T()) : p2.terms.front().first;
 T lci = one(T()) / lc;
 f *= lci;
```

```
  n = n - m + 1;
  r = 0; j = 1;
  while(j < n) { r++; j *= 2; }

  for(i=1,j=2;i<=r;i++,j*=2)
  {
   g *= (two - f*g);
   while(!g.terms.empty() && g.terms.front().second >= j)
    g.terms.pop_front();
  }
  g.variable = p2.variable;

  Polynomial<T> result =  p1.reverse()*g*lci;
  while(!result.terms.empty() && result.terms.front().second >= n)
   result.terms.pop_front();

  if(p1.constant()) result.variable = p2.variable;
  else              result.variable = p1.variable;
  return result.reverse();
}

template <class T>
Polynomial<T> Polynomial<T>::gcd(const Polynomial<T> &a,
                                 const Polynomial<T> &b) const
{
 Polynomial<T> c(a);
 Polynomial<T> d(b);

 if(c.terms.empty()) return Polynomial<T>(one(T()));
 if(d.terms.empty()) return Polynomial<T>(one(T()));

 c /= c.terms.front().first; d /= d.terms.front().first;
 while(d != Polynomial<T>(zero(T())))
 {
  Polynomial<T> r = c % d;
  c = d; d = r;
 }
 return c / c.terms.front().first;
}

template <class T>
Polynomial<T> Polynomial<T>::Diff(const string &x) const
{
 Polynomial<T> p2;
 typename list<pair<T,int> >::const_iterator i = terms.begin();

 // if(variable != x) return Polynomial<T>(zero(T()));
 if(variable != x)
 {
  p2.variable = variable;
  for(;i!=terms.end();i++)
    p2.terms.push_back(pair<T,int>(::Diff(i->first,x),i->second));
  p2.remove_zeros();
  return p2;
 }
 p2.variable = variable;
 for(;i!=terms.end();i++)
  if(i->second > 0)
```

```cpp
   p2.terms.push_back(pair<T,int>(i->first*T(i->second),i->second-1));

 return p2;
}


template <class T>
T Int(const T &t,const string &x)
{
 cerr << "Tried to integrate a datatype with respect to " << x
      << " when it is not supported." << endl;
 return t;
}

// partial template specialization for polynomials
template <class T>
Polynomial<T> Int(const Polynomial<T> &p,const string &x)
{ return p.Int(x); }

template <class T>
Polynomial<T> Polynomial<T>::Int(const string &x) const
{
 Polynomial<T> p2;
 typename list<pair<T,int> >::const_iterator i = terms.begin();

 if(variable != x)
 {
  p2.variable = variable;
  for(;i!=terms.end();i++)
    p2.terms.push_back(pair<T,int>(::Int(i->first,x),i->second));
 }

 p2.variable = variable;
 for(;i!=terms.end();i++)
   p2.terms.push_back(pair<T,int>(i->first/T(i->second+1),i->second+1));

 return p2;
}

template <class T>
Polynomial<T> Polynomial<T>::reverse() const
{
 int degree = terms.front().second;
 typename list<pair<T,int> >::const_reverse_iterator i;
 Polynomial<T> p(variable);

 p.terms.clear();
 for(i=terms.rbegin();i!=terms.rend();i++)
  p.terms.push_back(make_pair(i->first,degree - i->second));

 return p;
}

template <class T>
list<Polynomial<T> > Polynomial<T>::squarefree() const
{
 list<Polynomial<T> > l;
 T lc = (terms.empty()) ? one(T()) : terms.front().first;
 Polynomial<T> a = *this/lc;
```

```
  Polynomial<T> c = gcd(a,Diff(variable));
  Polynomial<T> w = a/c;
  while(c != Polynomial<T>(one(T())))
  {
   Polynomial<T> y = gcd(w,c);
   Polynomial<T> z = w/y;
   l.push_back(z);
   w = y; c /= y;
  }
  l.push_back(w);
  l.push_front(Polynomial<T>(lc));
  return l;
}

template <class T>
int Polynomial<T>::operator==(const Polynomial<T> &p) const
{ return ((*this-p).terms.size() == 0); }

template <class T>
int Polynomial<T>::operator!=(const Polynomial<T> &p) const
{ return !(*this == p); }

template <class T>
int Polynomial<T>::operator==(const T &c) const
{ return *this == Polynomial<T>(c); }

template <class T>
int Polynomial<T>::operator!=(const T &c) const
{ return !(*this == c); }

template <class T>
T Polynomial<T>::operator()(const T &c) const
{
 T factor = one(T()), result = zero(T());
 typename list<pair<T,int> >::const_reverse_iterator i = terms.rbegin();

 for(int j=0;i!=terms.rend();i++)
 {
  for(int k=j;k<i->second;k++) factor *= c;
  j = i->second;
  result += i->first * factor;
 }

 return result;
}

template <class T>
ostream &Polynomial<T>::output1(ostream &o) const
{
 typename list<pair<T,int> >::const_iterator i = terms.begin();

 if(i == terms.end()) return o << zero(T());
 o << ((i->first >= zero(T())) ? "" : "-");
 while(i != terms.end())
 {
  if(i->first != one(T()) || i->second == zero(T()))
   o << ((i->first >= zero(T())) ? i->first : -(i->first));
  if(i->second > zero(T())) o << variable;
```

```
 if(i->second > one(T())) o << "^" << i->second;
 if(++i != terms.end()) o << ((i->first >= zero(T())) ? "+" : "-");
}
return o;
}


template <class T>
ostream &Polynomial<T>::output2(ostream &o) const
{
 typename list<pair<T,int> >::const_iterator i = terms.begin();

 if(i == terms.end()) return o << zero(T());
 while(i != terms.end())
 {
  if(i->first != one(T()) || i->second == 0)
   o << "(" << i->first << ")";
  if(i->second > 0) o << variable;
  if(i->second > 1) o << "^" << i->second;
  if(++i != terms.end()) o << "+";
 }
 return o;
}


template <class T>
ostream &operator<<(ostream &o,const Polynomial<T> &p)
{ return p.output1(o); }

// partial template specialization for polynomial coefficients
template <class T>
ostream &operator<<(ostream &o,const Polynomial<Polynomial<T> > &p)
{ return p.output2(o); }

template <class T>
int Polynomial<T>::constant(void) const
{
 return terms.size() == 0 ||
             (terms.size() == 1 && terms.front().second == 0);
}


template <class T>
void Polynomial<T>::remove_zeros(void)
{
 typename list<pair<T,int> >::iterator i, j;

 for(i=j=terms.begin();i!=terms.end();)
 {
  j++;
  if(i->first == zero(T())) terms.erase(i);
  i=j;
 }
}

// additional functions that are not members of the class
template <class T>
Polynomial<T> operator+(const T &c,const Polynomial<T> &p)
{ return Polynomial<T>(c) + p; }

template <class T>
```

```
Polynomial<T> operator-(const T &c,const Polynomial<T> &p)
{ return Polynomial<T>(c) - p; }

template <class T>
Polynomial<T> operator*(const T &c,const Polynomial<T> &p)
{ return Polynomial<T>(c) * p; }

template <class T>
Polynomial<T> operator/(const T &c,const Polynomial<T> &p)
{ return Polynomial<T>(c) / p; }

template <class T>
Polynomial<T> operator%(const T &c,const Polynomial<T> &p)
{ return Polynomial<T>(c) % p; }

template <class T>
int operator==(const T &c,const Polynomial<T> &p) { return p == c; }

template <class T>
int operator!=(const T &c,const Polynomial<T> &p) { return p != c; }

template <class T>
Polynomial<T> zero(Polynomial<T>) { return Polynomial<T>(zero(T())); }

template <class T>
Polynomial<T> one(Polynomial<T>) { return Polynomial<T>(one(T())); }

#endif
```

13.10 Multinomial Class

The public interface of the `Multinomial` class:

- `Multinomial()` : Default Constructor.

- `Multinomial(string)` : Constructor giving the variable name.

- `Multinomial(T)` : Constant polynomial constructor.

- `Polynomial<T> Diff(const string &x)` : Differentiation with respect to the variable named x.

- Arithmetic operators : +(unary), -(unary), +, -, *, ^, =, +=, -=, *=

- Relational operators : ==, !=

- Stream operator : <<

For a detailed description of the class structure and each member function, please refer to Section 7.10.

```cpp
// multinomial.h

#ifndef _MULTINOMIAL
#define _MULTINOMIAL

#include <cassert>
#include <iostream>
#include <list>
#include <sstream>
#include <string>
#include <utility>  // for pair
#include <vector>
#include "identity.h"
using namespace std;

// Multinomial class

template <class T>
class Multinomial
{
 public:
    Multinomial(): type(number),n(zero(T())) {}
    Multinomial(T);
    Multinomial(string x);
    Multinomial(const Multinomial<T> &p):
    variable(p.variable),type(p.type),n(p.n),u(p.u),m(p.m) {}

    Multinomial<T>& operator=(const T&);
    Multinomial<T>& operator=(const Multinomial<T>&);

    Multinomial<T> operator+() const;
    Multinomial<T> operator-() const;

    Multinomial<T> operator+(const Multinomial<T>&) const;
    Multinomial<T> operator-(const Multinomial<T>&) const;
    Multinomial<T> operator*(const Multinomial<T>&) const;
    Multinomial<T> operator+(const T&) const;
    Multinomial<T> operator-(const T&) const;
    Multinomial<T> operator*(const T&) const;

    Multinomial<T> operator^(unsigned int) const;

    Multinomial<T>& operator+=(const Multinomial<T>&);
    Multinomial<T>& operator-=(const Multinomial<T>&);
    Multinomial<T>& operator*=(const Multinomial<T>&);
    Multinomial<T>& operator+=(const T&);
    Multinomial<T>& operator-=(const T&);
    Multinomial<T>& operator*=(const T&);

    int operator==(const Multinomial<T>&) const;
    int operator!=(const Multinomial<T>&) const;
    int operator==(const T&) const;
    int operator!=(const T&) const;

    Multinomial<T> Diff(const string &) const;
    ostream &output(ostream &) const;
```

```
  protected:
    void remove_zeros(void);
    pair<Multinomial<T>,Multinomial<T> >
     reconcile(const Multinomial<T>&,const Multinomial<T>&) const;
    vector<string> toarray(void) const;
    string variable;
    enum { number, univariate, multivariate } type;
    T n;                            //number
    list<pair<T,int> > u;           //univariate
    list<pair<Multinomial<T>,int> > m; //multivariate
};


// additional functions that are not members of the class
template <class T>
Multinomial<T> operator+(const T&,const Multinomial<T>&);
template <class T>
Multinomial<T> operator-(const T&,const Multinomial<T>&);
template <class T>
Multinomial<T> operator*(const T&,const Multinomial<T>&);
template <class T>
int operator==(T,const Multinomial<T>&);
template <class T>
int operator!=(T,const Multinomial<T>&);


// implementation

template <class T>
Multinomial<T>::Multinomial(T c) :
variable(""), type(number), n(c) {}

template <class T>
Multinomial<T>::Multinomial(string x): variable(x),type(univariate)
{u.push_back(pair<T,int>(one(T()),1));}

template <class T>
Multinomial<T>& Multinomial<T>::operator=(const T &c)
{return *this = Multinomial<T>(c);}

template <class T>
Multinomial<T>& Multinomial<T>::operator=(const Multinomial<T> &p)
{
 if(this == &p) return *this;
 variable = p.variable;
 type = p.type; u = p.u; m = p.m;
 return *this;
}

template <class T>
Multinomial<T> Multinomial<T>::operator+() const
{return *this;}

template <class T>
Multinomial<T> Multinomial<T>::operator-() const
{
 Multinomial<T> p2(*this);
 typename list<pair<T,int> >::iterator i = p2.u.begin();
```

```cpp
 typename list<pair<T,int> >::iterator j = p2.m.begin();
 for(;i!= p2.u.end();i++) i->first = -(i->first);
 for(;j!= p2.v.end();j++) j->first = -(j->first);
 return p2;
}


template <class T>
list <pair<T,int> >
merge(const list <pair<T,int> > &l1,const list <pair<T,int> > &l2)
{
 list <pair<T,int> > l;
 typename list<pair<T,int> >::const_iterator i = l1.begin();
 typename list<pair<T,int> >::const_iterator j = l2.begin();
 while(i!=l1.end() && j!=l2.end())
 {
  if      (i->second<j->second) l.push_back(*(j++));
  else if(i->second>j->second) l.push_back(*(i++));
  else // i->second==j->second
  {
   l.push_back(make_pair(i->first + j->first,i->second));
   i++; j++;
  }
 }
 while(i!=l1.end()) l.push_back(*(i++));
 while(j!=l2.end()) l.push_back(*(j++));
 return l;
}


template <class T>
Multinomial<T>
Multinomial<T>::operator+(const Multinomial<T> &p) const
{
 Multinomial<T> p2;
 pair<Multinomial<T>,Multinomial<T> > xy = reconcile(*this,p);

 switch(xy.first.type)
 {
  case number:       p2.type = number; p2.n = xy.first.n + xy.second.n;
                     break;
  case univariate:   p2.type = univariate;
                     p2.variable = xy.first.variable;
                     p2.u = merge(xy.first.u,xy.second.u);
                     break;
  case multivariate: p2.type = multivariate;
                     p2.variable = xy.first.variable;
                     p2.m = merge(xy.first.m,xy.second.m);
                     break;
 }
 p2.remove_zeros();
 return p2;
}


template <class T>
Multinomial<T>
Multinomial<T>::operator-(const Multinomial<T> &p) const
{ return (*this) + (-p); }


template <class T>
```

```
list <pair<T,int> >
distribute(const list <pair<T,int> > &l1,const list <pair<T,int> > &l2)
{
 list <pair<T,int> > l;
 typename list<pair<T,int> >::const_iterator i;
 typename list<pair<T,int> >::const_iterator j;

 for(i=l1.begin();i!=l1.end();i++)
 {
  list <pair<T,int> > t;
  for(j=l2.begin();j!=l2.end();j++)
   t.push_back(make_pair(i->first * j->first,i->second + j->second));
  l = merge(l,t);
 }
 return l;
}

template <class T>
Multinomial<T> Multinomial<T>::operator*(const Multinomial<T> &p) const
{
 Multinomial<T> p2;
 typename list<pair<T,int> >::const_iterator i, j;
 pair<Multinomial<T>,Multinomial<T> > xy = reconcile(*this,p);

 switch(xy.first.type)
 {
  case number:       p2.type = number; p2.n = xy.first.n * xy.second.n;
                     break;
  case univariate:   p2.type = univariate;
                     p2.variable = xy.first.variable;
                     p2.u = distribute(xy.first.u,xy.second.u);
                     break;
  case multivariate: p2.type = multivariate;
                     p2.variable = xy.first.variable;
                     p2.m = distribute(xy.first.m,xy.second.m);
                     break;
 }
 p2.remove_zeros();
 return p2;
}

template <class T>
Multinomial<T> Multinomial<T>::operator^(unsigned int n) const
{
 Multinomial<T> result(one(T())), factor(*this);
 while(n>0)
 {
  if(n%2 == 1) result *= factor;
  factor *= factor;
  n /= 2;
 }
 return result;
}

template <class T>
Multinomial<T> Multinomial<T>::operator+(const T &c) const
{return (*this) + Multinomial<T>(c);}
```

```
template <class T>
Multinomial<T> Multinomial<T>::operator-(const T &c) const
{return (*this) - Multinomial<T>(c);}

template <class T>
Multinomial<T> Multinomial<T>::operator*(const T &c) const
{return (*this) * Multinomial<T>(c);}

template <class T>
Multinomial<T>& Multinomial<T>::operator+=(const Multinomial<T> &p)
{return *this = *this + p;}

template <class T>
Multinomial<T>& Multinomial<T>::operator-=(const Multinomial<T> &p)
{return *this = *this - p;}

template <class T>
Multinomial<T>& Multinomial<T>::operator*=(const Multinomial<T> &p)
{return *this = *this * p;}

template <class T>
Multinomial<T> &Multinomial<T>::operator+=(const T &c)
{*this += Multinomial<T>(c);}

template <class T>
Multinomial<T> &Multinomial<T>::operator-=(const T &c)
{*this -= Multinomial<T>(c);}

template <class T>
Multinomial<T> &Multinomial<T>::operator*=(const T &c)
{*this *= Multinomial<T>(c);}

template <class T>
T Diff(const T &t,const string &x)
{return zero(T());}

// partial template specialization for polynomials
template <class T>
Multinomial<T> Diff(const Multinomial<T>&p,const string &x)
{return p.Diff(x);}

template <class T>
Multinomial<T> Multinomial<T>::Diff(const string &x) const
{
 Multinomial<T> p2;
 typename list<pair<T,int> >::const_iterator i = u.begin();
 typename list<pair<Multinomial<T>,int> >::const_iterator j = m.begin();

 if(type==number) return Multinomial<T>(zero(T()));
 if(type==univariate)
 {
  if(variable!=x) return Multinomial<T>(zero(T()));
  p2.variable = variable;
  p2.type = univariate;
  for(;i!=u.end();i++)
    p2.u.push_back(make_pair(i->first * T(i->second),i->second-1));
 }
 if(type == multivariate)
```

```
    {
      p2.variable = variable;
      p2.type = multivariate;
      if(variable==x)
        for(;j!=m.end();j++)
          p2.m.push_back(make_pair(j->first * T(j->second),j->second-1));
      else
        for(;j!=m.end();j++)
          p2.m.push_back(make_pair(j->first.Diff(x),j->second));
    }
    p2.remove_zeros();
    return p2;
}

template <class T>
int Multinomial<T>::operator==(const Multinomial<T> &p) const
{
  pair<Multinomial<T>,Multinomial<T> > xy = reconcile(*this,p);

  xy.first.remove_zeros();
  xy.second.remove_zeros();
  switch(xy.first.type)
  {
    case number:        return xy.first.n == xy.second.n;
                        break;
    case univariate:    return xy.first.u == xy.second.u;
                        break;
    case multivariate:  return xy.first.m == xy.second.m;
                        break;
  }
  return 0;
}

template <class T>
int Multinomial<T>::operator!=(const Multinomial<T> &p) const
{return !(*this == p);}

template <class T>
int Multinomial<T>::operator==(const T &c) const
{return *this == Multinomial<T>(c);}

template <class T>
int Multinomial<T>::operator!=(const T &c) const
{return !(*this == c);}

template <class T>
vector<string>
Multinomial<T>::toarray(void) const
{
  ostringstream o;
  vector<string> v;
  typename list<pair<T,int> >::const_iterator i = u.begin();
  typename list<pair<Multinomial<T>,int> >::const_iterator j = m.begin();

  switch(type)
  {
    case number:        o << n; v.push_back(o.str()); break;
    case univariate:    if(i==u.end()) v.push_back("0");
```

```
                        while(i!=u.end())
                        {
                         if(i->first!=one(T()) || i->second==0)
                          o << "(" << i->first << ")";
                         if(i->second>0) o << variable;
                         if(i->second>1) o << "^" << i->second;
                         v.push_back(o.str());
                         o.str("");  i++;
                        }
                        break;
  case multivariate: if(j==m.end()) v.push_back("0");
                        while(j!=m.end())
                        {
                         if(j->second>0) o << variable;
                         if(j->second>1) o << "^" << j->second;
                         if(j->first!=one(T()) || j->second==0)
                         {
                          vector<string> v1 = j->first.toarray();
                          for(int k=0;k<int(v1.size());k++)
                            v.push_back(v1[k] + o.str());
                         }
                         o.str("");  j++;
                        }
                        break;
 }
 return v;
}

template <class T>
ostream &Multinomial<T>::output(ostream &o) const
{
 int k;
 typename list<pair<T,int> >::const_iterator i = u.begin();

 switch(type)
 {
  case number:       o << n; break;
  case univariate:   if(i==u.end()) o << "0";
                        while(i!=u.end())
                        {
                         if(i->first!=one(T()) || i->second==0)
                          o << "(" << i->first << ")";
                         if(i->second>0) o << variable;
                         if(i->second>1) o << "^" << i->second;
                         if(!(++i==u.end())) o << " + ";
                        }
                        break;
  case multivariate: vector<string> v = toarray();
                        for(k=0;k<int(v.size())-1;k++)
                         o << v[k] << " + ";
                        if(k<int(v.size())) o << v[k];
                        else o << "0";
                        break;
 }
 return o;
}

template <class T>
```

```
ostream &operator<<(ostream &o,const Multinomial<T> &p)
{return p.output(o);}

template <class T>
void Multinomial<T>::remove_zeros(void)
{
 {
  typename list<pair<T,int> >::iterator i, j;
  for(i=j=u.begin();i!=u.end();)
  {
   j++;
   if(i->first==zero(T())) u.erase(i);
   i=j;
  }
 }
 {
  typename list<pair<Multinomial<T>,int> >::iterator i, j;
  for(i=j=m.begin();i!=m.end();)
  {
   j++;
   i->first.remove_zeros();
   if(i->first==zero(T())) m.erase(i);
   i=j;
  }
 }
}

template <class T>
pair<Multinomial<T>,Multinomial<T> >
Multinomial<T>::reconcile(const Multinomial<T> &x,
                          const Multinomial<T> &y) const
{
 if(x.type==number && y.type==number) return make_pair(x,y);
 if(x.type==number && y.type==univariate)
 {
  if(y.u.empty()) return make_pair(x,Multinomial<T>(zero(T())));
  Multinomial<T> t(y.variable);
  t.u.clear();
  t.u.push_back(make_pair(x.n,0));
  return make_pair(t,y);
 }
 if(x.type==number && y.type==multivariate)
 {
  if(y.m.empty()) return make_pair(x,Multinomial<T>(zero(T())));
  Multinomial<T> t(y.variable);
  t.u.clear();
  t.u.push_back(make_pair(x.n,0));
  return reconcile(t,y);
 }
 if(x.type==univariate && y.type==univariate)
 {
  if(x.variable==y.variable) return make_pair(x,y);
  if(x.variable< y.variable)
  {
   Multinomial<T> t1(y.variable);
   Multinomial<T> t2(y.variable);
   t1.type = t2.type = multivariate;
   t1.u.clear();
```

```
 t1.m.push_back(make_pair(x,0));
 typename list<pair<T,int> >::const_iterator i = y.u.begin();
 for(;i!=y.u.end();i++)
 {
  Multinomial<T> t3(x.variable);
  t3.u.clear();
  t3.u.push_back(make_pair(i->first,0));
  t2.m.push_back(make_pair(t3,i->second));
 }
 return make_pair(t1,t2);
}
else
{
 Multinomial<T> t1(x.variable);
 Multinomial<T> t2(x.variable);
 t1.type = t2.type = multivariate;
 t2.u.clear();
 t2.m.push_back(make_pair(y,0));
 typename list<pair<T,int> >::const_iterator i = x.u.begin();
 for(;i!=x.u.end();i++)
 {
  Multinomial<T> t3(y.variable);
  t3.u.clear();
  t3.u.push_back(make_pair(i->first,0));
  t1.m.push_back(make_pair(t3,i->second));
 }
 return make_pair(t1,t2);
 }
}
if(x.type==univariate && y.type==multivariate)
{
 if(y.m.empty())
 {
  Multinomial<T> t(x.variable);
  t.u.clear();
  return make_pair(x,t);
 }
 if(x.variable==y.variable)
  return make_pair(reconcile(x,y.m.front().first).first,y);
 if(x.variable< y.variable)
 {
  Multinomial<T> t(y.variable);
  t.type = multivariate;
  t.u.clear();
  t.m.push_back(make_pair(x,0));
  return make_pair(t,y);
 }
 else
 {
  Multinomial<T> t(x.variable);
  t.type = multivariate;
  t.u.clear();
  typename list<pair<T,int> >::const_iterator i = x.u.begin();
  for(;i!=x.u.end();i++)
  {
   Multinomial<T> t3(y.m.front().first.variable);
   t3.u.clear();
   t3.u.push_back(make_pair(i->first,0));
```

```
      t.m.push_back(make_pair(t3,i->second));
     }
     return reconcile(t,y);
   }
 }
 if(x.type==multivariate && y.type==multivariate)
 {
  if(x.variable==y.variable) return make_pair(x,y);
  if(x.variable< y.variable)
  {
   Multinomial<T> t(y.variable);
   t.type = multivariate;
   t.u.clear();
   t.m.push_back(make_pair(x,0));
   return make_pair(t,y);
  }
  else
  {
   Multinomial<T> t(x.variable);
   t.type = multivariate;
   t.u.clear();
   t.m.push_back(make_pair(y,0));
   return make_pair(x,t);
  }
 }

 pair<Multinomial<T>,Multinomial<T> > p;
 p = reconcile(y,x);
 return make_pair(p.second,p.first);
}

// additional functions that are not members of the class
template <class T>
Multinomial<T> operator+(const T &c,const Multinomial<T> &p)
{return Multinomial<T>(c) + p;}

template <class T>
Multinomial<T> operator-(const T &c,const Multinomial<T> &p)
{return Multinomial<T>(c) - p;}

template <class T>
Multinomial<T> operator*(const T &c,const Multinomial<T> &p)
{return Multinomial<T>(c) * p;}

template <class T>
int operator==(const T &c,const Multinomial<T> &p) {return p == c;}

template <class T>
int operator!=(const T &c,const Multinomial<T> &p) {return p != c;}

template <class T>
Multinomial<T> zero(Multinomial<T>) {return Multinomial<T>(zero(T()));}

template <class T>
Multinomial<T> one(Multinomial<T>) {return Multinomial<T>(one(T()));}

#endif
```

13.11   Symbolic Class

The public interface of the `Symbolic` class:

- `Symbolic()` : Default constructor.

- `Symbolic(const int&)` : Constructor from integers.

- `Symbolic(const double&)` : Constructor from real numbers.

- `Symbolic(const string &x)` : Constructor for a symbolic variable named x.

- `Symbolic(const char *x)` : Constructor for a symbolic variable named x.

- `Symbolic(const string &x,int)` : Constructor for a vector of numbered symbolic variable named x.

- `Symbolic(const char *x,int)` : Constructor for a vector of numbered symbolic variable named x.

- `Symbolic(const Symbolic &x,int)` : Constructor for a vector with elements initialized to x.

- `Symbolic(const string &x,int,int)` : Constructor for a matrix of numbered symbolic variable named x.

- `Symbolic(const char *x,int,int)` : Constructor for a matrix of numbered symbolic variable named x.

- `Symbolic(const Symbolic &x,int,int)` : Constructor for a matrix with elements initialized to x.

- `Symbolic(const list<Symbolic> &l)` : Constructor for a vector consisting of each of the elements in l.

- `Symbolic(const list<list<Symbolic> > &l)` : Constructor for a matrix consisting of each of the elements in l.

- `Symbolic subst(const Symbolic&,const Symbolic&)` : Replaces an expression by another.

- `Symbolic subst(const Symbolic&,const int&)` : Replaces an expression by an integer.

- `Symbolic subst(const Symbolic&,const double&)` : Replaces an expression by a real number.

- `Symbolic subst(const Equation&)` : Replaces all instances of the left hand side of the equation with the right hand side of the equation.

- `Symbolic subst(const list<Equation>&)` : Replaces all instances of the left hand sides of the equation with the corresponding right hand sides of the equations in the given list.

- Symbolic subst_all(const Symbolic&,const Symbolic&) : Replaces an expression by another, until the first expression no longer appears in the expression.

- Symbolic subst_all(const Equation&) : Replaces all instances of the left hand side of the equation with the right hand side of the equation, until the first expression no longer appears in the expression.

- Symbolic subst_all(const list<Equation>&) : Replaces all instances of the left hand sides of the equation with the corresponding right hand sides of the equations in the given list, until the first expression no longer appears in the expression.

- Symbolic coeff(const Symbolic &s) : coefficient of the expression s.

- Symbolic coeff(const Symbolic &s,int n) : coefficient of the expression s^n, if n is 0 then the function returns the constant term with respect to s.

- Symbolic coeff(int n) : returns the constant term divided by n.

- Symbolic coeff(double n) : returns the constant term divided by n.

- Symbolic commutative(int) : returns a variable which is commutative or non commutative as specified by the argument.

- int rows() : returns the number of rows in a matrix.

- int columns() : returns the number of columns in a matrix.

- Symbolic row(int) : returns the specified row of a matrix.

- Symbolic column(int) : returns the specified column of a matrix.

- Symbolic identity() : returns the identity matrix with the same number of rows and columns as the underlying matrix.

- Symbolic transpose() : returns the transpose of the matrix.

- Symbolic trace() : returns the trace of the matrix.

- Symbolic determinant() : returns the determinant of the matrix.

- Symbolic vec() : returns the vectorized matrix.

- Symbolic kron(const Symbolic &m) : returns the Kronecker product of the matrix with the matrix m.

- Symbolic dsum(const Symbolic &m) : returns the direct sum of the matrix with the matrix m.

- Symbolic hadamard(const Symbolic &m) : returns the Hadamard product of the matrix with the matrix m.

- Symbolic inverse() : returns the inverse of the matrix.

- Indexing operator : [] (dependency and substitution)

- Function operator : () (indexing)

The functions operating on the Symbolic class:

- Symbolic sin(const Symbolic&) : creates the sin of the argument.

- Symbolic cos(const Symbolic&) : creates the cos of the argument.

- Symbolic tan(const Symbolic&) : creates the tan of the argument.

- Symbolic cot(const Symbolic&) : creates the cot of the argument.

- Symbolic sec(const Symbolic&) : creates the sec of the argument.

- Symbolic csc(const Symbolic&) : creates the csc of the argument.

- Symbolic sinh(const Symbolic&) : creates the sinh of the argument.

- Symbolic cosh(const Symbolic&) : creates the cosh of the argument.

- Symbolic ln(const Symbolic&) : creates the ln of the argument.

- Symbolic log(const Symbolic &a,const Symbolic &b) : creates the $\log_a b$ of the arguments.

- Symbolic pow(const Symbolic &x,const Symbolic &y) : creates the power x^y of the arguments.

- Symbolic exp(const Symbolic&) : creates the exponential e^x of the argument.

- Symbolic sqrt(const Symbolic&) : creates the square root $\sqrt{x}$ of the argument.

- Symbolic df(const Symbolic &y,const Symbolic &x) : determines the derivative of y with respect to x.

- Symbolic df(const Symbolic &y,const Symbolic &x,int n) : determines n-th derivative of y with respect to x.

- Symbolic integrate(const Symbolic &y,const Symbolic &x) : determines the integral of y with respect to x.

- Symbolic integrate(const Symbolic &y,const Symbolic &x,int n) : determines n-th integral of y with respect to x.

- Symbolic tr(const Symbolic&) : the trace of the matrix argument.

- Symbolic trace(const Symbolic&) : the trace of the matrix argument.

- Symbolic det(const Symbolic&) : the determinant of the matrix argument.

- Symbolic determinant(const Symbolic&) : the determinant of the matrix argument.

- Symbolic kron(const Symbolic&,const Symbolic&) : the Kronecker of the matrix arguments.

- Symbolic dsum(const Symbolic&,const Symbolic&) : the direct sum of the matrix arguments.

- Symbolic &rhs(list<Equations>&,const Symbolic &lhs) : find the right hand side (in a list of equations) of an equation where the left hand side is given by lhs.

- Symbolic &lhs(list<Equations>&,const Symbolic &rhs) : find the left hand side (in a list of equations) of an equation where the right hand side is given by lhs.

- Arithmetic operators : +(unary), -(unary), ++, --, +, -, *, /, ^, ~, =, +=, -=, *=, /=

- Relational operators : ==, !=

- Vector operators : | (scalar product), % (cross product)

- Stream operators : >>, <<

- Cast to int and double.

For a detailed description of the class structure and each member function, please refer to Chapter 8.

13.11.1   Main Header File

```
// symbolicc++.h

// normal include headers
#include <iostream>
#include <iterator>
#include <list>
#include "cloning.h"
#include "identity.h"

// phased include headers
// according to class hierarchy
#include "symbolic.h"  // SymbolicInterface, Symbolic ...
#include "equation.h"  //    Equation : CloningSymbolicInterface
#include "number.h"    //    Numeric  : CloningSymbolicInterface
#include "product.h"   //    Product  : CloningSymbolicInterface
#include "sum.h"       //    Sum      : CloningSymbolicInterface
#include "symbol.h"    //    Symbol   : CloningSymbolicInterface
```

```
#include "functions.h" //     Sin    : Symbol ...
#include "symmatrix.h" //  SymbolicMatrix : CloningSymbolicInterface
#include "symerror.h"  //  SymbolicError  : CloningSymbolicInterface
#include "constants.h"

#ifndef SYMBOLIC_CPLUSPLUS
#define SYMBOLIC_CPLUSPLUS

// Include the relevant classes in 3 phases
//   Phase 1 ensures that every class has a forward
//    declaration for use in phase 2.
//   Phase 2 ensures that every constructor and method
//    has a forward declaration for use in phase 3.
// 1. Forward declarations: class X;
// 2. Declaraions:          class X { ... };
// 3. Definitions:          X::X() ...

// This overcomes mutual recursion in dependencies,
// for example class Sum needs class Product
// and class Product needs class Sum.

// forward declarations of all classes first
#define SYMBOLIC_FORWARD
#include "symbolicc++.h"
#undef  SYMBOLIC_FORWARD

// declarations of classes without definitions
#define SYMBOLIC_DECLARE
#include "symbolicc++.h"
#undef  SYMBOLIC_DECLARE

// definitions for non-member functions
// also used in definition phase for clarity

typedef list<Equation> Equations;
typedef list<Symbolic> SymbolicList;

Symbolic expand(const SymbolicInterface &s)
{ return s.expand(); }

ostream &operator<<(ostream &o,const Symbolic &s)
{ s.print(o); return o; }

ostream &operator<<(ostream &o,const Equation &s)
{ s.print(o); return o; }

Symbolic operator+(const Symbolic &s)
{ return s; }

Symbolic operator+(const Symbolic &s1,const Symbolic &s2)
{ return Sum(s1,s2); }

Symbolic operator+(const int &s1,const Symbolic &s2)
{ return Symbolic(Number<int>(s1)) + s2; }

Symbolic operator+(const Symbolic &s1,const int &s2)
{ return s1 + Symbolic(Number<int>(s2)); }
```

```
Symbolic operator+(const double &s1,const Symbolic &s2)
{ return Symbolic(Number<double>(s1)) + s2; }

Symbolic operator+(const Symbolic &s1,const double &s2)
{ return s1 + Symbolic(Number<double>(s2)); }

Symbolic operator++(Symbolic &s)
{ return s = s + 1; }

Symbolic operator++(Symbolic &s,int)
{ Symbolic t = s; ++s; return t; }

Symbolic operator-(const Symbolic &s)
{ return Product(Number<int>(-1),s); }

Symbolic operator-(const Symbolic &s1,const Symbolic &s2)
{ return Sum(s1,-s2); }

Symbolic operator-(const int &s1,const Symbolic &s2)
{ return Symbolic(s1) - s2; }

Symbolic operator-(const Symbolic &s1,const int &s2)
{ return s1 - Symbolic(s2); }

Symbolic operator-(const double &s1,const Symbolic &s2)
{ return Symbolic(s1) - s2; }

Symbolic operator-(const Symbolic &s1,const double &s2)
{ return s1 - Symbolic(s2); }

Symbolic operator--(Symbolic &s)
{ return s = s - 1; }

Symbolic operator--(Symbolic &s,int)
{ Symbolic t = s; --s; return t; }

Symbolic operator*(const Symbolic &s1,const Symbolic &s2)
{ return Product(s1,s2); }

Symbolic operator*(const int &s1,const Symbolic &s2)
{ return Symbolic(s1) * s2; }

Symbolic operator*(const Symbolic &s1,const int &s2)
{ return s1 * Symbolic(s2); }

Symbolic operator*(const double &s1,const Symbolic &s2)
{ return Symbolic(s1) * s2; }

Symbolic operator*(const Symbolic &s1,const double &s2)
{ return s1 * Symbolic(s2); }

Symbolic operator/(const Symbolic &s1,const Symbolic &s2)
{ return Product(s1,Power(s2,Number<int>(-1))); }

Symbolic operator/(const int &s1,const Symbolic &s2)
{ return Symbolic(s1) / s2; }

Symbolic operator/(const Symbolic &s1,const int &s2)
```

```
{ return s1 / Symbolic(s2); }

Symbolic operator/(const double &s1,const Symbolic &s2)
{ return Symbolic(s1) / s2; }

Symbolic operator/(const Symbolic &s1,const double &s2)
{ return s1 / Symbolic(s2); }

Symbolic operator+=(Symbolic &s1,const Symbolic &s2)
{ return s1 = s1 + s2; }

Symbolic operator+=(Symbolic &s1,const int &s2)
{ return s1 = s1 + s2; }

Symbolic operator+=(Symbolic &s1,const double &s2)
{ return s1 = s1 + s2; }

Symbolic operator-=(Symbolic &s1,const Symbolic &s2)
{ return s1 = s1 - s2; }

Symbolic operator-=(Symbolic &s1,const int &s2)
{ return s1 = s1 - s2; }

Symbolic operator-=(Symbolic &s1,const double &s2)
{ return s1 = s1 - s2; }

Symbolic operator*=(Symbolic &s1,const Symbolic &s2)
{ return s1 = s1 * s2; }

Symbolic operator*=(Symbolic &s1,const int &s2)
{ return s1 = s1 * s2; }

Symbolic operator*=(Symbolic &s1,const double &s2)
{ return s1 = s1 * s2; }

Symbolic operator/=(Symbolic &s1,const Symbolic &s2)
{ return s1 = s1 / s2; }

Symbolic operator/=(Symbolic &s1,const int &s2)
{ return s1 = s1 / s2; }

Symbolic operator/=(Symbolic &s1,const double &s2)
{ return s1 = s1 / s2; }

Equation operator==(const Symbolic &s1,const Symbolic &s2)
{ return Equation(s1,s2); }

Equation operator==(const Symbolic &s1,int i)
{ return s1 == Symbolic(i); }

Equation operator==(int i,const Symbolic &s1)
{ return Symbolic(i) == s1; }

Equation operator==(const Symbolic &s1,double d)
{ return s1 == Symbolic(d); }

Equation operator==(double d,const Symbolic &s1)
{ return Symbolic(d) == s1; }
```

```
int operator!=(const Symbolic &s1,const Symbolic &s2)
{ return !s1.compare(s2); }

int operator!=(const Symbolic &s1,int i)
{ return s1 != Symbolic(i); }

int operator!=(int i,const Symbolic &s1)
{ return Symbolic(i) != s1; }

int operator!=(const Symbolic &s1,double d)
{ return s1 != Symbolic(d); }

int operator!=(double d,const Symbolic &s1)
{ return Symbolic(d) != s1; }

Symbolic sin(const Symbolic &s)
{ return Sin(s); }

Symbolic cos(const Symbolic &s)
{ return Cos(s); }

Symbolic tan(const Symbolic &s)
{ return sin(s) / cos(s); }

Symbolic cot(const Symbolic &s)
{ return cos(s) / sin(s); }

Symbolic sec(const Symbolic &s)
{ return 1 / cos(s); }

Symbolic csc(const Symbolic &s)
{ return 1 / sin(s); }

Symbolic sinh(const Symbolic &s)
{ return Sinh(s); }

Symbolic cosh(const Symbolic &s)
{ return Cosh(s); }

Symbolic ln(const Symbolic &s)
{ return Log(SymbolicConstant::e,s); }

Symbolic log(const Symbolic &a,const Symbolic &b)
{ return Log(a,b); }

Symbolic pow(const Symbolic &s,const Symbolic &n)
{ return Power(s,n); }

Symbolic operator^(const Symbolic &s,const Symbolic &n)
{ return Power(s,n); }

Symbolic operator^(const Symbolic &s,int i)
{ return Power(s,Symbolic(i)); }

Symbolic operator^(int i,const Symbolic &s)
{ return Power(Symbolic(i),i); }
```

```
Symbolic operator^(const Symbolic &s,double d)
{ return Power(s,Symbolic(d)); }

Symbolic operator^(double d,const Symbolic &s)
{ return Power(Symbolic(d),s); }

Symbolic exp(const Symbolic &s)
{ return SymbolicConstant::e ^ s; }

Symbolic sqrt(const Symbolic &s)
{ return s ^ (Number<int>(1) / 2); }

Symbolic df(const Symbolic &s,const Symbolic &x)
{ return s.df(x); }

Symbolic integrate(const Symbolic &s,const Symbolic &x)
{ return s.integrate(x); }

Symbolic df(const Symbolic &s,const Symbolic &x,unsigned int i)
{
 Symbolic r = s;
 while(i-- > 0) r = r.df(x);
 return r;
}

Symbolic integrate(const Symbolic &s,const Symbolic &x,unsigned int i)
{
 Symbolic r = s;
 while(i-- > 0) r = r.integrate(x);
 return r;
}

Symbolic &rhs(list<Equation> &l,const Symbolic &lhs)
{
 list<Equation>::iterator i = l.begin();
 for(i=l.begin();i!=l.end();i++)
  if(i->lhs == lhs) return i->rhs;
 cerr << "Equation list does not contain lhs " << lhs << endl;
 throw SymbolicError(SymbolicError::NoMatch);
 return i->rhs;
}

Symbolic &lhs(list<Equation> &l,const Symbolic &rhs)
{
 list<Equation>::iterator i = l.begin();
 for(i=l.begin();i!=l.end();i++)
  if(i->rhs == rhs) return i->lhs;
 cerr << "Equation list does not contain rhs " << rhs << endl;
 throw SymbolicError(SymbolicError::NoMatch);
 return i->lhs;
}

template<> Symbolic zero(Symbolic) { return Number<int>(0); }

template<> Symbolic one(Symbolic) { return Number<int>(1); }

list<Equation>
operator,(const Equation &x,const Equation &y)
```

```
{
 list<Equation> l;
 l.push_back(x); l.push_back(y);
 return l;
}

list<Equation>
operator,(const list<Equation> &x,const Equation &y)
{
 list<Equation> l(x);
 l.push_back(y);
 return l;
}

list<Equation>
operator,(const Equation &x,const list<Equation> &y)
{
 list<Equation> l(y);
 l.push_front(x);
 return l;
}

list<Symbolic>
operator,(const Symbolic &x,const Symbolic &y)
{
 list<Symbolic> l;
 l.push_back(x);
 l.push_back(y);
 return l;
}

list<Symbolic> operator,(const int &x,const Symbolic &y)
{ return (Symbolic(x), y); }

list<Symbolic> operator,(const double &x,const Symbolic &y)
{ return (Symbolic(x), y); }

list<Symbolic> operator,(const Symbolic &x,const int &y)
{ return (x,Symbolic(y)); }

list<Symbolic> operator,(const Symbolic &x,const double &y)
{ return (x,Symbolic(y)); }

list<Symbolic>
operator,(const list<Symbolic> &x,const Symbolic &y)
{
 list<Symbolic> l(x);
 l.push_back(y);
 return l;
}

list<Symbolic> operator,(const list<Symbolic> &x,const int &y)
{ return (x, Symbolic(y)); }

list<Symbolic> operator,(const list<Symbolic> &x,const double &y)
{ return (x, Symbolic(y)); }

list<Symbolic>
```

```cpp
operator,(const Symbolic &x,const list<Symbolic> &y)
{
 list<Symbolic> l(y);
 l.push_front(x);
 return l;
}

list<Symbolic> operator,(const int &x,const list<Symbolic> &y)
{ return (Symbolic(x), y); }

list<Symbolic> operator,(const double &x,const list<Symbolic> &y)
{ return (Symbolic(x), y); }

list<list<Symbolic> >
operator,(const list<Symbolic> &x,const list<Symbolic> &y)
{
 list<list<Symbolic> > l;
 l.push_back(x); l.push_back(y);
 return l;
}

list<list<Symbolic> >
operator,(const list<list<Symbolic> > &x,const list<Symbolic> &y)
{
 list<list<Symbolic> > l(x);
 l.push_back(y);
 return l;
}

list<list<Symbolic> >
operator,(const list<Symbolic> &x,const list<list<Symbolic> > &y)
{
 list<list<Symbolic> > l(y);
 l.push_front(x);
 return l;
}

ostream &operator<<(ostream &o,const Equations &e)
{
 Equations::const_iterator i = e.begin();
 o << "[ ";
 while(i != e.end())
 {
  o << *(i++);
  if(i != e.end()) o << ",\n  ";
 }
 o << " ]" << endl;
 return o;
}

ostream &operator<<(ostream &o,const SymbolicList &e)
{
 SymbolicList::const_iterator i = e.begin();
 o << "[ ";
 while(i != e.end())
 {
  o << *(i++);
  if(i != e.end()) o << ", ";
```

```
      }
    o << " ]" << endl;
    return o;
    }

    Symbolic tr(const Symbolic &x) { return x.trace(); }

    Symbolic trace(const Symbolic &x) { return x.trace(); }

    Symbolic det(const Symbolic &x) { return x.determinant(); }

    Symbolic determinant(const Symbolic &x) { return x.determinant(); }

    Symbolic kron(const Symbolic &x,const Symbolic &y)
    { return x.kron(y); }

    Symbolic dsum(const Symbolic &x,const Symbolic &y)
    { return x.dsum(y); }

    Symbolic hadamard(const Symbolic &x,const Symbolic &y)
    { return x.hadamard(y); }

    // definitions for classes, member functions
    #define SYMBOLIC_DEFINE
    #include "symbolicc++.h"
    #undef   SYMBOLIC_DEFINE

    #endif
```

13.11.2   Memory Management

```
// cloning.h

#ifndef SYMBOLIC_CPLUSPLUS_CLONING
#define SYMBOLIC_CPLUSPLUS_CLONING

using namespace std;

class Cloning
{
 public: Cloning();
         Cloning(const Cloning&);
         virtual Cloning *clone() const = 0;
         template <class T> static Cloning *clone(const T&);
         virtual ~Cloning() {}
         virtual Cloning *reference(int) const;
         virtual int unreference() const;
         int refcount;
};

class CloningPtr
{
 private: Cloning *value;
 public:  CloningPtr();
          CloningPtr(const Cloning&,int=0);
          CloningPtr(const CloningPtr&,int=0);
```

```
        ~CloningPtr();

        CloningPtr &operator=(const CloningPtr&);
        CloningPtr &operator=(const Cloning&);

        Cloning *operator->() const;
        Cloning &operator*() const;
};

template <class T>
class CastPtr: public CloningPtr
{
 public:  CastPtr();
        CastPtr(const Cloning&,int=0);
        CastPtr(const CloningPtr&,int=0);
        ~CastPtr();

        T *operator->() const;
        T &operator*() const;
};

////////////////////////////////
// Cloning Implementation    //
////////////////////////////////

Cloning::Cloning() : refcount(0) {}

Cloning::Cloning(const Cloning &c) : refcount(0) {}

template <class T> Cloning *Cloning::clone(const T &t)
{ return new T(t); }

Cloning *Cloning::reference(int forceclone) const
{
 Cloning *c;
 if(refcount == 0 || forceclone)
 {
  c = clone();
  c->refcount = 1;
 }
 else
 {
  // we have to cast away the constness,
  // but we do know *this was dynamically allocated
  c = const_cast<Cloning*>(this);
  (c->refcount)++;
 }
 return c;
}

int Cloning::unreference() const
{
 if(refcount == 0) return 0;
 if(refcount == 1) return 1;
 // we have to cast away the constness,
 // but we do know *this was dynamically allocated
 (const_cast<Cloning*>(this)->refcount)--;
 return 0;
```

```
}

/////////////////////////////
// CloningPtr Implementation //
/////////////////////////////

CloningPtr::CloningPtr() : value(NULL) {}

CloningPtr::CloningPtr(const Cloning &p,int forceclone)
{ value = p.reference(forceclone); }

CloningPtr::CloningPtr(const CloningPtr &p,int forceclone)
{
 if(p.value == NULL) value = NULL;
 else value = p.value->reference(forceclone);
}

CloningPtr::~CloningPtr()
{
 if(value != NULL)
   if(value->unreference())
     delete value;
}

CloningPtr &CloningPtr::operator=(const Cloning &p)
{
 Cloning *v = value;
 if(value == &p) return *this;
 value = p.reference(0);
 if(v != NULL && v->unreference()) delete v;
 return *this;
}

CloningPtr &CloningPtr::operator=(const CloningPtr &p)
{
 Cloning *v = value;
 if(this == &p) return *this;
 if(value == p.value) return *this;
 if(p.value == NULL) value = NULL;
 else value = p.value->reference(0);
 if(v != NULL && v->unreference()) delete v;
 return *this;
}

Cloning *CloningPtr::operator->() const
{ return value; }

Cloning &CloningPtr::operator*() const
{ return *value; }

/////////////////////////////
// CastPtr Implementation    //
/////////////////////////////

template <class T> CastPtr<T>::CastPtr() : CloningPtr() {}

template <class T> CastPtr<T>::CastPtr(const Cloning &p,int forceclone)
 : CloningPtr(p,forceclone) {}
```

```
template <class T> CastPtr<T>::CastPtr(const CloningPtr &p,int forceclone)
 : CloningPtr(p,forceclone) {}

template <class T> CastPtr<T>::~CastPtr() {}

template <class T> T *CastPtr<T>::operator->() const
{ return dynamic_cast<T*>(CloningPtr::operator->()); }

template <class T> T &CastPtr<T>::operator*() const
{ return *operator->(); }

#endif
```

13.11.3 Constants

```
// constants.h

#include "symbolic.h"
#include "functions.h"

#ifndef SYMBOLIC_CPLUSPLUS_CONSTANTS

#ifdef  SYMBOLIC_DECLARE
#ifndef SYMBOLIC_CPLUSPLUS_CONSTANTS_DECLARE
#define SYMBOLIC_CPLUSPLUS_CONSTANTS_DECLARE

namespace SymbolicConstant
{
 Symbolic i = Symbolic(Power(Symbolic(-1), Power(Symbolic(2), Symbolic(-1))));
 Symbolic i_symbol("i");
 Symbolic e        ("e");
 Symbolic pi       ("pi");
}

#endif
#endif
#endif
```

13.11.4 Equations

```
// equation.h

#ifndef SYMBOLIC_CPLUSPLUS_EQUATION

using namespace std;

#ifdef  SYMBOLIC_FORWARD
#ifndef SYMBOLIC_CPLUSPLUS_EQUATION_FORWARD
#define SYMBOLIC_CPLUSPLUS_EQUATION_FORWARD

class Equation;

#endif
```

```
#endif

#ifdef   SYMBOLIC_DECLARE
#ifndef SYMBOLIC_CPLUSPLUS_EQUATION_DECLARE
#define SYMBOLIC_CPLUSPLUS_EQUATION_DECLARE

class Equation: public CloningSymbolicInterface
{
 public: Symbolic lhs, rhs;
         Equation(const Equation&);
         Equation(const Symbolic&,const Symbolic&);
         ~Equation();

         void print(ostream&) const;
         Symbolic subst(const Symbolic&,const Symbolic&,int &n) const;
         Simplified simplify() const;
         int compare(const Symbolic&) const;
         Symbolic df(const Symbolic&) const;
         Symbolic integrate(const Symbolic&) const;
         Symbolic coeff(const Symbolic&) const;
         Expanded expand() const;
         int commute(const Symbolic&) const;

         operator bool() const;
         operator int() const;

         Cloning *clone() const { return Cloning::clone(*this); }
};

#endif
#endif

#ifdef   SYMBOLIC_DEFINE
#ifndef SYMBOLIC_CPLUSPLUS_EQUATION_DEFINE
#define SYMBOLIC_CPLUSPLUS_EQUATION_DEFINE
#define SYMBOLIC_CPLUSPLUS_EQUATION

Equation::Equation(const Equation &s)
: CloningSymbolicInterface(s), lhs(s.lhs), rhs(s.rhs) {}

Equation::Equation(const Symbolic &s1,const Symbolic &s2)
: lhs(s1), rhs(s2) {}

Equation::~Equation() {}

void Equation::print(ostream &o) const
{ o << lhs << " == " << rhs; }

Symbolic Equation::subst(const Symbolic &x,const Symbolic &y,int &n) const
{ return Equation(lhs.subst(x,y,n),rhs.subst(x,y,n)); }

Simplified Equation::simplify() const
{ return Equation(lhs.simplify(),rhs.simplify()); }

int Equation::compare(const Symbolic &s) const
{
 if(s.type() != type()) return 0;
```

```
 CastPtr<const Equation> e = s;
 return (lhs.compare(e->lhs) && rhs.compare(e->rhs)) ||
        (lhs.compare(e->rhs) && rhs.compare(e->lhs));
}

Symbolic Equation::df(const Symbolic &s) const
{ return Equation(lhs.df(s),rhs.df(s)); }

Symbolic Equation::integrate(const Symbolic &s) const
{ return Equation(lhs.integrate(s),rhs.integrate(s)); }

Symbolic Equation::coeff(const Symbolic &s) const
{ return 0; }

Expanded Equation::expand() const
{ return Equation(lhs.expand(),rhs.expand()); }

int Equation::commute(const Symbolic &s) const
{ return 0; }

Equation::operator bool() const
{ return lhs.compare(rhs); }

Equation::operator int() const
{ return lhs.compare(rhs); }

#endif
#endif
#endif
```

13.11.5 Functions

```
// functions.h

#ifndef SYMBOLIC_CPLUSPLUS_FUNCTIONS

#include <cmath>
using namespace std;

#ifdef  SYMBOLIC_FORWARD
#ifndef SYMBOLIC_CPLUSPLUS_FUNCTIONS_FORWARD
#define SYMBOLIC_CPLUSPLUS_FUNCTIONS_FORWARD

class Sin;
class Cos;
class Sinh;
class Cosh;
class Log;
class Power;
class Derivative;
class Integral;

#endif
#endif

#ifdef  SYMBOLIC_DECLARE
```

```
#ifndef SYMBOLIC_CPLUSPLUS_FUNCTIONS_DECLARE
#define SYMBOLIC_CPLUSPLUS_FUNCTIONS_DECLARE

class Sin: public Symbol
{
 public: Sin(const Sin&);
         Sin(const Symbolic&);

         Simplified simplify() const;
         Symbolic df(const Symbolic&) const;
         Symbolic integrate(const Symbolic&) const;

         Cloning *clone() const { return Cloning::clone(*this); }
};

class Cos: public Symbol
{
 public: Cos(const Cos&);
         Cos(const Symbolic&);

         Simplified simplify() const;
         Symbolic df(const Symbolic&) const;
         Symbolic integrate(const Symbolic&) const;

         Cloning *clone() const { return Cloning::clone(*this); }
};

class Sinh: public Symbol
{
 public: Sinh(const Sinh&);
         Sinh(const Symbolic&);

         Simplified simplify() const;
         Symbolic df(const Symbolic&) const;
         Symbolic integrate(const Symbolic&) const;

         Cloning *clone() const { return Cloning::clone(*this); }
};

class Cosh: public Symbol
{
 public: Cosh(const Cosh&);
         Cosh(const Symbolic&);

         Simplified simplify() const;
         Symbolic df(const Symbolic&) const;
         Symbolic integrate(const Symbolic&) const;

         Cloning *clone() const { return Cloning::clone(*this); }
};

class Log: public Symbol
{
 public: Log(const Log&);
         Log(const Symbolic&, const Symbolic&);

         void print(ostream&) const;
         Simplified simplify() const;
```

```
        Symbolic df(const Symbolic&) const;
        Symbolic integrate(const Symbolic&) const;

        Cloning *clone() const { return Cloning::clone(*this); }
};

class Power: public Symbol
{
 public: Power(const Power&);
        Power(const Symbolic&,const Symbolic&);

        void print(ostream&) const;
        Simplified simplify() const;
        Expanded expand() const;
        Symbolic subst(const Symbolic &x,const Symbolic &y,int &n) const;
        Symbolic df(const Symbolic&) const;
        Symbolic integrate(const Symbolic&) const;

        Cloning *clone() const { return Cloning::clone(*this); }
};

class Derivative: public Symbol
{
 public: Derivative(const Derivative&);
        Derivative(const Symbolic&,const Symbolic&);

        Symbolic subst(const Symbolic &x,const Symbolic &y,int &n) const;
        Symbolic df(const Symbolic&) const;
        Symbolic integrate(const Symbolic&) const;
        int compare(const Symbolic&) const;

        Cloning *clone() const { return Cloning::clone(*this); }
};

class Integral: public Symbol
{
 public: Integral(const Integral&);
        Integral(const Symbolic&,const Symbolic&);

        Symbolic subst(const Symbolic &x,const Symbolic &y,int &n) const;
        Symbolic df(const Symbolic&) const;
        Symbolic integrate(const Symbolic&) const;

        Cloning *clone() const { return Cloning::clone(*this); }
};

#endif
#endif

#ifdef  SYMBOLIC_DEFINE
#ifndef SYMBOLIC_CPLUSPLUS_FUNCTIONS_DEFINE
#define SYMBOLIC_CPLUSPLUS_FUNCTIONS_DEFINE
#define SYMBOLIC_CPLUSPLUS_FUNCTIONS

////////////////////////////////////
// Implementation of Sin           //
////////////////////////////////////
```

```
Sin::Sin(const Sin &s) : Symbol(s) {}

Sin::Sin(const Symbolic &s) : Symbol(Symbol("sin")[s]) {}

Simplified Sin::simplify() const
{
 const Symbolic &s = parameters.front().simplify();
 if(s == 0) return Number<int>(0);
 if(s.type() == typeid(Product))
 {
  CastPtr<const Product> p(s);
  if(p->factors.front() == -1) return -Sin(-s);
 }
 if(s.type() == typeid(Numeric) &&
    Number<void>(s).numerictype() == typeid(double))
  return Number<double>(sin(CastPtr<const Number<double> >(s)->n));
 return *this;
}

Symbolic Sin::df(const Symbolic &s) const
{ return cos(parameters.front()) * parameters.front().df(s); }

Symbolic Sin::integrate(const Symbolic &s) const
{
 const Symbolic &x = parameters.front();
 if(x == s) return -cos(x);
 if(df(s) == 0) return *this * s;
 return Integral(*this,s);
}

//////////////////////////////////////
// Implementation of Cos            //
//////////////////////////////////////

Cos::Cos(const Cos &s) : Symbol(s) {}

Cos::Cos(const Symbolic &s) : Symbol(Symbol("cos")[s]) {}

Simplified Cos::simplify() const
{
 const Symbolic &s = parameters.front().simplify();
 if(s == 0) return Number<int>(1);
 if(s.type() == typeid(Product))
 {
  CastPtr<const Product> p(s);
  if(p->factors.front() == -1) return Cos(-s);
 }
 if(s.type() == typeid(Numeric) &&
    Number<void>(s).numerictype() == typeid(double))
  return Number<double>(cos(CastPtr<const Number<double> >(s)->n));
 return *this;
}

Symbolic Cos::df(const Symbolic &s) const
{ return -sin(parameters.front()) * parameters.front().df(s); }

Symbolic Cos::integrate(const Symbolic &s) const
{
```

```
 const Symbolic &x = parameters.front();
 if(x == s) return sin(x);
 if(df(s) == 0) return *this * s;
 return Integral(*this,s);
}


/////////////////////////////////////
// Implementation of Sinh            //
/////////////////////////////////////

Sinh::Sinh(const Sinh &s) : Symbol(s) {}

Sinh::Sinh(const Symbolic &s) : Symbol(Symbol("sinh")[s]) {}

Simplified Sinh::simplify() const
{
 const Symbolic &s = parameters.front().simplify();
 if(s == 0) return Number<int>(0);
 if(s.type() == typeid(Product))
 {
  CastPtr<const Product> p(s);
  if(p->factors.front() == -1) return -Sinh(-s);
 }
 if(s.type() == typeid(Numeric) &&
    Number<void>(s).numerictype() == typeid(double))
  return Number<double>(sinh(CastPtr<const Number<double> >(s)->n));
 return *this;
}

Symbolic Sinh::df(const Symbolic &s) const
{ return cosh(parameters.front()) * parameters.front().df(s); }

Symbolic Sinh::integrate(const Symbolic &s) const
{
 const Symbolic &x = parameters.front();
 if(x == s) return cosh(x);
 if(df(s) == 0) return *this * s;
 return Integral(*this,s);
}


/////////////////////////////////////
// Implementation of Cosh            //
/////////////////////////////////////

Cosh::Cosh(const Cosh &s) : Symbol(s) {}

Cosh::Cosh(const Symbolic &s) : Symbol(Symbol("cosh")[s]) {}

Simplified Cosh::simplify() const
{
 const Symbolic &s = parameters.front().simplify();
 if(s == 0) return Number<int>(1);
 if(s.type() == typeid(Product))
 {
  CastPtr<const Product> p(s);
  if(p->factors.front() == -1) return Cosh(-s);
 }
 if(s.type() == typeid(Numeric) &&
```

```
    Number<void>(s).numerictype() == typeid(double))
  return Number<double>(cosh(CastPtr<const Number<double> >(s)->n));
 return *this;
}

Symbolic Cosh::df(const Symbolic &s) const
{ return sinh(parameters.front()) * parameters.front().df(s); }

Symbolic Cosh::integrate(const Symbolic &s) const
{
 const Symbolic &x = parameters.front();
 if(x == s) return sinh(x);
 if(df(s) == 0) return *this * s;
 return Integral(*this,s);
}

////////////////////////////////////////
// Implementation of Log              //
////////////////////////////////////////

Log::Log(const Log &s) : Symbol(s) {}

Log::Log(const Symbolic &s1,const Symbolic &s2) : Symbol(Symbol("log")[s1,s2]) {}

void Log::print(ostream &o) const
{
 if(parameters.front() == SymbolicConstant::e)
 {
  Log l = *this;
  l.name = "ln";
  l.parameters.pop_front(),
  l.print(o);
 }
 else
  Symbol::print(o);
}

Simplified Log::simplify() const
{
 // log_a(b)
 const Symbolic &a = parameters.front().simplify();
 const Symbolic &b = parameters.back().simplify();
 if(b == 1) return Number<int>(0);
 if(b == a) return Number<int>(1);
 if(b.type() == typeid(Power))
 {
  CastPtr<const Power> p = b;
  if(p->parameters.front() == a)
   return p->parameters.back();
 }
 if(b.type() == typeid(Numeric)                          &&
    Number<void>(b).numerictype() == typeid(double) &&
    CastPtr<const Number<double> >(b)->n > 0.0)
  return Product(Number<double>(log(CastPtr<const Number<double> >(b)->n)),
                 Power(ln(a),-1)).simplify();
 return *this;
}
```

```
// d/ds log_a(b) = d/ds (ln(a) / ln(b))
//                = (1/b db/ds - log_a(b) / a da/ds) / ln(a)
//                = (a db/ds - b log_a(b) da/ds) / (a b ln(a))
Symbolic Log::df(const Symbolic &s) const
{
 const Symbolic &a = parameters.front();
 const Symbolic &b = parameters.back();
 return (a * b.df(s) - b * *this * a.df(s)) / (a * b * ln(a));
}

Symbolic Log::integrate(const Symbolic &s) const
{
 const Symbolic &x = parameters.back();
 const Symbolic &a = parameters.front();
 // int(log_a(x)) = (x ln(x) - x) / ln(a)
 if(x == s && a.df(s) == 0)
  return (x * *this - x) / ln(a);
 if(df(s) == 0) return *this * s;
 return Integral(*this,s);
}

/////////////////////////////////////
// Implementation of Power          //
/////////////////////////////////////

Power::Power(const Power &s) : Symbol(s) {}

Power::Power(const Symbolic &s,const Symbolic &p) : Symbol("pow")
{ parameters.push_back(s); parameters.push_back(p); }

void Power::print(ostream &o) const
{
  if(*this == SymbolicConstant::i)
  { o << SymbolicConstant::i_symbol; return; }

  int parens1 = parameters.front().type() == typeid(Symbol)
             || parameters.front().type() == typeid(Sin)
             || parameters.front().type() == typeid(Cos)
             || parameters.front().type() == typeid(Sinh)
             || parameters.front().type() == typeid(Cosh)
             || parameters.front().type() == typeid(Log)
             || parameters.front().type() == typeid(Derivative);
  int parens2 = parameters.back().type() == typeid(Symbol)
             || parameters.back().type() == typeid(Sin)
             || parameters.back().type() == typeid(Cos)
             || parameters.back().type() == typeid(Sinh)
             || parameters.back().type() == typeid(Cosh)
             || parameters.back().type() == typeid(Log)
             || parameters.back().type() == typeid(Derivative);
  parens1 = !parens1;
  parens2 = !parens2;
  if(parens1) o << "(";
  parameters.front().print(o);
  if(parens1) o << ")";
  o << "^";
  if(parens2) o << "(";
  parameters.back().print(o);
  if(parens2) o << ")";
```

```
}
Simplified Power::simplify() const
{
 list<Symbolic>::iterator i, j;
 const Symbolic &b = parameters.front().simplify();
 const Symbolic &n = parameters.back().simplify();
 if(n == 0) return Number<int>(1);
 if(n == 1) return b;
 if(b == 0) return Number<int>(0);
 if(b == 1) return Number<int>(1);
 if(b.type() == typeid(Power))
 {
  CastPtr<const Power> p = b;
  return (p->parameters.front() ^ (p->parameters.back() * n)).simplify();
 }
 if(b.type() == typeid(Numeric) &&
    n.type() == typeid(Numeric) &&
    Number<void>(n).numerictype() == typeid(int))
 {
  int i = CastPtr<const Number<int> >(n)->n;
  int inv = (i<0) ? 1 : 0;
  Number<void> r = Number<int>(1), x = b;
  i = (inv) ? -i : i;
  while(i != 0)
  {
   if(i & 1)  r = r * x;
   x = x * x;
   i >>= 1;
  }
  if(inv) return (Number<int>(1) / r);
  return r;
 }
 if(n.type() == typeid(Log))
 {
  CastPtr<const Log> l(n);
  if(l->parameters.front() == b)
   return l->parameters.back();
 }
 if(b.type() == typeid(Numeric) &&
    n.type() == typeid(Numeric) &&
    Number<void>(b).numerictype() == typeid(double))
 {
  double nd, bd = CastPtr<const Number<double> >(b)->n;
  if(Number<void>(n).numerictype() == typeid(int))
   nd = CastPtr<const Number<int> >(n)->n;
  else if(Number<void>(n).numerictype() == typeid(double))
   nd = CastPtr<const Number<double> >(n)->n;
  else if(Number<void>(n).numerictype() == typeid(Rational<Number<void> >))
   nd = double(CastPtr<const Number<Rational<Number<void> > > >(n)->n);
  else return Power(b,n);
  if(bd >= 0.0 || int(nd) == nd)
   return Number<double>(pow(bd,nd));
 }
 if(b.type() == typeid(Numeric) &&
    n.type() == typeid(Numeric) &&
    Number<void>(n).numerictype() == typeid(double))
 {
```

```
   double bd, nd = CastPtr<const Number<double> >(n)->n;
   if(Number<void>(b).numerictype() == typeid(int))
    bd = CastPtr<const Number<int> >(b)->n;
   else if(Number<void>(b).numerictype() == typeid(double))
    bd = CastPtr<const Number<double> >(b)->n;
   else if(Number<void>(b).numerictype() == typeid(Rational<Number<void> >))
    bd = double(CastPtr<const Number<Rational<Number<void> > > >(b)->n);
   else return Power(b,n);
   if(bd >= 0.0 || int(nd) == nd)
    return Number<double>(pow(bd,nd));
 }
 return Power(b,n);
}

Expanded Power::expand() const
{
 Symbolic b = parameters.front().expand();
 Symbolic n = parameters.back().expand();
 if(b.type() != typeid(Sum)      &&
    b.type() != typeid(Product) &&
    b.type() != typeid(Numeric))
  return Power(b,n);
 if(n.type() != typeid(Numeric) ||
    Number<void>(n).numerictype() != typeid(int))
  return Power(b,n);

 int ae = Symbolic::auto_expand;
 int i = CastPtr<const Number<int> >(n)->n;
 if(i<0) return *this;
 int sgn = (i>=0) ? 1 : -1;
 Symbolic r = 1, x = b;
 i *= sgn;

 Symbolic::auto_expand = 0;
 while(i != 0)
 {
  if(i & 1) r = Product(r,x).expand();
  i >>= 1;
  if(i != 0) x = Product(x,x).expand();
 }
 Symbolic::auto_expand = ae;

 return Power(r,Number<int>(sgn));
}

Symbolic Power::subst(const Symbolic &x,const Symbolic &y,int &n) const
{
 if(x.type() == typeid(Power))
 {
  CastPtr<const Power> p(x);
  if(parameters.front() == p->parameters.front()                    &&
     parameters.back().type() == typeid(Numeric)                    &&
     p->parameters.back().type() == typeid(Numeric)                 &&
     Number<void>(parameters.back()).numerictype() == typeid(int) &&
     Number<void>(p->parameters.back()).numerictype() == typeid(int))
  {
   int s = CastPtr<const Number<int> >(parameters.back())->n;
   int t = CastPtr<const Number<int> >(p->parameters.back())->n;
```

```
     if((s > 0) && (t > 0) && (s >= t))
     {
       n++;
       return Power(parameters.front(),s - t).subst(x,y,n) * y;
     }
   }
 }
 return Symbol::subst(x,y,n);
}


// d/ds (a^b) = d/ds exp(b ln(a))
//            = (ln(a) db/ds + b/a da/ds) a^b
Symbolic Power::df(const Symbolic &s) const
{
 const Symbolic &a = parameters.front();
 const Symbolic &b = parameters.back();
 return (ln(a)*b.df(s) + (b/a)*a.df(s)) * *this;
}


Symbolic Power::integrate(const Symbolic &s) const
{
 const Symbolic &a = parameters.front();
 const Symbolic &b = parameters.back();
 if(a == s && b.df(s) == 0)
 {
   if(b == -1) return ln(a);
   return (a^(b+1)) / (b+1);
 }
 if(a == SymbolicConstant::e && b == s)
   return *this;
 if(df(s) == 0) return *this * s;
 return Integral(*this,s);
}


///////////////////////////////////////
// Implementation of Derivative      //
///////////////////////////////////////

Derivative::Derivative(const Derivative &d) : Symbol(d) { }

Derivative::Derivative(const Symbolic &s1,const Symbolic &s2)
: Symbol("df")
{
 if(s1.type() == typeid(Derivative))
 {
   parameters = CastPtr<const Derivative>(s1)->parameters;
   parameters.push_back(s2);
 }
 else
 {
   parameters.push_back(s1);
   parameters.push_back(s2);
 }
}


Symbolic Derivative::subst(const Symbolic &x,const Symbolic &y,int &n) const
{
 if(*this == x) return y;
```

```cpp
list<Symbolic>::const_iterator i;
list<Symbolic>::iterator j;

if(x.type() == type() &&
   parameters.front() == CastPtr<const Derivative>(x)->parameters.front())
{
 // make a copy of this
 CastPtr<Derivative> l(*this,1);
 CastPtr<const Derivative> d(x);
 i = d->parameters.begin();
 for(i++;i!=d->parameters.end();i++)
 {
  j = l->parameters.begin();
  for(j++;j!=l->parameters.end();j++)
   if(*j == *i) break;
  if(j == l->parameters.end()) break;
  l->parameters.erase(j);
 }
 if(i == d->parameters.end())
 {
  n++;
  Symbolic newdf = y;
  j = l->parameters.begin();
  for(j++;j!=l->parameters.end();j++)
   newdf = ::df(newdf,*j);
  return newdf;
 }
}

i = parameters.begin();
Symbolic dy = i->subst(x,y,n);
for(i++;i!=parameters.end();i++)
 dy = dy.df(i->subst(x,y,n));
return dy;
}

Symbolic Derivative::df(const Symbolic &s) const
{
 list<Symbolic>::const_iterator i;

 if(parameters.front().type() == typeid(Symbol))
 {
  Symbolic result;
  CastPtr<const Symbol> sym(parameters.front());
  for(i=sym->parameters.begin();i!=sym->parameters.end();i++)
   result = result + Derivative(*this,*i) * i->df(s);

  return result;
 }

 if(parameters.front().df(s) != 0)
 {
  Derivative d(*this);
  d.parameters.push_back(s);
  return d;
 }
```

```
  return 0;
}

int Derivative::compare(const Symbolic &s) const
{
 list<Symbolic>::const_iterator i;
 list<Symbolic>::iterator j;

 if(s.type() != type()) return 0;
 // make a copy of s
 CastPtr<Derivative> d(s,1);
 if(d->parameters.size() != parameters.size()) return 0;
 if(d->parameters.front() != parameters.front()) return 0;
 for(i=parameters.begin(),i++;i!=parameters.end();i++)
 {
  for(j=d->parameters.begin(),j++;j!=d->parameters.end();j++)
   if(*i == *j) break;
  if(j == d->parameters.end()) return 0;
  d->parameters.erase(j);
 }
 return 1;
}

Symbolic Derivative::integrate(const Symbolic &s) const
{
 int n = 0, n1;
 list<Symbolic>::const_iterator i, i1 = parameters.cnd();
 list<Symbolic>::iterator j;

 for(i=parameters.begin();i!=parameters.end();i++,n++)
  if(*i == s) { i1 = i, n1 = n; }

 if(i1 != parameters.end())
 {
  // make a copy of *this
  CastPtr<Derivative> d(*this,1);
  for(j=d->parameters.begin();n1!=0;j++,n1--);
  d->parameters.erase(j);
  if(d->parameters.size() == 1) return d->parameters.front();
  return *d;
 }

 if(parameters.front().df(s) == 0) return *this * s;
 return Integral(*this,s);
}

/////////////////////////////////////
// Implementation of Integral       //
/////////////////////////////////////

Integral::Integral(const Integral &d) : Symbol(d) { }

Integral::Integral(const Symbolic &s1,const Symbolic &s2) : Symbol("int")
{
 if(s1.type() == typeid(Integral))
 {
  parameters = CastPtr<const Integral>(s1)->parameters;
  parameters.push_back(s2);
```

```
 }
 else
 {
  parameters.push_back(s1);
  parameters.push_back(s2);
 }
}


Symbolic Integral::subst(const Symbolic &x,
                         const Symbolic &y,int &n) const
{
 if(*this == x) { n++; return y; }

 list<Symbolic>::const_iterator i = parameters.begin();
 Symbolic dy = i->subst(x,y,n);
 for(i++;i!=parameters.end();i++)
  dy = dy.integrate(i->subst(x,y,n));
 return dy;
}


Symbolic Integral::df(const Symbolic &s) const
{
 int n = 0, n1;
 list<Symbolic>::const_iterator i, i1 = parameters.end();
 list<Symbolic>::iterator j;

 for(i=parameters.begin();i!=parameters.end();i++,n++)
  if(*i == s) { i1 = i; n1 = n; }

 if(i1 != parameters.end())
 {
  // make a copy of *this
  CastPtr<Integral> in(*this,1);
  for(j=in->parameters.begin();n1!=0;j++,n1--);
  in->parameters.erase(j);
  if(in->parameters.size() == 1) return in->parameters.front();
  return *in;
 }

 if(parameters.front().df(s) == 0) return 0;
 return Derivative(*this,s);
}


Symbolic Integral::integrate(const Symbolic &s) const
{
 if(parameters.front().df(s) != 0)
 {
  Integral i(*this);
  i.parameters.push_back(s);
  return i;
 }

 return *this * s;
}

#endif
#endif
#endif
```

13.11.6 Numbers

```
// number.h

#ifndef SYMBOLIC_CPLUSPLUS_NUMBER

#include <cmath>
#include <iostream>
#include <limits>
#include <typeinfo>
#include <utility>
#include "rational.h"
#include "verylong.h"
using namespace std;

#ifdef  SYMBOLIC_FORWARD
#ifndef SYMBOLIC_CPLUSPLUS_NUMBER_FORWARD
#define SYMBOLIC_CPLUSPLUS_NUMBER_FORWARD

class Numeric;
template <class T> class Number;

#endif
#endif

#ifdef  SYMBOLIC_DECLARE
#ifndef SYMBOLIC_CPLUSPLUS_NUMBER_DECLARE
#define SYMBOLIC_CPLUSPLUS_NUMBER_DECLARE

class Numeric: public CloningSymbolicInterface
{
 public: Numeric();
         Numeric(const Numeric&);
         virtual const type_info &numerictype() const = 0;
         virtual Number<void> add(const Numeric&) const = 0;
         virtual Number<void> mul(const Numeric&) const = 0;
         virtual Number<void> div(const Numeric&) const = 0;
         virtual Number<void> mod(const Numeric&) const = 0;
         virtual int isZero() const = 0;
         virtual int isOne() const = 0;
         virtual int isNegative() const = 0;
         virtual int cmp(const Numeric &) const = 0;
         pair<Number<void>,Number<void> >
             match(const Numeric&,const Numeric&) const;
         Symbolic subst(const Symbolic&,const Symbolic&,int &n) const;
         int compare(const Symbolic&) const;
         Symbolic df(const Symbolic&) const;
         Symbolic integrate(const Symbolic&) const;
         Symbolic coeff(const Symbolic&) const;
         Expanded expand() const;
         int commute(const Symbolic&) const;
};

template <class T>
class Number: public Numeric
{
 public: T n;
         Number();
```

```
        Number(const Number&);
        Number(const T&);
        ~Number();

        Number &operator=(const Number&);
        Number &operator=(const T&);

        void print(ostream&) const;
        const type_info &type() const;
        const type_info &numerictype() const;
        Simplified simplify() const;

        Number<void> add(const Numeric&) const;
        Number<void> mul(const Numeric&) const;
        Number<void> div(const Numeric&) const;
        Number<void> mod(const Numeric&) const;
        int isZero() const;
        int isOne() const;
        int isNegative() const;
        int cmp(const Numeric &) const;

        Cloning *clone() const { return Cloning::clone(*this); }
};

template<>
class Number<void>: public CastPtr<Numeric>
{
 public: Number();
        Number(const Number&);
        Number(const Numeric&);
        Number(const Symbolic&);
        ~Number();

        const type_info &numerictype() const;
        int isZero() const;
        int isOne() const;
        int isNegative() const;
        pair<Number<void>,Number<void> >
            match(const Numeric&,const Numeric&) const;
        pair<Number<void>,Number<void> >
            match(const Number<void>&,const Number<void>&) const;

        Number<void> operator+(const Numeric&) const;
        Number<void> operator-(const Numeric&) const;
        Number<void> operator*(const Numeric&) const;
        Number<void> operator/(const Numeric&) const;
        Number<void> operator%(const Numeric&) const;
        Number<void> &operator+=(const Numeric&);
        Number<void> &operator*=(const Numeric&);
        Number<void> &operator/=(const Numeric&);
        Number<void> &operator%=(const Numeric&);
        int operator==(const Numeric&) const;
        int operator<(const Numeric&) const;
        int operator>(const Numeric&) const;
        int operator<=(const Numeric&) const;
        int operator>=(const Numeric&) const;

        Number<void> operator+(const Number<void>&) const;
```

```
        Number<void> operator-(const Number<void>&) const;
        Number<void> operator*(const Number<void>&) const;
        Number<void> operator/(const Number<void>&) const;
        Number<void> operator%(const Number<void>&) const;
        Number<void> &operator+=(const Number<void>&);
        Number<void> &operator*=(const Number<void>&);
        Number<void> &operator/=(const Number<void>&);
        Number<void> &operator%=(const Number<void>&);
        int operator==(const Number<void>&) const;
        int operator<(const Number<void>&) const;
        int operator>(const Number<void>&) const;
        int operator<=(const Number<void>&) const;
        int operator>=(const Number<void>&) const;

        Numeric &operator*() const;
};

Number<void> operator+(const Numeric&,const Number<void>&);
Number<void> operator-(const Numeric&,const Number<void>&);
Number<void> operator*(const Numeric&,const Number<void>&);
Number<void> operator/(const Numeric&,const Number<void>&);
Number<void> operator%(const Numeric&,const Number<void>&);

#endif
#endif

#ifdef  SYMBOLIC_DEFINE
#ifndef SYMBOLIC_CPLUSPLUS_NUMBER_DEFINE
#define SYMBOLIC_CPLUSPLUS_NUMBER_DEFINE
#define SYMBOLIC_CPLUSPLUS_NUMBER

Numeric::Numeric() : CloningSymbolicInterface() {}

Numeric::Numeric(const Numeric &n) : CloningSymbolicInterface(n) {}

// Template specialization for Rational<Number<void> >
template <> Rational<Number<void> >::operator double() const
{
 pair<Number<void>,Number<void> > pr = p.match(p,q);
 if(pr.first.numerictype() == typeid(int))
 {
  CastPtr<const Number<int> > i1 = pr.first;
  CastPtr<const Number<int> > i2 = pr.second;
  return double(Rational<int>(i1->n,i2->n));
 }
 if(pr.first.numerictype() == typeid(Verylong))
 {
  CastPtr<const Number<Verylong> > v1 = pr.first;
  CastPtr<const Number<Verylong> > v2 = pr.second;
  return double(Rational<Verylong>(v1->n,v2->n));
 }
 cerr << "convert to double : " << pr.first.numerictype().name() << endl;
 throw SymbolicError(SymbolicError::NotDouble);
 return 0.0;
}

//////////////////////////////////////
// Implementation of Numeric        //
```

////////////////////////////////////

```
pair<Number<void>,Number<void> >
Numeric::match(const Numeric &n1,const Numeric &n2) const
{
 const type_info &t1 = n1.numerictype();
 const type_info &t2 = n2.numerictype();
 const type_info &type_int = typeid(int);
 const type_info &type_double = typeid(double);
 const type_info &type_verylong = typeid(Verylong);
 const type_info &type_rational = typeid(Rational<Number<void> >);

 if(t1 == type_int)
 {
  CastPtr<const Number<int> > i1 = n1;

  if(t2 == type_int)
   return pair<Number<void>,Number<void> >(n1,n2);
  if(t2 == type_double)
   return pair<Number<void>,Number<void> >(Number<double>(i1->n),n2);
  if(t2 == type_verylong)
   return pair<Number<void>,Number<void> >(Number<Verylong>(i1->n),n2);
  if(t2 == type_rational)
   return pair<Number<void>,Number<void> >
          (Number<Rational<Number<void> > >
              (Rational<Number<void> >(Number<void>(*i1)))
          ,n2);
  cerr << "Numeric cannot use " << t2.name() << endl;
  throw SymbolicError(SymbolicError::UnsupportedNumeric);
 }
 if(t1 == type_double)
 {
  CastPtr<const Number<double> > d1 = n1;

  if(t2 == type_int)
  {
   CastPtr<const Number<int> > i2 = n2;
   return pair<Number<void>,Number<void> >(n1,Number<double>(i2->n));
  }
  if(t2 == type_double)
   return pair<Number<void>,Number<void> >(n1,n2);
  if(t2 == type_verylong)
  {
   CastPtr<const Number<Verylong> > v2 = n2;
   return pair<Number<void>,Number<void> >(n1,Number<double>(v2->n));
  }
  if(t2 == type_rational)
  {
   CastPtr<const Number<Rational<Number<void> > > > r2 = n2;
   return pair<Number<void>,Number<void> >
          (n1,Number<double>(double(r2->n)));
  }
  cerr << "Numeric cannot use " << t2.name() << endl;
  throw SymbolicError(SymbolicError::UnsupportedNumeric);
 }
 if(t1 == type_verylong)
 {
  CastPtr<const Number<Verylong> > v1 = n1;
```

```
   if(t2 == type_int)
   {
    CastPtr<const Number<int> > i2 = n2;
    return pair<Number<void>,Number<void> >(n1,Number<Verylong>(i2->n));
   }
   if(t2 == type_double)
    return pair<Number<void>,Number<void> >(Number<double>(v1->n),n2);
   if(t2 == type_verylong)
    return pair<Number<void>,Number<void> >(n1,n2);
   if(t2 == type_rational)
    return pair<Number<void>,Number<void> >
          (Number<Rational<Number<void> > >
             (Rational<Number<void> >(Number<void>(*v1))),n2);
   cerr << "Numeric cannot use " << t2.name() << endl;
   throw SymbolicError(SymbolicError::UnsupportedNumeric);
 }
 if(t1 == type_rational)
 {
   CastPtr<const Number<Rational<Number<void> > > > r1 = n1;

   if(t2 == type_int)
   {
    CastPtr<const Number<int> > i2 = n2;
    return pair<Number<void>,Number<void> >
          (n1,
           Number<Rational<Number<void> > >
              (Rational<Number<void> >(Number<void>(*i2))));
   }
   if(t2 == type_double)
    return pair<Number<void>,Number<void> >
          (Number<double>(double(r1->n)),n2);
   if(t2 == type_verylong)
   {
    CastPtr<const Number<Verylong> > v2 = n2;
    return pair<Number<void>,Number<void> >
          (n1,
           Number<Rational<Number<void> > >
              (Rational<Number<void> >(Number<void>(*v2))));
   }
   if(t2 == type_rational)
    return pair<Number<void>,Number<void> >(n1,n2);
   cerr << "Numeric cannot use " << t2.name() << endl;
   throw SymbolicError(SymbolicError::UnsupportedNumeric);
 }
 cerr << "Numeric cannot use " << t1.name() << endl;
 throw SymbolicError(SymbolicError::UnsupportedNumeric);
 return pair<Number<void>,Number<void> >(n1,n2);
}

Symbolic Numeric::subst(const Symbolic &x,const Symbolic &y,int &n) const
{
 if(*this == x) { n++; return y; }
 return *this;
}

int Numeric::compare(const Symbolic &s) const
{
```

```
 if(s.type() != type()) return 0;
 pair<Number<void>,Number<void> > p = match(*this,*Number<void>(s));
 return p.first->cmp(*(p.second));
}

Symbolic Numeric::df(const Symbolic &s) const
{ return Number<int>(0); }

Symbolic Numeric::integrate(const Symbolic &s) const
{ return Symbolic(*this) * s; }

Symbolic Numeric::coeff(const Symbolic &s) const
{
 if(s.type() == typeid(Numeric)) return *this / Number<void>(s);
 return Number<int>(0);
}

Expanded Numeric::expand() const
{ return *this; }

int Numeric::commute(const Symbolic &s) const
{ return 1; }

/////////////////////////////////////
// Implementation of Number         //
/////////////////////////////////////

template <class T> Number<T>::Number() : n()
{ simplified = expanded = 1; }

template <class T> Number<T>::Number(const Number &n)
 : Numeric(n), n(n.n) {}

template <class T> Number<T>::Number(const T &t) : n(t)
{ simplified = 0; expanded = 1; }

template <class T> Number<T>::~Number() {}

template <class T> Number<T> &Number<T>::operator=(const Number &n)
{
 if(this != &n) n = n.n;
 return *this;
}

template <class T> Number<T> &Number<T>::operator=(const T &t)
{ n = t; return *this; }

template <class T> void Number<T>::print(ostream &o) const
{ o << n; }

template <class T> const type_info &Number<T>::type() const
{ return typeid(Numeric); }

template <class T> const type_info &Number<T>::numerictype() const
{ return typeid(T); }

template <class T>
Simplified Number<T>::simplify() const
```

```
{ return *this; }

template <>
Simplified Number<Verylong>::simplify() const
{
 if(n <= Verylong(numeric_limits<int>::max())
    && n > Verylong(numeric_limits<int>::min()))
  return Number<int>(n);
 return *this;
}

template <>
Simplified Number<Rational<Number<void> > >::simplify() const
{
 if(n.den().isOne()) return *(n.num());
 return *this;
}

template <class T> Number<void> Number<T>::add(const Numeric &x) const
{
 if(numerictype() != x.numerictype())
   throw SymbolicError(SymbolicError::IncompatibleNumeric);
 CastPtr<const Number<T> > p = x;
 return Number<T>(n + p->n);
}

template <class T> Number<void> Number<T>::mul(const Numeric &x) const
{
 if(numerictype() != x.numerictype())
   throw SymbolicError(SymbolicError::IncompatibleNumeric);
 CastPtr<const Number<T> > p = x;
 return Number<T>(n * p->n);
}

template <class T> Number<void> Number<T>::div(const Numeric &x) const
{
 if(numerictype() != x.numerictype())
   throw SymbolicError(SymbolicError::IncompatibleNumeric);
 CastPtr<const Number<T> > p = x;
 return Number<T>(n / p->n);
}

template <class T> Number<void> Number<T>::mod(const Numeric &x) const
{
 if(numerictype() != x.numerictype())
   throw SymbolicError(SymbolicError::IncompatibleNumeric);
 CastPtr<const Number<T> > p = x;
 return Number<T>(n - p->n * (n / p->n));
}

template <> Number<void> Number<int>::add(const Numeric &x) const
{
 if(numerictype() != x.numerictype())
   throw SymbolicError(SymbolicError::IncompatibleNumeric);
 CastPtr<const Number<int> > p = x;
 int sum = n + p->n;
 if((n < 0 && p->n < 0 && sum >= 0) ||
    (n > 0 && p->n > 0 && sum <= 0))
```

```
  return Number<Verylong>(Verylong(n) + Verylong(p->n));
  return Number<int>(n + p->n);
}

template <> Number<void> Number<int>::mul(const Numeric &x) const
{
 if(numerictype() != x.numerictype())
   throw SymbolicError(SymbolicError::IncompatibleNumeric);
 CastPtr<const Number<int> > p = x;
 int product = n * p->n;
 if(n != 0 && product / n != p->n)
  return Number<Verylong>(Verylong(n) * Verylong(p->n));
 return Number<int>(product);
}

template <> Number<void> Number<int>::div(const Numeric &x) const
{
 if(numerictype() != x.numerictype())
   throw SymbolicError(SymbolicError::IncompatibleNumeric);
 CastPtr<const Number<int> > p = x;
 if(n % p->n != 0)
  return Number<Rational<Number<void> > >
          (Rational<Number<void> >(Number<void>(*this),Number<void>(x)));
 return Number<int>(n / p->n);
}

template <> Number<void> Number<int>::mod(const Numeric &x) const
{
 if(numerictype() != x.numerictype())
   throw SymbolicError(SymbolicError::IncompatibleNumeric);
 return Number<int>(n % CastPtr<const Number<int> >(x)->n);
}

template <> Number<void> Number<double>::mod(const Numeric &x) const
{
 if(numerictype() != x.numerictype())
   throw SymbolicError(SymbolicError::IncompatibleNumeric);
 return Number<double>(fmod(n,CastPtr<const Number<double> >(x)->n));
}

template <> Number<void> Number<Verylong>::div(const Numeric &x) const
{
 if(numerictype() != x.numerictype())
   throw SymbolicError(SymbolicError::IncompatibleNumeric);
 CastPtr<const Number<Verylong> > p = x;
 if(n % p->n != Verylong(0))
  return Number<Rational<Number<void> > >
          (Rational<Number<void> >(Number<void>(*this),Number<void>(x)));
 return Number<Verylong>(n / p->n);
}

template <class T> int Number<T>::isZero() const
{ return (n == zero(T())); }

template <class T> int Number<T>::isOne() const
{ return (n == one(T())); }

template <class T> int Number<T>::isNegative() const
```

```
{ return (n < zero(T())); }

template <class T> int Number<T>::cmp(const Numeric &x) const
{
 if(numerictype() != x.numerictype())
   throw SymbolicError(SymbolicError::IncompatibleNumeric);
 return (numerictype() == x.numerictype()) &&
        (n == CastPtr<const Number<T> >(x)->n);
}

////////////////////////////////////
// Implementation of Number<void> //
////////////////////////////////////

Number<void>::Number() : CastPtr<Numeric>(Number<int>(0)) {}

Number<void>::Number(const Number &n) : CastPtr<Numeric>(n) {}

Number<void>::Number(const Numeric &n) : CastPtr<Numeric>(n) {}

Number<void>::Number(const Symbolic &n) : CastPtr<Numeric>(Number<int>(0))
{
 if(n.type() != typeid(Numeric))
   throw SymbolicError(SymbolicError::NotNumeric);
 CastPtr<Numeric>::operator=(n);
}

Number<void>::~Number() {}

const type_info &Number<void>::numerictype() const
{ return (*this)->numerictype(); }

int Number<void>::isZero() const
{ return (*this)->isZero(); }

int Number<void>::isOne() const
{ return (*this)->isOne(); }

int Number<void>::isNegative() const
{ return (*this)->isNegative(); }

pair<Number<void>,Number<void> >
Number<void>::match(const Numeric &n1,const Numeric &n2) const
{ return (*this)->match(n1,n2); }

pair<Number<void>,Number<void> >
Number<void>::match(const Number<void> &n1,const Number<void> &n2) const
{ return (*this)->match(*n1,*n2); }

Number<void> Number<void>::operator+(const Numeric &n) const
{
 pair<Number<void>,Number<void> > p = match(*this,n);
 return p.first->add(*(p.second));
}

Number<void> Number<void>::operator-(const Numeric &n) const
{
 pair<Number<void>,Number<void> > p = match(*this,n);
```

```
    return p.first->add(*(p.second));
}

Number<void> Number<void>::operator*(const Numeric &n) const
{
 pair<Number<void>,Number<void> > p = match(*this,n);
 return p.first->mul(*(p.second));
}

Number<void> Number<void>::operator/(const Numeric &n) const
{
 pair<Number<void>,Number<void> > p = match(*this,n);
 return p.first->div(*(p.second));
}

Number<void> Number<void>::operator%(const Numeric &n) const
{
 pair<Number<void>,Number<void> > p = match(*this,n);
 return p.first->mod(*(p.second));
}

Number<void> &Number<void>::operator+=(const Numeric &n)
{ return *this = *this + n; }

Number<void> &Number<void>::operator*=(const Numeric &n)
{ return *this = *this * n; }

Number<void> &Number<void>::operator/=(const Numeric &n)
{ return *this = *this / n; }

Number<void> &Number<void>::operator%=(const Numeric &n)
{ return *this = *this % n; }

int Number<void>::operator==(const Numeric &n) const
{
 pair<Number<void>,Number<void> > p = match(*(*this),n);
 return p.first->compare(*(p.second));
}

int Number<void>::operator<(const Numeric &n) const
{
 pair<Number<void>,Number<void> > p = match(*(*this),n);
 return (p.first - p.second).isNegative();
}

int Number<void>::operator>(const Numeric &n) const
{ return !(*this < n) && !(*this == n); }

int Number<void>::operator<=(const Numeric &n) const
{ return (*this < n) || (*this == n); }

int Number<void>::operator>=(const Numeric &n) const
{ return !(*this < n); }

Number<void> Number<void>::operator+(const Number<void> &n) const
{ return operator+(*n); }

Number<void> Number<void>::operator-(const Number<void> &n) const
```

```
{ return operator-(*n); }

Number<void> Number<void>::operator*(const Number<void> &n) const
{ return operator*(*n); }

Number<void> Number<void>::operator/(const Number<void> &n) const
{ return operator/(*n); }

Number<void> Number<void>::operator%(const Number<void> &n) const
{ return operator%(*n); }

Number<void> &Number<void>::operator+=(const Number<void> &n)
{ return *this = *this + n; }

Number<void> &Number<void>::operator*=(const Number<void> &n)
{ return *this = *this * n; }

Number<void> &Number<void>::operator/=(const Number<void> &n)
{ return *this = *this / n; }

Number<void> &Number<void>::operator%=(const Number<void> &n)
{ return *this = *this % n; }

int Number<void>::operator==(const Number<void> &n) const
{ return operator==(*n); }

int Number<void>::operator<(const Number<void> &n) const
{ return operator<(*n); }

int Number<void>::operator>(const Number<void> &n) const
{ return operator>(*n); }

int Number<void>::operator<=(const Number<void> &n) const
{ return operator<=(*n); }

int Number<void>::operator>=(const Number<void> &n) const
{ return operator>=(*n); }

Numeric &Number<void>::operator*() const
{ return CastPtr<Numeric>::operator*(); }

template <> Number<void> zero(Number<void>)
{ return Number<int>(0); }

template <> Number<void> one(Number<void>)
{ return Number<int>(1); }

Number<void> operator+(const Numeric &n1,const Number<void> &n2)
{ return Number<void>(n1) + n2; }

Number<void> operator-(const Numeric &n1,const Number<void> &n2)
{ return Number<void>(n1) - n2; }

Number<void> operator*(const Numeric &n1,const Number<void> &n2)
{ return Number<void>(n1) * n2; }

Number<void> operator/(const Numeric &n1,const Number<void> &n2)
{ return Number<void>(n1) / n2; }
```

```
Number<void> operator%(const Numeric &n1,const Number<void> &n2)
{ return Number<void>(n1) % n2; }

#endif
#endif
#endif
```

13.11.7   Products

```
// product.h

#ifndef SYMBOLIC_CPLUSPLUS_PRODUCT

#include <algorithm>
#include <list>
#include <vector>
using namespace std;

#ifdef  SYMBOLIC_FORWARD
#ifndef SYMBOLIC_CPLUSPLUS_PRODUCT_FORWARD
#define SYMBOLIC_CPLUSPLUS_PRODUCT_FORWARD

class Product;

#endif
#endif

#ifdef  SYMBOLIC_DECLARE
#ifndef SYMBOLIC_CPLUSPLUS_PRODUCT_DECLARE
#define SYMBOLIC_CPLUSPLUS_PRODUCT_DECLARE

class Product: public CloningSymbolicInterface
{
 public: list<Symbolic> factors;
         Product();
         Product(const Product&);
         Product(const Symbolic&,const Symbolic&);
         ~Product();

         Product &operator=(const Product&);
         int printsNegative() const;

         void print(ostream&) const;
         Symbolic subst(const Symbolic&,const Symbolic&,int &n) const;
         Simplified simplify() const;
         int compare(const Symbolic&) const;
         Symbolic df(const Symbolic&) const;
         Symbolic integrate(const Symbolic&) const;
         Symbolic coeff(const Symbolic&) const;
         Expanded expand() const;
         int commute(const Symbolic&) const;

         Cloning *clone() const { return Cloning::clone(*this); }
};
```

```
#endif
#endif

#ifdef  SYMBOLIC_DEFINE
#ifndef SYMBOLIC_CPLUSPLUS_PRODUCT_DEFINE
#define SYMBOLIC_CPLUSPLUS_PRODUCT_DEFINE
#define SYMBOLIC_CPLUSPLUS_PRODUCT

Product::Product() {}

Product::Product(const Product &s)
 : CloningSymbolicInterface(s), factors(s.factors) {}

Product::Product(const Symbolic &s1,const Symbolic &s2)
{
 if(s1.type() == typeid(Product))
  factors = CastPtr<const Product>(s1)->factors;
 else factors.push_back(s1);
 if(s2.type() == typeid(Product))
 {
  CastPtr<const Product> p(s2);
  factors.insert(factors.end(),p->factors.begin(),p->factors.end());
 }
 else factors.push_back(s2);
}

Product::~Product() {}

Product &Product::operator=(const Product &p)
{
 if(this != &p) factors = p.factors;
 return *this;
}

int Product::printsNegative() const
{
 return factors.size() > 1                          &&
        factors.front().type() == typeid(Numeric) &&
        CastPtr<const Numeric>(factors.front())->isNegative();
}

void Product::print(ostream &o) const
{
 if(factors.empty()) o << 1;
 if(factors.size() == 1) factors.begin()->print(o);
 else
  for(list<Symbolic>::const_iterator i=factors.begin();i!=factors.end();i++)
  {
   o << ((i==factors.begin()) ? "":"*");
   if(*i == -1) { o << "-"; i++; }
   if(i->type() != typeid(Product)    &&
      i->type() != typeid(Power)      &&
      i->type() != typeid(Sin)        &&
      i->type() != typeid(Cos)        &&
      i->type() != typeid(Log)        &&
      i->type() != typeid(Derivative) &&
      i->type() != typeid(Symbol)     &&
      i->type() != typeid(Numeric))
```

```
        { o << "("; i->print(o); o << ")"; }
      else i->print(o);
    }
}


Symbolic Product::subst(const Symbolic &x,const Symbolic &y,int &n) const
{
 if(x.type() == type())
  {
   CastPtr<const Product> p(x);
   // vector<T>::iterator has ordering comparisons
   // while list<T>::iterator does not
   list<Symbolic> u;
   vector<Symbolic> v;
   list<Symbolic>::const_iterator i;
   list<Symbolic>::const_iterator i1;
   vector<Symbolic>::iterator j, insert;
   list< vector<Symbolic>::iterator >::iterator k;
   // we store lists of locations (iterators) in v in the list l,
   // each list in l should describe a path of unique locations through v
   list<list< vector<Symbolic>::iterator > > l;
   list<list< vector<Symbolic>::iterator > >::iterator li;

   // expand positive integer powers to match each factor individually
   for(i=p->factors.begin();i!=p->factors.end();i++)
    {
     if(i->type() == typeid(Power))
      {
       CastPtr<const Power> p(*i);
       if(p->parameters.back().type() == typeid(Numeric) &&
          Number<void>(p->parameters.back()).numerictype() == typeid(int))
        {
         int n = CastPtr<const Number<int> >(p->parameters.back())->n;
         if(n>0) for(int m=0;m<n;m++) u.push_back(p->parameters.front());
         else u.push_back(*i);
        }
       else u.push_back(*i);
      }
     else u.push_back(*i);
    }

   // expand positive integer powers to match each factor individually
   for(i1=factors.begin();i1!=factors.end();i1++)
    {
     if(i1->type() == typeid(Power))
      {
       CastPtr<const Power> p(*i1);
       if(p->parameters.back().type() == typeid(Numeric) &&
          Number<void>(p->parameters.back()).numerictype() == typeid(int))
        {
         int n - CastPtr<const Number<int> >(p->parameters.back())->n;
         if(n>0) for(int m=0;m<n;m++) v.push_back(p->parameters.front());
         else v.push_back(*i1);
        }
       else v.push_back(*i1);
      }
     else v.push_back(*i1);
    }
```

```
    i = u.begin();

    // initialize each path in l to begin with the location
    // of the first factor of x (copied in p)
    // there could be none, one or many such paths
    for(j=v.begin();j!=v.end();j++)
     if(*j == *i)
     {
      l.push_back(list< vector<Symbolic>::iterator >());
      l.back().push_back(j);
     }

    // build each path by considering all possible
    // locations of subsequent factors of x and creating
    // all possible paths
    for(i++;i!=u.end();i++)
    {
     list<list< vector<Symbolic>::iterator > > l2;
     for(j=v.begin();j!=v.end();j++)
      if(*j == *i)
       for(li=l.begin();li!=l.end();li++)
        if(find(li->begin(),li->end(),j) == li->end())
        {
         l2.push_back(*li);
         l2.back().push_back(j);
        }
     l = l2;
    }

    // erase paths that are too short
    for(li=l.begin();li!=l.end();)
     if(li->size() != u.size()) l.erase(li++);
     else li++;

    // search for a path which may be substituted
    for(li=l.begin();li!=l.end();li++)
    {
     for(k=li->begin();k!=li->end();k++)
     {
      list< vector<Symbolic>::iterator >::iterator k1 = k;
      // when consecutive values are in the wrong order
      // check that they commute
      for((k1=k)++;k1!=li->end();k1++)
       if(*k1 < *k && !(*k)->commute(**k1)) break;
      if(k1!=li->end()) break;
     }
     // values cannot commute to the right order
     if(k != li->end()) continue;

     // try to find a position that all values
     // can commute to, and set that as the place for substitution
     for(insert = v.begin();insert!=v.end();insert++)
     {
      for(k=li->begin();k!=li->end();k++)
      {
       vector<Symbolic>::iterator beg, end;
       if(insert <= *k) { beg = insert; end = *k; }
```

```
      else { beg = *k + 1; end = insert; }
     for(j=beg;j!=end;j++)
      if(find(li->begin(),li->end(),j) == li->end() &&
         !j->commute(**k)) break;
     if(j != end) break;
     }
    if(k == li->end()) break;
   }
  if(insert != v.end()) break;
 }

 // found a match
 if(li != l.end() && insert != v.end())
 {
  Product resultl, resultr;

  // if the term did not play a role in the substitution just copy it
  for(j=v.begin();j!=insert;j++)
   if(find(li->begin(),li->end(),j) == li->end())
    if(j->commute(x)) resultr.factors.push_back(*j);
    else              resultl.factors.push_back(*j);

  // perform the substitution
  resultl.factors.push_back(y);
  n++;

  for(;j!=v.end();j++)
   if(find(li->begin(),li->end(),j) == li->end())
    resultr.factors.push_back(*j);

  return resultl*resultr.subst(x,y,n);
 }
}

// product does not contain expression for substitution
// try to substitute in each factor
Product p;
for(list<Symbolic>::const_iterator i=factors.begin();i!=factors.end();i++)
 p.factors.push_back(i->subst(x,y,n));
return p;
}

Simplified Product::simplify() const
{
 list<Symbolic>::const_iterator i;
 list<Symbolic>::iterator j, k, k1;
 Product r;

 // 1-element product:  (a) -> a
 if(factors.size() == 1) return factors.front().simplify();

 for(i=factors.begin();i!=factors.end();i++)
 {
  // absorb product of product: a * (a * a) * a -> a * a * a * a
  Simplified s = i->simplify();
  if(s.type() == typeid(Product))
  {
   CastPtr<const Product> product(s);
```

```
       r.factors.insert(r.factors.end(),product->factors.begin(),
                        product->factors.end());
   }
  else r.factors.push_back(s);
 }

 // if any matrices appear in the product,
 // pull everything (commutative) into the matrix
 SymbolicMatrix m(1,1,1);
 for(j=r.factors.begin();j!=r.factors.end();)
 {
  // found a matrix
  if(j->type() == typeid(SymbolicMatrix))
  {
   int br = 0;
   list<Symbolic>::iterator k;
   m = *CastPtr<SymbolicMatrix>(*j);
   // some terms preceding the matrix must be brought in from the left
   k = j;
   if(j!=r.factors.begin())
    for(k--;1;)
    {
     // this is the last element so end the loop
     if(k == r.factors.begin()) br = 1;
     // only multiply with elements that commute, i.e. "scalars"
     if(k->commute(m))
     {
      m = *k * m;
      k1 = k;
      if(k != r.factors.begin()) k--;
      r.factors.erase(k1);
     }
     else br = 1;
     if(br) break;
    }

   // some terms following the matrix must be brought in from the right
   k = j;
   for(k++;k!=r.factors.end();)
    // multiply matrices with matrices
    if(k->type() == typeid(SymbolicMatrix))
    {
     m = m * *CastPtr<SymbolicMatrix>(*k);
     r.factors.erase(k++);
    }
    // only multiply with elements that commute, i.e. "scalars"
    else if(k->commute(m))
    {
     m = m * *k;
     r.factors.erase(k++);
    }
    else break;

  // set *j to the resulting matrix
  *j = m.simplify();
  // set j to the next element after the last non-commuting element
  j = k;
 }
```

```
 else j++;
}

// group common terms
for(j=r.factors.begin();j!=r.factors.end();j++)
{
 // numbers will be grouped later
 if(j->type() == typeid(Numeric)) continue;

 Symbolic n = 1;
 Symbolic j1 = *j;

 // the exponent in products must be ignored in grouping comparisons
 if(j1.type() == typeid(Power))
 {
  CastPtr<Power> j2 = j1;
  n = j2->parameters.back();
  j1 = j2->parameters.front();
 }

 for((k=j)++;k!=r.factors.end() && j->commute(*k);)
 {
  // numbers will be grouped later
  if(k->type() == typeid(Numeric)) { k++; continue; }

  Symbolic k1 = *k;
  Symbolic power = 1;
  // the exponent in products must be ignored in grouping comparisons
  if(k1.type() == typeid(Power))
  {
   CastPtr<Power> k2 = k1;
   power = k2->parameters.back();
   k1 = k2->parameters.front();
  }

  if(j1 == k1)
  {
   n = n + power;
   r.factors.erase(k++);
  }
  else k++;
 }
 if(n == 0)
  *j = 1;
 else
  if(n == 1)
   *j = j1;
  else
   *j = (j1 ^ n).simplify();
}

// move numbers to the front
Number<void> n = Number<int>(1);
for(j=r.factors.begin();j!=r.factors.end();)
{
 if(j->type() == typeid(Numeric))
 {
  n = n * Number<void>(*j);
```

```
   if(n.isZero()) return n;
    r.factors.erase(j++);
   }
   else j++;
  }

  if(!n.isOne()) r.factors.push_front(n->simplify());
  if(r.factors.size()==0) return Number<int>(1);
  if(r.factors.size()==1) return r.factors.front();
  return r;
}

int Product::compare(const Symbolic &s) const
{
  int c = 0;
  if(type() != s.type()) return 0;

  return (subst(s,Symbolic(1),c) == 1 && c == 1);
}

Symbolic Product::df(const Symbolic &s) const
{
  list<Symbolic>::iterator i;
  Product p(*this);
  Sum r;

  for(i=p.factors.begin();i!=p.factors.end();i++)
  {
   Symbolic t = *i;
   *i = i->df(s);
   r.summands.push_back(p);
   *i = t;
  }
  return r;
}

Symbolic Product::integrate(const Symbolic &s) const
{
  int count = 0;
  list<Symbolic>::const_iterator i, i1;

  for(i=factors.begin();i!=factors.end();i++)
   if(i->df(s) != 0) { count++; i1 = i; }

  if(count == 1)
  {
   Product p;
   for(i=factors.begin();i!=factors.end();i++)
    if(i == i1) p.factors.push_back(i->integrate(s));
    else        p.factors.push_back(*i);
   return p;
  }

  if(count == 0) return *this * s;
  return Integral(*this,s);
}

Symbolic Product::coeff(const Symbolic &s) const
```

```
{
 int c = 0;
 Symbolic result = subst(s,1,c);

 if(c != 1) return 0;

 if(s.type() == typeid(Product))
 {
  CastPtr<const Product> p(s);
  list<Symbolic>::const_iterator i;
  for(i=p->factors.begin();i!=p->factors.end();i++)
  {
   result.subst(*i,1,c);
   if(c != 1) return 0;
  }
 }

 if(s.type() == typeid(Power))
 {
  CastPtr<const Power> p(s);
  result.subst(p->parameters.front(),1,c);
  if(c != 1) return 0;
 }

 if(s * result == *this) return result;

 return 0;
}

Expanded Product::expand() const
{
 list<Symbolic>::const_iterator i, k;
 list<Symbolic>::iterator j;
 Product r;

 for(i=factors.begin();i!=factors.end();i++)
 {
  Symbolic s = i->expand();
  if(s.type() == typeid(Sum))
  {
   // make a copy of s
   CastPtr<Sum> sum(s,1);
   k = i; k++;
   for(j=sum->summands.begin();j!=sum->summands.end();j++)
   {
    Product p = r;
    p.factors.push_back(*j);
    p.factors.insert(p.factors.end(),k,factors.end());
    *j = p;
   }
   return sum->expand();
  }
  else r.factors.push_back(s);
 }
 return r;
}

int Product::commute(const Symbolic &s) const
```

```
{
 // Optimize the case for numbers
 if(s.type() == typeid(Numeric)) return 1;

 list<Symbolic>::const_iterator i;
 for(i=factors.begin();i!=factors.end();i++)
  if(!i->commute(s)) return 0;
 return 1;
}

#endif
#endif
#endif
```

13.11.8 Sums

```
// sum.h

#ifndef SYMBOLIC_CPLUSPLUS_SUM

#include <list>
using namespace std;

#ifdef  SYMBOLIC_FORWARD
#ifndef SYMBOLIC_CPLUSPLUS_SUM_FORWARD
#define SYMBOLIC_CPLUSPLUS_SUM_FORWARD

class Sum;

#endif
#endif

#ifdef  SYMBOLIC_DECLARE
#ifndef SYMBOLIC_CPLUSPLUS_SUM_DECLARE
#define SYMBOLIC_CPLUSPLUS_SUM_DECLARE

class Sum: public CloningSymbolicInterface
{
 public: list<Symbolic> summands;
        Sum();
        Sum(const Sum&);
        Sum(const Symbolic&,const Symbolic&);
        ~Sum();

        Sum &operator=(const Sum&);

        void print(ostream&) const;
        Symbolic subst(const Symbolic&,const Symbolic&,int &n) const;
        Simplified simplify() const;
        int compare(const Symbolic&) const;
        Symbolic df(const Symbolic&) const;
        Symbolic integrate(const Symbolic&) const;
        Symbolic coeff(const Symbolic&) const;
        Expanded expand() const;
        int commute(const Symbolic&) const;
```

```
          Cloning *clone() const { return Cloning::clone(*this); }
};

#endif
#endif

#ifdef  SYMBOLIC_DEFINE
#ifndef SYMBOLIC_CPLUSPLUS_SUM_DEFINE
#define SYMBOLIC_CPLUSPLUS_SUM_DEFINE
#define SYMBOLIC_CPLUSPLUS_SUM

Sum::Sum() {}

Sum::Sum(const Sum &s)
 : CloningSymbolicInterface(s), summands(s.summands) {}

Sum::Sum(const Symbolic &s1,const Symbolic &s2)
{
 if(s1.type() == typeid(Sum)) summands = CastPtr<const Sum>(s1)->summands;
 else summands.push_back(s1);
 if(s2.type() == typeid(Sum))
 {
  CastPtr<const Sum> p(s2);
  summands.insert(summands.end(),p->summands.begin(),p->summands.end());
 }
 else summands.push_back(s2);
}

Sum::~Sum() {}

Sum &Sum::operator=(const Sum &s)
{
 if(this != &s) summands = s.summands;
 return *this;
}

void Sum::print(ostream &o) const
{
 if(summands.empty()) o << 0;
 for(list<Symbolic>::const_iterator i=summands.begin();i!=summands.end();
     i++)
 {
  if((i->type() != typeid(Numeric)
      || !CastPtr<const Numeric>(*i)->isNegative()) &&
     (i->type() != typeid(Product)
      || !CastPtr<const Product>(*i)->printsNegative()))
  o << ((i==summands.begin()) ? "":"+");
  i->print(o);
 }
}

Symbolic Sum::subst(const Symbolic &x,const Symbolic &y,int &n) const
{
 if(x.type() == type())
 {
  list<Symbolic>::const_iterator i;
  list<Symbolic>::iterator j;
  // make a copy of *this
```

```
  CastPtr<Sum> s1(*this,1);
  CastPtr<const Sum> s2(x);
  for(i=s2->summands.begin();i!=s2->summands.end();i++)
  {
   for(j=s1->summands.begin();j!=s1->summands.end() && *i != *j;j++);
   if(j == s1->summands.end()) break;
   s1->summands.erase(j);
  }
  if(i == s2->summands.end())
  {
   n++;
   // reset the simplified and expanded flags
   // since substitution may have changed this
   s1->simplified = s1->expanded = 0;
   return s1->subst(x,y,n) + y;
  }
 }

 // sum does not contain expression for substitution
 // try to substitute in each summand
 Sum s;
 for(list<Symbolic>::const_iterator i=summands.begin();i!=summands.end();
     i++)
  s.summands.push_back(i->subst(x,y,n));
 return s;
}

Simplified Sum::simplify() const
{
 list<Symbolic>::const_iterator i;
 list<Symbolic>::iterator j, k;
 Sum r;

 // 1-element sum:  (a) -> a
 if(summands.size() == 1) return summands.front().simplify();

 // absorb sum of sums: a + (a + a) + a -> a + a + a + a
 for(i=summands.begin();i!=summands.end();i++)
 {
  Simplified s = i->simplify();
  if(s.type() == typeid(Sum))
  {
   CastPtr<Sum> sum(s);
   r.summands.insert(r.summands.end(),sum->summands.begin(),
                     sum->summands.end());
  }
  else r.summands.push_back(s);
 }

 // collect matrices
 int firstm = 1;
 SymbolicMatrix m(1,1);
 for(j=r.summands.begin();j!=r.summands.end();)
 {
  if(j->type() == typeid(SymbolicMatrix))
  {
   if(firstm)
   {
```

```
    m = *CastPtr<SymbolicMatrix>(*j);
    firstm = 1 - firstm;
   }
   else
    m = m + *CastPtr<SymbolicMatrix>(*j);
   r.summands.erase(j++);
  }
  else j++;
 }
 if(!firstm) r.summands.push_back(m.simplify());

 // group common terms
 for(j=r.summands.begin();j!=r.summands.end();)
 {
  // numbers will be grouped later
  if(j->type() == typeid(Numeric)) { j++; continue; }

  Number<void> n = Number<int>(1);
  Symbolic j1 = *j;

  // the leading coefficient of products must be ignored in grouping comparisons
  if(j1.type() == typeid(Product))
  {
   // make a copy of j1
   CastPtr<Product> j2(j1,1);
   if(!j2->factors.empty() && j2->factors.front().type() == typeid(Numeric))
   {
    n = Number<void>(j2->factors.front());
    j2->factors.pop_front();
    j1 = *j2;
   }
  }

  for((k=j)++;k!=r.summands.end();)
  {
   // numbers will be grouped later
   if(k->type() == typeid(Numeric)) { k++; continue; }

   Symbolic k1 = *k;
   Number<void> coeff = Number<int>(1);
   // the leading coefficient of products must be ignored
   // in grouping comparisons

   if(k1.type() == typeid(Product))
   {
    // make a copy of k1
    CastPtr<Product> k2(k1,1);
    if(!k2->factors.empty() && k2->factors.front().type() == typeid(Numeric))
    {
     coeff = Number<void>(k2->factors.front());
     k2->factors.pop_front();
     k1 = *k2;
    }
   }

   if(j1 == k1)
   {
    n = n + coeff;
```

```
      r.summands.erase(k++);
     }
     else k++;
    }
    if(n.isZero()) r.summands.erase(j++);
    else *(j++) = (Symbolic(n) * j1).simplify();
  }

  // move numbers to the back
  Number<void> n = Number<int>(0);
  for(j=r.summands.begin();j!=r.summands.end();)
  {
    if(j->type() == typeid(Numeric))
    {
      n = n + Number<void>(*j);
      r.summands.erase(j++);
    }
    else j++;
  }

  if(!n.isZero()) r.summands.push_back(n->simplify());
  if(r.summands.size()==0) return Number<int>(0);
  if(r.summands.size()==1) return r.summands.front();

  return r;
}

int Sum::compare(const Symbolic &s) const
{
  if(type() != s.type()) return 0;
  // make a copy of s
  CastPtr<Sum> p(s,1);

  list<Symbolic>::const_iterator i;
  list<Symbolic>::iterator j;

  if(summands.size() != p->summands.size()) return 0;
  for(i=summands.begin();i!=summands.end();i++)
  {
    for(j=p->summands.begin();j!=p->summands.end() && *i != *j;j++);
    if(j == p->summands.end()) return 0;
    p->summands.erase(j);
  }

  return 1;
}

Symbolic Sum::df(const Symbolic &s) const
{
  list<Symbolic>::const_iterator i;
  Sum r;
  for(i=summands.begin();i!=summands.end();i++)
    r.summands.push_back(i->df(s));
  return r;
}

Symbolic Sum::integrate(const Symbolic &s) const
{
```

```cpp
 list<Symbolic>::const_iterator i;
 Sum r;
 for(i=summands.begin();i!=summands.end();i++)
  r.summands.push_back(i->integrate(s));
 return r;
}


Symbolic Sum::coeff(const Symbolic &s) const
{
 list<Symbolic>::const_iterator i;
 Sum r;
 for(i=summands.begin();i!=summands.end();i++)
  r.summands.push_back(i->coeff(s));
 return r;
}


Expanded Sum::expand() const
{
 list<Symbolic>::const_iterator i;
 Sum r;
 for(i=summands.begin();i!=summands.end();i++)
  r.summands.push_back(i->expand());
 return r;
}

int Sum::commute(const Symbolic &s) const
{
 list<Symbolic>::const_iterator i;

 // Optimize the case for numbers
 if(s.type() == typeid(Numeric)) return 1;

 // Optimize the case for a single symbol
 if(s.type() == typeid(Symbol))
 {
  list<Symbolic>::const_iterator i;
  for(i=summands.begin();i!=summands.end();i++)
   if(!i->commute(s)) return 0;
  return 1;
 }

 // if every term in the sum commutes with s
 // then the sum commutes with s
 for(i=summands.begin();i!=summands.end();i++)
  if(!i->commute(s)) break;
 if(i == summands.end()) return 1;

 // calculate the commutator [A, B] = A*B - B*A
 Expanded p1 = Product(*this,s).expand();
 Expanded p2 = Product(-s,*this).expand();
 Symbolic sum = Sum(p1,p2);
 // [A, B] == 0 implies A*B = B*A
 return sum == 0;
}

#endif
#endif
#endif
```

13.11.9 Symbols

```
// symbol.h

#ifndef SYMBOLIC_CPLUSPLUS_SYMBOL

#include <list>
#include <string>
using namespace std;

#ifdef  SYMBOLIC_FORWARD
#ifndef SYMBOLIC_CPLUSPLUS_SYMBOL_FORWARD
#define SYMBOLIC_CPLUSPLUS_SYMBOL_FORWARD

class Symbol;

#endif
#endif


#ifdef  SYMBOLIC_DECLARE
#ifndef SYMBOLIC_CPLUSPLUS_SYMBOL_DECLARE
#define SYMBOLIC_CPLUSPLUS_SYMBOL_DECLARE

class Symbol: public CloningSymbolicInterface
{
 public: string name;
         list<Symbolic> parameters;
         int commutes;
         Symbol(const Symbol&);
         Symbol(const string&,int = 1);
         Symbol(const char*,int = 1);
         ~Symbol();

         void print(ostream&) const;
         Symbolic subst(const Symbolic&,const Symbolic&,int &n) const;
         Simplified simplify() const;
         int compare(const Symbolic&) const;
         Symbolic df(const Symbolic&) const;
         Symbolic integrate(const Symbolic&) const;
         Symbolic coeff(const Symbolic&) const;
         Expanded expand() const;
         int commute(const Symbolic&) const;

         Symbol operator[](const Symbolic&) const;
         Symbol operator[](const list<Symbolic> &l) const;
         Symbol commutative(int=0) const;
         Symbol operator~() const;

         Cloning *clone() const { return Cloning::clone(*this); }
};

#endif
#endif


#ifdef  SYMBOLIC_DEFINE
#ifndef SYMBOLIC_CPLUSPLUS_SYMBOL_DEFINE
#define SYMBOLIC_CPLUSPLUS_SYMBOL_DEFINE
#define SYMBOLIC_CPLUSPLUS_SYMBOL
```

```
Symbol::Symbol(const Symbol &s)
: CloningSymbolicInterface(s),
  name(s.name), parameters(s.parameters), commutes(s.commutes) {}

Symbol::Symbol(const string &s,int c) : name(s), commutes(c) {}

Symbol::Symbol(const char *s,int c)   : name(s), commutes(c) {}

Symbol::~Symbol() {}

void Symbol::print(ostream &o) const
{
 o << name;
 if(!parameters.empty())
 {
  o << ((type() == typeid(Symbol)) ? "[":"(");
  parameters.front().print(o);
  for(list<Symbolic>::const_iterator i=++parameters.begin();
      i!=parameters.end();i++)
  {
   o << ((i==parameters.begin()) ? "":",");
   i->print(o);
  }
  o << ((type() == typeid(Symbol)) ? "]":")");
 }
}

Symbolic Symbol::subst(const Symbolic &x,const Symbolic &y,int &n) const
{
 if(*this == x)
 {
  n++;
  return y;
 }
 list<Symbolic>::iterator i;
 // make a copy of *this
 CastPtr<Symbol> s(*this,1);
 for(i=s->parameters.begin();i!=s->parameters.end();i++)
  *i = i->subst(x,y,n);
 // reset the simplified and expanded flags
 // since substitution may have changed this
 s->simplified = s->expanded = 0;
 return *s;
}

Simplified Symbol::simplify() const
{
 list<Symbolic>::iterator i;
 // make a copy of *this
 CastPtr<Symbol> sym(*this,1);

 for(i=sym->parameters.begin();i!=sym->parameters.end();i++)
  *i = i->simplify();

 return *this;
}
```

```
int Symbol::compare(const Symbolic &s) const
{
  list<Symbolic>::const_iterator i;
  list<Symbolic>::const_iterator j;

  if(s.type() != type()) return 0;
  CastPtr<const Symbol> sym = s;
  if(sym->name != name) return 0;
  if(sym->parameters.size() != parameters.size()) return 0;
  for(i=parameters.begin(),j=sym->parameters.begin();
      i!=parameters.end();
      i++,j++)
   if(*i != *j) return 0;
  return 1;
}

Symbolic Symbol::df(const Symbolic &s) const
{
 list<Symbolic>::const_iterator i;

 if(*this == s) return 1;

 Symbolic result = 0;
 for(i=parameters.begin();i!=parameters.end();i++)
  result = result + Derivative(*this,*i) * i->df(s);

 return result;
}

Symbolic Symbol::integrate(const Symbolic &s) const
{
 list<Symbolic>::const_iterator i;

 if(*this == s) return (s^2)/2;

 for(i=parameters.begin();i!=parameters.end();i++)
  if(i->df(s) != 0) return Integral(*this,s);

 return *this * s;
}

Symbolic Symbol::coeff(const Symbolic &s) const
{
 if(*this == s) return 1;
 return 0;
}

Expanded Symbol::expand() const
{
 // make a copy of *this
 CastPtr<Symbol> r(*this,1);
 list<Symbolic>::iterator i;
 for(i=r->parameters.begin();i!=r->parameters.end();i++)
  *i = i->expand();
 return *r;
}

int Symbol::commute(const Symbolic &s) const
```

```
{
 list<Symbolic>::const_iterator i;

 if(*this == s) return 1;
 for(i=parameters.begin();i!=parameters.end();i++)
  if(!i->commute(s)) return 0;

 // only Symbols should be non-commutative
 if(type() != typeid(Symbol)) return 1;

 if(s.type() == typeid(Symbol))
  return commutes || CastPtr<const Symbol>(s)->commutes;

 if(s.type() == typeid(SymbolicMatrix))
  return commutes;

 return s.commute(*this);
}

// this should not be used with the functions cos, exp etc. since
// the overrides on simplification and other methods will be lost
Symbol Symbol::operator[](const Symbolic &s) const
{
 if(type() != typeid(Symbol))
  cerr << "Warning: " << *this << " [" << s
       << "] discards any methods which have been overridden." << endl;
 Symbol r(*this);
 r.parameters.push_back(s);
 return r;
}

// this should not be used with the functions cos, exp etc.
// since the overide on simplification and other methods
// will be lost
Symbol Symbol::operator[](const list<Symbolic> &l) const
{
 if(type() != typeid(Symbol))
  cerr << "Warning: " << *this << " [..."
       << "] discards any methods which have been overridden." << endl;
 Symbol r(*this);
 r.parameters.insert(r.parameters.end(),l.begin(),l.end());
 return r;
}

// this should not be used with the functions cos, exp etc.
// since the overide on simplification and other methods
// will be lost
Symbol Symbol::commutative(int c) const
{
 if(type() != typeid(Symbol))
  cerr << "Warning: " << *this << " .commutative(" << c
       << ") discards any methods which have been overridden." << endl;
 Symbol r(*this);
 r.commutes = c;
 return r;
}

Symbol Symbol::operator~() const
```

```
{ return commutative(!commutes); }

#endif
#endif
#endif
```

13.11.10   Symbolic Expressions

```
// symbolic.h

#ifndef SYMBOLIC_CPLUSPLUS_SYMBOLIC

#include <iostream>
#include <typeinfo>
#include <list>
using namespace std;

#ifdef  SYMBOLIC_FORWARD
#ifndef SYMBOLIC_CPLUSPLUS_SYMBOLIC_FORWARD
#define SYMBOLIC_CPLUSPLUS_SYMBOLIC_FORWARD

class CloningSymbolicInterface;
class Expanded;
class Simplified;
class Symbolic;
class SymbolicInterface;
class SymbolicProxy;

#endif
#endif

#ifdef  SYMBOLIC_DECLARE
#ifndef SYMBOLIC_CPLUSPLUS_SYMBOLIC_DECLARE
#define SYMBOLIC_CPLUSPLUS_SYMBOLIC_DECLARE

class SymbolicInterface
{
 public: int simplified, expanded;
         SymbolicInterface();
         SymbolicInterface(const SymbolicInterface&);
         virtual ~SymbolicInterface();

         virtual void print(ostream&) const = 0;
         virtual const type_info &type() const;
         virtual Symbolic subst(const Symbolic&,
                                const Symbolic&,int &n) const = 0;
         virtual Simplified simplify() const = 0;
         virtual int compare(const Symbolic&) const = 0;
         virtual Symbolic df(const Symbolic&) const = 0;
         virtual Symbolic integrate(const Symbolic&) const = 0;
         virtual Symbolic coeff(const Symbolic&) const = 0;
         virtual Expanded expand() const = 0;
         virtual int commute(const Symbolic&) const = 0;
};

class CloningSymbolicInterface : public SymbolicInterface, public Cloning
```

```
{
 public: CloningSymbolicInterface();
         CloningSymbolicInterface(const CloningSymbolicInterface &);
};

class SymbolicProxy: public SymbolicInterface,
                     public CastPtr<CloningSymbolicInterface>
{
 public: SymbolicProxy(const CloningSymbolicInterface&);
         SymbolicProxy(const SymbolicProxy&);
         SymbolicProxy(const Number<void>&);

         void print(ostream&) const;
         const type_info &type() const;
         Symbolic subst(const Symbolic&,const Symbolic&,int &n) const;
         Simplified simplify() const;
         int compare(const Symbolic&) const;
         Symbolic df(const Symbolic&) const;
         Symbolic integrate(const Symbolic&) const;
         Symbolic coeff(const Symbolic&) const;
         Expanded expand() const;
         int commute(const Symbolic&) const;

         SymbolicProxy &operator=(const CloningSymbolicInterface&);
         SymbolicProxy &operator=(const SymbolicProxy&);
};

class Simplified: public SymbolicProxy
{
 public: Simplified(const CloningSymbolicInterface&);
         Simplified(const SymbolicProxy&);
         Simplified(const Number<void>&);
};

class Expanded: public SymbolicProxy
{
 public: Expanded(const CloningSymbolicInterface&);
         Expanded(const SymbolicProxy&);
         Expanded(const Number<void>&);
};

class Symbolic: public SymbolicProxy
{
 public: static int auto_expand;
         static int subst_count;

         Symbolic();
         Symbolic(const Symbolic&);
         Symbolic(const CloningSymbolicInterface&);
         Symbolic(const SymbolicProxy&);
         Symbolic(const Number<void>&);
         Symbolic(const int&);
         Symbolic(const double&);
         Symbolic(const string&);
         Symbolic(const char*);
         Symbolic(const string&,int);
         Symbolic(const char*,int);
         Symbolic(const Symbolic&,int);
```

```
     Symbolic(const string&,int,int);
     Symbolic(const char*,int,int);
     Symbolic(const Symbolic&,int,int);
     Symbolic(const list<Symbolic>&);
     Symbolic(const list<list<Symbolic> >&);
     ~Symbolic();

     SymbolicProxy &operator=(const CloningSymbolicInterface&);
     SymbolicProxy &operator=(const SymbolicProxy&);
     SymbolicProxy &operator=(const int&);
     SymbolicProxy &operator=(const double&);
     SymbolicProxy &operator=(const string&);
     SymbolicProxy &operator=(const char*);
     SymbolicProxy &operator=(const list<Symbolic>&);
     SymbolicProxy &operator=(const list<list<Symbolic> >&);
     Symbolic operator[](const Equation&) const;
     Symbolic operator[](const list<Equation>&) const;
     Symbolic operator[](const Symbolic&) const;
     Symbolic operator[](const list<Symbolic>&) const;
     Symbolic &operator()(int);
     Symbolic &operator()(int,int);
     const Symbolic &operator()(int) const;
     const Symbolic &operator()(int,int) const;
     Symbolic subst(const Symbolic&,
                    const Symbolic&,int &n=subst_count) const;
     Symbolic subst(const Symbolic&,
                    const int&,int &n=subst_count) const;
     Symbolic subst(const Symbolic&,
                    const double&,int &n=subst_count) const;
     Symbolic subst(const Equation&,int &n=subst_count) const;
     Symbolic subst(const list<Equation>&,int &n=subst_count) const;
     Symbolic subst_all(const Symbolic&,
                        const Symbolic&,int &n=subst_count) const;
     Symbolic subst_all(const Equation&,int &n=subst_count) const;
     Symbolic subst_all(const list<Equation>&,int &n=subst_count) const;
     Symbolic coeff(const Symbolic&) const;
     Symbolic coeff(const Symbolic&,int) const;
     Symbolic coeff(const int&) const;
     Symbolic coeff(const double&) const;

     Symbolic commutative(int) const;
     Symbolic operator~() const;
     operator int() const;
     operator double() const;

     Symbolic operator|(const Symbolic&) const;
     Symbolic operator%(const Symbolic&) const;

     int rows() const;
     int columns() const;
     Symbolic row(int) const;
     Symbolic column(int) const;
     Symbolic identity() const;
     Symbolic transpose() const;
     Symbolic trace() const;
     Symbolic determinant() const;
     Symbolic vec() const;
     Symbolic kron(const Symbolic&) const;
```

```
          Symbolic dsum(const Symbolic&) const;
          Symbolic hadamard(const Symbolic&) const;
          Symbolic inverse() const;
};

#endif
#endif

#ifdef  SYMBOLIC_DEFINE
#ifndef SYMBOLIC_CPLUSPLUS_SYMBOLIC_DEFINE
#define SYMBOLIC_CPLUSPLUS_SYMBOLIC_DEFINE
#define SYMBOLIC_CPLUSPLUS_SYMBOLIC

//////////////////////////////////////////////
// Implementation for SymbolicInterface        //
//////////////////////////////////////////////

SymbolicInterface::SymbolicInterface()
{ simplified = expanded = 0; }

SymbolicInterface::SymbolicInterface(const SymbolicInterface &s)
{ simplified = s.simplified; expanded = s.expanded; }

SymbolicInterface::~SymbolicInterface() {}

const type_info &SymbolicInterface::type() const
{ return typeid(*this); }

//////////////////////////////////////////////
// Implementation for CloningSymbolicInterface   //
//////////////////////////////////////////////

CloningSymbolicInterface::CloningSymbolicInterface()
 : SymbolicInterface(), Cloning() {}

CloningSymbolicInterface::CloningSymbolicInterface(
                                   const CloningSymbolicInterface &s)
 : SymbolicInterface(s), Cloning(s) {}

//////////////////////////////////////////////
// Implementation for SymbolicProxy            //
//////////////////////////////////////////////

SymbolicProxy::SymbolicProxy(const CloningSymbolicInterface &s)
 : CastPtr<CloningSymbolicInterface>(s) {}

SymbolicProxy::SymbolicProxy(const SymbolicProxy &s)
 : CastPtr<CloningSymbolicInterface>(s) {}

SymbolicProxy::SymbolicProxy(const Number<void> &n)
 : CastPtr<CloningSymbolicInterface>(n) {}

void SymbolicProxy::print(ostream &o) const
{ (*this)->print(o); }

const type_info &SymbolicProxy::type() const
{ return (*this)->type(); }
```

```
Symbolic SymbolicProxy::subst(const Symbolic &x,
                             const Symbolic &y,int &n) const
{ return (*this)->subst(x,y,n); }

Simplified SymbolicProxy::simplify() const
{
 if((*this)->simplified) return *this;
 return (*this)->simplify();
}

int SymbolicProxy::compare(const Symbolic &s) const
{ return (*this)->compare(s); }

Symbolic SymbolicProxy::df(const Symbolic &s) const
{ return (*this)->df(s); }

Symbolic SymbolicProxy::integrate(const Symbolic &s) const
{ return (*this)->integrate(s); }

Symbolic SymbolicProxy::coeff(const Symbolic &s) const
{ return (*this)->coeff(s); }

Expanded SymbolicProxy::expand() const
{
 if((*this)->expanded) return *this;
 return (*this)->expand();
}

int SymbolicProxy::commute(const Symbolic &s) const
{ return (*this)->commute(s); }

SymbolicProxy &SymbolicProxy::operator=(const CloningSymbolicInterface &s)
{
 CastPtr<CloningSymbolicInterface>::operator=(s);
 return *this;
}

SymbolicProxy &SymbolicProxy::operator=(const SymbolicProxy &s)
{
 CastPtr<CloningSymbolicInterface>::operator=(s);
 return *this;
}

//////////////////////////////////////////////////
// Implementation for Simplified                 //
//////////////////////////////////////////////////

// these constructors should only be used by
// SymbolicInterface::simplify()

Simplified::Simplified(const CloningSymbolicInterface &s) : SymbolicProxy(s)
{ (*this)->simplified = 1; }

Simplified::Simplified(const SymbolicProxy &s) : SymbolicProxy(s)
{ (*this)->simplified = 1; }

Simplified::Simplified(const Number<void> &n) : SymbolicProxy(n)
{ (*this)->simplified = 1; }
```

```
///////////////////////////////////////////////
// Implementation for Expanded                 //
///////////////////////////////////////////////

// these constructors should only be used by
// SymbolicInterface::expand()

Expanded::Expanded(const CloningSymbolicInterface &s) : SymbolicProxy(s)
{ (*this)->expanded = 1; }

Expanded::Expanded(const SymbolicProxy &s) : SymbolicProxy(s)
{ (*this)->expanded = 1; }

Expanded::Expanded(const Number<void> &n) : SymbolicProxy(n)
{ (*this)->expanded = 1; }

///////////////////////////////////////////////
// Implementation for Symbolic                 //
///////////////////////////////////////////////

int Symbolic::auto_expand = 1;
int Symbolic::subst_count = 0;

Symbolic::Symbolic() : SymbolicProxy(Number<int>(0)) {}

Symbolic::Symbolic(const Symbolic &s) : SymbolicProxy(s)
{
 if(s.type() == typeid(SymbolicMatrix))
  // s is presumed const, so indexing via operator()
  // should access a copy
  {
   CastPtr<SymbolicMatrix> csm(*s,1);
   SymbolicProxy::operator=(*csm);
  }
}

Symbolic::Symbolic(const CloningSymbolicInterface &s)
 : SymbolicProxy(Number<int>(0))
{
 if(auto_expand)
  SymbolicProxy::operator=(s.expand().simplify());
 else
  SymbolicProxy::operator=(s.simplify());
}

Symbolic::Symbolic(const SymbolicProxy &s) : SymbolicProxy(s)
{
 if(s.type() == typeid(SymbolicMatrix))
  // s is presumed const, so indexing via operator()
  // should access a copy
  {
   CastPtr<SymbolicMatrix> csm(*s,1);
   SymbolicProxy::operator=(*csm);
  }
}

Symbolic::Symbolic(const Number<void> &n) : SymbolicProxy(n) {}
```

```
Symbolic::Symbolic(const int &i)
  : SymbolicProxy(Number<int>(i).simplify()) {}

Symbolic::Symbolic(const double &d)
  : SymbolicProxy(Number<double>(d).simplify()) {}

Symbolic::Symbolic(const string &s)
  : SymbolicProxy(Symbol(s).simplify()) {}

Symbolic::Symbolic(const char *s)
  : SymbolicProxy(Symbol(s).simplify()) {}

Symbolic::Symbolic(const string &s,int n)
  : SymbolicProxy(SymbolicMatrix(s,n,1)) {}

Symbolic::Symbolic(const char *s,int n)
  : SymbolicProxy(SymbolicMatrix(s,n,1)) {}

Symbolic::Symbolic(const Symbolic &s,int n)
  : SymbolicProxy(SymbolicMatrix(s,n,1)) {}

Symbolic::Symbolic(const string &s,int n,int m)
  : SymbolicProxy(SymbolicMatrix(s,n,m)) {}

Symbolic::Symbolic(const char *s,int n,int m)
  : SymbolicProxy(SymbolicMatrix(s,n,m)) {}

Symbolic::Symbolic(const Symbolic &s,int n,int m)
  : SymbolicProxy(SymbolicMatrix(s,n,m)) {}

Symbolic::Symbolic(const list<Symbolic> &l) : SymbolicProxy(Number<int>(0))
{
 list<list<Symbolic> > ll;
 ll.push_back(l);
 (*this) = SymbolicMatrix(ll);
}

Symbolic::Symbolic(const list<list<Symbolic> > &l)
  : SymbolicProxy(SymbolicMatrix(l)) {}

Symbolic::~Symbolic() {}

SymbolicProxy &Symbolic::operator=(const CloningSymbolicInterface &s)
{
 if(auto_expand)
  SymbolicProxy::operator=(s.expand().simplify());
 else
  SymbolicProxy::operator=(s.simplify());
 return *this;
}

SymbolicProxy &Symbolic::operator=(const SymbolicProxy &s)
{ return SymbolicProxy::operator=(s); }

SymbolicProxy &Symbolic::operator=(const int &i)
{ return *this = Number<int>(i); }
```

```
SymbolicProxy &Symbolic::operator=(const double &d)
{ return *this = Number<double>(d); }

SymbolicProxy &Symbolic::operator=(const string &s)
{ return *this = Symbolic(s); }

SymbolicProxy &Symbolic::operator=(const char *s)
{ return *this = Symbolic(s); }

SymbolicProxy &Symbolic::operator=(const list<Symbolic> &l)
{ return *this = Symbolic(l); }

SymbolicProxy &Symbolic::operator=(const list<list<Symbolic> > &l)
{ return *this = Symbolic(l); }

Symbolic Symbolic::operator[](const Equation &p) const
{ return subst(p); }

Symbolic Symbolic::operator[](const list<Equation> &l) const
{ return subst(l); }

Symbolic Symbolic::operator[](const Symbolic &p) const
{
 if(type() == typeid(Symbol))
  return CastPtr<const Symbol>(*this)->operator[](p);
 if(type() == typeid(SymbolicMatrix))
 {
  // make a copy of *this
  CastPtr<SymbolicMatrix> m(*this,1);
  int i, j;
  for(i=m->rows()-1;i>=0;i--)
   for(j=m->cols()-1;j>=0;j--)
    (*m)[i][j] = (*m)[i][j][p];
  return (*m);
 }
 return *this;
}

Symbolic Symbolic::operator[](const list<Symbolic> &l) const
{
 Symbolic result(*this);
 for(list<Symbolic>::const_iterator i=l.begin();i!=l.end();i++)
  result = result[*i];
 return result;
}

Symbolic &Symbolic::operator()(int i)
{
 if(type() != typeid(SymbolicMatrix))
 {
  if(i != 0)
  {
   cerr << "Attempted to cast " << *this
        << " in operator() to SymbolicMatrix failed." << endl;
   throw SymbolicError(SymbolicError::NotMatrix);
  }
  return *this;
 }
```

```
 CastPtr<SymbolicMatrix> m(*this);
 if(m->cols() == 1) return (*m)[i][0];
 if(m->rows() == 1) return (*m)[0][i];

 cerr << "Attempted to cast " << *this
      << " in operator() to SymbolicMatrix (vector) failed." << endl;
 throw SymbolicError(SymbolicError::NotVector);

 return *this;
}

Symbolic &Symbolic::operator()(int i,int j)
{
 if(type() != typeid(SymbolicMatrix))
 {
  if(i != 0 || j != 0)
  {
   cerr << "Attempted to cast " << *this
        << " in operator() to SymbolicMatrix failed." << endl;
   throw SymbolicError(SymbolicError::NotMatrix);
  }
  return *this;
 }

 CastPtr<SymbolicMatrix> m(*this);
 return (*m)[i][j];
}

const Symbolic &Symbolic::operator()(int i) const
{
 if(type() != typeid(SymbolicMatrix))
 {
  if(i != 0)
  {
   cerr << "Attempted to cast " << *this
        << " in operator() to SymbolicMatrix failed." << endl;
   throw SymbolicError(SymbolicError::NotMatrix);
  }
  return *this;
 }
 CastPtr<const SymbolicMatrix> m(*this);
 if(m->cols() == 1) return (*m)[i][0];
 if(m->rows() == 1) return (*m)[0][i];

 cerr << "Attempted to cast " << *this
      << " in operator() to SymbolicMatrix (vector) failed." << endl;
 throw SymbolicError(SymbolicError::NotVector);

 return *this;
}

const Symbolic &Symbolic::operator()(int i,int j) const
{
 if(type() != typeid(SymbolicMatrix))
 {
  if(i != 0 || j != 0)
  {
   cerr << "Attempted to cast " << *this
```

```
        << " in operator() to SymbolicMatrix failed." << endl;
   throw SymbolicError(SymbolicError::NotMatrix);
 }
 return *this;
}


 CastPtr<const SymbolicMatrix> m(*this);
 return (*m)[i][j];
}

Symbolic Symbolic::subst(const Symbolic &x,const Symbolic &y,int &n) const
{ return SymbolicProxy::subst(x,y,n); }

Symbolic Symbolic::subst(const Symbolic &x,const int &j,int &n) const
{ return subst(x,Number<int>(j),n); }

Symbolic Symbolic::subst(const Symbolic &x,const double &d,int &n) const
{ return subst(x,Number<double>(d),n); }

Symbolic Symbolic::subst(const Equation &e,int &n) const
{ return subst(e.lhs,e.rhs,n); }

Symbolic Symbolic::subst(const list<Equation> &l,int &n) const
{
 Symbolic result(*this);
 for(list<Equation>::const_iterator i=l.begin();i!=l.end();i++)
  result = result.subst(*i,n);
 return result;
}


Symbolic Symbolic::subst_all(const Symbolic &x,
                             const Symbolic &y,int &n) const
{
 int n1 = n;
 Symbolic r = subst(x,y,n);
 while(n != n1)
 {
  n1 = n;
  r = r.subst(x,y,n);
 }
 return r;
}

Symbolic Symbolic::subst_all(const Equation &e,int &n) const
{ return subst_all(e.lhs,e.rhs,n); }

Symbolic Symbolic::subst_all(const list<Equation> &l,int &n) const
{
 int n1;
 Symbolic result(*this);
 do
 {
  n1 = n;
  for(list<Equation>::const_iterator i=l.begin();i!=l.end();i++)
   result = result.subst(*i,n);
 } while(n != n1);

 return result;
```

```
}

Symbolic Symbolic::coeff(const Symbolic &s) const
{ return SymbolicProxy::coeff(s); }

Symbolic Symbolic::coeff(const Symbolic &s,int i) const
{
 if(i == 0) return subst(s,Number<int>(0));
 return SymbolicProxy::coeff(s^i);
}

Symbolic Symbolic::coeff(const int &i) const
{ return coeff(Number<int>(i)); }

Symbolic Symbolic::coeff(const double &d) const
{ return coeff(Number<double>(d)); }

Symbolic Symbolic::commutative(int c) const
{
 if(type() == typeid(Symbol))
  return CastPtr<const Symbol>(*this)->commutative(c);
 if(type() == typeid(SymbolicMatrix))
 {
  // make a copy *this
  CastPtr<SymbolicMatrix> m(*this,1);
  int i, j;
  for(i=m->rows()-1;i>=0;i--)
   for(j=m->cols()-1;j>=0;j--)
    (*m)[i][j] = ~ (*m)[i][j];
  return (*m);
 }
 return *this;
}

Symbolic Symbolic::operator~() const
{
 if(type() == typeid(Symbol))
  return CastPtr<const Symbol>(*this)->operator~();
 if(type() == typeid(SymbolicMatrix))
 {
  // make a copy of *this
  CastPtr<SymbolicMatrix> m(*this,1);
  int i, j;
  for(i=m->rows()-1;i>=0;i--)
   for(j=m->cols()-1;j>=0;j--)
    (*m)[i][j] = ~ (*m)[i][j];
  return (*m);
 }
 return *this;
}

Symbolic::operator int(void) const
{
 if(type() == typeid(Numeric) &&
    Number<void>(*this).numerictype() == typeid(int))
  return CastPtr<const Number<int> >(*this)->n;
 cerr << "Attempted to cast " << *this << " to int failed." << endl;
 throw SymbolicError(SymbolicError::NotInt);
```

```
 return 0;
}

Symbolic::operator double(void) const
{
 if(type() == typeid(Numeric) &&
    Number<void>(*this).numerictype() == typeid(double))
  return CastPtr<const Number<double> >(*this)->n;
 if(type() == typeid(Numeric) &&
    Number<void>(*this).numerictype() == typeid(int))
  return double(CastPtr<const Number<int> >(*this)->n);
 if(type() == typeid(Numeric) &&
    Number<void>(*this).numerictype() == typeid(Rational<Number<void> >))
 {
  CastPtr<const Number<Rational<Number<void> > > > n(*this);
  Symbolic num = n->n.num();
  Symbolic den = n->n.den();
  return double(num)/double(den);
 }
 cerr << "Attempted to cast " << *this << " to double failed." << endl;
 throw SymbolicError(SymbolicError::NotDouble);
 return 0.0;
}

Symbolic Symbolic::operator|(const Symbolic &s) const
{
 if(rows() != s.rows() || columns() != s.columns() ||
    (rows() != 1 && columns() != 1))
 {
  cerr << "Attempt to dot product " << *this
       << " and " << s << " failed." << endl;
  throw SymbolicError(SymbolicError::IncompatibleVector);
 }

 CastPtr<const SymbolicMatrix> m1(*this);
 CastPtr<const SymbolicMatrix> m2(s);

 if(m1->rows() == 1)
  return ((*m1)[0] | (*m2)[0]);
 else
  return ((*m1)(0) | (*m2)(0));
}

Symbolic Symbolic::operator%(const Symbolic &s) const
{
 if(rows() != s.rows() || columns() != s.columns() ||
    (rows() != 1 && columns() != 1))
 {
  cerr << "Attempt to cross product " << *this
       << " and " << s << " failed." << endl;
  throw SymbolicError(SymbolicError::IncompatibleVector);
 }

 CastPtr<const SymbolicMatrix> m1(*this);
 CastPtr<const SymbolicMatrix> m2(s);

 if(m1->rows() == 1)
  return SymbolicMatrix((*m1)(0) % (*m2)(0));
```

```
    else
      return SymbolicMatrix((*m1)[0] % (*m2)[0]);
}


int Symbolic::rows() const
{
  if(type() != typeid(SymbolicMatrix)) return 1;
  return CastPtr<const SymbolicMatrix>(*this)->rows();
}


int Symbolic::columns() const
{
  if(type() != typeid(SymbolicMatrix)) return 1;
  return CastPtr<const SymbolicMatrix>(*this)->cols();
}


Symbolic Symbolic::row(int i) const
{
  if(type() != typeid(SymbolicMatrix))
  {
    if(i == 0) return *this;
    cerr << "Attempted to cast " << *this
         << " in row() to SymbolicMatrix failed." << endl;
    throw SymbolicError(SymbolicError::NotMatrix);
  }
  Matrix<Symbolic> m = (*CastPtr<const SymbolicMatrix>(*this))[i];
  return SymbolicMatrix(m.transpose());
}


Symbolic Symbolic::column(int i) const
{
  if(type() != typeid(SymbolicMatrix))
  {
    if(i == 0) return *this;
    cerr << "Attempted to cast " << *this
         << " in column() to SymbolicMatrix failed." << endl;
    throw SymbolicError(SymbolicError::NotMatrix);
  }
  return SymbolicMatrix((*CastPtr<const SymbolicMatrix>(*this))(i));
}


Symbolic Symbolic::identity() const
{
  if(type() != typeid(SymbolicMatrix)) return Symbolic(1);
  return SymbolicMatrix(CastPtr<SymbolicMatrix>(*this,1)->identity());
}


Symbolic Symbolic::transpose() const
{
  if(type() != typeid(SymbolicMatrix)) return *this;
  return SymbolicMatrix(CastPtr<const SymbolicMatrix>(*this)->transpose());
}


Symbolic Symbolic::trace() const
{
  if(type() != typeid(SymbolicMatrix)) return *this;
  return CastPtr<const SymbolicMatrix>(*this)->trace();
}
```

```
Symbolic Symbolic::determinant() const
{
 if(type() != typeid(SymbolicMatrix)) return *this;
 return CastPtr<const SymbolicMatrix>(*this)->determinant();
}


Symbolic Symbolic::vec() const
{
 if(type() != typeid(SymbolicMatrix))
 {
   cerr << "Attempted to cast " << *this
        << " in vec() to SymbolicMatrix failed." << endl;
   throw SymbolicError(SymbolicError::NotMatrix);
 }
 return SymbolicMatrix(CastPtr<const SymbolicMatrix>(*this)->vec());
}


Symbolic Symbolic::kron(const Symbolic &s) const
{
 if(type() != typeid(SymbolicMatrix) ||
    s.type() != typeid(SymbolicMatrix))

 {
   cerr << "Attempted to cast " << *this << " or " << s
        << " in kron() to SymbolicMatrix failed." << endl;
   throw SymbolicError(SymbolicError::NotMatrix);
 }
 CastPtr<const SymbolicMatrix> m1(*this), m2(s);
 return SymbolicMatrix(m1->kron(*m2));
}


Symbolic Symbolic::dsum(const Symbolic &s) const
{
 if(type() != typeid(SymbolicMatrix) ||
    s.type() != typeid(SymbolicMatrix))
 {
   cerr << "Attempted to cast " << *this << " or " << s
        << " in dsum() to SymbolicMatrix failed." << endl;
   throw SymbolicError(SymbolicError::NotMatrix);
 }
 CastPtr<const SymbolicMatrix> m1(*this), m2(s);
 return SymbolicMatrix(m1->dsum(*m2));
}


Symbolic Symbolic::hadamard(const Symbolic &s) const
{
 if(type() != typeid(SymbolicMatrix) ||
    s.type() != typeid(SymbolicMatrix))
 {
   cerr << "Attempted to cast " << *this << " or " << s
        << " in hadamard() to SymbolicMatrix failed." << endl;
   throw SymbolicError(SymbolicError::NotMatrix);
 }
 CastPtr<const SymbolicMatrix> m1(*this), m2(s);
 return SymbolicMatrix(m1->hadamard(*m2));
}

Symbolic Symbolic::inverse() const
```

```
{
  if(type() != typeid(SymbolicMatrix)) return 1/(*this);
  CastPtr<const SymbolicMatrix> m(*this);
  return SymbolicMatrix(m->inverse());
}

#endif
#endif
#endif
```

13.11.11   Symbolic Matrices

```
// symmatrix.h

#ifndef SYMBOLIC_CPLUSPLUS_SYMBOLICMATRIX

#include "matrix.h"
using namespace std;

#ifdef  SYMBOLIC_FORWARD
#ifndef SYMBOLIC_CPLUSPLUS_SYMBOLICMATRIX_FORWARD
#define SYMBOLIC_CPLUSPLUS_SYMBOLICMATRIX_FORWARD

class SymbolicMatrix;

#endif
#endif

#ifdef  SYMBOLIC_DECLARE
#ifndef SYMBOLIC_CPLUSPLUS_SYMBOLICMATRIX_DECLARE
#define SYMBOLIC_CPLUSPLUS_SYMBOLICMATRIX_DECLARE

class SymbolicMatrix
: public CloningSymbolicInterface, public Matrix<Symbolic>
{
 public: SymbolicMatrix(const SymbolicMatrix&);
         SymbolicMatrix(const Matrix<Symbolic>&);
         SymbolicMatrix(const list<list<Symbolic> >&);
         SymbolicMatrix(const string&,int,int);
         SymbolicMatrix(const char*,int,int);
         SymbolicMatrix(const Symbolic&,int,int);
         SymbolicMatrix(int,int);
         ~SymbolicMatrix();

         void print(ostream&) const;
         Symbolic subst(const Symbolic&,const Symbolic&,int &n) const;
         Simplified simplify() const;
         int compare(const Symbolic&) const;
         Symbolic df(const Symbolic&) const;
         Symbolic integrate(const Symbolic&) const;
         Symbolic coeff(const Symbolic&) const;
         Expanded expand() const;
         int commute(const Symbolic&) const;

         Cloning *clone() const { return Cloning::clone(*this); }
};
```

```
#endif
#endif

#ifdef  SYMBOLIC_DEFINE
#ifndef SYMBOLIC_CPLUSPLUS_SYMBOLICMATRIX_DEFINE
#define SYMBOLIC_CPLUSPLUS_SYMBOLICMATRIX_DEFINE
#define SYMBOLIC_CPLUSPLUS_SYMBOLICMATRIX

SymbolicMatrix::SymbolicMatrix(const SymbolicMatrix &s)
: CloningSymbolicInterface(s), Matrix<Symbolic>(s) {}

SymbolicMatrix::SymbolicMatrix(const Matrix<Symbolic> &s)
: Matrix<Symbolic>(s)
{}

SymbolicMatrix::SymbolicMatrix(const list<list<Symbolic> > &sl)
{
 int cols = 0, k, l;
 list<Symbolic>::const_iterator j;
 list<list<Symbolic> >::const_iterator i;

 for(i=sl.begin();i!=sl.end();i++)
  if(int(i->size()) > cols) cols = i->size();

 Matrix<Symbolic>::resize(sl.size(),cols,Symbolic(0));
 for(k=0,i=sl.begin();i!=sl.end();k++,i++)
 {
  for(l=0,j=i->begin();j!=i->end();l++,j++)
   Matrix<Symbolic>::operator[](k)[l] = *j;
 }
}

SymbolicMatrix::SymbolicMatrix(const string &s,int n,int m)
 : Matrix<Symbolic>(n,m)
{
 for(int i=0;i<n;i++)
  for(int j=0;j<m;j++)
  {
   ostringstream os;
   if(n == 1 || m == 1)
    os << s << i + j;
   else
    os << s << "(" << i << "," << j << ")";
   Matrix<Symbolic>::operator[](i)[j] = Symbolic(os.str());
  }
}

SymbolicMatrix::SymbolicMatrix(const Symbolic &s,int n,int m)
 : Matrix<Symbolic>(n,m,s) {}

SymbolicMatrix::SymbolicMatrix(const char *s,int n,int m)
 : Matrix<Symbolic>(n,m)
{
 int i, j;
 for(i=0;i<n;i++)
  for(j=0;j<m;j++)
  {
```

```
    ostringstream os;
    if(n == 1 || m == 1)
     os << s << i + j;
    else
     os << s << "(" << i << "," << j << ")";
    Matrix<Symbolic>::operator[](i)[j] = Symbolic(os.str());
   }
}

SymbolicMatrix::SymbolicMatrix(int n,int m)
 : Matrix<Symbolic>(n,m,Symbolic(0)) {}

SymbolicMatrix::~SymbolicMatrix() {}

void SymbolicMatrix::print(ostream &o) const
{ o << endl << *this; }

Symbolic SymbolicMatrix::subst(const Symbolic &x,
                               const Symbolic &y,int &n) const
{
 if(*this == x) { n++; return y; }

 SymbolicMatrix m(rows(),cols());
 int r, c;
 for(r = rows()-1;r>=0;r--)
  for(c = cols()-1;c>=0;c--)
   m[r][c] = Matrix<Symbolic>::operator[](r)[c].subst(x,y,n);

 return m;
}

Simplified SymbolicMatrix::simplify() const
{
 // single element matrix -> number
 if(rows() == 1 && cols() == 1)
  return Matrix<Symbolic>::operator[](0)[0].simplify();

 SymbolicMatrix m(rows(),cols());
 for(int r = rows()-1;r>=0;r--)
  for(int c = cols()-1;c>=0;c--)
   m[r][c] = Matrix<Symbolic>::operator[](r)[c].simplify();

 return m;
}

int SymbolicMatrix::compare(const Symbolic &s) const
{
 if(s.type() != type()) return 0;
 CastPtr<const SymbolicMatrix> m = s;
 return (*this) == (*m);
}

Symbolic SymbolicMatrix::df(const Symbolic &s) const
{
 SymbolicMatrix m(rows(),cols());
 for(int r = rows()-1;r>=0;r--)
  for(int c = cols()-1;c>=0;c--)
   m[r][c] = Matrix<Symbolic>::operator[](r)[c].df(s);
```

```
 return m;
}

Symbolic SymbolicMatrix::integrate(const Symbolic &s) const
{
 SymbolicMatrix m(rows(),cols());
 for(int r = rows()-1;r>=0;r--)
  for(int c = cols()-1;c>=0;c--)
   m[r][c] = Matrix<Symbolic>::operator[](r)[c].integrate(s);

 return m;
}

Symbolic SymbolicMatrix::coeff(const Symbolic &s) const
{ return 0; }

Expanded SymbolicMatrix::expand() const
{
 SymbolicMatrix m(rows(),cols());
 for(int r = rows()-1;r>=0;r--)
  for(int c = cols()-1;c>=0;c--)
   m[r][c] = Matrix<Symbolic>::operator[](r)[c].expand();

 return m;
}

int SymbolicMatrix::commute(const Symbolic &s) const
{
 if(s.type() == typeid(SymbolicMatrix))
 {
  CastPtr<const SymbolicMatrix> m(s);
  return (*m) * (*this) - (*this) * (*m) == Matrix<Symbolic>(rows(),cols(),0);
 }
 return s.commute(*this);
}

#endif
#endif
#endif
```

13.11.12   Errors

```
// symerror.h

#ifndef SYMBOLIC_CPLUSPLUS_ERRORS

#include <string>

#ifdef  SYMBOLIC_FORWARD
#ifndef SYMBOLIC_CPLUSPLUS_ERRORS_FORWARD
#define SYMBOLIC_CPLUSPLUS_ERRORS_FORWARD

class SymbolicError;

#endif
```

```
#endif

#ifdef  SYMBOLIC_DECLARE
#ifndef SYMBOLIC_CPLUSPLUS_ERRORS_DECLARE
#define SYMBOLIC_CPLUSPLUS_ERRORS_DECLARE

class SymbolicError
{
 public: typedef enum {
                        IncompatibleNumeric,
                        IncompatibleVector,
                        NoMatch,
                        NotDouble,
                        NotInt,
                        NotMatrix,
                        NotNumeric,
                        NotVector,
                        UnsupportedNumeric
                       } error;

        error errno;
        SymbolicError(const error &e);
        string message() const;
};

#endif
#endif

#ifdef  SYMBOLIC_DEFINE
#ifndef SYMBOLIC_CPLUSPLUS_ERRORS_DEFINE
#define SYMBOLIC_CPLUSPLUS_ERRORS_DEFINE

SymbolicError::SymbolicError(const error &e) : errno(e) {}

string SymbolicError::message() const
{
 switch(errno)
 {
   case IncompatibleNumeric:
        return "Tried to use two incompatible number together";
   case IncompatibleVector:
        return "Tried to use two incompatible types in a vector operation";
   case NoMatch:
        return "No match found";
   case NotDouble:
        return "The value is not of type double";
   case NotInt:
        return "The value is not of type int";
   case NotMatrix:
        return "The value is not a matrix";
   case NotNumeric:
        return "The value is not numeric";
   case NotVector:
        return "The value is not a vector";
   case UnsupportedNumeric:
        return "The data type is not supprted by Numeric";
   default:
        return "Unknown error";
```

```
    }
}

#endif
#endif
#endif
```

Bibliography

[1] Ammeraal Leen, *STL for C++ Programmers*, John Wiley, Chichester (1997).

[2] Anderson J. R., Corbett A. T. and Reiser B. J., *Essential LISP*, Addison-Wesley (1987).

[3] Aslaksen H., *Multiple-valued Complex Functions and Computer Algebra*, SIGSAM Bull., **30**, 12–20 (1996).

[4] Ayres F., *Modern Algebra*, Schaum's Outline Series, McGraw-Hill, New York (1965).

[5] Barendregt H. P., *The Lambda Calculus, Its Syntax and Semantics, Revised Edition*, North-Holland, Amsterdam, 1984.

[6] Barendregt H. P. and Barendsen E., "Introduction to Lambda Calculus", http://citeseer.ist.psu.edu/barendregt94introduction.html, 1994.

[7] Barnsley M. F., *Fractals Everywhere*, 2nd ed., Academic Press Professional, Boston (1993).

[8] Bauer C., Frink A. and Kreckel R., "Introduction to the GiNaC Framework for Symbolic Computation within the C++ Programming Language", *Journal Symbolic Computation*, **33**, 1–12 (2002).

[9] Berry J. T., *The Waite Group's C++ Programming*, 2nd ed., SAMS, Carmel, Indiana (1992).

[10] Brackx F., *Computer Algebra with LISP and REDUCE : An Introduction to Computer-aided Pure Mathematics*, Kluwer Academic, Boston (1991).

[11] Budd T. A., *Classic Data Structures in C++*, Addison-Wesley (1994).

[12] Char B. W., *First Leaves – A Tutorial Introduction to MAPLE V*, Springer-Verlag, New York (1991).

[13] Cohen J. S., *Computer Algebra and Symbolic Computation, Mathematical Methods*, A K Peters, Natick, MA (2003).

[14] Dautcourt G., Jann K. P., Riemer E. and Riemer M., *Astronomische Nachrichten*, **102**, 1 (1981).

[15] Davenport J. H., Siret Y. and Tournier E., *Computer Algebra: Systems and Algorithms for Algebraic Computation*, 2 ed., Academic Press, London (1993).

[16] Ellis M. A. and Stroustrup B., *The Annotated C++ Reference Manual*, Addison-Wesley, (1990).

[17] Epstein R. L. and Carnielli W. A., *Computability*, Wadsworth & Brooks/Cole, Pacific Grove, California (1989).

[18] Ferreira C., "Gene Expression Programming: a New Adaptive Algorithm for Solving Problems", `http://xxx.lanl.gov`, `cs.AI/0102027` http://www.gene-expression-programming.com/

[19] Graham P., *ANSI Common Lisp*, Prentice Hall, Upper Saddle River, NJ (1996).

[20] Fröberg C. E., *Numerical Mathematics, Theory and Computer Applications*, Benjamin-Cummings, Menlo Park (1985).

[21] Geddes K. O., Czapor S. R. and Labahn G., *Algorithms for Computer Algebra*, Kluwer Academic Publishers, Boston (2003).

[22] Hardy Y. and W.-H. Steeb, "Gene expression programming and one-dimensional chaotic maps", *International Journal of Modern Physics C*, **13**, 13–24, 2002.

[23] The Haskell website: `http://www.haskell.org`

[24] Hearn A., *REDUCE User's Manual, Version 3.5*, RAND publication CP78 (Rev. 10/93).

[25] Hehl F. W., Winkelmann V. and Meyer H., *REDUCE*, 2 ed., Springer-Verlag (1993).

[26] Hekmatpour S., *An Introduction to LISP and Symbol Manipulation*, Prentice Hall, New York (1988).

[27] Howson A. G., *A Handbook of Terms used in Algebra and Analysis*, Cambridge University Press, Cambridge (1972).

[28] Jamsa K., *Success with C++*, Jamsa Press, Las Vegas (1994).

[29] Jenks R. D. and Sutor R. S., *Axiom: The Scientific Computation System*, Springer-Verlag, New York (1992).

[30] Jones S. P. (Ed), *Haskell 98 Language and Libraries*, Cambridge University Press, Cambridge (2003).

[31] Kowalski K. and Steeb W.-H., *Nonlinear Dynamical Systems and Carleman Linearization*, World-Scientific, Singapore (1991).

[32] Knuth D. E., *Seminumerical Algorithms*, 2nd ed., vol. 2 of *The Art of Computer Programming*, Addison-Wesley, Reading, MA (1981).

[33] Lang S., *Linear Algebra*, Addison-Wesley, Reading, MA (1968).

[34] Lippman S. B., *C++ Primer*, 2nd ed., Addison-Wesley, Reading, MA (1991).

[35] Marchisotto E. A. and Zakeri G-A., "An Invitation to Integration in Finite Terms", *The College Mathematics Journal*, **25**, 295 – 308, 1994.

[36] McCarthy J. et al., *LISP 1.5 Programmer's Manual*, 2 ed., MIT Press (1965).

[37] J. McCarthy, "History of Lisp", http://www-formal.stanford.edu/jmc/history/lisp/lisp.html, 1979.

[38] MacCallum M. A. H. and Wright F. J., *Algebraic Computing with Reduce*, Clarendon Press, Oxford (1991).

[39] Meyers S., *Effective C++*, Addison-Wesley, Reading, MA (1992).

[40] Norvig P., *Paradigms of Artificial Intelligence Programming: Case Studies in Common Lisp*, Morgan Kaufman, San Mateo (1991).

[41] Oevel W., Postel F., Rüscher G., and Wehmeier S., *Das MuPAD Tutorium*, SciFace Software, Paderborn (1998).

[42] Oltean M., "Multi Expression Programming", Technical Report, Babes-Bolyai University, Romania.

[43] Pinkus A. Z. and Winitzki S., "YACAS: a do-it-yourself computer algebra system", *Lecture Notes in Artificial Intelligence*, **2385**, 332 – 336 (2002).

[44] Press W. H., Teukolsky S. A., Vetterling W. T. and Flannery B. P., *Numerical Recipes in C, The Art of Scientific Computing*, Cambridge University Press (1992).

[45] Rayna G., *REDUCE: Software for Algebraic Computation*, Springer-Verlag, New York (1987).

[46] Risch R. H., *Transactions of the American Mathematical Society*, **139**, 167–189 (1969).

[47] Smith M. A., *Object-Oriented Software in C++*, Chapman and Hall, London (1993).

[48] Steeb W. H. and Lewien D., *Algorithms and Computation with Reduce*, BI-Wissenschaftsverlag, Mannheim (1992).

[49] Steeb W. H., Lewien D. and Boine-Frankenhein O., *Object-Oriented Programming in Science with C++*, BI-Wissenschaftsverlag, Mannheim (1993).

[50] Steeb W.-H., *Quantum Mechanics using Computer Algebra*, World Scientific, (1994).

[51] Steeb W.-H., Differential Equations and Computer Algebra, WSSIAA, **4**, 631–646 (1995).

[52] Steeb W. H., *Continuous Symmetries, Lie Algebras and Differential Equations*, second edition, World-Scientific, Singapore (2007).

[53] Steeb W.-H., *Problems and Solutions in Theoretical and Mathematical Physics*, second edition, World Scientific, Singapore (2003).

[54] Steeb W.-H., *The Nonlinear Workbook: Chaos, Fractals, Cellular Automata, Neural Networks, Genetic Algorithms, Gene Expression Programming, Support Vector Machine, Wavelets, Hidden Markov Models, Fuzzy Logic with C++, Java and SymbolicC++ Programs*, fourth edition World Scientific, Singapore (2008).

[55] Steeb W.-H., Hardy Y., Hardy A. and Stoop R., *Problems and Solutions in Scientific Computing with C++ and Java Simulations*, World Scientific, Singapore (2004).

[56] Steeb W.-H., *Problems and Solutions in Introductory and Advanced Matrix Calculus*, World Scientific, Singapore (2006).

[57] Steele G. Jr. *Common Lisp : The Language*, Second Edition, Digital Press (1990)

[58] Stoutemeyer D. R., *Notices of the American Mathematical Society*, Volume 38, pp. 778–785, (1991).

[59] Stroustrup B., *The C++ Programming Language*, 3rd ed., Addison-Wesley, Reading, MA (2005).

[60] Tan Kiat Shi, Steeb W.-H., and Hardy Y., *SymbolicC++ (2nd extended and revised edition)*, Springer Verlag, London (2000).

[61] Touretzky D. S., *Common Lisp: A Gentle Introduction to Symbolic Computation*, Benjamin/Cummings, Redwood City (1990).

[62] von zur Gathen J. and Gerhard J., *Modern Computer Algebra*, 2nd ed., Cambridge University Press, Cambridge (2003).

[63] Winston P. H., *Lisp*, Addison-Wesley, (1989).

[64] Wolfram S., *Mathematica: A System for Doing Mathematics by Computer*, 2nd ed., Addison-Wesley (1992).

Index